An Operator Theory Problem Book

An Operator Theory Problem Book

Mohammed Hichem Mortad

University of Oran 1, Algeria

NEW JERSEY • LONDON • SINGAPORE • BEIJING • SHANGHAI • HONG KONG • TAIPEI • CHENNAI • TOKYO

Published by

World Scientific Publishing Co. Pte. Ltd.
5 Toh Tuck Link, Singapore 596224
USA office: 27 Warren Street, Suite 401-402, Hackensack, NJ 07601
UK office: 57 Shelton Street, Covent Garden, London WC2H 9HE

Library of Congress Cataloging-in-Publication Data
Names: Mortad, Mohammed Hichem, 1978– author.
Title: An operator theory problem book / by Mohammed Hichem Mortad
(University of Oran 1, Algeria).
Description: New Jersey : World Scientific, 2018. |
Includes bibliographical references and index.
Identifiers: LCCN 2018011305 | ISBN 9789813236257 (hardcover : alk. paper)
Subjects: LCSH: Operator theory--Textbooks. | Functional analysis--Textbooks.
Classification: LCC QA329 .M67 2018 | DDC 515/.724--dc23
LC record available at https://lccn.loc.gov/2018011305

British Library Cataloguing-in-Publication Data
A catalogue record for this book is available from the British Library.

For any available supplementary material, please visit
https://www.worldscientific.com/worldscibooks/10.1142/10884#t=suppl

Printed in Singapore

Contents

Preface

Operator Theory is such a beautiful and elegant part of modern (pure and applied) Mathematics. It belongs to a larger domain which is Functional Analysis. It is also indispensable to Physics, in particular, Quantum Mechanics as well as some parts of Engineering and Statistics.

The main subject treated in this manuscript is linear operators on a Hilbert space (with a not negligible part on the foundations of Functional Analysis). The book is mainly intended for an undergraduate course: 3rd and 4th (even 5th in some cases) year students depending on each country and on each university. It is also a good source of interesting exercises, problems, examples and counterexamples for lectures and other researchers.

There are good books on Operator Theory and Functional Analysis in the existing literature but only a very few propose such a wide and varied range of exercises on many topics and with very detailed solutions. So, this is one main value of this book. Other features (among others) of this manuscript is the way positive operators and square roots are presented. So is the case with the absolute value of an operator. Also, a special treatment is given to the exponential of an operator.

In general, I have tried as much as possible to stick to the classical facts and notions of Operator Theory. At the same time, I have tried to keep all the material involved readily reachable by an undergraduate readership. This is why deeper notions on Spectral Theory or Operator Theory have not been included or have only been mentioned superficially. For instance, the spectral theorem and the spectral measures have been included but not too deeply. However, the way they are presented and the good amount of examples (such as the important properties of A^α, e^A and $\log(A)$ with some conditions on $A \in B(H)$) will hopefully help students and beginners to understand this beautiful and powerful theorem. Besides, due to the rarity of simple examples in famous books dealing with the spectral theorem, students and nonspecialists often find this theorem fairly hard to understand and they are usually struggling with applying it. In the beginning, they regularly wonder what is permissible? What is prohibited? This is not only my point of view. This opinion is shared by many other mathematicians including the "big names" as "The Halmos" who wrote in [**92**]: "*Another reason the spectral theorem is thought to be hard is that its proof is hard*".

Non-normal operators like subnormal, quasinormal and paranormal operators have not been included. Hyponormal operators, however, are presented in detail. Banach or C^*-algebras have not really been considered either. As regards unbounded operators, they have been covered without considering the concept of the adjoint. As a result, self-adjoint and normal unbounded operators have not been considered. There is, however, a good part on closed operators.

The prerequisites to use this book are basics of:

(1) Real Analysis.
(2) Linear Algebra.
(3) Metric Spaces.
(4) Basic Topology.
(5) Measure and Integration.
(6) Complex Analysis.

The topics covered in this book (split into 13 chapters) are:

(1) Banach Spaces.
(2) Hilbert Spaces.
(3) Bounded Linear Operators.
(4) Closed Operators.
(5) Spectral Theory.
(6) Spectral Theorem and Functional Calculi.
(7) Hyponormal Operators.

The manuscript is divided into two big parts: "Exercises et al." and "Solutions". It contains more than 720 problems (and just over 100 are without solutions) which are classified in four types and the book is structured as follows (as in [**153**]):

(1) `Basics:` In this part, we recall the essential of notions and results which are needed for the exercises. Some proofs are given as solved exercises but not all of them. So, readers may wish to consult the following standard references for further reading: [**2**], [**12**], [**13**], [**25**], [**48**], [**63**], [**64**], [**84**], [**96**], [**112**], [**114**], [**123**], [**124**], [**125**], [**138**], [**178**], [**190**], [**191**], [**199**], [**210**] and [**216**].
(2) `True or False:` In this part (liked by many), some interesting questions are proposed to the reader. Sometimes, the questions contain traps. Some of these are common errors which

appear with different students almost every year. Thanks to this section, students should hopefully assimilate well the presented material and should avoid making many silly mistakes. Moreover, this part is an important back-up for the "Basics" Section. Readers may even observe some redundancies in some cases, but this is mainly because it is meant to deepen their understanding. I am certainly not a big fan of pleonasm, however, I do believe that no detail is unnecessary!

(3) `Exercises with Solutions`: The major part and the core of each chapter where many exercises are given with detailed solutions, and as an illustration of this point, we notice that the solutions of Exercises 9.3.21 & 10.3.9 are over twenty page long!

The exercises are culled from different sources (some are classic, others are research papers and some are made by myself). I have cited most of the relevant references, and if there is some source which I have forgotten to mention, then I sincerely apologize for that. Finally, I have done my best to include some elegant proofs on many topics.

(4) `Tests`: This section usually contains short exercises given with just answers or simply hints.

(5) `More Exercises`: In this part, some exercises are proposed without solutions to the interested reader.

Writing a book in mathematics is not an easy task and most authors would certainly agree with me. I have written many papers and several books but this one has really taken much of my time and in many times I just wanted to abandon the whole project. Nonetheless, even with the hard times I have gone through, I can say that overall I did enjoy writing this book. I really hope that readers will enjoy reading it and that it will benefit them. To conclude, I can say that the day I thought would never come has finally and thankfully arrived! This manuscript was accomplished on January the 06th, 2018 in the city of Oran.

I would like to warmly thank all the staff of World Scientific Publishing Company for their patience and help (in particular, Dr Lim Swee Cheng and Ms Tan Rok Ting).

Lastly, even though I believe that I have done my best, as a human work, the book is certainly not perfect. So, I will be happy to hear from readers about any eventual errors, typos, suggestions or omissions at my personal email address: **mhmortad@gmail.com.**

Mohammed Hichem Mortad

Notation and Terminology

0.1. Notation

- $\mathbb{N}$ is the set of natural numbers, i.e. $\mathbb{N}$ is the set $\{1, 2, \cdots\}$ (note that the French $\mathbb{N}$ contains 0 too).
- $\mathbb{Z}$ is the set of all integers while $\mathbb{Z}^+$ (or $\mathbb{Z}_+$ sometimes) is the set of positive numbers.
- $\mathbb{Q}$ is the set of rational numbers.
- $\mathbb{R}$ is the set of real numbers.
- $\mathbb{R}^+ = [0, \infty)$.
- $\mathbb{C}$ is the set of complex numbers.
- In general, $\mathbb{F} = \mathbb{R}$ or $\mathbb{C}$.
- $[a, b]$ is the closed interval with endpoints a and b.
- (a, b) is the open interval with endpoints a and b.
- (a, b) also denotes an ordered pair.
- i the complex square root of -1.
- $\mathbb{1}_X$ denotes the indicator function of a set X.
- X^c is the complement of a set X.
- $X^\perp$ is the orthogonal set of X.
- The interior of a set A is denoted by $\overset{\circ}{A}$.
- The closure of a set A is denoted by $\overline{A}$.
- X' is the topological dual of X.
- $\mathfrak{B}(\mathbb{R})$ denotes the set of bounded Borel functions.
- The set of polynomials may be denoted by $\mathbb{C}[X]$, $\mathbb{C}(X)$, $\mathbb{R}_n[X]$,...
- sp$A$ is the span of A, i.e. the set of all linear combinations of all finite subsets of A. It is also denoted by span A.
- $\|\cdot\|$ denotes a norm.
- $< \cdot, \cdot >$ denotes an inner product.
- c is the vector space of convergent sequences.
- c_0 is the vector space of convergent sequences having the limit zero.
- c_{00} is the vector space of all finitely non-zero sequences.
- $B(a, r)$ is the open ball of center $a \in X$ and radius $r > 0$.
- $B_c(a, r)$ is the closed ball of center $a \in X$ and radius $r \geq 0$.
- $S(a, r)$ is the sphere of center $a \in X$ and radius $r \geq 0$.

- $L(X, Y)$ is the space of linear operators from X into Y.
- $L(X) = L(X, X)$.
- $B(X, Y)$ is the space of linear bounded operators from X into Y.
- $B(X) = B(X, X)$.
- S.O.T. stands for the Strong Operator Topology.
- W.O.T. designates the Weak Operator Topology.
- If $A \in L(X, Y)$, then $G(A)$ denotes the graph of A.
- Let $A \in B(X, Y)$. Then $\ker A$ denotes the kernel of A and $\operatorname{ran} A$ the range or image of A
- If $A \in B(X)$, then e^A is its exponential.
- If $A, B \in B(H)$, then $[A, B]$ is their commutator.
- $\operatorname{Lat} A$ denotes the collection of all invariant subspaces for $A \in B(X)$.
- If $A \in B(H)$ is positive, then $\sqrt{A}$ or $A^{\frac{1}{2}}$ denotes the (unique) positive square root of A.
- If $A \in B(H)$, then $\operatorname{Re} A$ denotes its real part and $\operatorname{Im} A$ denotes its imaginary part.
- If $A \in B(H)$, then $|A|$ denotes its absolute value.
- If $A \in B(H)$, then A^* is the adjoint of A.
- If $A \in L(H)$, then $D(A)$ is its domain.
- Let A be an operator. Then $\sigma(A)$ is the spectrum of A, $\sigma_p(A)$ is the point spectrum, $\sigma_r(A)$ is the residual spectrum, $\sigma_c(A)$ is the continuous spectrum and $\sigma_a(A)$ is the approximate point spectrum. $\rho(A)$ denotes the resolvent set.
- $R(\lambda, A)$ denotes the resolvent function of A.
- $r(A)$ denotes the spectral radius of $A \in B(H)$.
- $W(A)$ denotes the numerical range of $A \in L(H)$.
- $K(H)$ is the space of compact operators.

0.2. Terminology

- It will be clear from the context whether "$\subset$" denotes inclusions between two sets or two operators.
- From time to time the reader will see "(why?)". In such case, this means that this is a question whose answer should be known by the reader. This is used by other authors such as J. B. Conway (see [**48**]). Some question tags are also used for a similar purpose.
- In general, letters such as A and B are exclusively reserved to operators on Banach or Hilbert spaces. However, "famous" operators may be called by their initials. For example: S for

the shift, V for Volterra, M for the multiplication operator, P for a projection and so on.

- If we say the shift operator, then we mean the unilateral shift on ℓ^2.
- ℓ^2 is $\ell^2(\mathbb{N})$. Otherwise, we will write $\ell^2(\mathbb{Z})$.
- WLOG, as it pleases many, stands for "without loss of generality".
- As it is used almost everywhere, "iff" means if and only if (for the fun, the French use "ssi" for "si et seulement si". Even in Arabic, it has been sorted out by doubling a letter in the end!).

Part 1

EXERCISES ET AL.

CHAPTER 1

Normed Vector Spaces. Banach Spaces

1.1. Basics

1.1.1. Definitions and Examples. First, we assume readers are very familiar with definitions and results from Linear Algebra such as: vector spaces, (linear) subspaces, bases etc...

DEFINITION 1.1.1. *Let X be a vector space on $\mathbb{F}$ (where $\mathbb{F}$ stands for either $\mathbb{R}$ or $\mathbb{C}$). A* ***norm*** *on X is a function $\|\cdot\| : X \to \mathbb{R}^+$ satisfying:*

(1) $\|x\| = 0 \Longleftrightarrow x = 0$.
(2) $\forall x \in X, \forall \lambda \in \mathbb{F} : \|\lambda x\| = |\lambda|\|x\|$.
(3) $\forall x, y \in X : \|x + y\| \leq \|x\| + \|y\|$ *(called the* ***triangle inequality****).*

Then $(X, \|\cdot\|)$ is called a ***normed vector space*** *(or a* ***normed space****). Any element of X is called a* ***vector****.*

REMARK. Let's agree henceforth that we will not be usually showing that a given set is a vector space as this will be assumed most of the time.

EXAMPLES 1.1.2.

(1) *The* ***absolute value*** *function is a norm on $\mathbb{R}$.*
(2) *Similarly, the* ***modulus*** *of a complex number is a norm on $\mathbb{C}$.*
(3) *On $\mathbb{R}^n$, we may define different norms such as*

$$x \mapsto \|x\|_2 = \sqrt{|x_1|^2 + |x_2|^2 + \cdots + |x_n|^2}$$

where $x = (x_1, x_2, \cdots, x_n) \in \mathbb{R}^n$. This norm is called the ***standard norm*** *or* ***Euclidean norm*** *on $\mathbb{R}^n$. Other norms on $\mathbb{R}^n$ are*

$$\|x\|_1 = |x_1| + |x_2| + \cdots + |x_n| \text{ and } \|x\|_\infty = \max(|x_1|, |x_2|, \cdots, |x_n|).$$

(4) *The* ***supremum norm*** *on $C([a,b],\mathbb{F})$, where $\mathbb{F} = \mathbb{R}$ or $\mathbb{C}$, is defined by*

$$\|f\|_\infty = \sup_{x \in [a,b]} |f(x)|.$$

Remark. If $f \geq 0$, then for any $\alpha > 0$, we have

$$\|f^\alpha\|_\infty = \|f\|_\infty^\alpha$$

where $f^\alpha(x) = (f(x))^\alpha$ for all x. This observation could be used without further notice.

Recall also that if f is bounded on some set I with $\inf_{x \in I} f(x) > 0$, then

$$\sup_{x \in I} \frac{1}{f(x)} = \frac{1}{\inf\limits_{x \in I} f(x)}.$$

(5) *Let $X = C([a,b],\mathbb{F})$ (where $\mathbb{F} = \mathbb{R}$ or $\mathbb{C}$) and let $1 \leq p < \infty$. For all $f \in X$, define*

$$\|f\|_p = \sqrt[p]{\int_a^b |f(x)|^p dx} \text{ and } \|f\|_\infty = \sup_{x \in [a,b]} |f(x)|.$$

Then the previous are norms on X.

(6) *Let*

$$\ell^p = \{(x_n) : \mathbb{N} \to \mathbb{C} : \sum_{n=1}^{\infty} |x_n|^p < \infty\},$$

where $1 \leq p < \infty$. Then ℓ^p is a vector space on which the standard norm is defined by

$$\|(x_n)\|_p = \left(\sum_{n=1}^{\infty} |x_n|^p\right)^{\frac{1}{p}}$$

(7) *Let ℓ^∞ be the vector space of all bounded sequences in $\mathbb{C}$. The usual norm on ℓ^∞ is given by*

$$\|(x_n)\|_\infty = \sup_{n \in \mathbb{N}} |x_n|.$$

Remark. The following examples are important linear subspaces of ℓ^∞:

(1) c is the vector space of convergent sequences. $(c, \|\cdot\|_\infty)$ is a normed vector space.
(2) c_0 is the vector space of convergent sequences having the limit zero. $(c_0, \|\cdot\|_\infty)$ is a normed vector space (notice that $c_0 \subset c$).
(3) c_{00} is the vector space of all finitely non-zero sequences.

Having defined ℓ^p spaces, we may ask whether they are comparable? This is indeed the case.

PROPOSITION 1.1.3. *Let $1 \leq p \leq q \leq \infty$. Then*

$$\ell^p \subset \ell^q \text{ and } \|x\|_q \leq \|x\|_p$$

for all $x \in \ell^p$.

The restriction of a norm must obviously be defined w.r.t. to a linear subspace:

PROPOSITION 1.1.4. *Let $(X, \|\cdot\|)$ be a normed vector space and let Y be a linear subspace of X. If $\|\cdot\|_Y$ is the restriction of $\|\cdot\|$ to Y, then $\|\cdot\|_Y$ is a norm on Y.*

The following result can be quite handy in many situations (a proof may be found in Exercise 1.3.2).

THEOREM 1.1.5. *Let $(X, \|\cdot\|)$ be a normed vector space. Then, for all $x, y \in X$,*

$$|\|x\| - \|y\|| \leq \|x \pm y\|$$

*(these inequalities are also known as the "**Generalized Triangle Inequalities**").*

DEFINITION 1.1.6. *Let $(X, \|\cdot\|)$ be a normed vector space, and let $x \in X$. Then x is said to be a **unit vector** whenever $\|x\| = 1$.*

REMARK. It is clear that to each $x \neq 0$, there corresponds a unit vector. Indeed, since x does not vanish, we have

$$\left\|\frac{x}{\|x\|}\right\| = \frac{\|x\|}{\|x\|} = 1.$$

PROPOSITION 1.1.7. *Let X be normed vector space and let $d : X \times X \to \mathbb{R}^+$ be defined by $d(x, y) = \|x - y\|$ for all $x, y \in X$. Then d is a metric on X. In other words, every normed vector space is a metric space.*

Now, we turn to special subsets of normed vector spaces and to operations on normed vector spaces.

DEFINITION 1.1.8. *Let $(X, \|\cdot\|)$ be a normed vector space.*

(1) *An **open ball** of center $a \in X$ and radius $r > 0$, denoted by $B(a, r)$, is defined by:*

$$B(a, r) = \{x \in X : \|x - a\| < r\}.$$

(2) *A **closed ball** of center $a \in X$ and radius $r \geq 0$, denoted by $B_c(a, r)$, is defined by:*

$$B_c(a, r) = \{x \in X : \|x - a\| \leq r\}.$$

(3) *A **sphere** of center $a \in X$ and radius $r \geq 0$, and denoted by $S(a,r)$, is given by:*

$$S(a,r) = \{x \in X : \ \|x-a\| = r\}.$$

Definition 1.1.9. *Let X be a vector space and let $Y, Z \subset X$ be non empty. Then*

$$Y + Z = \{y + z : \ y \in Y, z \in Z\}$$

and

$$\alpha Y = \{\alpha y : \ y \in Y\}.$$

Remark. In case $Y = \{y\}$, then it is customary to write $y + Z$ instead of $\{y\} + Z$.

The following result is easy to prove (a proof may be found in the "True or False" Section).

Proposition 1.1.10. *Let $(X, \|\cdot\|)$ be a normed vector space. Then*

$$B(a,r) = a + B(0,r) = a + rB(0,1).$$

A similar result holds for $B_c(a,r)$ and $S(a,r)$.

There are examples (see e.g. **[153]**) in metric spaces showing that the closure of an open ball need not be the corresponding closed ball (of the same radius and center). This, however, cannot occur in a normed vector space as we have:

Proposition 1.1.11. *(see Exercise 1.3.9) Let $(X, \|\cdot\|)$ be a normed vector space. Let $r > 0$ and let $a \in X$. Show that*

$$B_c(a,r) = \overline{B(a,r)}.$$

Next, the convergence of a sequence in a normed vector space is introduced.

Definition 1.1.12. *Let $(X, \|\cdot\|)$ be a normed vector space. A sequence (x_n) in X is said to **converge** to $x \in X$ if:*

$$\lim_{n\to\infty} \|x_n - x\| = 0.$$

Remark. When there is no risk of confusion, we write $x_n \to x$ to mean that $\|x_n - x\| \to 0$.

The following result is elementary, but we recall it anyway.

Proposition 1.1.13. *In a normed vector space, convergent sequences cannot have more than one limit.*

PROPOSITION 1.1.14. *Let $(X, \|\cdot\|)$ be a normed vector space. Let (x_n) and (y_n) be two convergent sequences in X to x and y respectively. Let (α_n) be a sequence in $\mathbb{F}$ which converges to $\alpha \in \mathbb{F}$. Then:*

(1) $\lim_{n\to\infty} \|x_n\| = \|x\|$.
(2) $\lim_{n\to\infty}(x_n + y_n) = x + y$.
(3) $\lim_{n\to\infty}(\alpha_n x_n) = \alpha x$.

REMARK. Since every normed vector space can be regarded as a metric space, the reader should be aware of results of convergence (and others) in metric spaces. For instance, we have the next notion and result.

DEFINITION 1.1.15. *Let X be a normed vector space. A sequence (x_n) is **bounded** if $\|x_n\| \leq M$ for some $M \geq 0$ and all n.*

PROPOSITION 1.1.16. *In a normed vector space, a convergent sequence is bounded.*

The following concept helps us, when this applies, to work with an "easier" norm:

DEFINITION 1.1.17. *Let X be a vector space. Define on X two norms which we denote by $\|\cdot\|_1$ and $\|\cdot\|_2$. We say that $\|\cdot\|_1$ and $\|\cdot\|_2$ are **equivalent** if:*

$$\exists \alpha, \beta > 0,\ \forall x \in X:\ \alpha\|x\|_1 \leq \|x\|_2 \leq \beta\|x\|_1.$$

EXAMPLES 1.1.18. *Let $X = C([0,1], \mathbb{R})$. Then*

$$\|f\|_1 = \int_0^1 |f(x)|dx,\ \|f\|_2 = \left(\int_0^1 |f(x)|^2 dx\right)^{\frac{1}{2}},\ \|f\|_\infty = \sup_{0\leq x\leq 1} |f(x)|$$

are three non-equivalent norms on X (see Exercise 1.3.7).

REMARK. It is plain that $\|\cdot\|_1$ and $\|\cdot\|_2$ are equivalent iff $\|\cdot\|_2$ and $\|\cdot\|_1$ are equivalent.

The following is also trivial.

PROPOSITION 1.1.19. *Let X be a vector space. Define on X three norms which we denote by $\|\cdot\|_1$, $\|\cdot\|_2$ and $\|\cdot\|_3$. If $\|\cdot\|_1$ is equivalent to $\|\cdot\|_2$, and $\|\cdot\|_2$ is equivalent to $\|\cdot\|_3$, then $\|\cdot\|_1$ and $\|\cdot\|_3$ are equivalent.*

Before giving more results concerning equivalent norms and before introducing Banach spaces, we give the following definition.

DEFINITION 1.1.20. *Let $(X, \|\cdot\|)$ be a normed vector space. We say that a sequence (x_n) is **Cauchy** if*

$$\forall \varepsilon > 0,\ \exists N \in \mathbb{N},\ \forall n, m \in \mathbb{N}\ (n, m \geq N \Longrightarrow \|x_n - x_m\| < \varepsilon).$$

PROPOSITION 1.1.21. *Let $(X, \|\cdot\|)$ be a normed vector space. Every Cauchy sequence (x_n) is* **bounded**, *i.e. for some positive M and all n, $\|x_n\| \leq M$.*

PROPOSITION 1.1.22. *Let X be a normed vector space. Then, every convergent sequence is Cauchy.*

The converse of the previous result is not true in general. This is why we are introducing the following definition.

DEFINITION 1.1.23. *Let $(X, \|\cdot\|)$ be a normed vector space. We say that $(X, \|\cdot\|)$ is a* ***Banach space*** *if every Cauchy sequence (x_n) in $(X, \|\cdot\|)$ converges to a limit $x \in X$.*

REMARK. Banach spaces have been called so in honor to Stefan Banach, a Polish mathematician and one of the very big names in Functional Analysis. One of his important works is [**13**]. It is worth noticing that the concept of complete normed spaces has also independently been introduced by N. Wiener.

The following result is practical.

PROPOSITION 1.1.24. *Let $(X, \|\cdot\|)$ be a normed vector space. If a Cauchy sequence (x_n) has a convergent subsequence to some limit x, then (x_n) itself converges to x as well.*

COROLLARY 1.1.25. $(\mathbb{R}, |\cdot|)$ *is a Banach space.*

EXAMPLES 1.1.26.

(1) $(c_0, \|\cdot\|_\infty)$ *is a Banach space.*
(2) $(c_{00}, \|\cdot\|_\infty)$ *is not a Banach space as it is not closed (cf. Exercise 1.3.10).*
(3) ℓ^p *is a Banach space, with $1 \leq p \leq \infty$.*
(4) $(C[0,1], \|\cdot\|_\infty)$ *is a Banach space (Exercise 1.3.14).*

DEFINITION 1.1.27. *We say that a normed vector space X is* ***separable*** *if X has a countable and dense subset Y.*

EXAMPLES 1.1.28.

(1) ℓ^p, *with $1 \leq p < \infty$ is separable, and so is c_0. See Exercise 1.3.17.*
(2) ℓ^∞ *is not separable. See Exercise 1.3.19.*

The next result (which also holds for a general metric space with the adequate changes) is practical for establishing the non-separability of some spaces.

PROPOSITION 1.1.29. *Let* $(X, \|\cdot\|)$ *be a normed vector space. If* $Y \subset X$ *is uncountable, and for some* $r > 0$, *we have that*

$$\forall x, y \in Y,\ x \neq y:\ \|x - y\| \geq r,$$

then X *is not separable.*

The next fundamental result is surely known to readers. Why we are stating here is that it will be used to show that some normed vector spaces are not Banach (with respect to the supremum norm), e.g. see Exercise 1.3.15.

It is also a fundamental tool when passing to the continuous functional calculus of self-adjoint or normal operators. See Chapter 11.

THEOREM 1.1.30. *(**Stone-Weierstrass Theorem**) Let* f *be a real-valued continuous function on* $[a, b]$. *Then there exists a sequence of polynomials* (p_n) *defined on* $[a, b]$ *such that*

$$\lim_{n\to\infty} \|p_n - f\|_\infty = 0.$$

REMARK. In fact, the correct spelling should be Weierstraß.

COROLLARY 1.1.31. *(Exercise 1.3.15) The space* $(C^1(0, 1], \|\cdot\|_\infty)$ *is not Banach.*

We also give a complex version of Stone-Weierstrass Theorem in a basic form which will be needed elsewhere.

THEOREM 1.1.32. *(**Stone-Weierstrass Theorem**, see e.g.* **[105]***) Let* f *be a complex-valued continuous function on* $[a, b]$. *Then there exists a sequence of complex polynomials* (p_n) *defined on* $[a, b]$ *such that*

$$\lim_{n\to\infty} \|p_n - f\|_\infty = 0.$$

The next result is par excellence the most important application of equivalent norms.

THEOREM 1.1.33. *Let* X *be a normed vector space equipped with two equivalent norms* $\|\cdot\|_1$ *and* $\|\cdot\|_2$. *Let* (x_n) *be a sequence in* X. *Then*

1. (x_n) *converges to* x *in* $(X, \|\cdot\|_1)$ *iff* (x_n) *converges to* x *in* $(X, \|\cdot\|_2)$.
2. (x_n) *is Cauchy in* $(X, \|\cdot\|_1)$ *iff* (x_n) *is Cauchy in* $(X, \|\cdot\|_2)$.
3. $(X, \|\cdot\|_1)$ *is a Banach space iff* $(X, \|\cdot\|_2)$ *is a Banach space.*

REMARK. Halves of the first two statements in the previous theorem may be obtained just by considering either $\|x\|_1 \leq \alpha\|x\|_2$ or $\|x\|_2 \leq \beta\|x\|_1$ for some $\alpha, \beta > 0$ and for all $x \in X$ (why?).

Finite dimensional spaces have interesting properties.

PROPOSITION 1.1.34. *Let X be a normed vector space such that* $\dim X < \infty$. ***Any*** *two norms on X are equivalent.*

EXAMPLE 1.1.35. *Since* $\dim \mathbb{R}^n = n$, *the following norms are equivalent*

$$\|x\|_2 = \sqrt{|x_1|^2 + |x_2|^2 + \cdots + |x_n|^2}, \ \|x\|_1 = |x_1| + |x_2| + \cdots + |x_n|$$

and

$$\|x\|_\infty = \max(|x_1|, |x_2|, \cdots, |x_n|).$$

COROLLARY 1.1.36. *If on a normed vector space X, two norms are not equivalent, then* $\dim X = \infty$.

The next result is also remarkable.

PROPOSITION 1.1.37. *If $(X, \|\cdot\|)$ is a normed vector space such that* $\dim X < \infty$, *then $(X, \|\cdot\|)$ is a Banach space.*

COROLLARY 1.1.38. *Let X be a normed vector space and let Y be a linear subspace of X such that either* $\dim X < \infty$ *or* $\dim Y < \infty$. *Then Y is closed.*

In case we have a non closed subspace Y, say, one may ask why not work with its closure $\overline{Y}$? Is $\overline{Y}$ a linear subspace anyway? The answer is yes as we have:

PROPOSITION 1.1.39. *(see Exercise 1.3.11) If X is a normed vector space and if Y is a linear subspace of X, then $\overline{Y}$ is a linear subspace of X.*

We know from a Topology Course that closed and bounded sets are compact in usual $\mathbb{R}^n$ or $\mathbb{C}^n$. In infinite dimensional spaces, however, a closed and bounded set need not be compact. This a consequence of the next famous result.

THEOREM 1.1.40. *(**Riesz**) Let X be a normed vector space such that* $\dim X \geq 1$, *and let Y be a closed linear subspace of X such that $X \neq Y$. Then*

$$\forall \alpha \in (0,1), \ \exists x_\alpha \in X : \ \|x_\alpha\| = 1 \text{ and } \|x_\alpha - y\| > \alpha, \ \forall y \in Y.$$

COROLLARY 1.1.41. *If X is a normed vector space such that* $\dim X = \infty$, *then neither the unit sphere nor the closed unit ball is compact.*

COROLLARY 1.1.42. *Let X be a normed vector space. Then* $\dim X$ *is finite iff the closed unit ball is compact.*

1.1.2. Operations on Banach Spaces. We start with products or direct sums. First, we have an introductory example.

EXAMPLE 1.1.43. *If X and Y are two normed vector spaces with respective norms $\|\cdot\|_1$ and $\|\cdot\|_2$, then the Cartesian product $X \times Y$ may be normed with the "norm" given by*

$$\|(x,y)\| = \|x\|_1 + \|y\|_2.$$

We may also use as another example, the equivalent norm:

$$\|(x,y)\| = \max(\|x\|_1, \|y\|_2).$$

REMARKS.

(1) The previous definition extends naturally to a finite number of normed vector spaces.

(2) We can define other norms on the Cartesian product $X \times Y$. For example, for $p > 1$,

$$\|(x,y)\| = (\|x\|_1^p + \|y\|_2^p)^{\frac{1}{p}}.$$

Are these equivalent norms? Are they equivalent with the two norms given just above? (Cf. Exercise 2.3.13).

DEFINITION 1.1.44. *The space $X \times Y$ is called the* **product** *of X and Y. It may also be called the* **direct sum** *of X and Y and it is denoted by $X \oplus Y$.*

PROPOSITION 1.1.45. *In the previous example, $(x,y) \mapsto \|(x,y)\|$ is indeed a norm on $X \times Y$ (or $X \oplus Y$, isn't it?).*

The previous example may easily be extended to the case of a finite number of normed vector spaces.

The case of an infinite number of spaces is a little delicate. Let I be a non necessarily countable set. Let $(X_i)_{i\in I}$ be a collection of Banach spaces. Define

$$X = \left\{ (x_i)_{i\in I}, x_i \in X_i : \sum_{i\in I} \|x_i\|_{X_i} < \infty \right\}$$

and

$$\|(x_i)\|_X = \sum_{i\in I} \|x_i\|_{X_i}.$$

THEOREM 1.1.46. *Let I be a non necessarily countable set and let $(X_i)_{i\in I}$ be a family of Banach spaces. If X and $\|\cdot\|_X$ are defined as before, then $(X, \|\cdot\|_X)$ is a complete normed vector space, i.e. a Banach space.*

Remark. To give a sense to the previous construction, we need the concept of a summable family which is assumed to be known by the reader.

Definition 1.1.47. *The set X defined above is called the* ***direct sum*** *of the spaces X_i, and it is denoted by*

$$X = \bigoplus_{i \in I} X_i.$$

Example 1.1.48. *If $I = \mathbb{N}$, then the direct sum of $\mathbb{C}$ a countable number of times gives us $\ell^1(\mathbb{N})$.*

In the end, we introduce the normed vector quotient space as well as some of its basic properties.

Let X be a normed vector space and let Y be a linear subspace of X. On X, we define a relation $\mathcal{R}$ by

$$x\mathcal{R}y \Longleftrightarrow x - y \in Y.$$

Proposition 1.1.49. *The relation $\mathcal{R}$ just defined is an equivalence relation.*

Remark. As usual, the set of equivalence classes is denoted by X/Y (called the **quotient space**). The equivalence class containing x is denoted by $\dot{x}$. We may write $\dot{x} = x + Y$.

We want to give the structure of a vector space to X/Y, that is, we need to equip it with an addition and a scalar multiplication.

Definition 1.1.50. *Let X be a normed vector space and let Y be a linear subspace of X. On X/Y, define*

$$\dot{x} + \dot{y} = \dot{\overbrace{x + y}} \text{ and } \alpha\dot{x} = \dot{\overbrace{\alpha x}}.$$

Now, we norm it.

Proposition 1.1.51. *Let X be a normed vector space and let Y be a* ***closed*** *linear subspace of X. For each $\dot{x}$, set*

$$\|\dot{x}\|_{X/Y} = \inf_{y \in \dot{x}} \|y\| = \inf_{y \in Y} \|x + y\|.$$

Then $x \mapsto \|\dot{x}\|_{X/Y}$ defines a norm on X/Y.

Remark. The assumption "Y closed" is crucial for proving that $\|\cdot\|_{X/Y}$ is a norm.

Remark. Since Y is a linear subspace, we may re-write $\|\dot{x}\|_{X/Y}$ as follows

$$\|\dot{x}\|_{X/Y} = \inf_{y \in Y} \|x - y\|$$

which allows us to get back a known formula, namely

$$\|\dot{x}\|_{X/Y} = d(x, Y).$$

The final result is an answer to when X/Y is a Banach space.

Theorem 1.1.52. *Let X be a Banach space, and let Y be a closed linear subspace of X. Then X/Y is a Banach space.*

1.1.3. Convex Sets. The next notion is important in the sense that it may replace vector spaces in some situations. It is also important in its own.

Definition 1.1.53. *Let X be a vector space and let Y be a subset of X. We say that Y is* ***convex*** *if:*

$$\forall x, y \in Y, \forall t \in [0, 1] : \ (1 - t)x + ty \in Y.$$

Examples 1.1.54.

(1) *A linear subspace is convex.*
(2) *$\mathbb{R}$, and intervals in general, are convex.*
(3) *The closed unit ball is convex.*

Remark. We could have instead written $\alpha x + \beta y$, with $\alpha + \beta = 1$ and $\alpha, \beta \geq 0$. More generally, we have

Proposition 1.1.55. *If Y is a convex set, and if $x_1, x_2, \cdots, x_n \in Y$, then all elements of the type $\alpha_1 x_1 + \alpha_2 x_2 + \cdots + \alpha_n x_n$ are in Y, whenever $\alpha_1 + \alpha_2 + \cdots + \alpha_n = 1$, with $\alpha_1, \alpha_2 \cdots, \alpha_n \geq 0$.*

Proposition 1.1.56. *Let X be a vector space. The intersection of* ***any*** *number of convex subsets of X is a convex subset.*

We finish this subsection with the notion of a convex hull.

Definition 1.1.57. *Let $Y \subset X$, where X is vector space. The* ***convex hull*** *(also known as the* ***convex envelope****) of Y is the intersection of all convex sets containing Y. The convex hull of Y is denoted by* conv$Y$.

The following theorem gives us a characterization of the convex hull.

Theorem 1.1.58. *Let X be a vector space, and let $Y \subset X$. Then*

$$\mathit{conv}Y = \left\{ \sum_{k=1}^{n} \alpha_k x_k : \ x_k \in Y, \alpha_k \geq 0, \ k = 1, \cdots, n, \ \sum_{k=1}^{n} \alpha_k = 1 \right\}.$$

1.1.4. L^p-Spaces. First, we assume the reader is familiar with results on Measure and Integration Theory such as: the monotonic convergence theorem, dominated convergence theorem, the Fatou's lemma, the Fubini theorem etc... A good reference (among many) on this is [**211**].

In this subsection, we give some properties of L^p-spaces ($1 \leq p \leq \infty$). These are some of the most important spaces in applications.

Definition 1.1.59. *Define $L^p(\mathbb{R}^n)$ (the* ***Lebesgue space****), with $1 \leq p < \infty$, to be:*

$$L^p(\mathbb{R}^n) = \left\{ f : \mathbb{R}^n \to \mathbb{C} \text{ measurable} : \int_{\mathbb{R}^n} |f(x)|^p dx < \infty \right\}.$$

The natural "norm" on $L^p(\mathbb{R}^n)$ is defined by

$$\|f\|_p = \left(\int_{\mathbb{R}^n} |f(x)|^p dx \right)^{\frac{1}{p}}.$$

Also, define

$$L^\infty(\mathbb{R}^n) = \{ f : \mathbb{R}^n \to \mathbb{C} \text{ measurable s.t. } \exists M \geq 0 : |f(x)| \leq M \text{ a.e. } x \in \mathbb{R}^n \}$$

The associated "norm" is

$$\|f\|_\infty = \inf\{M : \ |f(x)| \leq M \text{ a.e. } x \in \mathbb{R}^n\}.$$

The previous quantity is more known as the ***essential supremum*** *of f and it is denoted by* $\operatorname{esssup}_{x \in \mathbb{R}}$.

Remark. It is worth recalling that, in general, the essential supremum does not coincide with the usual supremum (of bounded functions!). Indeed, let

$$f(x) = \begin{cases} 100, & x \in \mathbb{Q}, \\ 0, & x \notin \mathbb{Q}. \end{cases}$$

If we equip $\mathbb{R}$ with the Lebesgue's measure, then

$$\operatorname*{esssup}_{x \in \mathbb{R}} f(x) = 0 \neq \sup_{x \in \mathbb{R}} f(x) = 100.$$

Remark. In the previous definition, we may replace $\mathbb{R}^n$ by any (non empty) subset of $\mathbb{R}^n$. Also, $\mathbb{C}$ can be replaced by $\mathbb{R}$.

Remark. It is legitimate to wonder why we have excluded $0 < p < 1$? A word on this may be found in the "True or False" section.

Remark. In Measure Theory, if we equip $L^p(X)$ with the counting measure and set $X = \mathbb{N}$, then L^p and ℓ^p coincide (for $1 \leq p \leq \infty$).

PROPOSITION 1.1.60. *Let $1 \leq p \leq \infty$. Then $L^p(\mathbb{R}^n)$ is a vector space.*

PROPOSITION 1.1.61. *Let $p \in [1, \infty]$. Then for all $f, g \in L^p(\mathbb{R}^n)$*

$$\|f + g\|_p \leq \|f\|_p + \|g\|_p$$

*(more known as the **Minkowski Inequality**).*

COROLLARY 1.1.62. *Let $p \in [1, \infty]$. Then $f \mapsto \|f\|_p$ and (defined above) is a norm on $L^p(\mathbb{R}^n)$.*

Another tremendous inequality is the following.

PROPOSITION 1.1.63. *(**Hölder Inequality**) Let $1 \leq p \leq \infty$ and let $\frac{1}{p} + \frac{1}{q} = 1$. If $f \in L^p(\mathbb{R}^n)$ and $g \in L^q(\mathbb{R}^n)$, then $fg \in L^1(\mathbb{R}^n)$ and*

$$\|fg\|_1 \leq \|f\|_p \|g\|_q.$$

REMARK. Here $p = 1$ gives $q = \infty$, and $p = \infty$ gives $q = 1$.

COROLLARY 1.1.64. *(**Cauchy-Schwarz Inequality**) Let $f, g \in L^2(\mathbb{R}^n)$, then $fg \in L^1(\mathbb{R}^n)$ and*

$$\|fg\|_1 \leq \|f\|_2 \|g\|_2.$$

REMARK. Throughout this book, we will see each time a more general form of the Cauchy-Schwarz Inequality.

COROLLARY 1.1.65. *If $1 \leq p \leq \infty$ and $f \in L^p(\mathbb{R}^n)$, then*

$$\|f\|_p = \sup_{\|g\|_q = 1} \|fg\|_1.$$

REMARK. The previous corollary may be utilized to supply another proof of Minkowski's inequality (do it!).

The following result is important too.

PROPOSITION 1.1.66. *Let $p, q \geq 1$ be such that $\frac{1}{p} + \frac{1}{q} = 1$. If $f \notin L^p(\mathbb{R}^n)$, then there is a function $g \in L^q(\mathbb{R}^n)$ (which can be taken nonnegative) such that*

$$fg \notin L^1(\mathbb{R}^n).$$

Based on the Hölder Inequality, we may prove a more general inequality (called the **Generalized Hölder Inequality**). A proof is given in Exercise 1.3.25.

COROLLARY 1.1.67. *Let $1 \leq p, q \leq \infty$ and let $\frac{1}{p} + \frac{1}{q} = \frac{1}{r}$ and $\frac{1}{p} + \frac{1}{q} \leq 1$. If $f \in L^p(\mathbb{R}^n)$ and $g \in L^q(\mathbb{R}^n)$, then $fg \in L^r(\mathbb{R}^n)$ and*

$$\|fg\|_r \leq \|f\|_p \|g\|_q.$$

We know that the ℓ^p spaces are comparable. But, in general, the $L^p(\mathbb{R}^n)$ spaces are not comparable (cf. Exercise 1.3.22). However, we have the following interesting interpolation result:

Proposition 1.1.68. *Let $1 \leq q \leq r \leq \infty$. If $f \in L^r(\mathbb{R}^n) \cap L^q(\mathbb{R}^n)$, then $f \in L^p(\mathbb{R}^n)$ for any $q \leq p \leq r$.*

Now, we introduce the important concept of convolution:

Definition 1.1.69. *Let f and g be two measurable functions on $\mathbb{R}^n$. The **convolution** of f and g, denoted by $f * g$, is defined by*

$$(f * g)(x) = \int_{\mathbb{R}^n} f(t)g(x-t)dt, \ x \in \mathbb{R}^n.$$

Remark. We have to be careful when $f * g$ makes sense. It makes sense if e.g. $f \in L^p(\mathbb{R}^n)$ and $g \in L^q(\mathbb{R}^n)$ with $\frac{1}{p} + \frac{1}{q} = 1$ (by the Hölder Inequality). If f and g are both in $L^1(\mathbb{R}^n)$, then we can no more apply the Hölder Inequality to guarantee that $f * g$ is well defined. We can however prove by other means that $f * g$ makes sense in this case. More classes can be included and this is one of the virtues of the convolution product. This is a consequence of the following result:

Theorem 1.1.70. *(**Young Convolution Inequality**) Consider $1 \leq p, q \leq \infty$ such that $\frac{1}{p} + \frac{1}{q} \geq 1$. Consider an r such that $\frac{1}{r} = \frac{1}{p} + \frac{1}{q} - 1$. If $f \in L^p(\mathbb{R}^n)$ and $g \in L^q(\mathbb{R}^n)$, then $f * g \in L^r(\mathbb{R}^n)$ and*

$$\|f * g\|_r \leq \|f\|_p \|g\|_q.$$

Remark. Another powerful property of the convolution product is that $f * g$ is smooth if only one of the two functions is smooth so that we may differentiate $f * g$. We shall not consider this in the present book.

Related to the previous, we now list some important dense subsets of $L^p(\mathbb{R}^n)$. Before that, we give two definitions:

Definition 1.1.71.

(1) *(See e.g. **[52]** or **[178]**) A function f is in the **Schwartz space**, denoted by $\mathcal{S}(\mathbb{R}^n)$, if it is infinitely differentiable and for all $a, b \in \mathbb{N}^n$*

$$\sup_{x \in \mathbb{R}^n} |x^a D^b f(x)| < \infty$$

where

$$D^b = \frac{\partial^{|b|}}{\partial x_1^{b_1} \cdots \partial x_n^{b_n}}.$$

(2) $C_0^\infty(\mathbb{R}^n)$ *is the space of infinitely often differentiable functions* f *whose* ***support***

$$\text{supp} f = \overline{\{x \in \mathbb{R}^n : \ f(x) \neq 0\}}$$

is a compact subset of $\mathbb{R}^n$.

THEOREM 1.1.72. *Let* $1 \leq p < \infty$. *Then*

(1) *The space of simple functions which belong to* $L^p(\mathbb{R}^n)$ *is dense in* $L^p(\mathbb{R}^n)$ *(here we may even allow* $p = \infty$*).*
(2) $C_0^\infty(\mathbb{R}^n)$ *is dense in* $L^p(\mathbb{R}^n)$.
(3) $\mathcal{S}(\mathbb{R}^n)$ *is dense in* $L^p(\mathbb{R}^n)$.

REMARK. The proof that $\mathcal{S}(\mathbb{R}^n)$ is dense in $L^p(\mathbb{R}^n)$ can be deduced from that of $C_0^\infty(\mathbb{R}^n)$ because $C_0^\infty(\mathbb{R}^n) \subset \mathcal{S}(\mathbb{R}^n) \subset L^p(\mathbb{R}^n)$. Using a similar idea, we may show e.g. that $L^1(\mathbb{R}^n) \cap L^2(\mathbb{R}^n)$ is dense in $L^2(\mathbb{R}^n)$.

The next natural question is: Is $L^p(\mathbb{R}^n)$ Banach? The next theorem gives an answer (a proof may be found in Exercise 1.3.21).

THEOREM 1.1.73. ***(Riesz-Fischer)*** *Let* $1 \leq p \leq \infty$. *Then* $L^p(\mathbb{R}^n)$ *is a Banach space. Besides, when a sequence* (f_n) *converges to* f *in the* L^p*-norm, then there exists a subsequence* (f_{n_k}) *of* (f_n) *such that*

$$\lim_{k\to\infty} f_{n_k}(x) = f(x) \textit{ almost everywhere.}$$

We finish this subsection with the concept of weak-L^p spaces.

DEFINITION 1.1.74. *A function* f *on* $\mathbb{R}^n$ *is said to be in* **weak-**L^p, *written* $f \in L^p_w$, *if there is a constant* $C < \infty$ *such that*

$$|\{x : |f(x)| > t\}| \leq Ct^{-p} \textit{ for all } t > 0.$$

If $f \in L^p_w$, *then we write*

$$\|f\|_{p,w} = \sup_t (t^p |\{x : |f(x)| > t\}|)^{\frac{1}{p}}.$$

REMARK. Notice that $\|\cdot\|_{p,w}$ is not a norm (why?). However, when $p > 1$, L^p_w carries the structure of a Banach space with a norm which is equivalent to $\|\cdot\|_{p,w}$.

REMARK. Any function in L^p is in L^p_w and we have:

$$\|f\|_{p,w} \leq \|f\|_p$$

(so that the appellation "weak" is justified).

In fact for any $t > 0$,

$$\|f\|_p^p \geq \int_{|f|>t} |f(x)|^p dx \geq |\{x : |f(x)| > t\}| t^p.$$

The inequality $t^p|\{x : |f(x)| > t\}| \leq \|f\|_p^p$ is called **Chebyshev's inequality**.

The next is an example showing that $L^p_w(\mathbb{R}^n) \not\subset L^p(\mathbb{R}^n)$.

EXAMPLE 1.1.75. *A typical example is the function* $|x|^{-\frac{n}{p}}$. *Then* $|\{x : |f(x)| > t\}| = c_n t^{-p}$ *where* c_n *is the volume of the unit ball in* $\mathbb{R}^n$. *Thus* $f \in L^p_w(\mathbb{R}^n)$ *but* f *is not in* $L^p(\mathbb{R}^n)$.

We now come to the major result on interpolation between weak-L^p spaces. Its proof may be found in Exercise 1.3.27.

THEOREM 1.1.76. *Let* $r \geq 1$. *If* $r < p < s$ *and* $f \in L^r_w \cap L^s_w$, *then* $f \in L^p$ *and*

$$\|f\|_p \leq a\|f\|_{r,w} + b\|f\|_{s,w},$$

where the constants a *and* b *depend on* p, r *and* s.

1.2. True or False: Questions

QUESTIONS. Comment on the following questions/statements and indicate those which are false and those which are true when this applies. Justify your answers.

(1) We know from Section 1 that a norm defines a metric. What about the converse?

(2) Let $(X, \|\cdot\|)$ be a normed vector space and let $x \in X$. Let $B(x,r)$ be the open ball of center x and radius $r > 0$. Denote here $\{x\}$ by x. Then we have

$$B(x,r) = x + B(0_X, r) = x + rB(0_X, 1).$$

(3) Let X be a (normed) vector space and let $A \subset X$. Then

$$A + \{-A\} = \{0\} \text{ and } A + A = 2A.$$

(4) Let $0 \leq p < 1$. Let

$$X = \left\{ f : [0,1] \to \mathbb{C} : f \text{ measurable and } \int_0^1 |f(x)|^p dx < \infty \right\}.$$

Define for all $f \in X$,

$$\|f\|_p = \sqrt[p]{\int_0^1 |f(x)|^p dx}.$$

Then $\|\cdot\|_p$ is a norm on X.

1.3. Exercises with Solutions

Exercise 1.3.1. We define on $\mathbb{R}^2$ the mapping

$$(x, y) \mapsto \|(x, y)\| = |3x + 2y|.$$

Is $\|\cdot\|$ a norm on $\mathbb{R}^2$?

Exercise 1.3.2. Let $(X, \|\cdot\|)$ be a normed vector space. Show that

$$|\|x\| - \|y\|| \leq \|x \pm y\|, \ \forall x, y \in X.$$

Exercise 1.3.3. Let (X, d) be a metric space where X is also equipped with a structure of a *vector space* on some field $\mathbb{F}$. Assume that d verifies the properties

(1) $\forall x, y, a \in X : \ d(x + a, y + a) = d(x, y)$.
(2) $\forall x, y \in X, \ \forall \alpha \in \mathbb{F} : \ d(\alpha x, \alpha y) = |\alpha| d(x, y)$.

Show that $x \mapsto \|x\| = d(x, 0)$ defines a norm on X.

Exercise 1.3.4. Show that every distance induced by a norm satisfies the listed properties in the foregoing exercise.

Exercise 1.3.5. Let X be a vector space (not reduced to $\{0\}$). Equip X with the discrete metric. Can X be normed?

Exercise 1.3.6. Let X be a normed vector space. Assume that $x, y \in X$ obey

$$\|x + y\| = \|x\| + \|y\|.$$

Show that for any $\alpha, \beta \geq 0$, we have

$$\|\alpha x + \beta y\| = \alpha\|x\| + \beta\|y\|.$$

Exercise 1.3.7. Let $X = C([0, 1], \mathbb{R})$.

(1) Show that

$$\|f\|_1 = \int_0^1 |f(x)| dx, \ \|f\|_2 = \left(\int_0^1 |f(x)|^2 dx\right)^{\frac{1}{2}}, \ \|f\|_\infty = \sup_{0 \leq x \leq 1} |f(x)|$$

define three norms on X.

(2) Show that

$$\|f\|_1 \leq \|f\|_2 \leq \|f\|_\infty, \ \forall f \in X.$$

(3) Show that these three norms are not equivalent.
(4) What is $\dim X$?

Exercise 1.3.8. Let $X = C^1([0, 1], \mathbb{R})$. Show that

$$\|f\| = |f(0)| + \|f'\|_\infty$$

is a norm on X, which is equivalent to $\|f\|_\infty + \|f'\|_\infty$.

Exercise 1.3.9. Let $(X, \|\cdot\|)$ be a normed vector space. Let $r > 0$ and let $a \in X$. Show that

$$B_c(a,r) = \overline{B(a,r)}$$

Hint: use the sequence $x_n = \frac{1}{n}a + \left(1 - \frac{1}{n}\right)x$ where $x \in X$.

Exercise 1.3.10. In $(\ell^1, \|\cdot\|_1)$, let

$$c_{00} = \{(x_n)_n \in \ell^1 : \exists N \in \mathbb{N}^*, \; x_n = 0, \forall n \geq N\}.$$

Show that c_{00} is not closed.

Exercise 1.3.11. Let X be a normed vector space. Show that if Y is a linear subspace of X, then so is $\overline{Y}$.

Exercise 1.3.12. Let X be a normed vector space. Let $Y, Z \subset X$. Set

$$Y + Z = \{y + z : \; y \in Y, z \in Z\}.$$

(1) Show that $Y + Z$ is open if Y is open.
(2) Show that $Y + Z$ is compact if Y and Z are so.
(3) Show that $Y + Z$ is connected if Y and Z are so.
(4) Is the sum of two closed sets always closed?
(5) Show that $Y + Z$ is closed if Y is compact and Z is closed.

Exercise 1.3.13. Let $X = C([0,1], \mathbb{R})$. We equip it with the *norm* defined, for all $f \in X$, by

$$\|f\|_1 = \int_0^1 |f(x)| dx.$$

Consider the sequence of *continuous* functions

$$f_n(x) = \begin{cases} 0, & 0 \leq x \leq \frac{1}{2} - \frac{1}{n} \\ nx + (1 - \frac{1}{2}n), & \frac{1}{2} - \frac{1}{n} \leq x \leq \frac{1}{2} \\ 1, & \frac{1}{2} \leq x \leq 1. \end{cases}$$

(1) Show that (f_n) is a Cauchy sequence.
(2) Is $(X, \|\cdot\|_1)$ complete?

Exercise 1.3.14. Let $X = C([0,1])$. Show that $(X, \|\cdot\|_\infty)$ is complete with respect to the norm (for $f \in X$)

$$\|f\|_\infty = \sup_{x \in [0,1]} |f(x)|.$$

Exercise 1.3.15. Set $X = C^1([0,1], \mathbb{R})$. We endow X with the *norm* defined by

$$\|f\|_\infty = \sup_{0 \leq x \leq 1} |f(x)|.$$

Is $(X, \|\cdot\|_\infty)$ a Banach space?

Exercise 1.3.16. Let X be the set of all polynomials (of any degree) defined on $[0,1]$ and equip it with the norm $\|\cdot\|_\infty$.

(1) Show that the function $x \mapsto e^x$ is not a polynomial.
(2) Using the sequence $P_n(x) = \left(1+\frac{x}{n}\right)^n$, $n \geq 1$, demonstrate that $(X, \|\cdot\|_\infty)$ is not complete.
(3) Give another proof of the non-completeness of $(X, \|\cdot\|_\infty)$ using the Weierstrass Theorem.

Exercise 1.3.17.

(1) Show that ℓ^p is separable when $p \in [1,\infty)$.
(2) Show that c_0 is also separable (with respect to the topology of $\|\cdot\|_\infty$).

Exercise 1.3.18.

(1) Show that $(c_0, \|\cdot\|_\infty)$ is a Banach space.
(2) Show that $\ell^p \neq c_0$.
(3) Show that $\ell^p \subset c_0$.
(4) Prove that ℓ^p is dense in $(c_0, \|\cdot\|_\infty)$.
(5) Show that c_{00} is dense (and not closed) in ℓ^p, $1 \leq p < \infty$.
(6) Is c_{00} dense in ℓ^∞? If not, what is its closure in ℓ^∞?

Exercise 1.3.19. Prove that ℓ^∞ is not separable.

Exercise 1.3.20. Show that $L^\infty(0,1)$ is not separable.

Exercise 1.3.21. Let $1 \leq p \leq \infty$. Prove that $L^p(\mathbb{R}^n)$ is a Banach space.

Exercise 1.3.22.

(1) Compare (with respect to "$\subset$") $L^1(\mathbb{R})$ with $L^2(\mathbb{R})$.
(2) If $\Omega \subset \mathbb{R}$ is such that $|\Omega| < \infty$, then $L^2(\Omega) \subset L^1(\Omega)$ and for all $f \in L^2(\Omega)$,

$$\|f\|_{L^1(\Omega)} \leq |\Omega|^{\frac{1}{2}} \|f\|_{L^2(\Omega)}.$$

(3) If $f \in L^\infty(\mathbb{R}) \cap L^1(\mathbb{R})$ (with essential supremum $K > 0$, say), then $f \in L^2(\mathbb{R})$ and

$$\|f\|_{L^2(\mathbb{R})} \leq K^{\frac{1}{2}} \|f\|_{L^1(\mathbb{R})}^{\frac{1}{2}}.$$

Exercise 1.3.23. Give an example of a measurable function f such that $f \notin L^2(0,\infty)$ and $f \notin L^\infty(0,\infty)$ but $f \in L^2(0,\infty) + L^\infty(0,\infty)$.

Exercise 1.3.24. Find a measurable $\varphi : \mathbb{R} \mapsto \mathbb{R}$ such that $\varphi \notin L^\infty(\mathbb{R})$,

$$\sup_{k \in \mathbb{Z}} \int_k^{k+1} |\varphi(t)|^2 dt < \infty \text{ and } \sum_{k=-\infty}^{\infty} \int_k^{k+1} |\varphi(t)|^2 dt = \infty$$

(the condition on the right hand side means that $\varphi \notin L^2(\mathbb{R})$).

Exercise 1.3.25. Give a proof of the Generalized Hölder Inequality.

Exercise 1.3.26. Let $f \in L^p(\mathbb{R}^n)$, where $p \geq 1$. Let E_λ denote the **distribution function** of f, i.e.

$$E_\lambda = \{x \in \mathbb{R}^n : |f(x)| \geq \lambda\}.$$

Show that

$$\|f\|_p^p = p \int_0^\infty \lambda^{p-1} |E_\lambda| d\lambda.$$

Exercise 1.3.27. Let $r \geq 1$. If $r < p < s$ and $f \in L_w^r \cap L_w^s$, then $f \in L^p$ and

$$\|f\|_p \leq a\|f\|_{r,w} + b\|f\|_{s,w} \tag{1}$$

where the constants a and b depend on p, r and s.

Exercise 1.3.28. Let $e_n = (0, \cdots, 0, 1, 0 \cdots)$ where 1 lies in n^{th} component of e_n.

(1) Show that (e_n) is a sequence in $B_c(0,1)$ (the closed ball in ℓ^1) from which we cannot extract any Cauchy subsequence.
(2) Deduce that ℓ^1 is infinite-dimensional.
(3) Deduce that ℓ^2 and ℓ^∞ are also infinite-dimensional.

Exercise 1.3.29. First, recall the following definition: *A normed vector space X is said to be **locally compact** at $x \in X$ if there exists a compact subspace Y of X such that $U \subset Y$ where U is a neighborhood of x. We say that X is locally compact if it is so at each of its points.*

(1) Show that X is locally compact iff the closed unit ball is compact.
(2) Assume that X is infinite dimensional on $\mathbb{F}$. Show that the closed unit ball is not compact in X.

1.4. Tests

Test 1.4.1. Is $[0, 1]$ a Banach space?

Test 1.4.2. Let $\mathcal{P}$ be the vector space of all polynomials on $[-1, 1]$. Is $\mathcal{P}$ open in $(C[-1, 1], \|\cdot\|_\infty)$?

Test 1.4.3. Let $X = C([0, 1], \mathbb{R})$. Define on X the following *norm*

$$\|f\|_2 = \sqrt{\int_0^1 |f(x)|^2 dx}.$$

Let

$$A = \{f \in X : f(0) = 0\}.$$

Is A closed in X?

1.5. More Exercises

Exercise 1.5.1. Let $X = \mathcal{M}_n(\mathbb{R})$ be the space of square matrices with real entries (it is a vector space on $\mathbb{R}$). Define for each $A = (a_{ij}) \in X$ the mapping

$$\|A\| = n \max_{i,j} |a_{ij}|.$$

(1) Show that $\|\cdot\|$ is a norm on X.
(2) Show that $\forall A, B \in X$: $\|AB\| \leq \|A\|\|B\|$.

Exercise 1.5.2. Let $X = C([a,b],\mathbb{R})$. For all $f \in X$, set

$$\|f\|_1 = \int_a^b e^x |f(x)| dx \text{ and } \|f\|_2 = \sqrt{\int_a^b e^x |f(x)|^2 dx}.$$

Show that $\|\cdot\|_1$ and $\|\cdot\|_2$ define two norms on X.

Exercise 1.5.3. Let $a, b > 0$. Define on $\mathbb{R}^2$ the function $\|\cdot\|$ by

$$\|(x,y)\| = \sqrt{a^2x^2 + b^2y^2}.$$

(1) Show that $\|\cdot\|$ is a norm on $\mathbb{R}^2$.
(2) Draw the closed ball of center $(0,0)$ and radius 1.

Exercise 1.5.4. Let X be a normed vector space. Show that if A is a convex subset of X, then so are $\overset{\circ}{A}$ and $\overline{A}$.

Exercise 1.5.5. Show that in a normed vector space $(X, \|\cdot\|)$, the only subsets which are open and closed at the same time are $\varnothing$ and X, that is, every normed vector space is connected.

Exercise 1.5.6. Let $1 \leq p \leq \infty$. Show that ℓ^p is a Banach space.

Exercise 1.5.7.

(1) Let $\mathbb{D} = \{z \in \mathbb{C} : |z| < 1\}$ and let $1 \leq p < \infty$. Let

$$H^p(\mathbb{D}) = \left\{ f : \mathbb{D} \to \mathbb{C} \text{ analytic}, \sup_{0 \leq r < 1} \left(\frac{1}{2\pi} \int_0^{2\pi} |f(re^{i\theta})|^p d\theta \right)^{\frac{1}{p}} < \infty \right\}.$$

Show that $(H^p(\mathbb{D}), \|\cdot\|)$ is a complete normed vector space where

$$\|f\| = \sup_{0 \leq r < 1} \left(\frac{1}{2\pi} \int_0^{2\pi} |f(re^{i\theta})|^p d\theta \right)^{\frac{1}{p}}.$$

(2) We may also define $H^\infty(\mathbb{D})$ similarly. The "norm" in this case is given by

$$\|f\|_\infty = \sup_{0 \le r < 1} |f(re^{i\theta})|$$

Show that $\|f\|_\infty$ is indeed a norm and that $(H^\infty(\mathbb{D}), \|\cdot\|_\infty)$ is a Banach space.

*The Banach space $H^p(\mathbb{D})$ is called a **Hardy space** with $1 \le p \le \infty$.*

Exercise 1.5.8. Let Ω be a non empty open set in $\mathbb{C}$. Let $1 \le p < \infty$. Set

$$\mathcal{A}^p = \left\{ f : \Omega \to \mathbb{C} \text{ analytic } : \iint_\Omega |f(x+iy)|^p dxdy < \infty \right\}.$$

*We then call $\mathcal{A}^p$ a **Bergman space**.* For $f \in \mathcal{A}^p$, set

$$\|f\| = \left(\iint_\Omega |f(x+iy)|^p dxdy < \infty \right)^{\frac{1}{p}}.$$

(1) Show that $(\mathcal{A}^p, \|\cdot\|)$ is a normed vector space.
(2) Show that $(\mathcal{A}^p, \|\cdot\|)$ is complete.

Exercise 1.5.9. First, recall the following definition: *Let f be a measurable function on X. The **essential range** of f, denoted by R_f, is the set*

$$R_f = \{\lambda \in \mathbb{C} : \{x \in X : |f(x) - \lambda| < \varepsilon\} \text{ has a positive measure } \forall \varepsilon > 0\}.$$

(1) Can R_f be empty?
(2) Give an example showing that R_f may be unbounded.
(3) Prove that if $f \in L^\infty(X)$, then R_f is closed in $\mathbb{C}$ and bounded (*hence it is compact*) and that

$$\|f\|_\infty = \sup\{|\lambda| : \lambda \in R_f\}.$$

Remark. Another proof of the compactness of R_f may be found in Exercise 10.3.9.

CHAPTER 2

Bounded Linear Operators on Banach Spaces

2.1. Basics

2.1.1. Basic Definitions. First recall the next known definition.

Definition 2.1.1. *Let X and Y be two vector spaces on $\mathbb{F}$. We say that the transformation $A : X \to Y$ is* ***linear*** *if*

$$\forall x, y \in X, \forall \alpha, \beta \in \mathbb{F} : \ A(\alpha x + \beta y) = \alpha A(x) + \beta A(y).$$

The linear transformation A will be called from now on a ***linear operator****.*

The set of linear operators from X into Y is denoted by $L(X, Y)$.

Remarks.

(1) All operators considered in the present manuscript are linear, unless otherwise indicated.
(2) It is customary to write $A(x)$ as Ax.

Definition 2.1.2. *Let $(X, \|\cdot\|_1)$ and $(Y, \|\cdot\|_2)$ be two normed vector spaces. Let $A : (X, \|\cdot\|_1) \to (Y, \|\cdot\|_2)$ be linear. We say that A is* ***bounded*** *if*

$$\exists M \geq 0, \forall x \in X : \ \|Ax\|_2 \leq M\|x\|_1.$$

If A is not bounded, then we say that A is ***unbounded****.*

Remark. The set of bounded operators from X into Y is denoted by $B(X, Y)$, and as it is customary, simply by $B(X)$ when $X = Y$. In some textbooks, they use $\mathcal{L}(X, Y)$ instead of $B(X, Y)$.

Remark. You should be careful with some textbooks in which linear operators mean *bounded* linear operators! We do not adopt this convention merely because there are many linear operators which are *unbounded*.

Remark. If we want to show that a given linear operator A on X is not bounded, then it is usually practical to carry this out using sequences. For example, if A is not bounded, then for each $n \in \mathbb{N}$, there exists a unit vector x_n such that $\|Ax_n\| \geq n$. The latter also implies the unboundedness of A. Alternatively, we may use the negation of sequential continuity.

EXAMPLES 2.1.3.

(1) *Let $I : X \to X$ be the* ***identity operator*** *on X, i.e. $Ix = x$ for all $x \in X$. Then clearly I is bounded.*
(2) *The linear operator $A : \ell^2 \to \ell^2$ defined by*

$$A(x_1, x_2, \cdots) = (0, 2x_1, x_2, 2x_3, x_4, \cdots)$$

is bounded.
(3) *If $A : X \to \mathbb{C}$, where X is the vector space of all polynomials on $[0, 1]$, then*

$$A(p) = p'(1)$$

is unbounded.

Thanks to the linearity of an operator A, we have the following characterization of bounded operators.

THEOREM 2.1.4. *Let $(X, \|\cdot\|_1)$ and $(Y, \|\cdot\|_2)$ be two normed vector spaces. Let $A : (X, \|\cdot\|_1) \to (Y, \|\cdot\|_2)$ be linear. Then the following are equivalent:*

(1) *A is uniformly continuous.*
(2) *A is continuous on X.*
(3) *A is continuous at 0.*
(4) *A is bounded.*

REMARK. Some mathematicians choose to say "continuous operators" in lieu of "bounded operators". The previous theorem justifies this choice. We also use the term continuous from time to time especially when using it as sequential continuity.

REMARK. In this book, bounded linear operators are usually defined in the whole Banach space X. We notice that some bounded operators may only be defined on a proper subspace of X and that unbounded operators (which are usually not defined on the entire space) may also be defined on the whole Banach space. We shall not give more detail on this particular point.

A proof of the next result may be found in the "True or False" section.

PROPOSITION 2.1.5. *If X is a normed vector space and $A \in B(X)$, then the function $f : X \to \mathbb{R}^+$ defined by $x \mapsto f(x) = \|Ax\|$ is continuous.*

Finite dimensional spaces also play a good role when it comes to boundedness of linear operators. The next result is a perfect illustration of this role (a proof is prescribed in Exercise 2.3.14).

THEOREM 2.1.6. *Let $A \in L(X,Y)$, where X and Y are two normed vector spaces such that* $\dim X < \infty$. *Then $A \in B(X,Y)$.*

REMARK. It is important to emphasize that this concerns the domain only. The fact that $\dim Y < \infty$ (obviously with $\dim X = \infty$!) is not sufficient to transform $A \in L(X,Y)$ into $A \in B(X,Y)$. A counterexample has already appeared above in Examples 2.1.3.

The definitions of the kernel and the range of a linear operator remain unchanged, yet we recall them here. We also include the definition of a graph.

DEFINITION 2.1.7. *Let $A \in L(X,Y)$ where X and Y are two normed vector spaces.*

(1) *The **kernel** of A (or the **null-space** of A) is the (linear) subspace*
$$\ker A = \{x \in X : \ Ax = 0_Y\}.$$
*Its dimension is called the **nullity** of A, that is, the number* $n(A) = \dim(\ker A)$.

(2) *The **range** of A (or **image** of A) is the subspace*
$$\mathrm{ran}A = A(X) = \{Ax : \ x \in X\}.$$
*Its dimension is called the **rank** of A, that is, the number* $\dim(\mathrm{ran}A)$.

(3) *The **graph** of A, denoted by $G(A)$, is defined by*
$$G(A) = \{(x, Ax) : \ x \in X\}$$
(which is clearly a subset of $X \times Y$).

Related to the notion of a graph, we have the following two results:

PROPOSITION 2.1.8. *(see Exercise 2.3.13) Let $(X, \|\cdot\|_X)$ and $(Y, \|\cdot\|_Y)$ be two normed vector spaces and let $A : X \to Y$ be linear. Then*
$$\|x\|_{G(A)} = \sqrt{\|x\|_X^2 + \|Ax\|_Y^2}$$
*defines a norm on X (we shall call it the **graph norm** of A).*

THEOREM 2.1.9. *(a proof may be found in Exercise 2.3.46) Let X and Y be two normed vector spaces. Let $A : X \to Y$ be a continuous map. Then $G(A)$ is closed in $X \times Y$.*

Remark. The converse is not always true. It holds under added assumptions and this is the Closed Graph Theorem (Theorem 2.1.111).

The fact that a linear operator takes its values in $\mathbb{R}$ or $\mathbb{C}$ has a particular name.

Definition 2.1.10. *Let X be a vector space. A linear operator $A : X \to \mathbb{K}$ is called a* ***linear functional****.*

Examples 2.1.11.

(1) *Let $X = C([0,1], \mathbb{C})$. The transformation $A : X \to \mathbb{C}$ defined by*
$$Af = f(0)$$
is a bounded linear functional.

(2) *Let $X = C([0,1], \mathbb{R})$. The transformation $A : X \to \mathbb{R}$ defined by*
$$Af = \int_0^1 f(x)dx$$
is a bounded linear functional.

Definition 2.1.12. *Let X be a normed vector space and let f be a linear functional on X which is not identically zero. The kernel of f, $\ker f$, is called a* ***hyperplane****.*

In general, if $A \in B(X,Y)$, then $\text{ran}A$ may be closed or dense or none of the latter. However, it is easy to see that if $A \in B(X,Y)$, then $\ker A$ is always closed; and that if $A \in L(X,Y)$ and $\ker A$ is closed, then A need not be continuous (find an example!). The following result tells us more about the particular case of linear functionals.

Theorem 2.1.13. *(a proof is given in Exercise 2.3.20) Let X be a normed vector space. Let f be a non-identically null linear functional on X. Then*

(1) *$\ker f$ is closed iff f is continuous.*
(2) *f is (everywhere) discontinuous iff $\ker f$ is dense in X.*

Having defined bounded linear operators, natural questions are: Is $B(X,Y)$ a vector space? Can it be normed?

We already know from a Linear Algebra Course that $L(X,Y)$ is a vector space.

Since $B(X,Y) \subset L(X,Y)$, the next result tells us that $B(X,Y)$ is actually a vector space.

Proposition 2.1.14. *Let X, Y be normed vector spaces over $\mathbb{K}$. For any $A, B \in B(X,Y)$, we have*

$$A + B \in B(X,Y) \text{ and } \alpha A \in B(X,Y)$$

whenever $\alpha \in \mathbb{K}$, that is, $B(X,Y)$ is a linear subspace of $L(X,Y)$.

Next, we show that in fact $B(X,Y)$ can also be normed. But first, we choose a candidate to be a norm on $B(X,Y)$.

Definition 2.1.15. *Let $(X, \|\cdot\|_X)$ and $(Y, \|\cdot\|_Y)$ be normed vector spaces and let $A \in B(X,Y)$. Set*

$$\|A\| = \inf\{\alpha \geq 0 : \ \|Ax\|_Y \leq \alpha \|x\|_X, \ x \in X\}.$$

*Then $\|A\|$ is called the "**norm**" of A.*

We wrote a norm in quotation marks since we did not have the following result yet:

Proposition 2.1.16. *Let X and Y be normed vector spaces and let $A \in B(X,Y)$. Then $A \mapsto \|A\|$ (just defined) satisfies:*

1. *$\|A\| \geq 0$ for all $A \in B(X,Y)$,*
2. *$\|\alpha A\| = |\alpha| \|A\|$ for all $A \in B(X,Y)$ and for all $\alpha \in \mathbb{K}$,*
3. *$\|A + B\| \leq \|A\| + \|B\|$ for all $A, B \in B(X,Y)$.*

*In other words, $A \mapsto \|A\|$ is a **norm** on $B(X,Y)$.*

The following result gives us a wide choice of *equal* norms to use on $B(X,Y)$.

Theorem 2.1.17. *Let $A \in B(X,Y)$. Then the following are all **equal norms** (all denoted by $\|A\|$) on $B(X,Y)$:*

$$\inf\{\alpha \geq 0 : \ \|Ax\| \leq \alpha \|x\|, \ x \in X\},$$

$$\sup\left\{\frac{\|Ax\|}{\|x\|} : \ x \in X, \ x \neq 0\right\},$$

$$\sup\{\|Ax\| : \ x \in X, \ \|x\| \leq 1\}$$

and

$$\sup\{\|Ax\| : \ x \in X, \ \|x\| = 1\}.$$

Remark. It is a common practise to write $A \in B(X,Y)$ iff $\|A\| < \infty$, and $\|A\| = \infty$ in case of unboundedness.

Remark. When dealing with topological concepts on $B(X)$ like closedness and openness among others, the norm by default is the operator norm unless otherwise specified.

Examples 2.1.18.

(1) *Let $I : X \to X$ be the identity operator on X. Then clearly* $\|I\| = 1$.
(2) *Let $X = C([0,1], \mathbb{C})$ be equipped with the supremum norm. The functional $A : X \to \mathbb{C}$ defined by*

$$Af = f(0)$$

has as a norm $\|A\| = 1$.

One of the immediate results is the following:

COROLLARY 2.1.19. *Let $(X, \|\cdot\|_X)$ and $(Y, \|\cdot\|_Y)$ be normed vector spaces and let $A \in B(X,Y)$. Then*

$$\|Ax\|_Y \leq \|A\|\|x\|_X, \ \forall x \in X.$$

It is known that the composition "$\circ$" of two linear and bounded operators stays linear and bounded. The following result gives an estimation of the norm of $A \circ B$:

PROPOSITION 2.1.20. *(see Exercise 2.3.6) Let $A, B \in B(X)$. Then $A \circ B \in B(X)$ and*

$$\|A \circ B\| \leq \|A\|\|B\|.$$

In Operator Theory, we usually make use of the following convention:

DEFINITION 2.1.21. *Let $A, B \in B(X)$. The composition $A \circ B$ is called the* ***product*** *of A with B and may be denoted just by AB.*

As readers are already wary: $AB \neq BA$ in general. So we have the fundamental concept:

DEFINITION 2.1.22. *Let $A, B \in B(X)$. We say that A* ***commutes*** *with B (or that A and B* ***commute****) if*

$$AB = BA, \text{ i.e. } ABx = BAx$$

for all $x \in X$.

REMARK. *Forewarned is forearmed:* For example,

$$(A - B)(A + B) \neq A^2 - B^2$$

unless $AB = BA$.

We have the following notion related to commutativity:

DEFINITION 2.1.23. *Let $A, B \in B(X)$. The* ***commutator*** *of A, B is defined by*

$$[A, B] = AB - BA.$$

Remarks. It is clear that:

(1) A commutes with B iff $[A, B] = 0$.
(2) $[A, B] = -[B, A]$ for any $A, B \in B(X)$.

We give a closely related notion.

Definition 2.1.24. *Let $\mathcal{C} \subset B(X)$. The **commutant** of $\mathcal{C}$, denoted by $\mathcal{C}'$ is defined by*

$$\mathcal{C}' = \{A \in B(X) : \ AC = CA, \ \forall C \in \mathcal{C}\}.$$

*The **double commutant** of $\mathcal{C}$ is defined by $\mathcal{C}'' = \{\mathcal{C}'\}'$.*

Remark. If we say the commutant of a certain $B \in B(X)$ (also denoted by $\{B\}'$), then we mean

$$\{B\}' = \{A \in B(X) : \ AB = BA\}.$$

A proof of the following result may be found in Exercises 2.3.11 (or Exercise 7.3.21).

Proposition 2.1.25. *If $A, B \in B(X)$, then $[A, B] \neq I$.*

Remark. In general,

$$\|AB\| \neq \|A\|\|B\|$$

even when $A = B$. There are, however, situations in which one has $\|A^2\| = \|A\|^2$ or more generally $\|A^n\| = \|A\|^n$ ($n \in \mathbb{N}$) (see Chapter 8), where A^n is defined next:

Definition 2.1.26. *Let $A \in B(X)$. We define the **iterated operator**, denoted by A^n, by composing A with itself n times. In other words, A^n is defined as follows:*

$$A^0 = I, \ A^{n+1} = A^n A$$

where I is the identity operator on X.

Proposition 2.1.27. *(see Exercise 2.3.7) Let $A \in B(X)$. Then $A^n \in B(X)$ for any $n \in \mathbb{N}$. Moreover,*

$$\|A^n\| \leq \|A\|^n$$

for all n.

Example 2.1.28. *Let $A \in B(X)$ and let $p, n \in \mathbb{N}$. Then*

$$(A^p)^n = A^{pn} \text{ and } A^{p+n} = A^p A^n.$$

Remark. Let $A \in B(X)$. If $p(x) = a_0 + a_1 x + \cdots + a_n x^n$ is a polynomial, then we can define symbolically $p(A)$ as

$$p(A) = a_0 I + a_1 A + \cdots + a_n A^n$$

where I is the identity operator on X. We content ourselves with this at the moment. More can be found in later chapters.

Corollary 2.1.29. *Let $A \in B(X)$. If $p(x) = a_0 + a_1 x + \cdots + a_n x^n$ is a polynomial on $\mathbb{C}$ say, then $p(A) \in B(X)$.*

Remark. In the same spirit, there is a so-called composition operator which we prefer to state here so that any eventual ambiguity or confusion should disappear right away.

Let X and Y be two compact Hausdorff spaces and let $C(X)$ (resp. $C(Y)$) be the space of continuous functions on X (resp. Y), and let $\varphi : X \to Y$ be a continuous function. The operator $C_\varphi : C(X) \to C(Y)$ defined by

$$C_\varphi f = \varphi \circ f, \text{ i.e. } \forall x \in X,\ C_\varphi f(x) = \varphi \circ f(x) = \varphi(f(x))$$

*is called the **composition operator** induced by φ.*

The previous definition is not special to $C(X)$. We may define C_φ on other spaces like L^p. See e.g. [**39**].

Having normed $B(X,Y)$, another natural question is: When is $B(X,Y)$ a Banach space? One may conjecture that this is the case if both X and Y are Banach spaces. In fact, less than that suffices.

But, when we talk about Banach spaces, we will be using Cauchy sequences and convergence. So, if (A_n) is a sequence in $B(X)$, how we define its convergence? Using $\|A_n - A\|_{B(X)} \to 0$ or $\|(A_n - A)x\|_X \to 0$ $(x \in X)$?

These are different concepts as we have:

Definition 2.1.30. *Let X be a Banach space and let (A_n) be a sequence in $B(X)$.*

(1) *We say that (A_n) **converges in norm (or uniformly)** to $A \in B(X)$ if*

$$\lim_{n\to\infty} \|A_n - A\|_{B(X)} = 0.$$

(2) *We say that (A_n) **strongly converges** to $A \in B(X)$ if*

$$\lim_{n\to\infty} \|(A_n - A)x\|_X = 0$$

for each $x \in X$.

Remark. More on this can be found in Subsection 3.1.8 of the next chapter. In particular, the fact that the convergence in norm implies the strong convergence, but not vice versa. We content ourselves with this at the moment and we give the following result:

Proposition 2.1.31. *(see Exercise 2.3.4) Let X be a Banach space and let $A, B \in B(X)$. Then $(A, B) \mapsto AB$ is continuous with respect to the convergence in norm as well as with respect to the strong convergence.*

Now, we give the result on the completeness of $B(X, Y)$:

Theorem 2.1.32. *(see Exercise 2.3.5) Let X be a normed vector space. If Y is a Banach space, then so is $B(X, Y)$.*

Remark. What about the converse? This is another natural and interesting question to ask whether $B(X, Y)$ being Banach entails that Y is Banach? See the "True or False" section for a discussion on this.

We have already talked about interpolation among L^p spaces (Proposition 1.1.68). Now, we give one striking result on interpolation of operators. Its proof and more results may be found in **[179]** or **[204]**.

Theorem 2.1.33. *(**Riesz-Thorin**) Let $1 \le p_0, p_1, q_0, q_1 \le \infty$ and let $t \in (0, 1)$ be such that*

$$\frac{1}{p} = \frac{1-t}{p_0} + \frac{t}{p_1} \text{ and } \frac{1}{q} = \frac{1-t}{q_0} + \frac{t}{q_1}.$$

If $A \in B(L^{p_0}, L^{q_0})$ and $A \in B(L^{p_1}, L^{q_1})$, then $A \in B(L^p, L^q)$. More precisely, if

$$\|Af\|_{q_0} \le M_0 \|f\|_{p_0},\ \forall f \in L^{p_0}$$

and

$$\|Af\|_{q_1} \le M_1 \|f\|_{p_1},\ \forall f \in L^{p_1},$$

then

$$\|Af\|_q \le M_0^{1-t} M_1^t \|f\|_p,\ \forall f \in L^p.$$

The last stop is at the definition of a bilinear operator.

Definition 2.1.34. *Let X, Y and Z be three normed vector spaces over the same field $\mathbb{F}$. Let $A : X \times Y \to Z$ be a mapping. We say that A is **bilinear** if for all $x, y \in X$, $u, v \in Y$, $\alpha, \beta \in \mathbb{F}$:*

$$A(\lambda x + \mu y, u) = \lambda A(x, u) + \mu A(y, u)$$

and

$$A(x, \lambda u + \mu v) = \lambda A(x, u) + \mu A(x, v),$$

that is, if A is linear in each variable.

If on the right hand side of the second equation, λ and μ comes out as $\overline{\lambda}$ and $\overline{\mu}$ (when $\mathbb{F} = \mathbb{C}$), then A will be called ***sesquilinear****.*

If $Z = \mathbb{R}$ or $Z = \mathbb{C}$, then we may speak of a ***bilinear form (or functional)****.*

If $Z = \mathbb{C}$, then each sesqulinear mapping is called a ***sesquilinear form (or functional)****.*

Finally, we say that a bilinear (or sesquilinear) mapping A is ***bounded*** *if*

$$\|A(x,y)\|_Z \le M\|x\|_X\|y\|_Y$$

for some $M \ge 0$ and all $(x,y) \in X \times Y$.

The set of all bounded bilinear mappings from $X \times Y$ into Z is denoted by $B(X,Y;Z)$.

EXAMPLE 2.1.35. *The matrix multiplication $\mathcal{M}_{m\times n} \times \mathcal{M}_{n\times p} \to \mathcal{M}_{m\times p}$ is bilinear.*

We finish with the norm of a bounded bilinear mapping.

DEFINITION 2.1.36. *Let A be a bounded bilinear mapping from $X \times Y$ into Z, i.e.*

$$\|A(x,y)\|_Z \le M\|x\|_X\|y\|_Y$$

for some $M \ge 0$ and all $(x,y) \in X \times Y$. The infimum of all M satisfying the previous inequality is called the ***norm of the bounded bilinear mapping*** *A. It is denoted by $\|A\|$.*

THEOREM 2.1.37. *Let $A \in B(X,Y;Z)$. Then*

$$\begin{aligned}\|A\| &= \sup\left\{\frac{\|A(x,y)\|}{\|x\|\|y\|} : x \in X, x \ne 0,\ y \in Y, y \ne 0\right\}\\ &= \sup\{\|A(x,y)\| : x \in X, \|x\| \le 1,\ y \in Y, \|y\| \le 1\}\\ &= \sup\{\|A(x,y)\| : x \in X, \|x\| = 1,\ y \in Y, \|y\| = 1\}.\end{aligned}$$

Also,

$$\forall (x,y) \in X \times Y : \ \|A(x,y)\| \le \|A\|\|x\|\|y\|.$$

2.1.2. Dual Spaces.

DEFINITION 2.1.38. *Let X, Y be two normed vector spaces, and let $A \in B(X,Y)$. We say that A is an* ***isometry*** *if*

$$\|Ax\| = \|x\|,\ \forall x \in X.$$

EXAMPLE 2.1.39. *The identity operator is an isometry.*

REMARK. It is clear that an isometry is one-to-one. Is it onto? The answer is no in general. But if it is so, then we have the next definition.

DEFINITION 2.1.40. *Let X, Y be two normed vector spaces, and let $A \in B(X,Y)$. If A is a surjective isometry, then we call it an* ***isometric isomorphism****. In such case, X and Y are said to be* ***isometrically isomorphic****. We also say that X may be* ***identified*** *with Y.*

DEFINITION 2.1.41. *Let $\mathbb{F} = \mathbb{R}$ or $\mathbb{C}$, and let X be a normed vector space over $\mathbb{F}$. The space $B(X, \mathbb{F})$ is called the* ***dual space*** *of X. It is denoted by X', that is, $B(X, \mathbb{F}) = X'$.*

REMARK. The dual in the previous definition may also be called the topological dual as it involves continuity. The reader is already aware of the algebraic dual X^*, that is, $X^* = L(X, \mathbb{K})$. The reader should also be wary of the American literature, in which they usually use X^* for the (topological) dual.

REMARK. Dual spaces have many applications in Mathematics and Physics. See [**178**].

The following corollary is a consequence of Theorem 2.1.32.

COROLLARY 2.1.42. *If X is a normed vector space, then X' is a Banach space.*

Now, we turn to examples. The following example will be treated in Exercise 2.3.24.

EXAMPLE 2.1.43. *$(\ell^1)'$ may be identified with ℓ^∞, written $(\ell^1)' = \ell^\infty$.*

REMARK. $(\ell^\infty)' \neq \ell^1$.

THEOREM 2.1.44. *If $1 < p < \infty$, then $(\ell^p)' = \ell^q$, where $\frac{1}{p} + \frac{1}{q} = 1$.*

Now we turn to L^p spaces. First, the following is just a consequence of the Hölder Inequality.

PROPOSITION 2.1.45. *Let $1 \leq p \leq \infty$ and let q be its conjugate exponent, i.e. $\frac{1}{p} + \frac{1}{q} = 1$. If $g \in L^q(\mathbb{R}^n)$, then*

$$\ell(f) = \int_{\mathbb{R}^n} f(x)g(x)dx$$

defines a bounded linear functional on $L^p(\mathbb{R}^n)$, i.e. $\ell \in [L^p(\mathbb{R}^n)]'$ and such that

$$\|\ell\| \leq \|g\|_{L^q}.$$

COROLLARY 2.1.46. *If $1 \leq p \leq \infty$, then $L^q \subset (L^p)'$.*

THEOREM 2.1.47. *Let $1 \leq p < \infty$, $\frac{1}{p} + \frac{1}{q} = 1$. Then $(L^p)' = L^q$, i.e. for each bounded linear function on L^p, there exists a unique $g \in L^q$ such that*

$$\ell(f) = \int_{\mathbb{R}^n} f(x)g(x)dx$$

for all $f \in L^p$. Moreover,

$$\|\ell\| = \|g\|_{L^q}.$$

REMARK. It is important to emphasize that $(L^\infty)' \neq L^1$, that is, not every bounded linear functional on L^∞ may be represented as $\ell(f) = \int fg$ for some $g \in L^1$. A counterexample may be found in Exercise 18 (Page 191) of [**211**]. By Corollary 2.1.46, we only have $L^1 \subset (L^\infty)'$.

REMARK. Duality is not very much within the scope of the present manuscript. There are books treating it with great detail as [**34**], [**178**], [**190**] and [**211**].

2.1.3. The Fourier Transform. The Fourier transform is a very useful tool which is appreciated by mathematicians and other scientists. One of its many great features is that it "converts differentiation into multiplication".

DEFINITION 2.1.48. *Let f be in $L^1(\mathbb{R})$. The **Fourier transform** of f is defined by*

$$\hat{f}(t) = \frac{1}{\sqrt{2\pi}} \int_{-\infty}^{\infty} f(x)e^{-itx}dx = \int_{-\infty}^{\infty} f(x)e^{-itx}d\mu(x)$$

where $i = \sqrt{-1}$.

REMARK. The listed results in this subsection are with respect to $d\mu(x)$.

REMARK. In case of $f \in L^1(\mathbb{R}^n)$, we define $\hat{f}$ as

$$\hat{f}(t) = \frac{1}{(2\pi)^{\frac{n}{2}}} \int_{\mathbb{R}^n} f(x)e^{-it \cdot x}dx$$

where $x \cdot t = x_1t_1 + \cdots + x_nt_n$.

The next theorem gathers a few basic properties of the Fourier transform:

THEOREM 2.1.49. *(see [**189**], [**203**] and [**204**])*

(1) *$\hat{f}$ is linear.*

(2) *$\hat{f}$ is continuous from $L^1(\mathbb{R})$ into $L^\infty(\mathbb{R})$, i.e.*

$$\exists M > 0: \ \|\hat{f}\|_{L^\infty(\mathbb{R})} \le M\|f\|_{L^1(\mathbb{R})}$$

for all $f \in L^1(\mathbb{R})$.

(3) *If $f \ge 0$, then*

$$\|\hat{f}\|_{L^\infty(\mathbb{R})} = \|f\|_{L^1(\mathbb{R})}.$$

(4) *If $g(x) = -ixf(x)$ and $g \in L^1(\mathbb{R})$, then $\hat{f}$ is differentiable and $\hat{f}'(t) = \hat{g}(t)$. Also, if $f, f' \in L^1(\mathbb{R})$, then $\widehat{f'} = it\hat{f}$.*

Proposition 2.1.50. *(**Inversion Formula,** [**189**]) Let $f, \hat{f} \in L^1(\mathbb{R})$. If*

$$g(x) = \int_{\mathbb{R}} \hat{f}(t)e^{itx}d\mu(t),$$

then $f = g$ a.e.

Proposition 2.1.51. *(**Uniqueness Theorem,** [**189**]) Let $f \in L^1(\mathbb{R})$ and $\hat{f}(t) = 0$ for all $t \in \mathbb{R}$, then $f(x) = 0$ a.e.*

Having defined the Fourier transform on $L^1(\mathbb{R})$, we want to extend this definition to $L^2(\mathbb{R})$ functions (can we? Why L^2?). The idea in general is to work on dense subsets (as $C_0^\infty(\mathbb{R})$ or $\mathcal{S}(\mathbb{R})$) of $L^2(\mathbb{R})$, then consider an extension to $L^2(\mathbb{R})$. Here, we choose the dense subset $L^1(\mathbb{R}) \cap L^2(\mathbb{R})$ in $L^2(\mathbb{R})$. Then, it may be shown that $f \mapsto \hat{f}$ is an isometry on $L^1(\mathbb{R}) \cap L^2(\mathbb{R})$. Therefore, using Theorem 2.1.104 below, we extend it to the whole of $L^2(\mathbb{R})$.

Theorem 2.1.52. *(**Plancherel Theorem**) Let $f \in L^1(\mathbb{R}) \cap L^2(\mathbb{R})$. Then $\hat{f} \in L^2(\mathbb{R})$ and we have $\|\hat{f}\|_{L^2(\mathbb{R})} = \|f\|_{L^2(\mathbb{R})}$. Then $f \mapsto \hat{f}$ has a unique extension to $L^2(\mathbb{R}) \to L^2(\mathbb{R})$ such that for all $f \in L^2(\mathbb{R})$*

$$\|\hat{f}\|_{L^2(\mathbb{R})} = \|f\|_{L^2(\mathbb{R})}.$$

Moreover, $f \mapsto \hat{f}$ is surjective isometry from $L^2(\mathbb{R})$ onto $L^2(\mathbb{R})$.

Remarks.

(1) We continue to denote the Fourier transform by $\hat{f}$ when $f \in L^1(\mathbb{R}) \cup L^2(\mathbb{R})$.

(2) If $f \in L^2(\mathbb{R})$, and $f \notin L^1(\mathbb{R})$, then $\hat{f}$ is to be understood as an $L^2(\mathbb{R})$-limit as follows ([**189**]): If

$$g_k(t) = \int_{-k}^{k} f(x)e^{-ixt}d\mu(x) \text{ and } h_k(t) = \int_{-k}^{k} \hat{f}(x)e^{ixt}d\mu(x),$$

then as k tends to ∞, we have

$$\|g_k - \hat{f}\|_2 \to 0 \text{ and } \|h_k - f\|_2 \to 0.$$

(3) All the previous results remain true in $\mathbb{R}^n$.
(4) We still have on $L^2(\mathbb{R})$ the important property that the Fourier transform of f' becomes a product by $t\hat{f}$ up to a factor. This also applies to partial derivatives.

PROPOSITION 2.1.53. *Let $n \geq 1$. If $f \in L^2(\mathbb{R}^n)$ and $\hat{f} \notin L^1(\mathbb{R}^n)$ with $\hat{f} \geq 0$, then $f \notin L^\infty(\mathbb{R}^n)$.*

One last word on the Fourier transform. We have defined it on $L^1(\mathbb{R})$ as well as on $L^2(\mathbb{R})$. What happens if we want to define it on other L^p spaces? The next result gives us an answer. Its proof is very simple using the Riesz-Thorin Theorem (Theorem 2.1.33).

COROLLARY 2.1.54. ***(Hausdorff-Young Inequality)*** *If $f \in L^p(\mathbb{R}^n)$, $1 \leq p \leq 2$, then $\hat{f} \in L^q(\mathbb{R}^n)$ (with $\frac{1}{p} + \frac{1}{q} = 1$) and*

$$\|\hat{f}\|_q \leq C\|f\|_p.$$

REMARK. Mathematicians are usually interested in best constants in inequalities. The best constant C in the previous result was found in [**18**].

2.1.4. Invertibility. Invertibility is first defined in the usual fashion. Then we will see how the particular features of $B(X)$ help us to facilitate the concept of invertibility.

DEFINITION 2.1.55. *Let $A \in B(X)$ where X is normed vector space. We say that A is* ***invertible*** *if there exists a $B \in B(X)$ such that*

$$AB = BA = I,$$

where I is the identity operator on X. B is then called the ***inverse*** *of A.*

REMARK. What the previous definition is telling us is that A is invertible if it is bijective and B is linear and bounded! The bijectivity alone of A is, in general, not sufficient to guarantee its invertibility. See Exercise 2.3.50 (cf. the Banach Isomorphism Theorem below).

For more practical results, see Theorem 2.1.64 and Corollary 2.1.65 (see also Exercise 2.3.25 for an application of the latter results).

The next proposition gives us straightforward properties of invertible elements in $B(X)$.

PROPOSITION 2.1.56. *Let $A, B \in B(X)$ be invertible. Then*

(1) *A^{-1} is invertible and $(A^{-1})^{-1} = A$.*
(2) *AB is invertible and*

$$(AB)^{-1} = B^{-1}A^{-1}.$$

(3)

$$AB = BA \Longleftrightarrow A^{-1}B^{-1} = B^{-1}A^{-1}.$$

In some cases, it may be quite practical to use the following definition:

Definition 2.1.57. *Let $A, B, C \in B(X)$. We say that B is the* ***left inverse*** *of A if $BA = I$. We then say that A is* ***left invertible****.*

We say that C is the ***right inverse*** *of A if $AC = I$. We then say that A is* ***right invertible****.*

The following proposition tells us why the previous definition is so important.

Proposition 2.1.58. *(see Test 2.4.3) If $A \in B(X)$ possesses a left inverse B and a right inverse C, then $B = C$ and hence A is invertible.*

Corollary 2.1.59. *If $A \in B(X)$ is invertible, then it has a unique inverse.*

Remark. For more related results, see Exercise 4.3.24, Exercise 6.3.26 and Exercise 9.5.2.

The following definition is known to readers in Matrix Theory:

Definition 2.1.60. *Let $A, B \in B(X)$. We say that A et B are* ***similar*** *if there exists an invertible operator $P \in B(X)$ such that*

$$P^{-1}AP = B.$$

Next, we introduce the notion of boundedness below:

Definition 2.1.61. *Let $A \in B(X)$ where X is normed vector space. We say that A is* ***bounded below*** *if there exists an $\alpha > 0$ such that*

$$\|Ax\| \geq \alpha\|x\|$$

for all $x \in X$.

It is clear that non-zero multiples of the identity are bounded below. More generally,

PROPOSITION 2.1.62. *Let X be a normed vector space and let $A \in B(X)$ be invertible. Then for all $x \in X$,*

$$\|Ax\| \geq \|A^{-1}\|^{-1}\|x\|,$$

that is, A is bounded below.

We have the following interesting result (a proof is given in Exercise 2.3.12).

PROPOSITION 2.1.63. *Let X be a Banach space and let $A \in B(X)$ be bounded below. Then* $\operatorname{ran}(A)$ *is closed.*

We may now state a characterization of invertibility of bounded operators on a Banach space.

THEOREM 2.1.64. *Let X be a Banach space and let $A \in B(X)$. Then A is invertible iff* $\operatorname{ran}(A)$ *is dense and A is bounded below.*

COROLLARY 2.1.65. *Let X be a Banach space and let $A \in B(X)$. Then A is not invertible if* ***either*** $\operatorname{ran}(A)$ *is not dense in X* ***or*** *if there exists a sequence (x_n) in X such that $\|x_n\| = 1$ for all n but $\lim\limits_{n\to\infty} Ax_n = 0$.*

For more on invertibility readers should consult the Neumann Series Subsection or the Banach Isomorphism Theorem (Corollary 2.1.109.). In particular, the reader should pay attention to the remark below Corollary 2.1.109.

2.1.5. Series in Banach Spaces. If X is a normed vector space, and if $x_1, x_2, \cdots, x_n \in X$, then it is clear what we mean by $\sum_{k=1}^{n} x_k$. The natural question is how to pass to infinity?

DEFINITION 2.1.66. *If X is a normed vector space and $x_1, x_2, \cdots, x_n$ are in X, then*

$$S_n := \sum_{k=1}^{n} x_k$$

is called the ***sequence of partial sums****.*

DEFINITION 2.1.67. *Let $(X, \|\cdot\|)$ be a normed vector space. We say that the series $\sum_{n=1}^{\infty} x_n$ converges in $(X, \|\cdot\|)$ if the sequence of partial sums converges in $(X, \|\cdot\|)$, that is, if there exists an $x \in X$ such that*

$$\lim_{n\to\infty} \left\| \sum_{k=1}^{n} x_k - x \right\| = 0.$$

Then x is called the ***sum*** *of the series $\sum_{n=1}^{\infty} x_n$.*

If the positive series $\sum_{n=1}^{\infty}\|x_n\|$ converges, then we say that the series $\sum_{n=1}^{\infty}x_n$ is ***absolutely convergent****.*

Remark. The good thing about absolutely convergent series is that although the normed vector space X may be as general as it can be, the norm is a positive quantity, and hence to see whether a given series is (or is not) absolutely convergent, we may use any of the known tests for positive series such as: comparison, ratio or root tests.

So far, everything seems to be just as in the case of real or complex numbers. Here is a little surprise: Does every absolutely convergent series converge? The answer is no in general! A counterexample may be found in Exercise 2.3.29. However, we have the following result (whose proof is supplied in Exercise 2.3.28).

Theorem 2.1.68. *(cf. Theorem 3.1.21) Let X be a normed vector space. Then X is a Banach space iff every absolutely convergent series in X converges.*

Besides, when $\sum_{n=1}^{\infty}\|x_n\|$ converges we have

$$\left\|\sum_{n=1}^{\infty}x_n\right\|\leq\sum_{n=1}^{\infty}\|x_n\| \text{ (\textbf{\textit{Generalized Triangle Inequality}})}.$$

Let us now give some examples of series in Banach spaces.

Example 2.1.69. *(see Exercise 2.3.32 for more details) Let X be a Banach space and let $A\in B(X)$. The series*

$$\sum_{n=0}^{\infty}\frac{A^n}{n!}$$

converges absolutely, and hence it converges. Its sum is denoted by e^A and it is called the ***exponential of*** *A.*

Example 2.1.70. *Let I be the identity on X and let $\alpha\in\mathbb{C}$. Then*

$$e^{\alpha I}=e^{\alpha}I.$$

Therefore,

$$e^{i\pi I}+I=0.$$

Remark. The function $t\mapsto e^{tA}$, where $t\in\mathbb{R}$ and $A\in B(X)$, is of great importance in the theory of Linear Differential Equations.

Remark. One of the remarkable results in Basic Real Analysis is the following limit

$$\lim_{n\to\infty}\left(I+\frac{1}{n}x\right)^n=e^x$$

where $x \in \mathbb{R}$. This limit can be carried over to $B(X)$, i.e. with A (in $B(X)$) instead of x. See Exercise 2.3.34.

Here are other similar examples.

EXAMPLES 2.1.71. *Let $A \in B(X)$, where X is a Banach space. Then*

$$\sin A = \sum_{n=0}^{\infty}(-1)^n \frac{A^{2n+1}}{(2n+1)!} \text{ and } \cos A = \sum_{n=0}^{\infty}(-1)^n \frac{A^{2n}}{(2n)!},$$

*called the **sine** and **cosine** of the operator A respectively.*

PROPOSITION 2.1.72. *Let $A \in B(X)$. Then*

$$e^{iA} = \cos A + i \sin A.$$

Also,

$$\cos^2 A + \sin^2 A = I.$$

REMARK. Similarly, we may also define $\sinh A$, $\cosh A$, etc...See also Exercise 2.5.4 or Exercise 11.5.2.

REMARK. The natural question is whether we can define $f(A)$ for any f and any A? The answer is that we can do a lot but not always. For more details, see Chapter 11.

It is quite handy when computing the exponential of an operator to meet an operator A whose power vanishes at some point. This concept, already known to readers from a Linear Algebra Course, is recalled here:

DEFINITION 2.1.73. *Let $A \in B(X)$. We say that A is **nilpotent** if $A^n = 0$ for some $n \in \mathbb{N}$. The smallest (which is the most important) $k \in \mathbb{N}$ satisfying $A^k = 0$ is called the **index of nilpotence** of A.*

A related definition is the following:

DEFINITION 2.1.74. *We say that the operator $A \in B(X)$ is **algebraic** if $p(A) = 0$ for some nonzero polynomial p.*

REMARK. Obviously, in case A is algebraic, then there is always a polynomial p of minimum degree such that $p(A) = 0$.

When working with series in $B(X)$, the following result will be so useful (it will be used without further notice):

PROPOSITION 2.1.75. *Let $A, B \in B(X)$. Let (A_n) be a sequence in $B(X)$ such that $\sum_{n=1}^{\infty} A_n = A$. Then*

$$\sum_{n=1}^{\infty} A_n B = AB \text{ and } \sum_{n=1}^{\infty} BA_n = BA.$$

Before finishing this subsection, we give a word on the Schauder basis and this seems to be the right place for it. It is well known that any vector x in $\mathbb{R}^n$ say, may be expressed uniquely as $x = \alpha_1 e_1 + \cdots \alpha_n e_n$, where $\alpha_1, \cdots, \alpha_n \in \mathbb{R}$ and $\{e_n : n = 1, \cdots, n\}$ is a basis in $\mathbb{R}^n$. Can we have this in an infinite dimensional space?

DEFINITION 2.1.76. *Let X be a Banach space on $\mathbb{F}$. A sequence (e_n) in X is said to be a **Schauder basis** if for any $x \in X$, there exists a unique sequence (α_n) (which may depend on x) in $\mathbb{F}$ such that*

$$x = \sum_{n=1}^{\infty} \alpha_n e_n, \text{ that is, } \lim_{n\to\infty} \left\| x - \sum_{k=1}^{n} \alpha_k e_k \right\| = 0.$$

EXAMPLE 2.1.77. *The set $\{e_n : n \in \mathbb{N}\}$, where*

$$e_1 = (1, 0, 0, \cdots),\ e_2 = (0, 1, 0, \cdots), \cdots,\ e_n = (0, \cdots, 0, \underset{nth}{1}, 0, \cdots), \cdots$$

constitutes a Schauder basis for ℓ^p, with $1 \leq p < \infty$.

Here are some properties related to Schauder bases.

THEOREM 2.1.78. *Every Banach space X with a Schauder basis is separable.*

REMARK. What about the converse? Banach himself asked whether every separable Banach space has a Schauder basis? This conjecture remained open for four decades. Then Enflo (remember this name!) in [**70**] answered this negatively by supplying a counterexample.

Having defined a Schauder basis, we would like to add another notion of a basis which generalizes naturally the finite-dimensional notion.

DEFINITION 2.1.79. *Let X be a vector space with a not necessarily finite dimension. A subset Y of X is called a **Hamel** basis if every $x \in X$ may be expressed uniquely as a finite linear combination of elements of Y.*

REMARK. The notion of a Hamel basis is purely algebraic whereas that of a Schauder basis needs Topology. This makes the concept of a Hamel basis a little insignificant in the analysis of an infinite dimensional setting. It is, however, used for instance for counterexamples.

EXAMPLE 2.1.80. *Let X be the vector space of all polynomials of with coefficients in $\mathbb{R}$. Then*

$$Y = \{x^n : n = 0, 1, 2, \cdots\}$$

constitutes a Hamel basis for X.

THEOREM 2.1.81. *Every nontrivial vector space has a Hamel basis.*

We finish this subsection with some interesting results.
First, recall the following definition:

DEFINITION 2.1.82. *Let X be a topological space and let $Y \subset X$. We say that Y is **nowhere dense** in X if $\overset{\circ}{\overline{Y}} = \varnothing$.*

Proofs of the following four results may be found in Exercise 2.3.44.

PROPOSITION 2.1.83. *Let X be a normed vector space. Every proper linear subspace of X has an empty interior.*

COROLLARY 2.1.84. *Let X be a normed vector space. Any proper and closed linear subspace of X is nowhere dense.*

COROLLARY 2.1.85. *Let X be a normed vector space. A linear subspace Y of X is either dense or nowhere dense.*

COROLLARY 2.1.86. *Let X be an infinite dimensional normed vector space which has a Hamel basis. Then this basis is uncountable.*

2.1.6. Neumann Series. We devote this subsection to an important series in Banach spaces.

We have already noted that invertibility in infinite dimensional spaces is not as easy as it used to be in their finite dimensional counterparts. However, there is a nice way for producing inverses (which also intervenes frequently in Integral Equations). It is based on the the following elementary result in real series:

We know that the series

$$1 + x + x^2 + \cdots + x^n + \cdots$$

converges if $|x| < 1$, and that its sum is given by $(1-x)^{-1}$. This basic result may be carried over to $B(X)$. First, we have a definition.

DEFINITION 2.1.87. *Let X be a Banach space and let $A \in B(X)$. We say that A is a **contraction** if $\|A\| \leq 1$; and a **pure (or strict) contraction** if $\|A\| < 1$.*

Here is the promised generalization:

THEOREM 2.1.88. *(see Exercise 2.3.30) Let X be a Banach space and let $A \in B(X)$ be a pure contraction, i.e. $\|A\| < 1$. Then $I - A$ is invertible and*

$$(I - A)^{-1} = \sum_{n=0}^{\infty} A^n \text{ (called the \textbf{Neumann Series})}$$

(the limit in the series being in $\|\cdot\|_{B(X)}$). Hence

$$\|(I-A)^{-1}\| \leq \frac{1}{1-\|A\|}.$$

Remark. The assumption "X being a Banach" is fundamental to ensure the convergence of the Neumann series.

Corollary 2.1.89. *Let X be a Banach space and let $A \in B(X)$ be such that $\|I-A\| < 1$. Then A is invertible and*

$$A^{-1} = I + \sum_{n=1}^{\infty} (I-A)^n.$$

Corollary 2.1.90. *Let X be a Banach space and let $A \in B(X)$. Let $\lambda \in \mathbb{C}$ be such that $|\lambda| > \|A\|$. Then $\lambda I - A$ is invertible and*

$$(\lambda I - A)^{-1} = \sum_{n=0}^{\infty} \lambda^{-n-1} A^n.$$

Remark. Obviously, if $A \in B(X)$ with $\|A\| < 1$, then $I + A$ too is invertible and

$$(I+A)^{-1} = \sum_{n=0}^{\infty} (-1)^n A^n.$$

This may be generalized to the following perturbation result:

Proposition 2.1.91. *(cf. Exercise 5.3.22) Let $A, B \in B(X)$. If A is invertible, then $A+B$ is invertible whenever $\|B\| < \|A^{-1}\|^{-1}$ and the inverse is given by*

$$(A+B)^{-1} = \sum_{n=0}^{\infty} (A^{-1}B)^n A^{-1}.$$

The following essential result will be needed later on (it is also important in e.g. Differential Calculus).

Theorem 2.1.92. *Let X be a Banach space and let $A \in B(X)$. Let $B_i(X)$ be the (group) of invertible elements of $B(X)$. Then*

1. *$B_i(X)$ is open in $B(X)$ (a proof is given in Exercise 2.3.39).*
2. *The map $A \mapsto A^{-1}$ from $B_i(X)$ into $B_i(X)$ is continuous (for a proof, see Exercise 2.3.40).*

2.1.7. Operator-Valued Analytic Functions. We saw in the definition of the Neumann Series, that we are working with functions from $\mathbb{C}$ into $B(X)$. More generally, we call them **operator-valued** (or **vector-valued**) **functions**. In this subsection, we go quickly through some essential background in the theory of vector-valued analytic functions. For details see e.g. [**104**] or [**114**]. See also [**84**].

A fundamental example of an operator-valued function is the exponential of an operator (already met in Example 2.1.69). We recall it for convenience.

EXAMPLE 2.1.93. *Let $z \in \mathbb{C}$. If X is a Banach space and $A \in B(X)$, then the series*

$$e^{zA} = \sum_{n=0}^{\infty} \frac{z^n}{n!} A^n$$

defines the exponential of zA.

We will need at some points to differentiate operator-valued functions. How do we do that? What about continuity?

DEFINITION 2.1.94. *Let X be a Banach space. Let $A(t)$ be defined for $t \in D$ (where D is a subset of either $\mathbb{R}$ or $\mathbb{C}$) and assume that it takes its values in $B(X)$. We say that $A(t)$ is **continuous** at $t = a$ if*

$$\lim_{t \to a} \|A(t) - A(a)\| = 0$$

and we shall write in such case $\lim_{t \to a} A(t) = A(a)$.

*The **derivative** of $A(t)$, denoted by $A'(t)$, is the limit*

$$\lim_{h \to 0} \frac{A(t+h) - A(t)}{h}$$

in $\| \cdot \|_{B(X)}$ when this limit exists.

*If this limit exists for all $t \in D$, then we say that $A(t)$ is **analytic** (or **holomorphic**) .*

*If $A(t)$ is analytic on $\mathbb{C}$, then we call it **entire**.*

EXAMPLE 2.1.95. *Define $A(t)$ from $\mathbb{R}$ say, to some normed vector space X by $A(t) = bI$ for some fixed element $b \in X$. Then $A'(t) = 0$.*

Here are some basic properties:

PROPOSITION 2.1.96. *Let X be a Banach space. Let $A(t)$ and $B(t)$ be defined from some domain D into $B(X)$ and let $x(t)$ be defined from D into X. Then*

$$\frac{d}{dt}[A(t)x(t)] = A'(t)x(t) + A(t)x'(t)$$

and

$$\frac{d}{dt}[A(t)B(t)] = A'(t)B(t) + A(t)B'(t)$$

(whenever everything is well-defined and the derivatives on the right exist).

PROPOSITION 2.1.97. *(a proof is included in the solution of Exercise 2.3.36) Let* $t \in \mathbb{C}$. *If* X *is a Banach space and* $A \in B(X)$*, then* e^{tA} *is entire and*

$$\frac{d}{dt}e^{tA} = Ae^{tA} = e^{tA}A.$$

REMARK. The previous exponential appears in the solution of certain differential equations on a Banach space. See Exercise 2.3.36

The Liouville Theorem has a vector-valued version.

THEOREM 2.1.98. ***(Liouville Theorem**, see e.g.* **[160]***) Let* X *be a normed vector space on* $\mathbb{C}$. *If* $f : \mathbb{C} \to X$, $z \mapsto f(z)$ *is entire and bounded, i.e. for some* $M > 0$ *and all* z, $\|f(z)\| \leq M$*, then* f *is contant.*

REMARK. The Liouville Theorem will be of great interest when proving that the spectrum of a bounded operator is not empty. It also helps in due time for proving the Fuglede-Putnam(-Rosenblum) Theorem.

2.1.8. The Four Pillars of Functional Analysis. In this subsection, we give some of the most fundamental theorems in Functional Analysis (most of them are due to Banach) and which have tremendous applications. They are four main results:

(1) Hahn-Banach Theorem.
(2) Banach-Steinhaus Theorem.
(3) Open Mapping Theorem.
(4) Closed Graph Theorem.

THEOREM 2.1.99. ***(Hahn-Banach, analytic form)*** *Let* X *be a vector space and let* $p : X \to \mathbb{R}$ *be a transformation satisfying:*

$$p(\alpha x) = \alpha p(x),\ \forall x \in X,\ \forall \alpha > 0,$$

and

$$p(x + y) \leq p(x) + p(y),\ \forall x, y \in X.$$

Now, let Y *be a linear subspace of* X *and let* $g : Y \to \mathbb{R}$ *be a linear transformation such that*

$$g(x) \leq p(x),\ \forall x \in Y.$$

Then, there exists a linear functional f, defined on X, which extends g, that is, $f(x) = g(x)$ for all $x \in X$ and such that

$$f(x) \leq p(x), \; \forall x \in X.$$

Remark. There is another form of the Hahn-Banach Theorem called the geometric form. We content ourselves to the analytic form. See e.g. [**34**].

The following consequence works for real and complex Banach spaces and that for the complex case an additional argument is needed.

Corollary 2.1.100. *Let Y be a linear subspace of a normed vector space X and let g be a bounded linear functional on Y. Then g may be extended to a bounded linear functional f on the whole of X such that*

$$\|f\|_{X'} = \|g\|_{Y'},$$

where X' and Y' are the topological duals of X and Y respectively.

The next result is also fundamental, in particular, it is a crucial tool for the converse of Theorem 2.1.32.

Corollary 2.1.101. *Let X be a normed vector space and let $x_0 \in X$. Then there exists $f_0 \in X'$ such that*

$$\|f_0\| = \|x_0\| \text{ and } f_0(x_0) = \|x_0\|^2.$$

Proposition 2.1.102. *Let Y be a linear subspace of a normed vector space X such that $\overline{Y} \neq X$. Then there exists $f \in Y'$, $f \neq 0$ such that*

$$f(x) = 0, \; \forall x \in Y.$$

Remark. The power of the previous result lies in the fact that it allows us to prove a given subspace Y is dense. To use it, we take a linear functional f on X such that $f = 0$ on Y. Then we show that $f = 0$ on all of X.

Example 2.1.103. *Let $1 < p < \infty$. Consider*

$$X = \left\{ x = (x_n) \in \ell^p : \sum_{n=0}^{\infty} x_n = 0 \right\}.$$

Then X is dense in ℓ^p (this question and more will be studied in Exercise 2.3.42).

We would like to give a small idea on the numerous and the tremendous applications of the Hahn-Banach Theorem. This is mainly quoted from [**34**]. In this reference, Brezis presented many many varied applications of the Hahn-Banach theorem. We just cite a few from Brezis book which, in his turn, only cited a few of the existing ones. Indeed, the Hahn-Banach Theorem has applications to:

(1) The Krein-Milman Theorem about compact convex sets.
(2) Partial Differential Equations, in particular, the existence of a fundamental solution of a differential operator $P(D)$ with constant coefficients (non identically null).
(3) Convex functions and applications to: Game Theory, Optimization, Economics, Mechanics and Plasma Theory, etc...
(4) Variational problems related to periodic solutions of Hamiltonian systems and non-linear equations of the vibrating strings.

We finish with an important discussion on how/when to extend bounded linear operators. Here is the problem: *Let X and Z be two Banach spaces, and let Y be a closed linear subspace of X. Let $B \in B(Y, Z)$. When do we have an extension of B to X, that is, when do we have the existence of $A \in B(X, Z)$ such that $A = B$ on Y?*

Here are some answers:

(1) Corollary 2.1.100 gave us an answer *only* in the case $Z = \mathbb{R}$.
(2) If $\dim Y < \infty$, then we may apply Corollary 2.1.100 to each component of the basis of Y.
(3) If X is a "Hilbert space", then we may use "$Y^{\perp}$" to complement X (to be defined in the next chapter).

As observed in **[34]**, a hard problem is to see when $\|A\|_{B(X,Z)} = \|B\|_{B(Y,Z)}$. The next result is closely related to what has just been said. It is very important and it is so used in Functional Analysis that it is not explicitly mentioned sometimes.

THEOREM 2.1.104. *Let X be a normed vector space, and let Z be a Banach space, and let Y be a* **dense** *linear subspace of X. If $A \in B(Y, Z)$, then there exists a* **unique** *$B \in B(X, Z)$ such that B coincides with A on Y and*

$$\|A\|_{B(X,Z)} = \|B\|_{B(Y,Z)}.$$

Another extension result is the following:

COROLLARY 2.1.105. *Let X and Y be two Banach spaces and let X' and Y' be two dense linear subspaces of X and Y respectively. If $A : X' \to Y'$ is a surjective isometry, then there exists a unique surjective isometry $\tilde{A} : X \to Y$ which extends A over X.*

The second pillar is more known as the **Principle of Uniform Boundedness** and it is due to Banach-Steinhaus. It is worth noticing that its proof as well as the proof of the Open Mapping Theorem (stated below), rely strongly on the Baire Category Theorem (cf. [**34**]) which we recall here for convenience (in one of its versions):

Theorem 2.1.106. ***(Baire)*** *If (X,d) is a non-empty complete metric space, and if $(X_n)_n$ is a sequence of closed subsets of X such that $\cup_{n=1}^{\infty} X_n = X$, then*

$$\exists n_0 \in \mathbb{N} : \overset{\circ}{X}_{n_0} \neq \varnothing.$$

Theorem 2.1.107. ***(Banach-Steinhaus)*** *Let X and Y be two Banach spaces and let $(A_i)_{i\in I}$ be a family (not necessarily countable) in $B(X,Y)$. Assume that*

$$\sup_{i\in I} \|A_i x\| < \infty, \ \forall x \in X.$$

Then

$$\sup_{i\in I} \|A_i\| < \infty,$$

that is, there is a constant $\alpha \geq 0$ such that

$$\|A_i x\| \leq \alpha \|x\|, \ \forall x \in X, \ \forall i \in I.$$

Next, we have the **Open Mapping Theorem**:

Theorem 2.1.108. *Let X and Y be two Banach spaces and let $A \in B(X,Y)$. If A is surjective, then there exists $\alpha > 0$ such that*

$$B'(0,\alpha) \subset A(B(0,1)),$$

where $B'(0,\alpha)$ is the open ball of center 0 and radius α in Y, and $B(0,1)$ is the unit ball in X.

Remark. A priori, there seems to be no apparent link between the result of the previous theorem and its name. The reader is asked in Exercise 2.3.47 to prove that any operator with the hypotheses of Theorem 2.1.108 is an open map.

Remark. We notice that the conclusion of the Open Mapping Theorem need not hold for bilinear mappings. In particular, it was asked by Rudin in [**187**] whether a continuous bilinear map from the product of two Banach spaces onto a Banach space must be open at the origin? The first counterexample was found by Cohen in [**46**] and it was a fairly non-trivial construction on an infinite-dimensional space. Much simpler was the counterexample found by Horowitz [**108**]. Indeed, what is striking about this example is that spaces are finite-dimensional, and usually interesting questions about Banach spaces are rather trivial in finite dimensions.

As an immediate consequence of the previous theorem, we have the following result (more commonly known as the "**Banach Isomorphism Theorem**")

COROLLARY 2.1.109. *Let X and Y be two Banach spaces and let $A \in B(X,Y)$ be bijective. Then $A^{-1} : Y \to X$ is also bounded.*

REMARK. Some may think what is so special with the previous result? Since A is bounded and bijective, it is certainly invertible! We have already observed that this is not necessarily true! There are many counterexamples (see e.g. Exercise 2.3.50). In fact, we do not say A is invertible unless we know that A^{-1} is also bounded.

Some authors, before checking that A^{-1} is bounded, call it an "algebraic inverse" (they use the word "algebraic" because the continuity of the inverse involves topology). I personally do not use this since even algebraically an inverse must belong to the group in which we are working. I propose, and certainly prefer, to call A^{-1} the **formal inverse** of A as soon as A is bijective (this is equivalent to saying that A is invertible in $L(X)$). Once one checks that A^{-1} is in effect bounded, then this formal inverse becomes the actual inverse (in $B(X)$).

A direct application of Corollary 2.1.109 is the following result concerning equivalent norms (a proof is given in Exercise 2.3.45).

COROLLARY 2.1.110. *Let X be a Banach space with respect to* ***two*** *norms $\|\cdot\|_1$ and $\|\cdot\|_2$. If*

$$\exists \alpha > 0,\ \|x\|_1 \leq \alpha \|x\|_2,\ \forall x \in X,$$

then these two norms are equivalent.

We close this saga with the **Closed Graph Theorem** which is an application of the Open Mapping Theorem (or the a direct application of the previous corollary). It is also a converse of Theorem 2.1.9.

THEOREM 2.1.111. *(see Exercise 2.3.46, cf. Exercise 2.3.49) Let X and Y be two Banach spaces. Let $A : X \to Y$ be linear such that its graph $G(A)$ is closed in $X \times Y$. Then $A \in B(X,Y)$.*

2.2. True or False: Questions

QUESTIONS. Comment on the following questions/statements and indicate those which are false and those which are true when this applies. Justify your answers.

(1) ℓ^∞ does not possess a Schauder basis...
(2) Let $A : \mathbb{R} \to \mathbb{R}$ (both $\mathbb{R}$ are endowed with the usual $|\cdot|$) be such that $A(x) = 2x$. Then is A bounded or not bounded?
(3) Let A be a bounded and invertible operator on a Banach space. Then

$$\|A^{-1}\| = \frac{1}{\|A\|}.$$

(4) A linear operator is either continuous everywhere or discontinuous everywhere.
(5) We saw in the definition of the operator norm that e.g. $\|A\| = \sup\{\|Ax\| : \|x\| = 1\}$, if $A \in B(X)$. Surely, we can replace "sup" by "max", can't we?
(6) Let X and Y be two normed vector spaces. Let $A \in L(X,Y)$. Then

$$\ker A \text{ is closed} \iff A \text{ is continuous.}$$

(7) Let X and Y be two normed vector spaces. If $B(X,Y)$ is a Banach space, then so is Y.
(8) Let $A \in B(X)$ such that $A^2 = I$. Then A is invertible.
(9) Let X be a Banach space. Let A be a bounded linear operator on X. Then

$$\|A\| < 1 \iff (I - A) \text{ is invertible.}$$

(10) Let X be a Banach space and let $A \in B(X)$. Then

$$\|A\| \leq 1 \iff \|Ax\| \leq \|x\|, \ \forall x \in X.$$

2.3. Exercises with Solutions

Exercise 2.3.1. Let $X = C([-2,3],\mathbb{R})$ equipped with the norm of uniform convergence. Let

$$A : f \mapsto Af = f(-2) - f(-1) + 3f(0) + 2f(1) + 3f(2) - 2f(3).$$

(1) Show that A is in $B(X,\mathbb{R})$.
(2) Find $\|A\|$.

Exercise 2.3.2. Let $\alpha = (\alpha_n) \in \ell^\infty$. Let $1 \leq p < \infty$.

(1) Show that $A : \ell^p \to \ell^p$ defined by

$$Ax = A(x_n) = (\alpha_n x_n)$$

is well-defined, linear, bounded and $\|A\| = \|\alpha\|_\infty$. *The operator A is called a "**Multiplication Operator**" (cf. Exercises 2.3.3, 3.3.43 & 10.3.9).*

(2) Show that $B : \ell^1 \to \mathbb{C}$ defined by

$$Bx = B(x_n) = \sum_{n=1}^{\infty} \alpha_n x_n$$

is well-defined, linear, bounded and $\|B\| = \|\alpha\|_\infty$.

Exercise 2.3.3. (cf. Exercises 2.5.1 & 10.3.9) Let φ be a continuous function on $[0,1]$ taking its values in $\mathbb{C}$. For all $f \in L^2([0,1])$ (works as well for $[a,b]$), set

$$(Af)(x) = \varphi(x)f(x).$$

We also call the operator A a "**Multiplication Operator**".

(1) Prove that A is linear and bounded.
(2) Find $\|A\|$.

Exercise 2.3.4. Let X be a Banach space and let $A, B \in B(X)$. Show that $(A, B) \mapsto AB$ is continuous (with respect to the topology of the operator norm).

What about continuity with respect to the strong operator topology (you may use for this question Theorem 2.1.107)?

Exercise 2.3.5. Show that if X is a normed vector space, and Y is a Banach space, then $B(X,Y)$ is a Banach space.

Exercise 2.3.6. Let A and B be in $B(X)$. Show that

$$\|BA\| \leq \|B\|\|A\|.$$

Exercise 2.3.7. Let $A \in B(X)$. Show that

$$\|A^n\| \leq \|A\|^n$$

for all $n \in \mathbb{N}$.

Exercise 2.3.8. Let $(A_k)_{k=1,\cdots,n}$ and $(B_k)_{k=1,\cdots,n}$ be both in $B(X)$. Let $p, q \geq 1$ be such that $\frac{1}{p} + \frac{1}{q} = 1$. Show that

$$\left\|\sum_{k=1}^{n} A_k B_k\right\| \leq \left(\sum_{k=1}^{n} \|A_k\|^p\right)^{\frac{1}{p}} \left(\sum_{k=1}^{n} \|B_k\|^q\right)^{\frac{1}{q}}.$$

Remark. A better estimate in the Cauchy-Schwarz case, i.e. when $p = q = 2$, may be found in Exercise 4.3.7.

Exercise 2.3.9. Let $A \in B(X)$. Show that $\lim_{n\to\infty} \|A^n\|^{\frac{1}{n}}$ exists and that

$$\lim_{n\to\infty} \|A^n\|^{\frac{1}{n}} = \inf_{n\in\mathbb{N}} \|A^n\|^{\frac{1}{n}}.$$

Exercise 2.3.10. Let X be a Banach space and let $A, B \in B(X)$ be such that $A^2 = A$, $B^2 = B$ and $AB = BA$.

(1) Show that for all $n \geq 0$: $(A - B)^{2n+1} = A - B$.
(2) Deduce that either $A = B$ or $\|A - B\| \geq 1$.

Exercise 2.3.11. (see also Exercise 7.3.21) Let X be a Banach space and let I be the identity operator on X. Do there exist $A, B \in B(X)$ such that $AB - BA = I$?

Hint: Show that $AB - BA = I$ implies that for any $n \in \mathbb{N}$: $A^n B - BA^n = nA^{n-1} \neq 0$.

Exercise 2.3.12. Let X be a Banach space and let Y be a normed vector space. Assume that $A \in B(X, Y)$ and that

$$\exists \alpha > 0,\ \forall x \in X:\ \|Ax\| \geq \alpha \|x\|.$$

Show that ran$A$ is closed.

Exercise 2.3.13. Let $(X, \|\cdot\|_X)$ and $(Y, \|\cdot\|_Y)$ be two normed vector spaces and let $A: X \to Y$ be linear.

(1) Show that

$$\|x\|_{G(A)} = (\|x\|_X^2 + \|Ax\|_Y^2)^{\frac{1}{2}}$$

defines a norm on X.

(2) Show that also

$$\|x\|_1 = \|x\|_X + \|Ax\|_Y$$

defines a norm on X.

Remark. The norm $\|\cdot\|_1$ is also called a **graph norm**.

(3) Are $\|\cdot\|_{G(A)}$ and $\|\cdot\|_1$ equivalent?

Exercise 2.3.14.

(1) Let $A \in L(X, Y)$, where X and Y are two normed vector spaces. Show that $A \in B(X, Y)$ whenever $\dim X < \infty$.
(2) Let X be the vector space of all polynomials on $[0, 1]$ with real coefficients. Let $A: X \to \mathbb{R}$ be defined by

$$Ap = p'(1),$$

where p' is the derivative of p. Show that A is unbounded (if X is equipped with the supremum norm).

Exercise 2.3.15. Let $X = C([0,1],\mathbb{R})$ be equipped with the supremum norm. Define $A : X \to \mathbb{R}$ and $B : X \to X$ by

$$Af = \int_0^1 f(x)dx \text{ and } [Bf](x) = \int_0^x f(y)dy$$

respectively, where $x \in [0,1]$ (B is called the **Volterra** operator and it will be studied in detail on $L^2[0,1]$ in Exercise 9.3.21).

(1) Show that $A \in B(X,\mathbb{R})$ and $B \in B(X)$.
(2) Show that

$$\|A\| = \|B\| = 1$$

Exercise 2.3.16. Let $X = C([0,1],\mathbb{R})$. Define on X the following two *norms*

$$\|f\|_1 = \int_0^1 |f(x)|dx, \ \|f\|_\infty = \sup_{0\le x\le 1} |f(x)|.$$

(1) Is the functional $A : (X, \|\cdot\|_\infty) \to \mathbb{R}$ defined by $Af = f(0)$ continuous?
(2) Is the functional $B : (X, \|\cdot\|_1) \to \mathbb{R}$ defined by $Bf = f(0)$ continuous?

Exercise 2.3.17. Let $X = C([0,1],\mathbb{R})$ be endowed with the supremum norm. Let w be a continuous function of a constant sign on $[0,1]$. Define $\varphi : X \to \mathbb{R}$ by

$$\forall f \in X: \ \varphi(f) = \int_0^1 f(x)w(x)dx.$$

(1) Show that $\varphi \in B(X,\mathbb{R})$.
(2) Show that $\|\varphi\| = \int_0^1 |w(x)|dx$.
(3) Does the result remain true if w does not have a constant sign on $[0,1]$?

Exercise 2.3.18. Let $X = C([-1,1],\mathbb{R})$ be endowed with the supremum norm. Let

$$\varphi(f) = \int_{-1}^0 f(x)dx - \int_0^1 f(x)dx.$$

(1) Show that φ is a linear and continuous functional on X.

(2) Show that $\|\varphi\| = 2$. **Hint:** you may use the sequence of functions (g_n) defined by

$$g_n(x) = \begin{cases} 1, & -1 \le x \le \frac{-1}{2n}, \\ -2nx, & \frac{-1}{2n} \le x \le \frac{1}{2n}, \\ -1, & \frac{1}{2n} \le x \le 1. \end{cases}$$

(3) Show that the norm is not attained, i.e. there is no $g \in X$ such that $\|g\|_\infty = 1$ and $\|\varphi\| = \varphi(g)$.

Exercise 2.3.19. Define the following *linear functional*

$$\varphi(f) = \int_0^1 xf(x)dx$$

on $X = C([0,1], \mathbb{R})$ equipped with the supremum norm.

(1) Find the norm of φ.
(2) Is the norm attained? i.e. is there an $f \in X$ (with supremum norm equal to 1) such that $|\varphi(f)| = \|\varphi\|$.
(3) Set

$$Y = \{f \in X : \ f(1) = 0\}.$$

Let ψ be the restriction of φ to Y.
  (a) Show that $\|\varphi\| = \|\psi\|$
  (b) Show that the norm of ψ is not attained on $(Y, \|\cdot\|_\infty)$.

Exercise 2.3.20. Let X be a normed vector space. Let f be a non-identically null linear functional on X.

(1) Show that $\ker f$ is closed iff f is continuous.
(2) Show that f is (everywhere) discontinuous iff $\ker f$ is dense in X.
(3) Let $X = C([0,1], \mathbb{R})$. Define on X the following two *norms*

$$\|f\|_1 = \int_0^1 |f(x)|dx, \ \ \|f\|_\infty = \sup_{0 \le x \le 1} |f(x)|.$$

Let

$$A = \{f \in X : \ f(0) = 0\}.$$

Is A closed in X with respect to either of the two norms?

Exercise 2.3.21. Let $X = C([-1,1], \mathbb{R})$. Endow X with the *norm*

$$\|f\|_1 = \int_{-1}^1 |f(x)|dx.$$

Let

$$Y = \left\{ f \in X : \ \int_{-1}^1 f(x)dx = 0 \right\}.$$

Is Y closed in X?

Exercise 2.3.22. Let $x_n : \mathbb{N} \to \mathbb{C}$ be a sequence. Let

$$X = \left\{ x = (x_n) : \sum_{n \in \mathbb{N}} n^2 |x_n| < \infty \right\}.$$

Set $\|x\| = \sum_{n=1}^{\infty} n^2 |x_n|$.

(1) Show that X is a vector space.
(2) Show that $\| \cdot \|$ is a norm on X.
(3) Prove that $(X, \| \cdot \|)$ is complete. **Hint:** show first that the *bounded linear* mapping $f : \ell^1 \to X$ defined by, $f(x) = \left(\frac{x_n}{n^2}\right)_{n \geq 1}$, is a surjective isometry.

Exercise 2.3.23. Show that a normed vector space $(X, \| \cdot \|)$ is homeomorphic to the open unit ball $B(0,1)$.

Exercise 2.3.24.

(1) Prove that $(\ell^1)' = \ell^\infty$.
(2) Show that $(c_0)' = \ell^1$.

Exercise 2.3.25. (cf. Exercise 2.3.26) Let $A, B \in B(L^2[0,1])$ be defined by

$$Af(x) = (1+x)f(x) \text{ and } Bf(x) = xf(x).$$

Is A invertible? Is B invertible?

Exercise 2.3.26. Consider the multiplication operator defined in Exercise 2.3.3 on $L^2([a,b])$ by

$$(Af)(x) = \varphi(x)f(x)$$

where φ is a continuous function on $[a,b]$ taking its values in $\mathbb{C}$. Give a necessary and sufficient condition for the invertibility of A.

Exercise 2.3.27. Define $A \in B(\ell^2)$ by

$$A(x_1, x_2, \cdots, x_n, \cdots) = \left(x_1, \frac{1}{2}x_2, \cdots, \frac{1}{n}x_n, \cdots \right).$$

Is A invertible?

Exercise 2.3.28. Let X be a normed vector space. Prove that X is Banach iff every absolutely convergent series is convergent.

Exercise 2.3.29. Let

$$c_{00} = \{(x_n)_n : \exists N \in \mathbb{N}^*, x_n = 0, \forall n \geq N\}$$

considered as a linear subspace of the space of real sequences tending to zero. We endow c_{00} with the *norm* $\|(x_n)_n\|_\infty = \sup_{n\in\mathbb{N}^*} |x_n|$.

(1) Show that c_{00} is not a Banach space (you may consider $S_n = (1, \frac{1}{2^2}, \frac{1}{3^2}, \cdots, \frac{1}{n^2}, 0, 0, \cdots))$.
(2) Based on the previous question, construct an example of an absolutely convergent series in c_{00} which is not convergent.
(3) Do you see any contradiction with the previous result?

Exercise 2.3.30. Let X be a Banach space and let $A \in B(X)$ be a pure contraction. Show that $I - A$ is invertible, that

$$(I-A)^{-1} = \sum_{n=0}^{\infty} A^n$$

and that

$$\|(I-A)^{-1}\| \le \frac{1}{1-\|A\|}.$$

Hint: If (S_n) is a sequence in $B(X)$ such that $\|S_n - S\| \to 0$ where $S \in B(X)$, then $\|(S_n - S)T\|, \|T(S_n - S)\| \to 0$. A proof can be found in the solution of Exercise 2.3.4

Exercise 2.3.31. Let A be a nilpotent operator on a Banach space X. Show that $I - A$ is invertible. What about $I + A$?

Exercise 2.3.32. Let $A, B \in B(X)$ where X is a *Banach* space. Let

$$x_n = \frac{A^n}{n!}, \quad n \in \mathbb{N}.$$

(1) Show that the series $\sum_{n\ge 0} x_n$ converges absolutely.
(2) Deduce from the previous question that $\sum_{n\ge 0} x_n$ converges. Denote its sum $\sum_{n=0}^{\infty} x_n$ by e^A. Infer that (cf. Exercise 11.3.5):
$$\|e^A\| \le e^{\|A\|}.$$
(3) Assume that A commutes with B. Show that
$$e^{A+B} = e^A e^B = e^B e^A.$$

Remark. For the converse, see the Functional Calculi Chapter.

(4) Supply an example showing the importance of the commutativity assumption.
(5) Is e^A invertible? If yes, what is its inverse?

Remark. See the fourth case of Examples 11.1.3 for another proof.

(6) **Application:** Solve the following differential system

$$\begin{cases} x'(t) = x(t) - y(t) - z(t), \\ y'(t) = 2x(t) + 5y(t) + 2z(t), \\ z'(t) = -2x(t) - 3y(t). \end{cases}$$

Exercise 2.3.33. Let $A, B \in B(X)$ where X is a Banach space. Do we have

$$e^A = e^B \Longrightarrow A = B?$$

Exercise 2.3.34. ([**32**]) Let $A \in B(X)$, where X is a Banach space. Show that

$$\lim_{n\to\infty} \left(I + \frac{1}{n}A\right)^n = e^A$$

(the limit is w.r.t. $\|\cdot\|_{B(X)}$).

Exercise 2.3.35. Let X be a Banach space. Let $A, B \in B(X)$ where A is invertible. Do we have

$$e^{A^{-1}BA} = A^{-1}e^B A?$$

Exercise 2.3.36. Let $A \in B(X)$ where X is a Banach space. Consider the following **initial value problem on a Banach space**:

$$y'(t) = A(t), \ -\infty < t < \infty, \ y(0) = x \in X$$

where $y : \mathbb{R} \to X$. Show that $y(t) = e^{tA}x$ is the only solution of the previous problem.

Exercise 2.3.37. Let X be a Banach space. Let $A, B, C \in B(X)$ such that $\|A\| < 1$ and $\|B\| < 1$. Find a solution $T \in B(X)$ of

$$T - ATB = C.$$

Exercise 2.3.38. Let X be a Banach space. Let $A, B, C \in B(X)$ such that A is invertible, $\|A^{-1}\| < 1$ and $\|B\| < 1$. Solve the equation

$$AT - TB = C$$

where $T \in B(X)$.

Exercise 2.3.39. Let X and Y be two Banach spaces. Let $B_i(X,Y)$ be the space of invertible operators from X onto Y.

(1) Can $B_i(X,Y)$ be empty?
(2) Show that $B_i(X,Y)$ is open in $B(X,Y)$. **Hint:** Use the Neumann series.
(3) Is $B_i(X,Y)$ closed in $B(X,Y)$?

Exercise 2.3.40. (cf. Exercise 7.3.21) Let X be a Banach space and let $A \in B(X)$. Let $B_i(X)$ be the (group) of invertible elements of $B(X)$.

(1) Show that the map $A \mapsto A^{-1}$ from $B_i(X)$ into $B_i(X)$ is continuous.
(2) Is $A \mapsto A^{-1}$ a homeomorphism?

Exercise 2.3.41. Let $x \in [0,1]$ and let $\lambda \in \mathbb{R}$. By finding a suitable Banach space, solve the following integral equation (**Fredholm's of the second kind**)

$$u(x) = f(x) + \lambda \int_0^1 e^{x-t} u(t) dt,$$

where f is given and u is the unknown.

Exercise 2.3.42. (mostly from [**50**]) Let $1 \leq p < \infty$. Consider

$$X_p = \left\{ x = (x_n) \in \ell^p : \sum_{n=0}^{\infty} x_n = 0 \right\}.$$

(1) Show that $X_p \neq \varnothing$.
(2) Show that X_p is a linear subspace of ℓ^p.
(3) Show that X_1 is closed, and that X_p is not closed in case $1 < p < \infty$.
(4) Using Proposition 2.1.102, show that X_p is dense in ℓ^p (in case $1 < p < \infty$).

Exercise 2.3.43. Let X and Y be two normed vector spaces with $X \neq \{0\}$. Show that if $B(X,Y)$ is a Banach space, then so is Y.

Exercise 2.3.44. Let X be a normed vector space.
(1) Show that any proper linear subspace of X has an empty interior (hence any proper and closed linear subspace of X is nowhere dense).
(2) Deduce that a linear subspace Y is either dense or nowhere dense.
(3) Using the Baire Category Theorem (Theorem 2.1.106), infer that an infinite dimensional *Banach* space X does not have a countable Hamel basis.

Exercise 2.3.45. Let X be a Banach space with respect to two norms $\|\cdot\|_1$ and $\|\cdot\|_2$. Assume that

$$\exists k > 0,\ \|x\|_1 \leq k\|x\|_2,\ \forall x \in X.$$

Show that these two norms are equivalent.

Exercise 2.3.46. Let X and Y be two *metric* spaces and let $A : X \to Y$ be a continuous transformation. Let $G(A)$ be its graph.
(1) Show that $G(A)$ is closed.

(2) Show that if X and Y are now two Banach spaces, and if A is *linear*, then A is bounded whenever $G(A)$ is closed.

Exercise 2.3.47. Show that any linear operator satisfying Theorem 2.1.108 is an open map.

Exercise 2.3.48. Let $X = C([0,1],\mathbb{R})$. We consider on X the following two norms

$$\|f\|_1 = \int_0^1 |f(x)|dx \text{ and } \|f\|_\infty = \sup_{x\in[0,1]} |f(x)|.$$

(1) Show that the identity operator $I : (X, \|\cdot\|_\infty) \to (X, \|\cdot\|_1)$ is bounded, onto but not open.
(2) Does the conclusion of the previous question violate the Open Mapping Theorem?

Exercise 2.3.49. Let $X = C^1([0,1],\mathbb{R})$ and let $Y = C([0,1],\mathbb{R})$. We endow both these spaces with the supremum norm and we define an operator $A : X \to Y$ by $Af(x) = f'(x)$ for all $f \in X$ and all $x \in [0,1]$.

(1) Show that A is linear and unbounded.
(2) Show that A has a closed graph.
(3) Does the conclusion of the previous question contradict the Closed Graph Theorem?

Exercise 2.3.50. Let

$$c_{00} = \{x = (x_n) \in \ell^\infty : \exists N \in \mathbb{N}, x_n = 0 \ \forall n \geq N\}$$

considered as a linear subspace endowed with the supremum norm. We define a mapping $A : c_{00} \to c_{00}$ by

$$A(x_n)_{n\geq 1} = \left(\frac{x_1}{2}, \frac{x_2}{3}, \cdots, \frac{x_n}{n+1}, \cdots\right).$$

(1) Show that A is linear and bounded.
(2) Show that A is bijective.
(3) Is A invertible?
(4) Why does the previous conclusion not contradict the Banach Isomorphism Theorem?

Exercise 2.3.51. Let $(X, \|\cdot\|)$ be a normed vector space. Show that for every $a \in X$, $a \neq 0$, there exists a non-zero linear functional F on X such that $F(a) = \|F\|\|a\|$.

Exercise 2.3.52. Let X and Y be two Banach spaces. Let (A_n) be a sequence of bounded linear operators on X into Y such that for

every $x \in X$, $\lim_{n\to\infty} A_n(x) = A(x)$. Show that A is linear and bounded.
Hint: Use the Banach-Steinhaus Theorem.

2.4. Tests

Test 2.4.1. Define an operator A from $L^2[0,1]$ into $\mathbb{R}$, say, by

$$Af = \int_0^1 f(x)dx.$$

Find $\|A\|$.

Test 2.4.2. Let $A \in B(X)$, where X is a normed vector space. Let

$$Y = \{B \in B(X) : \ BA = 0_{B(X)}\} \text{ and } Z = \{B \in B(X) : \ AB = BA\}.$$

Is Y closed in $B(X)$? Is Z closed in $B(X)$?

Test 2.4.3. Prove Proposition 2.1.58.

Test 2.4.4. Let $A, B \in B(X)$. Show that if BA is nilpotent, then so is AB.

Test 2.4.5. Are there two (square) matrices A and B on a finite dimensional space which verify the relation $AB - BA = I$ (use a different method from that of the solution of Exercise 2.3.11)?

Test 2.4.6. Let $1 \leq p \leq \infty$ and define $A \in B(L^p(\mathbb{R}))$ by $Af(x) = f(-x)$. Is A invertible?

Test 2.4.7. Let $1 \leq p < \infty$. Define a linear operator $V : \ell^p \to L^p[0,\infty)$ by

$$V(x) = \sum_{n=1}^{\infty} x_n \mathbb{1}_{[n-1,n)}$$

where $x = (x_1, x_2, \cdots, x_n, \dots)$. Is V an isometry?

Test 2.4.8. Using the method of exponentials of matrices, solve

$$\begin{cases} x'(t) = y(t), \\ y'(t) = x(t). \end{cases}$$

Test 2.4.9. Prove Corollary 2.1.100.

2.5. More Exercises

Exercise 2.5.1. (cf. Exercise 10.3.9) Let $p \geq 1$. Define $A : L^p[0,1] \to L^p[0,1]$ by

$$Af(x) = \varphi(x)f(x)$$

where $\varphi : [0,1] \to \mathbb{C}$ is continuous. Show that $A \in B(L^p[0,1])$ and that

$$\|A\|_{B(L^p[0,1])} = \sup_{0 \leq x \leq 1} |\varphi(x)|.$$

Exercise 2.5.2. Let X be a Banach space and let $A \in B(X)$. Show that for all $n \in \mathbb{N}$

$$(e^A)^n = e^{nA}.$$

Exercise 2.5.3. Let $A, B \in B(X)$. Show that

$$e^{A+B} = \lim_{n \to \infty} \left(e^{\frac{A}{n}} e^{\frac{B}{n}} \right)^n$$

w.r.t. the norm $\|\cdot\|_{B(X)}$ (this is more known as the **Lie Product Formula**).

Exercise 2.5.4. Let $A \in B(X)$. Guess the formula for "$\arctan A$" and prove it (you will need to assume that $\|A\| < 1$).

Exercise 2.5.5. Give an explicit example of a couple of two unbounded operators S and T that satisfy the relation $ST - TS = I$ in Exercise 2.3.11.

Hint: There is a famous example on functional spaces, but look for one couple as infinite *unbounded* matrices.

Exercise 2.5.6. Prove that if $f \in C_0^\infty(\mathbb{R})$ such that $\hat{f} \in C_0^\infty(\mathbb{R})$, then $f = 0$.

Exercise 2.5.7. ([**50**] or [**101**]) Let $1 < p < \infty$. Let $K : (0, \infty) \to \mathbb{R}$ be a Lebesgue measurable function satisfying

$$\int_0^\infty \frac{1}{x^{\frac{1}{p}}} |K(x)| dx < \infty$$

(K is then called a **Hardy kernel**). Now, let $I_K : L^p(0,\infty) \to L^p(0,\infty)$ and $M_K : L^p(0,\infty) \to L^p(0,\infty)$ be defined by

$$I_K f(x) = \frac{1}{x} \int_0^\infty K\left(\frac{y}{x}\right) f(y) dy \text{ and } M_K f(x) = x^{1-\frac{2}{p}} \int_0^\infty K(xy) f(y) dy$$

respectively.

(1) Prove that $I_K \in B(L^p(0,\infty))$ and $M_K \in B(L^p(0,\infty))$ and that they obey

$$\left|\int_0^\infty \frac{1}{y^{\frac{1}{p}}}K(y)dy\right| \leq \|I_K\| \leq \int_0^\infty \frac{1}{y^{\frac{1}{p}}}|K(y)|dy$$

and

$$\left|\int_0^\infty \frac{1}{y^{\frac{1}{p}}}K(y)dy\right| \leq \|M_K\| \leq \int_0^\infty \frac{1}{y^{\frac{1}{p}}}|K(y)|dy.$$

What happens when $K \geq 0$?

(2) By choosing a suitable kernel, apply the previous results to:

(a) Show that the norm of the **Hardy Operator**, defined by $H : L^p(0,\infty) \to L^p(0,\infty)$

$$Hf(x) = \frac{1}{x}\int_0^x f(y)dy,$$

is given by

$$\|H\| = \frac{p}{p-1}.$$

(b) Show that the norm of the **Hilbert Operator** defined by $J : L^p(0,\infty) \to L^p(0,\infty)$

$$Jf(x) = \int_0^\infty \frac{f(y)}{x+y}dy,$$

is given by

$$\|J\| = \frac{\pi}{\sin\frac{\pi}{p}}.$$

(c) Show that the norm of the **Schur Operator** defined by $S : L^p(0,\infty) \to L^p(0,\infty)$

$$Sf(x) = \int_0^\infty \frac{f(y)}{\max(x,y)}dy,$$

is given by

$$\|S\| = \frac{p^2}{p-1}.$$

Exercise 2.5.8. ([**83**]) Let $a, b \in \mathbb{C}^*$ be such that $0 < |b| < |a|$. Let

$$e_1 = (a,b,0,\cdots),\ e_2 = (0,a,b,0,\cdots), e_3 = (0,0,a,b,0,\cdots),\cdots.$$

Show that $\{e_n :\ n \in \mathbb{N}\}$ is a Schauder basis for ℓ^2.

Exercise 2.5.9. ([**48**], more can be found in [**33**]) First, recall a definition: An **algebra** $\mathcal{A}$ over $\mathbb{F}$ is a vector space over $\mathbb{F}$ which has a multiplication which makes $\mathcal{A}$ into a ring such that

$$\alpha(ab) = (\alpha a)b = a(\alpha b)$$

for all $\alpha \in \mathbb{F}$ and all $a, b \in \mathcal{A}$. If $\mathcal{A}$ has a norm $\|\cdot\|$ which satisfies

$$\|ab\| \leq \|a\|\|b\|$$

for all $a, b \in \mathcal{A}$, is called a **normed algebra**. If the normed algebra is complete, then we call it a **Banach algebra**. If $\mathcal{A}$ has an identity e, then we assume that $\|e\| = 1$. If the product law is commutative, then we say that $\mathcal{A}$ is commutative. For example, if X is a Banach space, then $B(X)$ is a (non-commutative) Banach algebra with an identity (where the multiplication is $\circ$ and the norm is the usual operator norm). Many known results on $B(X)$ remain valid on arbitrary Banach algebra. For instance, if $\mathcal{A}$ is a Banach algebra with identity e, then:

(1) Show that $x \in \mathcal{A}$ is invertible whenever $\|e - x\| < 1$. Give the expression of x^{-1}.
(2) Show that the subset of $\mathcal{A}$ constituted of invertible elements in $\mathcal{A}$ is open.

Exercise 2.5.10. Show that $(L^1(\mathbb{R}), +, *)$ is a commutative Banach algebra without an identity, where $*$ denotes the convolution product.

Exercise 2.5.11. Let $\mathcal{A}$ be a Banach algebra. Can we define e^a for $a \in \mathcal{A}$?

CHAPTER 3

Hilbert Spaces

3.1. Basics

3.1.1. Definitions and Examples.

Definition 3.1.1. *Let X be a vector space over $\mathbb{C}$. An* ***inner*** *(or* ***scalar****)* ***product*** *on X is a function $<\cdot,\cdot>: X \times X \to \mathbb{C}$ satisfying the following properties:*

(1) $<x,x>\in \mathbb{R}$ *and* $<x,x>\geq 0$ *whichever* $x \in X$;
(2) $<x,x>=0 \Leftrightarrow x=0$;
(3) *For all* $\alpha,\beta \in \mathbb{C}$ *and for all* $x,y,z \in X$:

$$<\alpha x+\beta y,z>=\alpha <x,z>+\beta <y,z>;$$

(4) *For any* $x,y \in X$: $<x,y>=\overline{<y,x>}$.

An ***inner product space*** *is a couple $(X,<\cdot,\cdot>)$, where X is a vector space and $<\cdot,\cdot>$ is an inner product.*

Remark. A real inner product is defined over a real vector space X by changing $\mathbb{C}$ by $\mathbb{R}$ above. In such case, the last property is replaced by

$$<x,y>=<y,x>$$

for all $x,y \in X$.

Remark. In some references (like [**189**]), an inner product space is called a **unitary space**. We may also say a **pre-Hilbert space**. We may also speak of a **Euclidean space** especially when $\dim X < \infty$.

Remark. The inner product is bilinear on $\mathbb{R}$, and is sesquilinear on $\mathbb{C}$.

Examples 3.1.2.

(1) *On $\mathbb{R}^n$, define*

$$<x,y>=\sum_{k=1}^{n} x_n y_n,$$

where $x=(x_1,\cdots,x_n)$ and $y=(y_1,\cdots,y_n)$. Then $<\cdot,\cdot>$ is a real inner product on $\mathbb{R}^n$.

(2) *Similarly, we may define an inner product on* $\mathbb{C}^n$ *by*

$$< x, y >= \sum_{k=1}^{n} x_n \overline{y_n},$$

with $x, y \in \mathbb{C}^n$.

(3) *For* $x = (x_n)$ *and* $y = (y_n)$ *in* ℓ^2, *define*

$$< x, y >= \sum_{n=1}^{\infty} x_n \overline{y_n}.$$

Then $< \cdot, \cdot >$ *is an inner product on* ℓ^2.

(4) *The twin space of* ℓ^2 *is the "bilateral* ℓ^2*" denoted by* $\ell^2(\mathbb{Z})$. *It is defined by*

$$\ell^2(\mathbb{Z}) = \{x = (x_n)_{n\in\mathbb{Z}}, x_n : \mathbb{Z} \to \mathbb{C} : \sum_{n=-\infty}^{\infty} |x_n|^2 < \infty\}.$$

Then the following defines an inner product on $\ell^2(\mathbb{Z})$

$$< x, y >= \sum_{n=-\infty}^{\infty} x_n \overline{y_n}.$$

(5) $C([0,1], \mathbb{R})$ *is an inner product space w.r.t.*

$$< f, g >= \int_0^1 f(x)\overline{g(x)}dx,$$

with $f, g \in C([0,1], \mathbb{R})$.

(6) $L^2(\mathbb{R})$ *is an inner product space w.r.t.*

$$< f, g >= \int_{\mathbb{R}} f(x)\overline{g(x)}dx,$$

where $f, g \in L^2(\mathbb{R})$.

Next, we give some elementary properties of inner products which are immediate consequences of the previous definition.

PROPOSITION 3.1.3. *Let* X *be an inner product space, and let* $x, y, z \in X$ *and* $\alpha, \beta \in \mathbb{F}$. *Then*

(1) $< 0, y >=< x, 0 >= 0$.
(2) $< x, \alpha y + \beta z >= \overline{\alpha} < x, y > + \overline{\beta} < x, z >$.

The following simple result is quite practical.

PROPOSITION 3.1.4. *(a proof is given in Exercise 3.3.6) Let X be an inner product space. Let $x \in X$. Then*

$$< x, y >= 0, \ \forall y \in X \Longleftrightarrow x = 0.$$

REMARK. This useful result must be remembered in its generality. However, the reader must not forget it when it comes to particular examples. For instance, if $f \in X = C([0,1])$ and

$$\forall g \in X : \int_0^1 f(x)\overline{g(x)}dx = 0,$$

then the previous result tells us that f must vanish on $[0,1]$.

COROLLARY 3.1.5. *Let X be an inner product space and let $y, z \in X$. Then*

$$\forall x \in X : < x, y >=< x, z >\Longrightarrow y = z.$$

The following result is very much used in Functional Analysis. This is the celebrated Cauchy-Schwarz Inequality. The reader must have already seen different versions of it. Here, we give a general version.

THEOREM 3.1.6. ***(Cauchy-Schwarz Inequality,** see Exercise 3.3.7) Let X be an inner product space and let $x, y \in X$. Then*

$$| < x, y > | \leq \sqrt{< x, x >}\sqrt{< y, y >}.$$

Equality holds iff $x = \lambda y$ for some $\lambda \in \mathbb{C}$.

REMARK. There are yet more general versions of the Cauchy-Schwarz Inequality. See Theorem 5.1.13. See also Exercises 6.5.7 & 11.5.17.

COROLLARY 3.1.7. *(see Exercise 3.3.7) Let X be an inner product space. Then the function $\|\cdot\| : X \to \mathbb{R}^+$, defined by $\|x\| = \sqrt{< x, x >}$, is a norm in X.*

REMARK. The Cauchy-Schwarz inequality can then be rewritten as

$$| < x, y > | \leq \|x\|\|y\|.$$

It is also plain that we can "(re)-rewrite" the Cauchy-Schwarz Inequality as the following determinant:

$$\begin{vmatrix} < x, x > & < x, y > \\ < y, x > & < y, y > \end{vmatrix} \geq 0.$$

We have just seen that each inner product space can be regarded as a normed vector space. What about the converse? The answer is no in general unless a certain condition is imposed on the norm.

THEOREM 3.1.8. *(see Exercise 3.3.9) Let X be an inner product space on $\mathbb{F}$. Then for all $x, y \in X$*

$$\|x+y\|^2 + \|x-y\|^2 = 2(\|x\|^2 + \|y\|^2)$$

(called the ***parallelogram law****). Moreover, if $\mathbb{F} = \mathbb{R}$, then*

$$< x, y >= \frac{1}{4}(\|x+y\|^2 - \|x-y\|^2);$$

and if $\mathbb{F} = \mathbb{C}$, then

$$< x, y >= \frac{1}{4}(\|x+y\|^2 - \|x-y\|^2 + i\|x+iy\|^2 - i\|x-iy\|^2)$$

(more known as the ***polarization identity****).*

REMARK. By the previous theorem we have the following observation: *A normed vector space is an inner product space if and only if it satisfies the parallelogram law.*

3.1.2. Modes of Convergence of Sequences. Since an inner product space is necessarily a normed vector space, we may keep on using the definition of a convergent sequence (see Definition 1.1.12). Since we will introduce another type of convergence, we rename the convergence in Definition 1.1.12 as follows:

DEFINITION 3.1.9. *Let X be an inner product space. We say that the sequence (x_n) in X* ***converges strongly*** *to $x \in X$ if*

$$\lim_{n\to\infty} \|x_n - x\| = 0.$$

Such an x is called the ***strong limit*** *of (x_n). We may denote the strong limit by "$s - \lim_{n\to\infty} x_n$" or $x_n \xrightarrow{s} x$.*

REMARK. The strong convergence is the convergence which we will be using by default.

Complete inner product spaces have a particular name.

DEFINITION 3.1.10. *A complete (w.r.t. to the induced norm) inner product space is called a* ***Hilbert space****.*

REMARK. In honor to Hilbert, we will be usually denoting a Hilbert space by H.

REMARK. By the remark below Theorem 3.1.8, we may say that a Banach space is a Hilbert space iff it satisfies the parallelogram law.

EXAMPLES 3.1.11.

(1) *Any finite-dimensional inner product space is a Hilbert space.*

(2) $X := C[0,1]$ *endowed with the inner product*

$$< f,g >= \int_0^1 f(x)\overline{g(x)}dx$$

$(f, g \in X)$ *is not a Hilbert space because it is not complete.*

(3) $L^2(\mathbb{R})$ *is a Hilbert space with the inner product*

$$< f,g >= \int_{\mathbb{R}} f(x)\overline{g(x)}dx$$

(4) ℓ^2 *is a Hilbert space, and so is* $\ell^2(\mathbb{Z})$.

(5) ℓ^p, *with* $p \neq 2$, *is not a Hilbert space for it does not satisfy the parallelogram law.*

As alluded to just above, we give the other mode of convergence in an inner product space.

DEFINITION 3.1.12. *Let H be a Hilbert space. We say that the sequence (x_n) in H* ***converges weakly*** *to $x \in H$ if*

$$\lim_{n\to\infty} < x_n, y >=< x, y >$$

for all $y \in H$. If x is as such, then it is called the ***weak limit*** *of (x_n). We may denote the weak limit by "$w - \lim_{n\to\infty} x_n$" or $x_n \overset{w}{\to} x$.*

The following result may easily be demonstrated.

PROPOSITION 3.1.13. *Weak limits are unique.*

Other properties which are equally simple to prove are given in the following proposition.

PROPOSITION 3.1.14. *Let H be a Hilbert space and let (x_n) and (y_n) be two weakly convergent sequences to x and y respectively. Let $\alpha \in \mathbb{F}$. Then*

$$x_n + y_n \overset{w}{\to} x + y \text{ and } \alpha x_n \overset{w}{\to} \alpha x.$$

The next result justifies the appellations "strong" and "weak".

PROPOSITION 3.1.15. *Every strongly convergent sequence is weakly convergent.*

REMARKS.

(1) See Exercise 3.3.11 for a proof of the previous result, and also for a counterexample to the backward implication (cf. Proposition 3.1.63).

(2) The weak convergence is purely an infinite-dimensional notion. Indeed, if the inner product space in question has a finite dimension, then the weak convergence and the strong convergence are equivalent (see Exercise 3.3.12).

We end this subsection with a result prescribing a relationship between weak and strong limits.

THEOREM 3.1.16. *Let (x_n) be a weakly convergent sequence to x in a given Hilbert space H. Then*

$$\|x\| \leq \liminf_{n\to\infty} \|x_n\|.$$

Besides, the following statements are equivalent

(1) $\lim_{n\to\infty} \|x_n\| = \|x\|$.
(2) $\|x\| \geq \limsup_{n\to\infty} \|x_n\|$.
(3) (x_n) *converges strongly to* x.

3.1.3. Orthogonality. One of the advantages of inner product spaces is their geometrical properties.

DEFINITION 3.1.17. *Let X be an inner product space and let Y be a linear subspace of X. Two vectors x and y in X are said to be* ***orthogonal (or perpendicular)*** *if*

$$< x, y >= 0.$$

We say that x is orthogonal to Y, and we write $x \perp Y$, if $< x, y >= 0$ for all $y \in Y$.

A set of vectors $\{x_i\}_i$ in X, where "i" runs through an index set I is said to be ***orthogonal*** *if*

$$< x_i, x_j >= 0, \ \forall i \neq j, \ \forall i, j \in I.$$

DEFINITION 3.1.18. *Let X be an inner product space and let Y be a subset of X. The* ***orthogonal complement*** *of Y, denoted by $Y^\perp$, is defined by*

$$Y^\perp = \{x \in X, \ < x, y >= 0, \ \forall y \in Y\}.$$

REMARK. It is important to emphasize that Y is not assumed to be a linear subspace of X.

One of the oldest theorems in Mathematics is the Pythagorean Theorem. Does it hold in inner product spaces? The answer is yes.

THEOREM 3.1.19. ***(Pythagorean Theorem)*** *Let X be an inner product space and let $x, y \in X$ be orthogonal. Then*

$$\|x + y\|^2 = \|x\|^2 + \|y\|^2.$$

The previous can easily be generalized to a finite orthogonal set.

COROLLARY 3.1.20. *Let X be an inner product space and let $\{x_k\}_{1\leq k\leq n}$ be an orthogonal set. Then*

$$\left\|\sum_{k=1}^{n} x_k\right\|^2 = \sum_{k=1}^{n} \|x_k\|^2$$

A natural question is whether Pythagorean Theorem holds for infinite series? See the next theorem (see Exercise 4.3.27 on how to prove at least one implication).

THEOREM 3.1.21. *Let $\{x_n : n \in \mathbb{N}\}$ be an orthogonal sequence in a Hilbert space H. Then the series $\sum_{n\in\mathbb{N}} x_n$ converges in H iff $\sum_{n\in\mathbb{N}} \|x_n\|^2$ converges. Moreover, when $\sum_{n\in\mathbb{N}} \|x_n\|^2$ converges, then*

$$\sum_{n=1}^{\infty} \|x_n\|^2 = \left\|\sum_{n=1}^{\infty} x_n\right\|^2$$

Related to the previous result, we have (see e.g. **[190]**):

PROPOSITION 3.1.22. *Let $\{x_n : n \in \mathbb{N}\}$ be an orthogonal sequence in a Hilbert space H. Then*

$$\sum_{n\in\mathbb{N}} x_n \text{ converges in } H \iff \sum_{n\in\mathbb{N}} < x_n, y > \text{ converges for all } y \in H.$$

Next, we give basic properties of the orthogonal complements (see also Exercise 3.5.7).

PROPOSITION 3.1.23. *Let X be an inner product space and let $Y, Z \subset X$. Then*

(1) $0 \in Y^{\perp}$.
(2) *If* $0 \in Y$, *then* $Y^{\perp} \cap Y = \{0\}$. *Otherwise,* $Y^{\perp} \cap Y = \varnothing$.
(3) $X^{\perp} = \{0\}$ *and* $\{0\}^{\perp} = X$.
(4) *If* $Y \subset Z$, *then* $Z^{\perp} \subset Y^{\perp}$.
(5) $Y^{\perp}$ *is always a closed linear subspace of* X.
(6) $Y \subset Y^{\perp\perp}$.

The following result is fundamental.

PROPOSITION 3.1.24. *Let Y be a linear subspace of an inner product space X. Then*

$$x \in Y^{\perp} \iff \|x - y\| \geq \|x\|, \ \forall y \in Y.$$

3.1.4. The Closest Point Property and the Projection Theorem. One of the tremendous results in Hilbert Space Theory is the closest point property:

THEOREM 3.1.25. *(**The closest point property**) Let H be a Hilbert space and let Y be a non-empty closed convex subset in H. Then for any $x \in H$, there is a unique point $y_0 \in Y$ such that*

$$\|x - y_0\| = \inf_{y \in Y} \|x - y\|.$$

Moreover, y_0 is characterized by the following property:

$$y_0 \in Y \text{ and } \mathrm{Re} < x - y_0, z - y_0 > \leq 0, \ \forall z \in Y.$$

DEFINITION 3.1.26. *The point y_0 in the preceding theorem is called the **projection of** x **onto** Y, and it is denoted by $P_Y(x)$.*

REMARK. Obviously, the previous theorem holds if H is only an inner product space and Y is complete.

It seems unnecessary to say that Theorem 3.1.25 holds if Y is a closed linear subspace. However, this stronger assumption does give us a quite interesting conclusion. Besides, this structure allows us to speak about linear operators.

THEOREM 3.1.27. *Let Y be a closed linear subspace of a Hilbert space H. Then $P_Y : H \to Y$ is a (surjective) linear operator. Moreover, if $x \in H$, then $P_Y(x)$ is the unique element $y_0 \in H$ which verifies: $y_0 \in Y$ such that $x - y_0 \in Y^\perp$.*

DEFINITION 3.1.28. *The point y_0 (i.e. $P_Y(x)$) in the preceding theorem is called the **orthogonal projection of** x **onto** Y.*

Another result of great importance is.

THEOREM 3.1.29. *(**The Orthogonal Decomposition Theorem**) Let Y be a closed linear subspace of a Hilbert space H. Any $x \in H$ may be uniquely expressed as $x = y + z$, where $y \in Y$ and $z \in Y^\perp$. The previous may be denoted by*

$$H = Y \oplus Y^\perp.$$

Besides,

$$\|x\|^2 = \|y\|^2 + \|z\|^2.$$

REMARK. The assumptions Y being closed and H being a Hilbert are indispensable. See Exercises 3.3.15, 3.3.17 & 3.3.18.

COROLLARY 3.1.30. *If Y is a closed linear subspace of a Hilbert space, then $Y^{\perp\perp} = Y$.*

COROLLARY 3.1.31. *If Y is a linear subspace of a Hilbert space, then $Y^{\perp\perp} = \overline{Y}$.*

REMARK. The previous result is particularly efficient if we want to show that a given subset is dense in a Hilbert space H. Indeed, to show that $\overline{Y} = H$, it then suffices to prove that $Y^{\perp} = \{0\}$ as this implies that

$$\overline{Y} = Y^{\perp\perp} = \{0\}^{\perp} = H.$$

3.1.5. Orthonormal Sets and Bases. As in Ordinary Geometry, we have the following concept.

DEFINITION 3.1.32. *Let X be an inner product space. A family of vectors $\{x_i : i \in I\}$ in X is said to be an* ***orthonormal set (or sequence if*** $I = \mathbb{N}$***)*** *if*

$$\|x_i\| = 1, \ \forall i \in I \ \text{and} \ < x_i, x_j >= 0, \ \forall i \neq j,$$

that is, if

$$< x_i, x_j >= \delta_{ij}$$

(where δ_{ij} denotes the usual ***Kronecker delta****).*

EXAMPLES 3.1.33.

(1) *Let*

$$e_1 = (1, 0, 0, \cdots), \ e_2 = (0, 1, 0, \cdots), ...etc.$$

Then $\{e_n : n \in \mathbb{N}\}$ is an orthonormal sequence in ℓ^2.

(2) *The sequence*

$$\{e^{2\pi inx} : n \in \mathbb{Z}\}$$

is orthonormal in the space $L^2[0,1]$.

(3) *The set $\{(0,1),(1,1)\}$ is not orthonormal in $\mathbb{R}^2$.*

(4) *The sequence $\{f_n : n \in \mathbb{Z}\}$ with $f_n : \mathbb{R} \to \mathbb{C}$ and where*

$$f_n(x) = \frac{(x-i)^n}{\sqrt{\pi}(x+i)^{n+1}}, \ x \in \mathbb{R}$$

is orthonormal in $L^2(\mathbb{R})$ (see e.g. **[220]***).*

The following is an easy exercise.

PROPOSITION 3.1.34. *The vectors of the orthonormal set $\{e_1, e_2, \cdots, e_n\}$ are linearly independent.*

The converse is not necessarily true, however, we have

PROPOSITION 3.1.35. *From a set $\{x_1, x_2, \cdots, x_n\}$ of linearly independent vectors, we can produce an orthonormal set $\{e_1, e_2, \cdots, e_n\}$.*

Remark. The method for producing such an orthonormal set in the foregoing proposition is commonly known as the **Gram-Schmidt Algorithm**. It goes as follows:

Set $e_1 = \frac{x_1}{\|x_1\|}$, and the remaining vectors are defined by:

$$e_n = \frac{x_n - \sum_{k=1}^{n-1} < x_n, e_k > e_k}{\left\| x_n - \sum_{k=1}^{n-1} < x_n, e_k > e_k \right\|},\ n \geq 2.$$

Remark. The previous algorithm also holds for a linear independent sequence of vectors. Notice that a Hamel basis $(x_i)_{i \in I}$ in a separable infinite dimensional Hilbert space cannot be made into an orthonormal collection $(e_i)_{i \in I}$ (why?).

Theorem 3.1.36. *An infinite dimensional inner product space always contains an orthonormal sequence.*

Proposition 3.1.37. *(**Bessel Inequality**) Let X be an inner product space and let $\{e_n : n \in \mathbb{N}\}$ be an orthonormal sequence in X. Then for any $x \in X$, the positive series $\sum_{n=1}^{\infty} | < x, e_n > |^2$ converges and*

$$\sum_{n=1}^{\infty} | < x, e_n > |^2 \leq \|x\|^2.$$

Corollary 3.1.38. *If $\{e_n : n \in \mathbb{N}\}$ is an orthonormal sequence in an inner product space X, then for any $x \in X$,*

$$\lim_{n \to \infty} < x, e_n >= 0.$$

Corollary 3.1.39. *Let H be a Hilbert space and let $\{e_n : n \in \mathbb{N}\}$ be an orthonormal sequence in H. If (x_n) is a sequence in $\mathbb{F}$, then $\sum_{n \in \mathbb{N}} x_n e_n$ converges iff $(x_n) \in \ell^2$.*

It is known from a Linear Algebra Course that if, $\{e_k\}_{k=1,\cdots,n}$ is a basis for some vector space X (with $\dim X = n$), then any $x \in X$ is uniquely expressible as $x = \sum_{k=1}^{n} \alpha_k e_k$ for some $\alpha_k \in \mathbb{F}$, $k = 1, \cdots, n$. Using inner products, we may reformulate this as

$$x = \sum_{k=1}^{n} < x, e_k > e_k.$$

Can we extend this to an infinite sum? By the Bessel Inequality we are sure that the series $\sum_{n=1}^{\infty} < x, e_n > e_n$ converges (why?). Another question: Does the previous series converges to x? If not, when does this occur? The following theorem answers this question.

Theorem 3.1.40. *Let H be a Hilbert space and let $\{e_n : n \in \mathbb{N}\}$ be an orthonormal sequence in H. Then the following statements are equivalent:*

(1) $\overline{\{e_n : n \in \mathbb{N}\}^{\perp} = \{0\}}$;
(2) $\overline{\operatorname{span}\{e_n : n \in \mathbb{N}\}} = H$;
(3) $\|x\|^2 = \sum_{n=1}^{\infty} | < x, e_n > |^2$ *for all* $x \in H$ ***(Parseval Identity)***;
(4) $x = \sum_{n=1}^{\infty} < x, e_n > e_n$ *for all* $x \in H$ ***(Fourier expansion)***.
(5) *For any* $x, y \in H$:

$$< x, y >= \sum_{n=1}^{\infty} < x, e_n > \overline{< y, e_n >}.$$

Remark. Each $< x, e_n >$ is called a **Fourier coefficient** of x.

Remark. If $\{e_n : n \in \mathbb{N}\}^{\perp} = \{0\}$, then we say that the sequence $\{e_n : n \in \mathbb{N}\}$ is **total (or complete)**.

Definition 3.1.41. *Let H be a Hilbert space and let $\{e_n\}_{n\in\mathbb{N}}$ be an orthonormal sequence in H. We say that $\{e_n\}_{n\in\mathbb{N}}$ is an **orthonormal basis** (or **Hilbert basis**) for H if one (hence all) of the five properties in the previous theorem holds.*

Examples 3.1.42.

(1) *If X is a finite dimensional vector space, then every orthonormal set in X is an orthonormal basis.*
(2) *Let*

$$e_1 = (1, 0, 0, \cdots), \ e_2 = (0, 1, 0, \cdots), ...etc.$$

*Then $\{e_n : n \in \mathbb{N}\}$ is an orthonormal basis in ℓ^2 (usually referred to as the **standard (or usual or natural) basis** in ℓ^2). See Exercise 3.3.20.*

(3) *The sequence*

$$\{e^{2\pi i n x} : n \in \mathbb{Z}\}$$

is an orthonormal basis in the space $L^2[0, 1]$ (Exercise 3.3.20).

Proposition 3.1.43. *Every orthonormal basis in a separable Hilbert space is a Schauder basis.*

Now, comparing with Theorem 3.1.36, we may wonder whether every Hilbert space has an orthonormal basis? It turns out that only separable Hilbert spaces have this property.

Theorem 3.1.44. *An infinite dimensional Hilbert space is separable iff it contains an orthonormal (countable) basis. If* $\dim H = n$, *then H is isomorphic to $\mathbb{C}^n$. If* $\dim H$ *is countably infinite, then H is isomorphic to ℓ^2.*

Remark. By the examples 3.1.42, we can say that ℓ^2 and $L^2[0,1]$ are separable.

Remark. For an example of a non separable Hilbert space, see Exercise 3.3.31.

Proposition 3.1.45. *Let H be a separable Hilbert space and let Y be an infinite dimensional closed subspace of H. Let $\{e_n : n \in \mathbb{N}\}$ be a Hilbert basis of Y. For each $x \in H$, the series $\sum_{n=1}^{\infty} < x, e_n > e_n$ is convergent in Y with a sum given by the projection of x onto Y:*

$$p_Y(x) = \sum_{n=1}^{\infty} < x, e_n > e_n.$$

3.1.6. Orthogonal Polynomials.

First, we give a definition.

Definition 3.1.46. *Let I be a closed interval in $\mathbb{R}$ and let $w : I \to \mathbb{R}$ be a continuous function. Assume further that w is strictly positive on $\mathring{I}$. We say that w is a* ***weight*** *if*

$$\forall n \in \mathbb{N} : \int_I |x|^n w(x)dx < \infty.$$

We give a few more or less known examples.

Examples 3.1.47.

(1) $w(x) = e^{-x^2}$, $I = \mathbb{R}$.
(2) $w(x) = e^{-x}$, $I = \mathbb{R}^+$.
(3) $w(x) = (1-x)^a(1+x)^b$, $I = [-1,1]$, $a, b \in (-1, \infty)$.

Using different weights we may produce different orthogonal polynomials. To this end, we define an inner product space. Let

$$W = \left\{ f : I \to \mathbb{R} \text{ continuous s.t. } \int_I (f(x))^2 w(x)dx < \infty \right\}.$$

Then it can be shown that W is a non-complete inner product space with respect to the inner product

$$< f, g >= \int_I f(x)g(x)w(x)dx,$$

where $f, g \in W$.

PROPOSITION 3.1.48. *The sequence* $\{1, x, x^2, \cdots, x^n, \cdots\}$ *is linearly independent in* W.

By the previous result, we may then apply the Gram-Schmidt algorithm to obtain orthonormal polynomials $p_0, p_1, \cdots, p_n, \cdots$. Based on what we have just said, here are some famous orthogonal polynomials.

EXAMPLES 3.1.49.

(1) *In the case,* $w(x) = e^{-x}$ *and* $I = \mathbb{R}^+$, *the produced polynomials are called the* ***Laguerre polynomials***.
(2) *In the case,* $w(x) = e^{-x^2}$ *and* $I = \mathbb{R}$, *the produced polynomials are called the* ***Hermite polynomials***.
(3) *In the case* $w(x) = (1-x)^a(1+x)^b$, $a, b > -1$ *and* $I = [-1, 1]$, *the produced polynomials are in general called the* ***Jacobi polynomials***. *Two important classes of Jacobi polynomials are the following:*
   (a) *In the case* $w(x) = (1-x)^a(1+x)^b$ *and* $I = [-1, 1]$, *with* $a = b = 0$, *the produced polynomials are called the* ***Legendre polynomials***.
   (b) *In the case* $w(x) = (1-x)^a(1+x)^b$ *and* $I = [-1, 1]$, *with* $a = b = -\frac{1}{2}$, *the produced polynomials are called the* ***Chebyshev polynomials***.

REMARK. There is a wide range of applications of orthogonal polynomials. They may be met when dealing with (among others):

(1) Continued fractions,
(2) Moment problem,
(3) Quadrature,
(4) Random matrices,
(5) Electrostatics,
(6) Polynomial solutions of eigenvalue problems,
(7) Harmonic analysis on spheres and balls,
(8) Approximation theory,
(9) etc...

More details may be found in e.g. **[85]**. The reference **[199]** is also of interest.

We finish with an interesting result.

THEOREM 3.1.50. *If* I *is a bounded interval, then the orthonormal sequence* $\{\frac{p_n}{\|p_n\|} : n \geq 0\}$ *is a Hilbert basis for* W.

3.1.7. Linear Operators on a Hilbert Space. In this subsection, we introduce very basic notions of linear operators and their adjoints. The next chapters contain much deeper material on this topic.

Since a Hilbert space is a Banach space, the definition of a bounded operator A on a Hilbert H remains the same as in the Banach case (using the induced norm by an inner product). As is customary, we write $B(H)$ instead of $B(H,H)$.

We have seen four equivalent norms of a bounded linear operator A on a Banach space (cf. Theorem 2.1.17). If $A \in B(H)$, there is still one more! (This is a simple consequence of Exercise 3.3.8):

Proposition 3.1.51. *Let $A \in B(H)$. Then*

$$\|A\| = \sup\{|< Ax, y >| : \ \|x\| = \|y\| = 1\}.$$

The next fundamental result on duality in indispensable for defining the important notion of the adjoint of an operator on a Hilbert space (see Theorem 3.1.53).

Theorem 3.1.52. ***(Riesz-Fréchet)*** *Let H be a Hilbert space and let $f \in H'$, i.e. let f be a continuous linear functional on H. Then there exists a unique $y \in H$ such that*

$$f(x) =< x, y >, \ \forall x \in H.$$

Furthermore, $\|f\|_{H'} = \|y\|_H$.

Remark. We insist on the space H being complete (cf. Exercise 3.3.35).

Theorem 3.1.53. *Let H and K be two complex Hilbert spaces with inner products $< \cdot, \cdot >_H$ and $< \cdot, \cdot >_K$ respectively. Let $A \in B(H, K)$. Then there exists a unique operator in $B(K, H)$, denoted by A^*, such that*

$$< Ax, y >_K =< x, A^*y >_H$$

for all $x \in H$ and all $y \in K$.

Definition 3.1.54. *Let $A \in B(H, K)$. Then the (unique) operator $A^* \in B(K, H)$, introduced in the previous theorem, is called the **adjoint** of A.*

Before giving some properties of the adjoint, we give a few examples.

Examples 3.1.55.

(1) *The adjoint of any identity operator is itself, i.e. $I^* = I$.*
(2) *In $\mathbb{C}^2$,*

$$\begin{pmatrix} a & b \\ c & d \end{pmatrix}^* = \begin{pmatrix} \overline{a} & \overline{c} \\ \overline{b} & \overline{d} \end{pmatrix}.$$

In $\mathbb{R}^2$, it is clear that

$$\begin{pmatrix} a & b \\ c & d \end{pmatrix}^* = \begin{pmatrix} a & c \\ b & d \end{pmatrix} = \begin{pmatrix} a & b \\ c & d \end{pmatrix}^t.$$

The next operator on ℓ^2 (or on other ℓ^p) is so important that we just can't imagine how would Operator Theory be without it!

DEFINITION 3.1.56. *Let $S \in B(\ell^2)$ be defined by*

$$S(x_1, x_2, \cdots) = (0, x_1, x_2, \cdots).$$

Then S is called the (unilateral) ***shift*** *operator.*

REMARK. In Definition 3.1.58, the appellation "unilateral" will be justified.

REMARK. The shift operator is a valuable source of counterexamples in Operator Theory.

REMARK. It is sometimes more appropriate to define an operator using its effect on the standard (or other) orthonormal basis. For instance, the shift acts on the standard basis $\{e_n : n \in \mathbb{N}\}$ as

$$Se_n = e_{n+1}, \ n \in \mathbb{N}.$$

We may also represent it as

$$S = \begin{pmatrix} 0 & 0 & 0 & 0 & \cdots \\ 1 & 0 & 0 & 0 & \cdots \\ 0 & 1 & 0 & 0 & \cdots \\ 0 & 0 & 1 & 0 & \cdots \\ \vdots & 0 & 0 & 1 & \ddots \\ \vdots & \vdots & \vdots & \ddots & \ddots \end{pmatrix}.$$

Related to it, we have already dealt with a multiplication operator on ℓ^2 (Exercise 2.3.2). Its action on the natural Hilbert basis of ℓ^2 is described by

$$Ae_n = \alpha_n e_n, \ n \in \mathbb{N}$$

(this expression is sometimes referred to as a **diagonal operator**).

PROPOSITION 3.1.57. *(a proof is supplied in Exercise 3.3.41) The adjoint of the shift operator is given by*

$$S^*(x_1, x_2, x_3, \cdots) = (x_2, x_3, \cdots).$$

Its action on the Hilbert basis $\{e_n : n \in \mathbb{N}\}$ is then:

$$S^* e_1 = 0_{\ell^2}, \ S^* e_n = e_{n-1}, \ n \geq 2.$$

Remark. Let H be a separable Hilbert space and let (e_n) be an orthonormal basis (not necessarily the standard one). Can you give a formula for the shift with respect to this basis?

Now, we introduce another "shift" operator:

Definition 3.1.58. *On $\ell^2(\mathbb{Z})$, define an operator R by*

$$R(x_n)_{n\in\mathbb{Z}} = R(\cdots, x_{-1}, \boldsymbol{x_0}, x_1, \cdots) = (\cdots, x_{-2}, \boldsymbol{x_{-1}}, x_0, \cdots),$$

where the "bold" indicates the zeros positions of the sequence. The operator R is called the ***bilateral (forward or right) shift****.*

Remark. Let us agree that whenever we say the bilateral shift, then we mean this operator R.

Proposition 3.1.59. *(see Exercise 3.3.42) Let R be the bilateral shift on $\ell^2(\mathbb{Z})$. Then $\|R\| = 1$ and its adjoint is given by*

$$R^*(\cdots, x_{-1}, \boldsymbol{x_0}, x_1, \cdots) = (\cdots, x_{-1}, x_0, \boldsymbol{x_1}, x_2, \cdots).$$

Definition 3.1.60. *The adjoint of R, i.e. R^*, is called the* ***bilateral backward (or left) shift****.*

Next, we pass to basic properties of the adjoint operation:

Proposition 3.1.61. *Let $A, B \in B(H)$ and let $\alpha, \beta \in \mathbb{C}$. Then*

(1) $(\alpha A + \beta B)^* = \overline{\alpha} A^* + \overline{\beta} B^*$.
(2) $(AB)^* = B^* A^*$.
(3) $(A^*)^* = A$.
(4) $\|A^*\| = \|A\|$.
(5) $\|A\|^2 = \|A^* A\|$.
(6) *If A is invertible, then its adjoint A^* too is invertible and*

$$(A^*)^{-1} = (A^{-1})^*.$$

Corollary 3.1.62. *Let $A \in B(H)$. Then*

$$\|A^* A\| = \|AA^*\| = \|A\|^2$$

so that $A^* A = 0$ iff $A = 0$.

We will see an example of a bounded sequence in a Hilbert space which does not have any strong convergent subsequence (cf. Exercise 9.3.1). But, as a consequence of the Riesz Representation Theorem we have the following result:

Proposition 3.1.63. *Let H be a Hilbert space. Any bounded sequence in H admits a weakly convergent subsequence.*

The following result generalizes the polarization identity already met in Theorem 3.1.8.

THEOREM 3.1.64. ***(Generalized Polarization Identity)*** *Let $A \in B(H)$ (where H is a $\mathbb{C}$-Hilbert). Then for any $x, y \in H$,*

$$< Ax, y >= \frac{1}{4}[< A(x+y), x+y > - < A(x-y), x-y >]$$
$$+ \frac{i}{4}[A(x+iy), x+iy > - < A(x-iy), x-iy >]$$

PROPOSITION 3.1.65. *Let H be a complex Hilbert space and let $A \in B(H)$. Then the following statements are equivalent:*

(1) $A = 0$,
(2) $< Ax, x >= 0$ *for all* $x \in H$,
(3) $< Ax, y >= 0$ *for all* $x, y \in H$.

The next theorem is fundamental.

THEOREM 3.1.66. *Let H be a Hilbert space and let $A \in B(H)$. Then*

(1) $\ker A = [\operatorname{ran}(A^*)]^\perp$.
(2) $\ker A^* = [\operatorname{ran}(A)]^\perp$.
(3) A^* *is injective iff* $\overline{\operatorname{ran}(A)} = H$.

PROPOSITION 3.1.67. *Let $A \in B(H)$. Then*

(1) $\ker A = \ker(A^*A)$.
(2) $\ker A^* = \ker(AA^*)$.

Operators on a Hilbert space with non-closed ranges do exist (see e.g. Exercise 3.3.44). The following result prescribes an interesting relation between the range of an operator and that of its adjoint (a proof is given in Exercise 3.3.48).

PROPOSITION 3.1.68. *(cf. Exercise 4.3.12) Let $A \in B(H)$. Then the range of A is closed if and only if the range of A^* is closed.*

THEOREM 3.1.69. *(see Exercise 3.3.46 for a proof, cf. Propositions 5.1.40 & 5.1.41) Let $A, B \in B(H)$. Then*

$$\forall x \in H: \ \|Ax\| \leq \|Bx\| \iff \exists K \in B(H) \text{ contraction } : A = KB.$$

The next theorem is very important in applications (the reader may consult its proof in Exercise 3.3.37):

THEOREM 3.1.70. ***(Hellinger-Toeplitz)*** *Let H a complex Hilbert space H. Let $A \in L(H)$ (i.e. A is only linear for the moment). Assume further that*

$$< Ax, y >=< x, Ay >, \ \forall x, y \in H.$$

Then $A \in B(H)$.

We finish this subsection with another important consequence (e.g. it could be used in the Borel Functional Calculus) of the Riesz-Fréchet Theorem to bilinear forms:

Proposition 3.1.71. *Let H be a Hilbert space and let B be a bounded sesquilinear form on H. Then*

$$\exists A \in B(H) \text{ unique}: \ B(x,y) = < Ax, y >$$

for all $x, y \in H$.

3.1.8. Operator Topologies.

Given a sequence (without any detail) involving elements of $B(H)$, then three natural sequences come to our minds, namely, $(A_n)_n$, $(A_n x)_n$ and $(< A_n x, y >)_n$. However, these act on different spaces as (A_n) is a sequence in $B(H)$, $(A_n x)$ is a sequence in H and $(< A_n x, y >)_n$ is a sequence in $\mathbb{C}$. This paragraph has prepared the ground for the next definition. Observe that the first two definitions have already been introduced on Banach spaces and we recall them here for readers convenience.

Definition 3.1.72. *Let H be a Hilbert space and let (A_n) be a sequence in $B(H)$.*

1. *We say that (A_n)* ***converges in norm (or uniformly)*** *to $A \in B(H)$ if*

   $$\lim_{n \to \infty} \|A_n - A\| = 0.$$

   The topology associated with this convergence is called the ***Topology of the Operator Norm.***
2. *We say that (A_n)* ***strongly converges*** *to $A \in B(H)$ if*

   $$\lim_{n \to \infty} \|(A_n - A)x\| = 0$$

   for each $x \in H$. The topology associated with this convergence is called the ***Strong Operator Topology*** *(denoted also by* ***S.O.T.****).*
3. *We say that (A_n)* ***weakly converges*** *to $A \in B(H)$ if*

   $$\lim_{n \to \infty} < A_n x, y > = < Ax, y >$$

   for each $x, y \in H$. The topology associated with this convergence is called the ***Weak Operator Topology*** *(denoted also by* ***W.O.T.****).*

Remark. We have already seen the concept of a weakly convergent sequence in a Hilbert space. The definition here is basically the same as (A_n) weakly converges to A iff $(A_n x)$ weakly converges to Ax according to Definition 3.1.12.

Remark. Proposition 2.1.31 is not valid w.r.t. weak convergence. See e.g. [**94**].

The following is easily seen to hold.

Proposition 3.1.73. *Let H be a Hilbert space and let (A_n) be a sequence in $B(H)$. Then uniform convergence implies strong convergence, and strong convergence entails weak convergence.*

Remark. None of the backward implications holds in the previous proposition. See Exercise 3.3.41 for counterexamples.

3.2. True or False: Questions

QUESTIONS. Comment on the following questions/statements and indicate those which are false and those which are true when this applies. Justify your answers.

(1) A complex inner product is linear with respect to the first vector and conjugate-linear with respect to the second vector.

(2) Let $f, g \in X = C([0,1], \mathbb{R})$. Let $f, g \geq 0$ be such that $fg \geq 1$. Then

$$\int_0^1 f(x)dx \int_0^1 g(x)dx \geq 1.$$

(3) In a *real* inner product space, the equation

$$\|x+y\|^2 = \|x\|^2 + \|y\|^2$$

implies the orthogonality of the vectors x and y.

(4) The same question in a complex inner product space this time.

(5) Let X be an inner product space. Let $Y \subset X$. Give one or two reasons why $Y = Y^{\perp\perp}$ need not hold.

(6) Let (e_n) be an orthonormal sequence in an inner product space X and let $x \in X$. Then

$$\lim_{n\to\infty} < x, e_n >= 0.$$

(7) A subsequence of an orthonormal basis is an orthnormal basis.

(8) Let H be a Hilbert space and let (e_n) be an orthonormal basis in H. Then the series $\sum_{n=1}^{\infty} \| < x, e_n > e_n \|$ is convergent.

(9) The adjoint of a linear operator is always unique.

(10) Let $A \in B(H)$. If $A \neq 0$, then $A^* \neq 0$.

(11) Let $A \in B(H)$. Then $\|A^2\| = \|A\|^2$.

(12) Let $A \in B(H)$. Say why a priori an equality of the type $(\ker A)^{\perp} = \operatorname{ran} A^*$ need not hold.

(13) Let H be a Hilbert space and let (A_n) be a sequence in $B(H)$. Let $A \in B(H)$. Then (A_n) converges in norm to A iff (A_n^*) converges in norm to A^*.

(14) Let H be a Hilbert space and let (A_n) be a sequence in $B(H)$. Then (A_n) strongly converges to A iff (A_n^*) strongly converges to A^*.

(15) Let H be a Hilbert space and let (A_n) be a sequence in $B(H)$. Then (A_n) weakly converges to A iff (A_n^*) weakly converges to A^*.

3.3. Exercises with Solutions

Exercise 3.3.1. On $\mathbb{R}^2 \times \mathbb{R}^2$, we define

$$< (x, y), (x', y') >= xx' - yy'.$$

Does this mapping define an inner product on $\mathbb{R}^2$?

Exercise 3.3.2. Is

$$< f, g >= \int_0^1 f'(x)\overline{g'(x)}dx$$

an inner product space on $C^1([0, 1], \mathbb{R})$?

Exercise 3.3.3. For all $x = (x_1, x_2, \cdots, x_n)$ and $y = (y_1, y_2, \cdots, y_n)$ in $\mathbb{R}^n$, set

$$< x, y >= \sum_{k=1}^{n} x_i y_i.$$

Show that this defines a real inner product on $\mathbb{R}^n$.

Exercise 3.3.4. Show that the following are inner products on X:

(1) $< f, g >= \int_0^1 f(x)\overline{g(x)}dx$, $X = C([0, 1], \mathbb{R})$.
(2) $< f, g >= \int_{-\infty}^{\infty} f(x)\overline{g(x)}dx$, $X = L^2(\mathbb{R})$.
(3) $< f, g >= \int_0^1 f'(x)\overline{g'(x)}dt + f(0)\overline{g(0)}$, $C^1([0, 1], \mathbb{R})$.
(4) $< f, g >= \int_0^1 f(x)\overline{g(x)}dx + \int_0^1 f'(x)\overline{g'(x)}dx$, $X = C^1([0, 1], \mathbb{R})$.
(5) $< f, g >= \int_{-1}^{1} x^2 f(x)g(x)dx$, $X = C([-1, 1], \mathbb{R})$.
(6) $< P, Q >= \int_{-\infty}^{\infty} e^{-x^2} P(x)Q(x)dx$, X is the vector space of all polynomials with real coefficients.

Exercise 3.3.5. Let $X = \mathcal{M}_{n,m}(\mathbb{R})$ be the vector space of matrices with n rows and m columns and with real entries. Let $A, B \in \mathcal{M}_{n,m}(\mathbb{R})$. Set

$$< A, B >= \operatorname{tr}(A^t B)$$

where "tr" denotes the trace of a square matrix (in $\mathbb{C}$, we use the conjugate transpose, denoted by $\overline{A}^t = \overline{A^t}$ or by A^*, instead of the transpose).

(1) Show that $< \cdot, \cdot >$ is an inner product on $\mathcal{M}_{n,m}(\mathbb{R})$.
(2) Deduce that $A \mapsto \|A\|_F = \sqrt{\operatorname{tr}(A^t A)}$ is a norm on X (this norm is more known as the **Frobenius Norm**).

Exercise 3.3.6. Let X be an inner product space. Let $x \in X$. Then

$$< x, y >= 0, \ \forall y \in X \iff x = 0.$$

Exercise 3.3.7. Let X be an inner product space and let $x, y \in X$.

(1) Show that
$$| < x, y > | \leq \sqrt{< x, x >}\sqrt{< y, y >}$$
for all $x, y \in X$.

(2) Show that $| < x, y > | = \sqrt{< x, x >}\sqrt{< y, y >}$ iff $x = \alpha y$ for some $\alpha \in \mathbb{C}^*$, that is, iff x and y are linearly dependent.

(3) Show that $\|x\| = \sqrt{< x, x >}$ defines a norm on X.

Exercise 3.3.8. Let X be an inner product space. Let $x, y \in X$. Show that
$$\|y\| = \sup_{\|x\|=1} | < x, y > |.$$

Exercise 3.3.9. Let X be an inner product space on $\mathbb{F}$.

(1) Show that for all $x, y \in X$
$$\|x + y\|^2 + \|x - y\|^2 = 2(\|x\|^2 + \|y\|^2)$$

(2) Show that if $\mathbb{F} = \mathbb{R}$, then
$$< x, y >= \frac{1}{4}(\|x + y\|^2 - \|x - y\|^2).$$

(3) Show that if $\mathbb{F} = \mathbb{C}$, then
$$< x, y >= \frac{1}{4}(\|x + y\|^2 - \|x - y\|^2 + i\|x + iy\|^2 - i\|x - iy\|^2).$$

Exercise 3.3.10. Show that the following norms cannot be induced by an inner product on X:

(1) $\|x\|_1 = \sum_{k=1}^{n} |x_k|$ and $\|x\|_\infty = \max_{1 \leq k \leq n} |x_k|$ both on $X = \mathbb{R}^n$.

(2) $\|f\|_1 = \int_0^1 |f(x)|dx$ and $\|f\|_\infty = \max_{x \in [0,1]} |f(x)|$ both on $X = C[0, 1]$.

(3) $\|A\| = \|(a_{ij})\| = \max_{1 \leq j \leq n} \left(\sum_{1 \leq i \leq n} |a_{ij}| \right)$ on X, the space of square matrices of order n having real entries.

(4) $\|f\| = \int_{\mathbb{R}} |f(x)|dx$ on $X = L^1(\mathbb{R})$.

Exercise 3.3.11. Let X be an inner product space.

(1) Show that every strongly convergent sequence in X converges weakly (to the same limit).

(2) Is the converse true?

(3) Show that if (x_n) is a sequence in X which converges weakly to $x \in X$ and if $\lim\limits_{n\to\infty} \|x_n\| = \|x\|$, then (x_n) converges strongly to x.

Exercise 3.3.12. Let X be a *finite dimensional* inner product space. Let $x \in X$ and let (x_n) be a sequence in X. Show that if (x_n) converges weakly to x, then (x_n) converges strongly to x. Conclude.

Exercise 3.3.13. Are the following Hilbert spaces?

(1) $X = C([0,1], \mathbb{R})$ with respect to the *inner product*

$$< f, g >= \int_0^1 f(x)g(x)dx$$

for all $f, g \in X$.

(2) $X = \mathbb{R}^n$ with respect to the *inner product*

$$< x, y >= \sum_{k=1}^{n} x_n y_n$$

for all $x = (x_n), y = (y_n) \in \mathbb{R}^n$.

(3) $X = L^\pi(\mathbb{R})$.

(4) ℓ^2 with respect to the *inner product*

$$< x, y >= \sum_{k=n}^{\infty} x_n \overline{y_n}$$

for all $x = (x_n), y = (y_n) \in \ell^2$.

Exercise 3.3.14. Show that ℓ^p is a Hilbert space iff $p = 2$.

Exercise 3.3.15. Let $X = C([0,1], \mathbb{R})$ be equipped with the *real inner product* $< f, g >= \int_0^1 f(x)g(x)dx$ and

$$Y = \left\{ f \in X : \ f(x) = 0 \text{ for all } 0 \le x \le \frac{1}{2} \right\}.$$

(1) Show that Y is a closed vector space.

(2) Find $Y^\perp$.

(3) Set $Z = \{f \in X : \ f(\frac{1}{2}) = 0\}$. Show that $Y \oplus Y^\perp = Z \neq X$.

(4) Why does this conclusion not contradict the Orthogonal Decomposition Theorem?

Exercise 3.3.16. Let X be an inner product space. Let y be a non-zero fixed element of X. Set

$$Y = \{x \in X : < x, y >= 0\}.$$

(1) Show that $Y = (\text{span}\{y\})^\perp$.

(2) Is Y closed in X?
(3) What is $Y^{\perp}$?
(4) Let $X = C([0,1],\mathbb{R})$ be equipped with its usual inner product. Let
$$Y = \left\{ f \in X : \int_0^1 f(x)dx = 0 \right\}.$$
Give $Y^{\perp}$.

Exercise 3.3.17. Let
$$Y = \{y \in c_{00} \subset \ell^2 : < y, x >= 0\}$$
where $x = (x_n) = \left(\frac{1}{n}\right)_n$ and where $< \cdot, \cdot >$ is the standard ℓ^2 inner product.

(1) Is Y closed?
(2) Show that $c_{00} \neq Y \oplus Y^{\perp}$.
(3) Why does the result of the previous question not contradict the Orthogonal Decomposition Theorem?

Exercise 3.3.18. Give an example of Hilbert space H and a non-closed subspace Y such that $Y \oplus Y^{\perp} \neq H$.

Exercise 3.3.19. ([**4**]) Show that an inner product space X, in which $Y = Y^{\perp\perp}$ for every closed linear subspace $Y \subset X$, is a Hilbert space.

Exercise 3.3.20. Show that the sequence $\{e_n\}$ is orthonormal in H in the following cases:

(1)
$$e_1 = (1, 0, 0, \cdots),\ e_2 = (0, 1, 0, \cdots), ...etc,\ H = \ell^2.$$
(2)
$$\left\{e_n = e^{2\pi i n x} : n \in \mathbb{Z}\right\},\ H = L^2[0,1].$$

Exercise 3.3.21. Let $\{e_n : n \in \mathbb{N}\}$ be the standard orthonormal basis in ℓ^2. Show that $\{e_{2n}\}$ is an orthonormal sequence which is not an orthonormal basis in ℓ^2.

Exercise 3.3.22. ([**50**]) Let $H^2(\mathbb{D})$ be as in Exercise 1.5.7.

(1) Demonstrate that if $f(z) = \sum_{n=0}^{\infty} a_n z^n$, then $(a_n) \in \ell^2$ iff $f \in H^2(\mathbb{D})$. Show also that the map $\varphi(f) : H^2(\mathbb{D}) \to \ell^2$, where $\varphi(f) = (a_n)_n$ iff $f(z) = \sum_{n=0}^{\infty} a_n z^n$, is an isometric isomorphism.
(2) Show that $H^2(\mathbb{D})$ is a Hilbert space.
(3) Show that $\{z^n : n \geq 0\}$ is an orthonormal basis for $H^2(\mathbb{D})$.

Remark. By Test 3.4.3, $H^2(\mathbb{D})$ is the only Hilbert space among the $H^p(\mathbb{D})$.

Exercise 3.3.23. Let X be a Hilbert space. We denote the distance from $x \in X$ to $Y \subset X$ by $d(x, Y)$.

(1) Let $a \in X$ be such that $a \neq 0$. Show that

$$d(x, \{a\}^{\perp}) = \frac{|< x, a >|}{\|a\|}, \ \forall x \in X.$$

(2) Set $X = L^2([0,1])$. Set

$$Y = \left\{ f \in X : \int_0^1 f(x)dx = 0 \right\}.$$

(a) Show that Y is closed vector subspace of X.
(b) Show that $Y^{\perp}$ consists of constant functions on $[0,1]$.
(c) Let $f(x) = x$. Find $d(f, Y)$.

Exercise 3.3.24. Let $X = C([0, 2\pi], \mathbb{C})$ be endowed with the *inner product*

$$< f, g >= \frac{1}{2\pi} \int_0^{2\pi} f(x)\overline{g(x)}dx.$$

(1) Show that the collection $(e_p)_{p \in \mathbb{Z}}$ of the functions defined by $e_p(x) = e^{ipx}$, $x \in \mathbb{R}$, is orthonormal.
(2) Let $p \in \mathbb{N}$. For every $f \in X$, show that

$$\lim_{p \to \infty} \frac{1}{2\pi} \int_0^{2\pi} f(x)e^{-ipx}dx = 0.$$

Exercise 3.3.25. Let X be an inner product space. For $x \in X$, define a function f_x from X into $\mathbb{C}$ by $f_x(y) =< x, y >$.

(1) Show that f_x is linear and continuous.
(2) Let $Y \subset X$. Deduce that $Y^{\perp}$ is a closed subspace of X.

Exercise 3.3.26. Using the Gram-Schmidt algorithm, find an orthonormal set out of $\{1, x, x^2\}$ in $L^2[-1,1]$.

Exercise 3.3.27. Find

$$\min_{\alpha,\beta,\gamma \in \mathbb{R}} \int_{-1}^1 |x^3 - \alpha x^2 - \beta x - \gamma|^2 dx.$$

Exercise 3.3.28. Evaluate

$$\inf_{a,b \in \mathbb{R}} \int_0^1 (x^2 - ax - b)^2 dx.$$

Exercise 3.3.29. Let $X = C([0,1],\mathbb{R})$ be endowed with the *inner product* defined by

$$< f, g >= \int_0^1 f(x)g(x)dx$$

for all $f, g \in X$. Let $Y = \mathbb{R}(1-x) = \{\alpha(1-x)\}_{\alpha\in\mathbb{R}}$.

(1) Explain briefly why Y is complete.
(2) Let $f(x) = 1$ for all $x \in [0,1]$. Find the orthogonal projection of f onto Y.

Exercise 3.3.30. Let H be an inner product space. Show that if H is separable, then every orthonormal set in H is countable.

Exercise 3.3.31. (A non-separable Hilbert space) Set $u_r(x) = e^{irx}$ where $x, r \in \mathbb{R}$. Let X be the complex vector space of all finite linear combinations of the functions u_r. For $f, g \in X$, set

$$< f, g >= \lim_{n\to\infty} \frac{1}{2n} \int_{-n}^{n} f(x)\overline{g(x)}dx.$$

(1) Show that $< \cdot, \cdot >$ is an inner product on X.
(2) Show that the completion of X is not separable.

Exercise 3.3.32. Let Ω be an open subset of the complex plane. Recall that (cf. Exercise 1.5.8)

$$\mathcal{A}^2 = \left\{ f : \Omega \to \mathbb{C} \text{ analytic } : \iint_{\Omega} |f(x+iy)|^2 dxdy < \infty \right\}$$

and define for $f, g \in \mathcal{A}^2$

$$< f, g >= \iint_{\Omega} f(x+iy)\overline{g(x+iy)}dxdy.$$

(1) Verify that $< \cdot, \cdot >$ is an inner product on $\mathcal{A}^2$.
(2) Show that if f is analytic and $B_c(a,r) \subset \Omega$, then

$$f(a) = \frac{1}{\pi r^2} \iint_{B_c(a,r)} f(x+iy)dxdy.$$

(3) Infer that if $f \in \mathcal{A}^2(\Omega)$, then

$$|f(a)| \le \frac{1}{\sqrt{\pi} r} \|f\|.$$

(4) Let K be a compact subset of Ω. Let $R > 0$ be such that

$$\bigcup_{a \in K} B_c(a, R) \subset \Omega$$

Show that for all $f \in \mathcal{A}^2(\Omega)$, we have

$$\sup_{z \in K} |f(z)| \leq \frac{1}{\sqrt{\pi} R} \|f\|.$$

(5) Show that $\mathcal{A}^2$ is a Hilbert space (you may use the fact that the space of analytic functions is closed in the space of continuous functions on compact subsets of Ω w.r.t. the supremum norm).

Exercise 3.3.33. Let H be an infinite dimensional Hilbert space. Equip $B(H)$ with the operator norm.

(1) Is $B(H)$ separable?
(2) Is $B(H)$ compact?
(3) Is $B(H)$ a Hilbert space?

Exercise 3.3.34. Let $A \in B(H)$ be defined by

$$Ax = < x, b > a + < x, a > b,$$

where a and b are in H, both non zero and orthogonal. Show that

$$\|A\| = \|a\| \|b\|.$$

Exercise 3.3.35. Let $c_{00} \subset \ell^2$ be endowed with the ℓ^2-*inner product*:

$$< x, y > = \sum_{n=1}^{\infty} x_n \overline{y_n}.$$

(1) Define $f : c_{00} \to \mathbb{C}$ by $f(x) = \sum_{n=1}^{\infty} \frac{x_n}{n}$. Prove that f is a continuous linear functional on c_{00} and that $\|f\| \leq \frac{\pi}{\sqrt{6}}$.
(2) Show that

$$\nexists y \in c_{00} : \ \forall x \in c_{00}, \ f(x) = < x, y > .$$

(3) Why does the conclusion of the previous question not violate the Riesz-Fréchet Theorem?

Exercise 3.3.36. For all $x = (x_n) \in \ell^2$, set

$$A(x_n) = (x_1, 2x_2, \cdots, nx_n, \cdots).$$

Is A bounded on ℓ^2?

Exercise 3.3.37. Let H be a complex Hilbert space H. Let $A \in L(H)$. Assume further that

$$< Ax, y >=< x, Ay >, \ \forall x, y \in H.$$

Show that $A \in B(H)$.

Exercise 3.3.38. Give the adjoint of the following matrices

$$A = \begin{pmatrix} i & 1 \\ -2i & 3 \end{pmatrix} \text{ and } B = \begin{pmatrix} 1 & -1 \\ 0 & 2 \end{pmatrix}.$$

Exercise 3.3.39. Let φ be a continuous function on $[0,1]$ taking its values in $\mathbb{C}$. For all $f \in L^2([0,1])$, consider the multiplication operator

$$(Af)(x) = \varphi(x)f(x).$$

Compute A^*.

Exercise 3.3.40. Let $H = L^2([0,1])$ and let $K : [0,1] \times [0,1] \to \mathbb{C}$ be a continuous function. Let A be defined by:

$$(Af)(x) = \int_0^1 K(x,y)f(y)dy.$$

(1) Show that A is a linear bounded operator on H (A is referred to as an "**Integral Operator**").
(2) Compute A^*.

Exercise 3.3.41. Let $S : \ell^2 \to \ell^2$ (the shift operator) be defined by

$$S(x) = S(x_1, x_2, \cdots, x_n, \cdots) = (0, x_1, x_2, \cdots, x_n, \cdots).$$

(1) Show that S is a bounded linear operator on ℓ^2. What is $\|S\|$?
(2) Is ran$S$ closed?
(3) Find the adjoint S^*.
(4) Find $\ker S^*$.
(5) Show that

$$\text{ran}S = \overline{\text{span}\{e_2, e_3, \cdots\}}$$

where (e_n) is the usual orthonormal basis.
(6) Set $A_n = (S^*)^n$. Show that as n tends to ∞, then $A_n x \to 0$ for all $x \in \ell^2$ but $\|A_n\| = 1$, for all $n \in \mathbb{N}$.
(7) Set $B_n = S^n$. Show that B_n converges weakly to 0 but not in the Strong Operator Topology (or that of the operator norm).

Exercise 3.3.42. On $\ell^2(\mathbb{Z})$, let R the bilateral shift defined by

$$R(x_n)_{n\in\mathbb{Z}} = R(\cdots, x_{-1}, \boldsymbol{x_0}, x_1, \cdots) = (\cdots, x_{-2}, \boldsymbol{x_{-1}}, x_0, \cdots),$$

where the "bold" indicates the zeros positions of the sequence.

(1) Show that $R \in B(\ell^2(\mathbb{Z}))$ and that $\|R\| = 1$.
(2) Find R^*.

Exercise 3.3.43. Let $\alpha = (\alpha_n)_n$ be a sequence of complex numbers. Define the multiplication operator A, on ℓ^2, by

$$A(x_1, x_2, \cdots, x_n, \cdots) = (\alpha_1 x_1, \alpha_2 x_2, \cdots, \alpha_n x_n, \cdots).$$

(1) Show that A is bounded on ℓ^2 iff (α_n) is bounded. In case of boundedness of A, find $\|A\|$.
Assume from now on that (α_n) *is bounded:*
(2) Give the expression of A^*.
(3) Show that A is invertible if and only if $\sup_{n\in\mathbb{N}} \frac{1}{|\alpha_n|} < \infty$.
(4) Do we have $\|A^{-1}\| = \frac{1}{\|A\|}$?

Exercise 3.3.44. Define a bounded operator $A : \ell^2 \to \ell^2$ by

$$A(x_1, x_2, \cdots, x_n, \cdots) = \left(\frac{1}{2}x_1, \frac{1}{3}x_2, \cdots, \frac{1}{n+1}x_n, \cdots\right).$$

Is the range of A closed?

Exercise 3.3.45. Let $A \in B(H)$ where H is a Hilbert space on $\mathbb{C}$.

(1) Show that if $< Ax, x >= 0$ for all $x \in H$, then $A = 0$. Deduce that if $B \in B(H)$, then

$$\forall x \in H, \ < Bx, x >=< Ax, x >\Longrightarrow B = A.$$

(2) Does this result remain true if H is an $\mathbb{R}$-Hilbert space? Justify your answer.

Exercise 3.3.46. Let $A, B \in B(H)$. Show that

$$\forall x \in H : \ \|Ax\| \leq \|Bx\| \Longleftrightarrow \exists K \in B(H) \text{ contraction } : A = KB.$$

Exercise 3.3.47. Let H be a Hilbert space and let (x_n) be a weakly convergent sequence to $x \in H$. Show that for all $A \in B(H)$, then (Ax_n) converges weakly to Ax.

Exercise 3.3.48. Let $A \in B(H)$. Show that $\text{ran}(A)$ is closed iff $\text{ran}(A^*)$ is closed. Deduce another proof of the closedness of $\text{ran}S$ where S is the shift operator.

Exercise 3.3.49. Let $A \in B(H)$. Show that if A is surjective, then A^* is bounded below.

3.4. Tests

Test 3.4.1. Let X be the space of Riemann integrable real-valued functions on $[0,1]$. Does

$$< f, g >= \int_0^1 f(x)g(x)dx,$$

where $f, g \in X$, define an inner product on X?

Test 3.4.2. Give a very simple proof of the Cauchy-Schwarz Inequality in an inner product space based on the Bessel Inequality.

Test 3.4.3. Show that the Hardy space $H^p(\mathbb{D})$ is a Hilbert space iff $p = 2$.

Test 3.4.4. Let H be a Hilbert space and let $Y \subset H$. Do we have

$$Y^{\perp\perp\perp} = Y^{\perp}?$$

Test 3.4.5. Let $Y = \{(x_n) \in \ell^2 : \ x_{2n} = 0, \ \forall n \in \mathbb{N}\}$. Find $Y^{\perp}$.

Test 3.4.6. Let H be an infinite dimensional Hilbert space. Equip $B(H)$ with the operator norm. Can $B(H)$ have a Schauder basis?

Test 3.4.7. Can the Frobenius Norm be regarded as an operator norm?

3.5. More Exercises

Exercise 3.5.1. Let $X = C([0,1], \mathbb{R})$. Show that the following *norms* cannot be induced by an inner product on X:

$$\|f\| = \int_0^1 e^x |f(x)| dx \text{ and } \|f\| = \int_0^1 x^2 |f(x)| dx.$$

Exercise 3.5.2. The same question with $X = C^1([0,1], \mathbb{R})$ and the *norm* is given by

$$\|f\|_\infty + \|f'\|_\infty.$$

Exercise 3.5.3. Show that the inner product space X, defined in Exercise 3.3.31, is not complete.

Exercise 3.5.4. Show that $L^p(\mathbb{R})$ is a Hilbert space if and only if $p = 2$.

Exercise 3.5.5. Find

$$\inf_{a,b,c \in \mathbb{R}} \int_0^\infty e^{-x} |x^3 - ax^2 - bx - c|^2 dx.$$

Exercise 3.5.6. Let $\mathcal{A}^2(\Omega)$ be the (Bergman) Hilbert space defined in Exercise 3.3.32. Let $\Omega = \mathbb{D} = \{z \in \mathbb{C} : |z| < 1\}$.

(1) Let
$$e_n(z) = \left(\frac{\sqrt{n+1}}{\sqrt{\pi}}\right) z^n, \ n \geq 0.$$
Show that $\{e_n(z) : n = 0, 1, \cdots\}$ is an orthonormal basis for $\mathcal{A}^2(\Omega)$.

(2) Let $\varphi : \mathcal{A}^2 \to \ell^2$ be defined by
$$\varphi(f) = \left(\sqrt{\frac{\pi}{1}}a_0, \sqrt{\frac{\pi}{2}}a_1, \cdots, \sqrt{\frac{\pi}{n+1}}a_n, \cdots\right)$$
iff $f(z) = \sum_{n=0}^{\infty} a_n z^n$. Prove that φ is an isometric isomorphism.

Exercise 3.5.7. Let X, Y be two linear subspaces of a Hilbert space H. Then

(1) $(X + Y)^{\perp} = X^{\perp} \cap Y^{\perp}$.
(2) $(X \cap Y)^{\perp} = \overline{X^{\perp} + Y^{\perp}}$.

Exercise 3.5.8. (**[136]**) Let $\mathbb{D}$ be the usual open unit disk. Set
$$Di(\mathbb{D}) = \left\{ f : \mathbb{D} \to \mathbb{C} \text{ analytic}, \iint_{\mathbb{D}} |f'(x+iy)|^2 dxdy < \infty \right\}.$$
Set for all $f, g \in Di(\mathbb{D})$
$$< f, g >= f(0)\overline{g(0)} + \frac{1}{\pi}\iint_{\mathbb{D}} f'(x+iy)\overline{g'(x+iy)}dxdy$$

(1) Show that $Di(\mathbb{D})$ is a Hilbert space (*the space* $Di(\mathbb{D})$ *is called the* ***Dirichlet space***).
(2) Find an orthonormal basis for the Dirichlet space.

Exercise 3.5.9. Let $X = C([0,1], \mathbb{R})$. Endow X with the two *norms*
$$\|f\|_{\infty} = \sup_{0 \leq x \leq 1} |f(x)| \text{ and } \|f\|_2 = \sqrt{\int_0^1 |f(x)|^2 dx}.$$
Define a *linear functional* φ on X by
$$\varphi(f) = f\left(\frac{1}{2}\right), \ f \in X.$$

(1) Is φ continuous on X with respect to $\|\cdot\|_{\infty}$?
(2) Is φ continuous on X with respect to $\|\cdot\|_2$?

(3) Is φ well-defined if $X = L^2(0,1)$?

Exercise 3.5.10. Let $\mathcal{A}$ be a Banach algebra. We give a definition: An **involution** is a map $a \to a^*$ from $\mathcal{A}$ into $\mathcal{A}$ such that for all $a, b \in \mathcal{A}$ and for all $\alpha \in \mathbb{C}$:

(1) $(a^*)^* = a$.
(2) $(ab)^* = b^*a^*$.
(3) $(\alpha a + b)^* = \overline{\alpha}a^* + b^*$.

If $\mathcal{A}$ has an identity e, then as a consequence of the first two properties: $e^* = e$.

A C^***-algebra** is a Banach algebra $\mathcal{A}$ with an involution which satisfies

$$\|a^*a\| = \|a\|^2$$

for all $a \in \mathcal{A}$.

The major example of a C^*-algebra is $B(H)$ where H is a Hilbert space and where the involution is given by the usual adjoint operation. Another example is $C(X)$, where X is compact and where

$$f^*(x) = \overline{f(x)}$$

($x \in X$ and $f \in C(X)$). Many results holding in $B(H)$ remains valid in an arbitrary C^*-algebra. For example, show that if $\mathcal{A}$ is a C^*-algebra, then

(1) $\|a\| = \|a^*\|$ for each $a \in \mathcal{A}$.
(2) a^* is invertible iff a is invertible.

See [9] for more on C^*-algebras.

CHAPTER 4

Classes of Linear Operators on Hilbert Spaces

4.1. Basics

4.1.1. Operator Theory Basics. We already introduced a few basic facts about linear operators in the Hilbert Space Chapter. In this chapter, we give deeper material on this.

DEFINITION 4.1.1. *Let H be a Hilbert space and let $A \in B(H)$. Let $I \in B(H)$ be the identity operator on H. We say that A is:*

(1) ***self-adjoint (or symmetric or Hermitian)*** *if $A = A^*$;*
(2) ***skew symmetric (or skew Hermitian)*** *if $A^* = -A$;*
(3) ***normal*** *if $AA^* = A^*A$;*
(4) ***unitary*** *if $AA^* = A^*A = I$;*
(5) *an **isometry** if $A^*A = I$;*
(6) *a **co-isometry** if $AA^* = I$;*
(7) *a **projection or idempotent** if $A^2 = A$; **tripotent** if $A^3 = A$.*
(8) *an **orthogonal projection** if $A^2 = A$ and $A^* = A$;*
(9) ***power bounded*** *if for some $M \geq 0$ and all $n \in \mathbb{N}$, we have $\|A^n\| \leq M$.*

EXAMPLES 4.1.2.

(1) *The matrix*

$$A = \begin{pmatrix} -1 & 1 \\ 1 & \pi \end{pmatrix}$$

is self-adjoint.

(2) *Let φ be a continuous function on $[0,1]$ taking its values in $\mathbb{C}$. For all $f \in L^2([0,1])$ set*

$$(Af)(x) = \varphi(x)f(x).$$

Then A is self-adjoint iff φ is real-valued. Also, A is normal for any φ.

(3) *The shift operator S is a non-unitary isometry, that is,*

$$S^*S = I \text{ and } SS^* \neq I.$$

Its adjoint S^* is therefore a co-isometry.

(4) *The bilateral shift is unitary (Test 4.4.2).*

(5) *The Fourier Transform is a unitary operator from $L^2(\mathbb{R})$ onto $L^2(\mathbb{R})$ (by Theorem 2.1.52).*

Here are some remarks and more examples:

REMARKS.

(1) Self-adjoint and unitary operators are obviously normal.
(2) It is clear that skew symmetric operators are not self-adjoint (except for the zero operator). However, if A is skew symmetric, then A^2 is always self-adjoint.
(3) Every operator $A \in B(H)$ may be expressed as the sum of one symmetric operator and another skew symmetric. More details may be found in the Cartesian Form Subsection.
(4) A skew symmetric operator is normal.
(5) A unitary operator is invertible, with the inverse given by its adjoint.
(6) If $A \in B(H)$ is self-adjoint, then clearly A^n remains self-adjoint ($n \in \mathbb{N}$).
(7) Let $A, B \in B(H)$. If B is self-adjoint, then A^*BA and ABA^* are self-adjoint.
(8) If $A, B \in B(H)$ are self-adjoint, then $AB - BA$ (the commutator of A and B) is skew symmetric.
(9) Let $A \in B(H)$. Then A is normal *if and only if* A^* is normal.
(10) Let $A \in B(H)$ be normal. Then (αA) remains normal for every $\alpha \in \mathbb{C}$. Nonetheless, if $\mathcal{N}$ denotes the collection of all normal operators in $B(H)$, then $\mathcal{N}$ is not necessarily a linear subspace of $B(H)$. The major obstacle is that the sum of two normal operators does not have to be normal. See Exercise 4.3.11 for a counterexample (cf. Theorem 4.1.16). See also Exercise 11.3.37 for further details.
(11) A normal isometry is unitary. So is a surjective isometry.
(12) If A is normal and invertible, then A^{-1} too is normal (Prove it!).
(13) If A is tripotent, then A^2 is idempotent. Notice, that the notion of tripotent matrices play a certain part in applications to digital image encryption as observed in **[30]**.

We start with the following (cf. Proposition 3.1.51):

PROPOSITION 4.1.3. *(cf. Test 6.4.5 & Exercise 6.5.8) Let $A \in B(H)$ be self-adjoint. Then*

$$\|A\| = \sup_{\|x\|=1} |< Ax, x >|.$$

REMARK. By the Cauchy-Schwarz Inequality,

$$\sup_{\|x\|=1} | < Ax, x > | \leq \|A\|.$$

To prove the other inequality, you may use the polarization identity as prescribed in [**178**]. You may also wish to look at (Theorem 9.2.2, [**123**]).

REMARK. The previous result also holds for normal operators (on an $\mathbb{R}$-Hilbert space as well? Or not?). There are different proofs. You may find one in Theorem 21.5 of [**12**] (and a different one in Theorem 12.25, [**190**]).

We have already introduced the concept of similarity of operators. The next is a stronger concept:

DEFINITION 4.1.4. *We say that $A, B \in B(H)$ are* ***unitarily equivalent*** *if there exists a unitary operator $U \in B(H)$ such that*

$$U^{-1}AU = B.$$

REMARK. It is clear that we can replace $U^{-1}AU = B$ by the equivalent condition $U^*AU = B$.

PROPOSITION 4.1.5. *(cf. Exercise 4.3.17) Let $U \in B(H)$ be a unitary operator and let $A, B \in B(H)$ be such that $U^{-1}AU = B$. Then A is self-adjoint (resp. normal) iff B is self-adjoint (resp. normal).*

REMARK. Normal matrices play an important role in Numerical Linear Algebra. In [**86**], the authors gave 70 conditions (sometimes with added hypotheses) equivalent to normality in the setting of $n \times n$ matrices. It is a quite interesting survey, even though the authors claimed it was not one.

Before passing to characterizations of some of the classes introduced above, we say a word on orthogonal projections. The eighth definition of Definition 4.1.1 is equivalent to the following:

DEFINITION 4.1.6. *Let H be a Hilbert space and let X be a closed linear subspace of H. A bounded operator P on H is called an* ***orthogonal projection*** *onto X if*

$$P(x + y) = x, \ \forall x \in X \text{ and } y \in X^{\perp}.$$

REMARK. It is fairly easy to see the following:

(1) P is linear.
(2) $\ker P = X^{\perp}$.
(3) $\operatorname{ran} P = X$.

(4) $Px = x$ for all $x \in X$.
(5) $I - P$ is an orthogonal projection onto $X^{\perp}$.

Here are some direct properties of orthogonal projections.

PROPOSITION 4.1.7. *Let $P \in B(H)$ be an orthogonal projection. Then*

(1) $\text{ran}P = \ker(I - P)$.
(2) $\text{ran}P$ *is closed.*
(3) *Each $x \in H$ may be written as $y + z$, where $Py = 0$ and $Pz = z$.*

THEOREM 4.1.8. *Let $P \in B(H)$ be a projection. Then the following statements are equivalent:*

(1) *P is normal (see Exercise 12.5.7 for a generalization).*
(2) *P is self-adjoint.*
(3) $\text{ran}P = (\ker P)^{\perp}$.
(4) $< Px, x >= \|Px\|^2$ *for all $x \in H$.*

Besides, if P and Q are two orthogonal projections, then

$$\text{ran}P \perp \text{ran}Q \iff PQ = 0.$$

The following contains equivalent definitions to some of the classes introduced above.

THEOREM 4.1.9. *Let $A \in B(H)$. Then*

(1) *A is self-adjoint iff $< Ax, x > \in \mathbb{R}$ for all $x \in H$.*
(2) *A is normal iff $\|Ax\| = \|A^*x\|$ for all $x \in H$.*
(3) *A is unitary iff $\|Ax\| = \|A^*x\| = \|x\|$ for all $x \in H$.*
(4) *A is an isometry iff $\|Ax\| = \|x\|$ for all $x \in H$, that is, iff $< Ax, Ay >=< x, y >$ for all $x, y \in H$.*

Related to the second point, we have a more general concept than normality (as defined in [**25**]):

DEFINITION 4.1.10. *We say that $A, B \in B(H)$ are **metrically equivalent** if $\|Ax\| = \|Bx\|$ for all $x \in H$.*

PROPOSITION 4.1.11. *Let $A, B \in B(H)$. Then*

$$A \text{ and } B \text{ are metrically equivalent} \iff A^*A = B^*B.$$

EXAMPLE 4.1.12. *It is clear that a normal operator and its adjoint are metrically equivalent.*

The following proposition tells us a little more about isometries.

PROPOSITION 4.1.13. *(see Exercise 4.3.6 for a proof) Let $A, B \in B(H)$. Let $V, W \in B(H)$ be isometries. Then*

(1) *VW is an isometry.*
(2) *Also,*

$$\|VA\| = \|A\| \text{ and } \|BV^*\| = \|B\|.$$

(3) $\|V^n\| = 1$ *for all* $n \in \mathbb{N}$.

One of the most powerful tools in the theory of normal operators is the **Fuglede Theorem**. We prefer to state here albeit without proof as it will help us deal with many situations. See Exercise 11.3.30 for its proof, and Chapter 11 for further details.

THEOREM 4.1.14. *Let $A, B \in B(H)$ be such that $AB = BA$. If A (or B) is normal, then $A^*B = BA^*$.*

REMARK. The Fuglede Theorem may be presented as follows: *If an operator commutes with a normal one, then it commutes with the normal's adjoint.*

COROLLARY 4.1.15. *(see Exercise 4.3.8) Let $A, B \in B(H)$ be such that A is normal. Then*

$$BA = AB \iff BA^* = A^*B \iff B^*A = AB^* \iff B^*A^* = A^*B^*.$$

As immediate applications of the Fuglede Theorem, we have:

THEOREM 4.1.16. *Let $A, B \in B(H)$ be both normal and such that $AB = BA$. Then*

(1) *AB, A^*B^*, AB^*, A^*B, BA, BA^*, B^*A and B^*A^* are all normal (see Exercise 4.3.9 for a proof, cf. Exercise 4.3.10).*
(2) *$A+B$, A^*+B^*, A^*+B and $A+B^*$ are all normal (see Exercise 4.3.11).*

REMARK. It is clear from Corollary 4.1.15 that once we prove AB is normal in the previous theorem, a similar argument leads to the normality of remaining products. The same applies to the sum $A + B$ and to the other sums.

4.1.2. Cartesian Form. We know that any complex number z may be expressed as $z = x+iy$, where $x, y \in \mathbb{R}$. This is more commonly known as the cartesian form of a complex number.

In $B(H)$, this has an analogue. Indeed, we may write any $T \in B(H)$ as $A + iB$, where A and B are two bounded self-adjoint operators on H (cf. Exercise 5.3.33). Thus, we may consider to some extent that self-adjoint operators play the role of real numbers.

DEFINITION 4.1.17. *Let $A \in B(H)$ and let "i" be the usual complex number. Then*

$$\operatorname{Re} A = \frac{A + A^*}{2} \text{ and } \operatorname{Im} A = \frac{A - A^*}{2i}$$

are called (respectively) the **real** *and the* ***imaginary parts*** *of* A.

The following is a simple exercise.

Lemma 4.1.18. *Let* $A \in B(H)$. *Then* $\operatorname{Re} A$ *and* $\operatorname{Im} A$ *are both self-adjoint.*

Theorem 4.1.19. *Let* $T \in B(H)$. *Then there exist two self-adjoint operators in* $B(H)$ *such that* $T = A + iB$. *Necessarily,* $A = \operatorname{Re} T$ *and* $B = \operatorname{Im} T$.

Definition 4.1.20. *The decomposition* $T = A + iB$ *is called the* ***Cartesian Form*** *of* $T \in B(H)$.

Like in the case of complex numbers, we have the following result:

Corollary 4.1.21. *Let* $A \in B(H)$. *Then* A *is self-adjoint iff* $\operatorname{Re} A = A$.

Normal operators have an interesting property as regards the Cartesian Form.

Proposition 4.1.22. *(a proof is given in 4.3.36) Let* $T \in B(H)$ *be such that* $T = A + iB$ *with* A *and* B *being self-adjoint. Then* T *is normal iff* A *commutes with* B.

4.1.3. Operator Matrices. We know from an Advanced Linear Algebra Course that (for example!) a matrix 4×4 e.g.

$$A = \begin{pmatrix} 0 & 0 & 1 & 1 \\ 1 & 1 & 2 & 2 \\ 2 & 2 & 3 & 3 \\ 3 & 3 & 4 & 4 \end{pmatrix}$$

may be written as

$$A = \begin{pmatrix} A_{11} & A_{12} \\ A_{21} & A_{22} \end{pmatrix},$$

where

$$A_{11} = \begin{pmatrix} 0 & 0 \\ 1 & 1 \end{pmatrix}, \ A_{12} = \begin{pmatrix} 1 & 1 \\ 2 & 2 \end{pmatrix}, \ A_{21} = \begin{pmatrix} 2 & 2 \\ 3 & 3 \end{pmatrix} \text{ and } A_{22} = \begin{pmatrix} 3 & 3 \\ 4 & 4 \end{pmatrix}.$$

The matrix A was then called a *block matrix* and the A_{ij} ($i, j \in \{1, 2\}$) were called its blocks.

The natural question is whether we can instead have operators on infinite dimensional spaces as blocks? i.e. can we have operators as entries? It turns out that we can in effect do that. One of the virtues of "matrices of operators" is that they will enable us to weaken hypotheses in some situations. This will be seen on several occasions throughout this manuscript.

Before giving the formal definition of this operator matrix, we "recall" the following:

Definition 4.1.23. *Let H_1 and H_2 be two Hilbert spaces. The set*

$$\{(x_1, x_2) : \ x_1 \in H_1, \ x_2 \in H_2\}$$

*is denoted by $H_1 \oplus H_2$, and it is called the **direct sum** of H_1 and H_2. It is a pre-Hilbert space with respect to the inner product*

$$< (x_1, x_2), (y_1, y_2) >_{H_1 \oplus H_2} = < x_1, y_1 >_{H_1} + < x_2, y_2 >_{H_2} .$$

Remark. There is a somehow related notion. If $H = X \oplus Y$ where X and Y are linear subspaces (not necessarily closed), then the **codimension** of X, denoted by $\operatorname{codim} X$, is $\dim Y$ (whether it is finite or infinite).

Proposition 4.1.24. *Let H_1 and H_2 be two Hilbert spaces. Then $(H_1 \oplus H_2, < \cdot, \cdot >_{H_1 \oplus H_2})$ is a Hilbert space.*

Remark. We may identify H_1 with $H_1 \oplus \{0\}$ and H_2 with $\{0\} \oplus H_2$. Hence, H_1 and H_2 are orthogonal subspaces of $H_1 \oplus H_2$, and $H_1 \oplus H_2$ is in fact the ***orthogonal*** direct sum of H_1 and H_2.

Definition 4.1.25. *Let $A_{ij} \in B(H_j, H_i)$, where $i, j = 1, 2$. The matrix*

$$A = \begin{pmatrix} A_{11} & A_{12} \\ A_{21} & A_{22} \end{pmatrix}$$

*is called an **operator matrix** and it defines a linear operator from $H_1 \oplus H_2$ into $H_1 \oplus H_2$ by*

$$A \begin{pmatrix} x_1 \\ x_2 \end{pmatrix} = \begin{pmatrix} A_{11}x_1 + A_{12}x_2 \\ A_{21}x_1 + A_{22}x_2 \end{pmatrix}$$

Example 4.1.26. *Let H and K be two Hilbert spaces. Let I_H be the identity operator on H and let I_K be the identity operator on K. The matrix of operators*

$$I_{H \oplus K} = \begin{pmatrix} I_H & 0 \\ 0 & I_K \end{pmatrix}$$

*is called the **identity** on $H \oplus K$ (where the upper 0 is the zero operator between K and H and the bottom one is the zero operator from H into K).*

Some natural questions arise as regards the A defined previously. Is A bounded? What is the adjoint of A? How do we define the product of two elements in $B(H_1 \oplus H_2)$? The following theorem answers these questions.

THEOREM 4.1.27. *Let $A_{ij} \in B(H_j, H_i)$, where $i, j = 1, 2$ and $B_{ij} \in B(H_j, H_i)$, where $i, j = 1, 2$. Set*

$$A = \begin{pmatrix} A_{11} & A_{12} \\ A_{21} & A_{22} \end{pmatrix} \text{ and } B = \begin{pmatrix} B_{11} & B_{12} \\ B_{21} & B_{22} \end{pmatrix}.$$

Then

(1) *$A \in B(H_1 \oplus H_2)$ and*

$$\max_{1 \le i,j \le 2} \|A_{ij}\| \le \|A\| \le 4 \max_{1 \le i,j \le 2} \|A_{ij}\|.$$

(2) *The sum $A + B$ is given by*

$$A + B = \begin{pmatrix} A_{11} + B_{11} & A_{12} + B_{12} \\ A_{21} + B_{21} & A_{22} + B_{22} \end{pmatrix}.$$

(3) *The product AB is given by*

$$AB = \begin{pmatrix} A_{11}B_{11} + A_{12}B_{21} & A_{11}B_{12} + A_{12}B_{22} \\ A_{21}B_{11} + A_{22}B_{21} & A_{21}B_{12} + A_{22}B_{22} \end{pmatrix}.$$

(4) *The adjoint A^* of A is given by*

$$A^* = \begin{pmatrix} A_{11} & A_{12} \\ A_{21} & A_{22} \end{pmatrix}^* = \begin{pmatrix} A_{11}^* & A_{21}^* \\ A_{12}^* & A_{22}^* \end{pmatrix}.$$

REMARK. In the third property of the preceding theorem, we must not alter the order of the "entries" A_{ij} and B_{ij} as they are operators!

As a special case, we have:

PROPOSITION 4.1.28. *If H and K are two Hilbert spaces, and if $A \in B(H)$ and $B \in B(K)$, then*

$$T(x, y) = (Ax, By)$$

defines a bounded operator on $H \oplus K$, which may be denoted by $A \oplus B$. We call it the ***direct sum of the operators*** *A and B. Moreover (cf. Example 1.1.43)*

$$\|T\| = \|A \oplus B\| = \max(\|A\|, \|B\|)$$

and

$$(A \oplus B)^* = A^* \oplus B^*.$$

Definition 4.1.29. *The operator T defined in the previous proposition which is represented as*

$$T = \begin{pmatrix} A & 0 \\ 0 & B \end{pmatrix}$$

is called a ***diagonal matrix of operators****.*

Remark. The extension to a finite sum is simple. As for the case of an infinite sum, the reader may wish to consult e.g. [**112**].

Corollary 4.1.30. *(see Exercise 4.3.39) Let $A, B \in B(H)$ and set*

$$T = \begin{pmatrix} 0 & A \\ B & 0 \end{pmatrix}.$$

Then

$$\|T\| = \max(\|A\|, \|B\|).$$

The last stop of this subsection is at when an operator matrix is invertible? First, we give a "natural definition".

Definition 4.1.31. *Let $A \in B(H \oplus H)$. Then A is said to be* ***invertible*** *if there exists a $B \in B(H \oplus H)$ such that*

$$AB = BA = I_{H \oplus H}.$$

Example 4.1.32. *Let $X \in B(H)$ and let I be the identity on H. Then $\begin{pmatrix} I & X \\ 0 & I \end{pmatrix}$ is invertible and*

$$\begin{pmatrix} I & X \\ 0 & I \end{pmatrix}^{-1} = \begin{pmatrix} I & -X \\ 0 & I \end{pmatrix}.$$

Proposition 4.1.33. *(see Exercise 4.3.40 for a proof) Let H and K be two Hilbert spaces and consider the operator $T = A \oplus B$ on $H \oplus K$. Then*

$$A \oplus B \text{ is invertible} \iff A \text{ and } B \text{ are invertible.}$$

In such case,

$$(A \oplus B)^{-1} = A^{-1} \oplus B^{-1}.$$

In other words,

$$\begin{pmatrix} A & 0 \\ 0 & B \end{pmatrix} \text{ is invertible} \iff A \text{ and } B \text{ are invertible}$$

and

$$\begin{pmatrix} A & 0 \\ 0 & B \end{pmatrix}^{-1} = \begin{pmatrix} A^{-1} & 0 \\ 0 & B^{-1} \end{pmatrix}.$$

A more general result on invertibility is the following (see e.g. [**94**]):

PROPOSITION 4.1.34. *Let*

$$A = \begin{pmatrix} A_{11} & A_{12} \\ A_{21} & A_{22} \end{pmatrix}.$$

If A_{21} commutes with A_{22}, and A_{22} is invertible, then A is invertible iff $A_{11}A_{22} - A_{12}A_{21}$ (more known as the ***formal determinant*** *of A) is invertible.*

REMARK. An example borrowed from [**94**] shows that we cannot merely drop the invertibility of A_{22}. Details may be found in Exercise 4.3.41.

4.1.4. A Word on the Invariant Subspace Problem. One of the very famous problems in Operator Theory, and in Functional Analysis in general, is the so-called "**The Invariant Subspace Problem**". First, we give a few definitions, examples and remarks:

DEFINITION 4.1.35. *Let $A \in B(H)$ and let M be a* ***closed*** *subspace of H.*

(1) *We say that M is an* ***invariant subspace*** *for (or under) A if $AM \subset M$, that is, $Ax \in M$, whenever $x \in M$.*
(2) *We say that M is a* ***reducing subspace*** *for A (or that M* ***reduces*** *A) if $AM \subset M$ and $AM^\perp \subset M^\perp$, that is, iff both M and $M^\perp$ are invariant subspaces for A.*
(3) *We say that M is a* ***hyperinvariant subspace*** *for B if $AM \subset M$ for all A in the commutant of B.*

The collection of all invariant subspaces for A is denoted by Lat$A$.

REMARKS.

(1) The definition of an *invariant* subspace is valid on Banach spaces as well.
(2) In old mathematical literature, the word "**stable**" was used instead of "invariant".
(3) It is clear that every hyperinvariant subspace for A is invariant.

EXAMPLES 4.1.36. *Let $A, B \in B(H)$.*

(1) *As is known, $\{0\}$ and H are closed in H. They are also clearly invariant under any $A \in B(H)$. We shall call $\{0\}$ and H* ***trivial invariant subspaces****.*
(2) *Let $A \in B(H)$ be a* ***scalar operator****, that is, $A = \alpha I$ for some α. Then clearly any subspace of H is invariant for any scalar operator.*
(3) *Since A is continuous,* $\ker A$ *is closed. Besides,* $\ker A \in \text{Lat}A$, *that is,* $\ker A$ *is invariant under any $A \in B(H)$. In fact,* $\ker A$ *is hyperinvariant for A as we have:*

(4) *If $AB = BA$, then* $\ker A$ *is invariant under B.*
(5) *Let V be the* **Volterra operator** *defined on $L^2[0,1]$, i.e.*

$$Vf(x) = \int_0^x f(t)dt$$

where $f \in L^2[0,1]$. Then V is bounded (cf. Exercise 2.3.15 or see Exercise 9.3.21). Now,

$$M_s = \{f \in L^2[0,1] : f(t) = 0 \text{ for } 0 \le t \le s\}$$

is invariant under V for each $s \in [0,1]$. In fact, Donoghue **[60]** *showed that on complex $L^2[0,1]$, we have*

$$\text{Lat}V = \{M_s : 0 \le t \le s\}.$$

More examples will be presented below. Let us now give a result concerning the shift operator separately:

Proposition 4.1.37. *(A proof can be found in Exercise 4.3.46) Let S be the shift operator on ℓ^2. Then:*
(1) *S has non-trivial invariant subspaces.*
(2) *S has no non-trivial reducing subspaces.*

Next, we give more properties about invariant subspaces.

Proposition 4.1.38. *(see Exercise 4.3.42) Let $A \in B(H)$. If a subspace M is invariant under A, then $M^\perp$ is invariant under A^*. In particular, if A is self-adjoint, then an invariant subspace is automatically reducing.*

Corollary 4.1.39. *Let $A \in B(H)$. Let M and M' be two linear subspaces of H. Then*

$$A(M) \subset M' \Longrightarrow A(\overline{M}) \subset \overline{M'}.$$

As another immediate consequence, we have the following:

Proposition 4.1.40. *Let $A \in B(H)$ and let M be a closed subspace of H. Then M reduces A iff M is invariant for both A and A^*.*

Since $A(\text{ran}A) \subset \text{ran}A$, the next result is trivial:

Corollary 4.1.41. *For any $A \in B(H)$, $\overline{\text{ran}A}$ is invariant (in fact, hyperinvariant) for A.*

Let M be a closed subspace of a Hilbert space H. Write $H = M \oplus M^\perp$ and hence write $A \in B(H)$ in the following form

$$A = \begin{pmatrix} A_{11} & A_{12} \\ A_{21} & A_{22} \end{pmatrix},$$

where $A_{11} \in B(M)$, $A_{12} \in B(M^\perp, M)$, $A_{21} \in B(M, M^\perp)$ and $A_{22} \in B(M^\perp)$.

With the previous notation, we have characterizations of invariant and reducing subspaces:

THEOREM 4.1.42. *(see e.g. [**48**]) Let $A \in B(H)$ and let M be a closed subspace of H. If P is the orthogonal projection onto M, then*

$$M \text{ is invariant for } A \iff A_{21} = 0 \iff PAP = AP.$$

$$M \text{ reduces } A \iff A_{21} = A_{12} = 0 \iff AP = PA.$$

DEFINITION 4.1.43. *If $M \subset H$ is an invariant subspace under $A \in B(H)$, then the* ***restriction*** *$A|_M : M \to M$ is defined by $(A|_M)x = Ax$ for each $x \in M$.*

REMARKS.

(1) As M is (closed) invariant, then $A|_M$ is a well-defined mapping between two Hilbert spaces.
(2) It is clear that $\|A|_M\| \leq \|A\|$.
(3) If $A, B \in B(H)$ and M is invariant under A and B, then M is invariant under $A + B$, AB and αA ($\alpha \in \mathbb{C}$). Besides,

$$(A + B)|_M = A|_M + B|_M.$$
$$(AB)|_M = A|_M B|_M.$$
$$(\alpha A)|_M = \alpha A|_M.$$

Here are some more properties of $A|_M$.

PROPOSITION 4.1.44. *(see Test 4.4.5) Let $A \in B(H)$. If M reduces A, then $(A|_M)^* = A^*|_M$ (where $A^*|_M$ is A^* restricted to M).*

COROLLARY 4.1.45. *Let $A \in B(H)$. If M reduces A, and A is normal (resp. unitary), then $A|_M$ is normal (resp. unitary).*

Propositions 4.1.38 & 4.1.44 yield:

COROLLARY 4.1.46. *Let $A \in B(H)$ be self-adjoint. If M is invariant under A, then $A|_M$ is self-adjoint.*

PROPOSITION 4.1.47. *Let $A \in B(H)$ be normal. If M is invariant under A, then $A|_M$ is normal iff M reduces A.*

Proposition 4.1.33 above implies the following:

COROLLARY 4.1.48. *Let $A \in B(H)$ and let M be a closed subspace of H such that $H = M \oplus M^\perp$. Assume that M is a reducing subspace of A. Denote the restriction of A to M and $M^\perp$ by $A|_M$ and $A|_{M^\perp}$ respectively. Then A is invertible iff $A|_M$ and $A|_{M^\perp}$ invertible.*

We have just observed that $\{0\}$ and H are invariant under any $A \in B(H)$, where H is a Hilbert (or a Banach space). Are there operators with only these two closed invariant subspaces? One may think of $\ker A$ as another (closed) invariant subspace for any $A \in B(H)$? But, $\ker A$ is not always non-trivial! This leads to:

The **Invariant Subspace Problem**: *Does every operator $A \in B(H)$ on a separable infinite dimensional complex Hilbert space H have a non-trivial invariant subspace M, i.e. M is closed, $M \neq \{0\}$ and $M \neq H$ and $AM \subset M$?* (We may also assume that the Hilbert space has a dimension greater than one).

In fact, this is the most important version of it and it is still an open problem, and so is the case with the **Hyperinvariant Subspace Problem** (just change "invariant" by "hyperinvariant"). This apparently innocent problem is one of the hardest in modern Functional Analysis, and if you are under forty and are interested in this problem, then solving it means that you will have a great chance to win the **Fields Medal**.

Let us speak briefly about some of the known cases. We will also see that each word in the statement of the Invariant Subspace Problem is important:

(1) If H is complex and $\dim H < \infty$, then every operator (matrix) A on H has an eigenvalue λ. Hence, the **eigenspace** $\ker(A - \lambda I)$ works just fine as a *non-trivial* invariant subspace for A. This is already known from Linear Algebra (see [**10**] for a determinant-free proof). However, it does not extend naturally to the *infinite dimensional* case as we will see that there are operators without any eigenvalue.

(2) In the previous observation, we insist on H being over $\mathbb{C}$. There are operators A on the vector space $\mathbb{R}^2$ (not on $\mathbb{R}^n$ with $n \geq 3$, see [**10**]) such that $\text{Lat}A$ is trivial. An example borrowed from [**48**] is:

$$A = \begin{pmatrix} 0 & -1 \\ 1 & 0 \end{pmatrix}.$$

Details may be found in Exercise 4.3.44.

(3) If H is not separable and $\dim H = \infty$, then the problem has an affirmative answer. This is Exercise 4.3.45.

(4) If the completeness assumption is dropped, then the problem is untrue. See [**173**], Example 6.1.

(5) If A is not necessarily bounded, then again the problem is false. See [**173**], Example 6.2.

(6) ***The problem is false in a Banach space:*** This took some time (but not as long as the Hilbert case has been doing). There are counterexamples on spaces as simple as ℓ^1. There is some known controversy around the counterexample. We just would like to say that Enflo [**71**] gave the main ideas of the counterexample and that Read [**175**] followed the ideas of Enflo and published two papers (the previous reference and [**176**]) before the one of Enflo [**71**]. Something Beauzamy [**16**] chose not to do as even if he found yet a simplified counterexample, he gave credit to Enflo. This does not call into question that the regretted C. J. Read was one of the big names related to the Invariant Subspace Problem.

Finally, readers should be aware that all these counterexamples are at least fairly complicated. In fact, Enflo's paper is basically unreadable. Maybe the simplest counterexample is Sirotkin's version of Read's counterexample (this appeared in [**201**]).

(7) Later, Read again constructed this time a bounded linear operator on ℓ^1 which has no non-trivial invariant closed *subsets*. So, when a solution to the Invariant Subspace Problem will be found (if this is ever going to happen!), another problem will follow immediately: *Does every operator $A \in B(H)$ on a separable infinite dimensional complex Hilbert space H have a non-trivial invariant* ***subset*** *M?*

(8) There are other classes of operators for which the invariant subspace holds in a Hilbert space. For example (this is a consequence of the spectral theorem which has not been stated yet)

THEOREM 4.1.49. *(see Exercise 11.3.22, cf. Theorem 9.1.15) Let $A \in B(H)$ be a non-scalar normal . Then A has a non-trivial hyperinvariant subspace.*

COROLLARY 4.1.50. *(see Test 11.4.10) Every isometry on H has a non-trivial invariant subspace.*

Other results will be given throughout this book, but so much for the Invariant Subspace Problem at present. For further reading, I recommend a recent textbook on this topic [**39**]. Other references (albeit a slightly less recent) may also be consulted as [**173**].

4.2. True or False: Questions

Questions. Comment on the following questions/statements and indicate those which are false and those which are true when this applies. Justify your answers.

(1) Let A be an operator in $B(H)$ where H is a Hilbert space. Then
$$A \text{ is self-adjoint } \iff < Ax, x > \in \mathbb{R},\ \forall x \in H.$$
(2) Let $A, B \in B(H)$ be self-adjoint. Then
$$\|AB\| = \|BA\|.$$
(3) Let $A, B \in B(H)$ be normal. Then
$$\|AB\| = \|BA\|.$$
(4) Let H be an $\mathbb{R}$-Hilbert space. Let $SA(H)$ be the set of self-adjoint operators on H. Then $SA(H)$ is a linear subspace of $B(H)$.
(5) The same question as before, but on a $\mathbb{C}$-Hilbert space this time.
(6) Let $A \in B(H)$ where H is a Hilbert space. Then A is unitary if and only if it is normal and invertible.
(7) Let $A \in B(H)$ where H is a Hilbert space. Then A is a contraction iff A is an isometry.
(8) Let $P \in B(H)$. If P is an orthogonal projection, then $\|P\| = 1$.
(9) A self-adjoint operator on a finite dimensional inner product space has always a symmetric associated matrix.

4.3. Exercises with Solutions

Exercise 4.3.1. Let $A \in B(H)$.

(1) Consider the operators, $A+A^*$, iA, $iA-iA^*$, A^*A, AA^*, A^2A^*, $A-A^*$ and $I+A$.
   (a) Which of the previous operators are self-adjoint?
   (b) Among the non self-adjoint operators above, which ones are normal?
(2) Assume now that A is self-adjoint. Are iA (i is the complex number) and $A+A^2$ self-adjoint? normal?.
(3) Now, assume that A is normal. Is A^2 normal? Is A^2A^* normal? Is iA normal?

Exercise 4.3.2. What is the general form of a 2×2 normal matrix with real entries?

Exercise 4.3.3. (cf. Exercise 5.3.39) Let H be a Hilbert space and let $A \in B(H)$ be such that $A^2 = 0$.

(1) Does it follow that $A = 0$?
(2) Show that $A = 0$ if one further assumes e.g. that A is self-adjoint.

Exercise 4.3.4. (cf. Exercise 3.3.45) Let $A \in B(H)$ where H is over $\mathbb{R}$. Show that if A is self-adjoint, then

$$< Ax, x >= 0, \ \forall x \in H \Longrightarrow A = 0.$$

Exercise 4.3.5. Let H be a $\mathbb{C}$-Hilbert space and let $A \in B(H)$. Show that

$$\forall x \in H, \ < Ax, x > \in \mathbb{R} \Longleftrightarrow A \text{ is self-adjoint.}$$

Exercise 4.3.6. Let $A, B \in B(H)$. Let $V, W \in B(H)$ be isometries and let $U \in B(H)$ be unitary.

(1) Show that (cf. Exercise 12.3.8)

$$\|VA\| = \|A\| \text{ and } \|BV^*\| = \|B\|.$$

Deduce that

$$\|VAV^*\| = \|A\| \text{ and } \|U^*BU\| = \|B\|.$$

REMARK. In particular, if A and B are unitarily equivalent, then

$$\|A\| = \|B\|.$$

(2) Do we always have $\|BV\| = \|B\|$?
(3) Show that VW is an isometry.
(4) $\|V^n x\| = \|x\|$ for all $n \in \mathbb{N}$ and all $x \in H$. What can you deduce from this last result?

Exercise 4.3.7. ([**50**]) Let $(A_k)_{k=1,\cdots,n}$ and $(B_k)_{k=1,\cdots,n}$ be both in $B(H)$. Show that

$$\left\|\sum_{k=1}^{n} A_k B_k\right\|^2 \le \left\|\sum_{k=1}^{n} A_k A_k^*\right\| \left\|\sum_{k=1}^{n} B_k^* B_k\right\|.$$

REMARK. Can we replace $A_k A_k^*$ above by $A_k^* A_k$?

Exercise 4.3.8. Let $A, B \in B(H)$ be such that A is normal. Show that

$$BA = AB \Longleftrightarrow BA^* = A^*B \Longleftrightarrow B^*A = AB^* \Longleftrightarrow B^*A^* = A^*B^*.$$

Exercise 4.3.9.

(1) Is the product of two self-adjoint operators always self-adjoint? Is it normal?
(2) Show that the product of two self-adjoint operators is self-adjoint iff these operators commute.
(3) Prove that the product of two commuting normal operators is normal (cf. Exercise 4.3.10). Is the converse always true?

Exercise 4.3.10. Let $A, B, C \in B(H)$ be all normal and such that $AC = BA$ and $AB = CA$. Show that (AB and $(AC)^*$) *and* ($(AB)^*$ and AC) are metrically equivalent.

Exercise 4.3.11. (cf. Exercise 4.5.11)

(1) Check that the sum of two self-adjoint operators is self-adjoint.
(2) Is the sum of two normal operators always normal?
(3) Show that the sum of two commuting normal operators is normal.
(4) Is the sum of two anti-commuting normal operators always normal? *Recall that $A, B \in B(H)$ are said to* ***anti-commute*** *if $AB = -BA$.*
(5) Let A and B be two normal operators such that $A + B$ is normal. Assume that A^*B and AB^* are self-adjoint. Show that A and B commute.
(6) Show, by an example, that the assumptions A^*B and AB^* being self-adjoint may not just be dropped.

Exercise 4.3.12. Let H be a complex Hilbert space, and let $A \in B(H)$.

(1) Show that if A is a normal operator, then $\ker A = \ker A^*$. What can you deduce?
(2) Is the converse true?
(3) Show that if A is normal then A is surjective iff A^* is surjective.

Exercise 4.3.13. (cf. Exercise 12.3.8) Let A and B be two bounded operators. Show that:

(1) If A is normal, then $\|AB\| = \|A^*B\|$.
(2) If B is normal, then $\|BA\| = \|A^*B\|$.

Exercise 4.3.14. Let $N \in B(H)$ be invertible and normal. Show that $N(N^*)^{-1}$ is unitary.

Exercise 4.3.15. Let $U, V \in B(H)$ be unitary.

(1) Show that UV is always unitary.
(2) Need $U + V$ be unitary?

Exercise 4.3.16. Let H be a Hilbert space and let $X \subset H$. Let $U \in B(H)$ be unitary. Show that

$$U(X^{\perp}) = (U(X))^{\perp}.$$

Exercise 4.3.17. Let $U \in B(H)$ be unitary and let $A, B \in B(H)$ be such that $U^{-1}AU = B$. Prove that A is normal iff B is normal.

Exercise 4.3.18. Let L be the Laplace Transform, that is the operator defined by

$$Lf(s) = \int_0^{\infty} f(t)e^{-st}dt,$$

where $f \in L^2(\mathbb{R}_+^*)$. Denote the $L^2(\mathbb{R}_+^*)$-norm by $\|\cdot\|_2$.

(1) Evaluate

$$I = \int_0^{\infty} e^{-x}x^{-\frac{1}{2}}dx.$$

(2) Show that L is bounded on $L^2(\mathbb{R}_+^*)$ such that

$$\|L\| \leq \sqrt{\pi}.$$

(3) Show that L is self-adjoint.
(4) Show that

$$\|L\| = \sqrt{\pi}.$$

Hint: You may use the Hilbert Operator and its norm which appeared in Exercise 2.5.7.

Exercise 4.3.19. Let $a \in \mathbb{R}$. Let U be a *bounded linear* operator defined on $L^2(\mathbb{R})$ by

$$(Uf)(x) = f(x+a), \ x \in \mathbb{R}.$$

Show that U is unitary.

Exercise 4.3.20. Show that $\sqrt{2}A$ is unitary where

$$Af(x) = \sqrt{x}f(1-x^2), \ f \in L^2(0,1), \ x \in (0,1).$$

Exercise 4.3.21. Define the bounded linear operator $A : L^2(0,1) \to L^2(0,1)$ by $Af(x) = xf(x)$ where $x \in (0,1)$. Show that A^2 is unitarily equivalent to A.

Exercise 4.3.22. Let R be the bilateral shift on $\ell^2(\mathbb{Z})$ and M the operator of multiplication on $\ell^2(\mathbb{Z})$ (with (α_n) is in ℓ^{∞})

$$M(\cdots, x_{-1}, \boldsymbol{x_0}, x_1, \cdots) = (\cdots, \alpha_{-1}x_{-1}, \boldsymbol{\alpha_0 x_0}, \alpha_1 x_1, \cdots),$$

where as usual the "bold" indicates the zeros positions of the sequence. Set $A = MR$.

(1) Is M bounded on $\ell^2(\mathbb{Z})$? Find M^*.

(2) Find explicitly A.
(3) When is A normal? unitary?

Exercise 4.3.23. Let $A, B \in B(H)$.

(1) Show that if $\dim H < \infty$, then AB is invertible iff A and B are invertible.
(2) Give an example which shows that the hypothesis "$\dim H < \infty$" may not just be dropped.
(3) Show that if A and B are invertible, then AB is invertible.
(4) Show that if $AB = BA$, then AB is invertible implies that A and B are invertible (even if $\dim H = \infty$).
(5) Deduce that if A is normal such that AA^* is invertible, then A (and A^*!) is invertible.
(6) Assume that A and B are self-adjoint. Show that if AB is invertible, then so are A and B (see Exercises 4.5.2 & 4.5.4 for a better result. Cf. Exercise 4.5.3).

Exercise 4.3.24. (see Exercise 6.3.26 and the remarks below it) Let $A, B \in B(H)$ where A is self-adjoint. If $AB = I$, show that A is invertible and that B is self-adjoint.

Exercise 4.3.25. On ℓ^2, define the following two operators

(1) $A(x_1, x_2, x_3, x_4, \cdots) = (x_1, 0, x_3, 0, x_5, \cdots)$ and
(2) $B(x_1, x_2, \cdots, x_n, \cdots) = (0, 0, \cdots, 0, x_n, x_{n+1} \cdots)$.

Show that both A and B are orthogonal projections on ℓ^2.

Exercise 4.3.26. Let $P, Q \in B(H)$ be two orthogonal projections.

(1) Show that PQ is an orthogonal projection iff $PQ = QP$.
(2) Prove that $PQ = 0$ iff $\text{ran}P \perp \text{ran}Q$.
(3) Show that $P + Q$ is an orthogonal projection iff $PQ = 0$ (see also Exercise 4.3.27).

Exercise 4.3.27. Let (P_n) be a sequence of orthogonal projections (on H) such that $P_n P_m = 0$ for all $n \neq m$. Show that the series $\sum_{n\in\mathbb{N}} P_n x$ converges for each $x \in H$.

Exercise 4.3.28. (cf. Test 6.4.6) Let $A, B \in B(H)$. Assume that $A + B = AB$.

(1) Show that A commutes with B if $\dim H < \infty$.
(2) Does the previous result necessarily hold if $\dim H = \infty$?
(3) In the coming questions we allow $\dim H = \infty$. Show that A commutes with B if one of the following occurs:
    (a) A (or B) is self-adjoint. In such case B (or A respectively) is necessarily self-adjoint.

(b) A and B^* are isometries. In this case, this implies that A and B are unitary.
(c) A and B are invertible.

Exercise 4.3.29. Let (P_n) be a sequence of projections in $B(H)$.

(1) If (P_n) converges *strongly* to P, then show that P is a projection.
(2) (cf. [**94**]) Show that if (P_n) is a sequence of *orthogonal* projections in $B(H)$ which converges *weakly* to an *orthogonal projection* $P \in B(H)$, then (P_n) converges to P strongly.

Exercise 4.3.30. Let H be a Hilbert space. Show that the set of self-adjoint operators is closed in $B(H)$ (w.r.t. the topology of the operator norm).

Exercise 4.3.31. Let H be a Hilbert space. Show that the set of normal operators is closed in $B(H)$ (w.r.t. the topology of the operator norm).

Exercise 4.3.32. Let $\mathcal{I}$ be the set of isometries on H.

(1) Show that $\mathcal{I}$ is closed in $B(H)$ (with respect to the strong operator topology).
(2) What about the set of unitary operators? (Cf. Exercise 4.5.9).

Exercise 4.3.33. (Returning to the exponential!, cf. Theorem 11.1.1) Let $A \in B(H)$ be self-adjoint.

(1) Show that e^{iA} is unitary.
(2) Show that e^{A} is self-adjoint.

Exercise 4.3.34. Let $A+iB$ be the cartesian decomposition of some $T \in B(H)$. Show that

$$T = 0 \Longleftrightarrow A = B = 0.$$

Exercise 4.3.35. Let $A + iB$ and $C + iD$ be two cartesian decompositions such that $A + iB = C + iD$. Does it follow that $A = B$ and $C = D$?

Exercise 4.3.36. Let $T = A + iB \in B(H)$. Show that

$$T \text{ is normal} \Longleftrightarrow AB = BA \Longleftrightarrow \|Tx\|^2 = \|Ax\|^2 + \|Bx\|^2$$

for all $x \in H$.

Exercise 4.3.37. Let $T = A + iB \in B(H)$ be normal. Show that T is invertible iff $A^2 + B^2$ is invertible.

Exercise 4.3.38. Let $N \in B(H)$ be normal. Show that

$$\|e^N\| \le e^{\|\operatorname{Re} N\|}.$$

Exercise 4.3.39. Let $A, B \in B(H)$ and set

$$T = \begin{pmatrix} 0 & A \\ B & 0 \end{pmatrix}.$$

Show that

$$\|T\| = \max(\|A\|, \|B\|).$$

Exercise 4.3.40. Let H and K be two Hilbert spaces and consider the operator $T = A \oplus B$ on $H \oplus K$. Show that

$$T = \begin{pmatrix} A & 0 \\ 0 & B \end{pmatrix} \text{ is invertible} \Longleftrightarrow A \text{ and } B \text{ are invertible}$$

and that

$$\begin{pmatrix} A & 0 \\ 0 & B \end{pmatrix}^{-1} = \begin{pmatrix} A^{-1} & 0 \\ 0 & B^{-1} \end{pmatrix}.$$

Exercise 4.3.41. Let S be the shift operator on ℓ^2 and let S^* be its adjoint. Define an operator matrix A on $\ell^2 \oplus \ell^2$ by

$$A = \begin{pmatrix} S^* & \mathbf{0} \\ \mathbf{0} & S \end{pmatrix}$$

where the zero entries represent the zero operator on ℓ^2.

(1) Check that the formal determinant of A is invertible.
(2) Show that A is not invertible.
(3) Why does this example not contradict Proposition 4.1.34?

Exercise 4.3.42. Let $A \in B(H)$. Let M and M' be two subspaces of H.

(1) Show that if M is invariant under A, then $M^\perp$ is invariant under A^*.
(2) Deduce that

$$A(M) \subset M' \Longrightarrow A(\overline{M}) \subset \overline{M'}.$$

Exercise 4.3.43. ([**124**]) Let $A, B \in B(H)$ be non-zero and satisfying $AB = 0$. Show that $\ker A$ and $\overline{\operatorname{ran} B}$ are nontrivial invariant subspaces for both A and B.

Exercise 4.3.44. Let

$$A = \begin{pmatrix} 0 & -1 \\ 1 & 0 \end{pmatrix}.$$

Show that $\operatorname{Lat} A$ is trivial.

Exercise 4.3.45. Let H be an infinite dimensional non-separable Hilbert space. Find an $A \in B(H)$ with a non-trivial invariant subspace.

Exercise 4.3.46. Let S be the shift operator on ℓ^2. Show that:
(1) S has non-trivial invariant subspaces.
(2) S has no non-trivial reducing subspaces.

4.4. Tests

Test 4.4.1. Let $f \in L^2[0,\infty)$ and define

$$Vf(x) = \int_0^x f(t)dt.$$

Is V bounded on $L^2[0,\infty)$?

Test 4.4.2. Show that the bilateral shift is unitary.

Test 4.4.3. Let $A, B \in B(H)$, where A is unitary and B is normal. Show that

$$AB \text{ is normal } \Longleftrightarrow BA \text{ is normal.}$$

Test 4.4.4. (cf. Exercise 12.3.8) Let $A, B \in B(H)$ be normal and satisfying $AB = 0$. Does it follow that $BA = 0$?

Test 4.4.5. Prove Proposition 4.1.44.

Test 4.4.6. Let S be the shift operator. It may be expressed as $S = A + iB$ where $A, B \in B(\ell^2)$ are self-adjoint. Then $S^* = A - iB$ and so

$$I + i0 = I = S^*S = A^2 + B^2 + i(AB - BA).$$

Therefore, we must have $A^2+B^2 = I$ and $AB-BA = 0$, i.e. $AB = BA$. Thus, S would become normal! What is wrong with this "proof"?

Test 4.4.7. Let $A \in B(H)$. Show that $\cos A$ and $\sin A$ are self-adjoint.

4.5. More Exercises

Exercise 4.5.1. Let $A, B, C \in B(H)$ be self-adjoint such that ABC is self-adjoint. Does it follow that at least two of A, B and C must commute?

Exercise 4.5.2. Let $A, B \in B(H)$ be such that AB is invertible and A is self-adjoint. Show that A is invertible and hence so is B.

Exercise 4.5.3. Let $A, B, C \in B(H)$ be such that ABC is invertible and A is self-adjoint. Show that A is invertible, however, we need not have the invertibility of B and C.

Exercise 4.5.4. Let $A_1, A_2, \cdots, A_n; B$ be all in $B(H)$ such that $A_1, A_2, \cdots, A_n$ are self-adjoint. Show that if the product $A_1A_2\cdots A_nB$ is right invertible, then each of $A_1, A_2, \cdots, A_n$ is invertible. Deduce that if we further assume that $A_1A_2\cdots A_nB$ is left invertible, then B is also invertible.

Remark. By Exercise 6.3.26, we may replace the self-adjointness of the operators by the normality in the three previous exercises. In fact, something weaker than normality suffices (see the remarks below Exercise 6.3.26.

Exercise 4.5.5. Show that a projection $A \in B(H)$ which is also a contraction is self-adjoint, that is, a contractive projection is an orthogonal projection.

Exercise 4.5.6. Let $A \in B(H)$ be self-adjoint and let $P \in B(H)$ be an orthogonal projection. Let $B \in B(H)$ be normal.

(1) Check that PAP is self-adjoint.
(2) Is PBP normal?
(3) Show that PBP is normal whenever $PB = BP$.

Exercise 4.5.7. Let $A \in B(H)$ be a contraction such that

$$A^n = A$$

for some $n \geq 2$. Show that A is normal. **Hint:** Use the previous exercise.

Exercise 4.5.8. Let $A \in B(H)$ be normal.

(1) Show that A^n is normal for any $n \in \mathbb{N}$.
(2) In fact, show that

$$A^n(A^*)^m = (A^*)^m A^n$$

for all $n, m \in \mathbb{N}$.

Exercise 4.5.9. Let $\mathcal{U}$ be the set of unitary operators on a Hilbert space H. Is $\mathcal{U}$ closed with respect to the operator norm?

Exercise 4.5.10. Under the same assumption as in Exercise 4.3.27, show that the sum $\sum_{n=1}^{\infty} P_n x$ is an orthogonal projection onto a space to be determined.

Exercise 4.5.11. Let $(A_n)_{n\in\mathbb{N}}$ be a sequence in $B(H)$ and let $(B_i)_{i=1,\cdots,k}$ be a family of elements of $B(H)$. Assume further that $\sum_{n\in\mathbb{N}} A_n$ converges.

(1) Show that if all B_i are self-adjoint, then so is $\sum_{i=1}^{k} B_i$.

(2) Show that if B_i are pairwise commuting normal operators, then $\sum_{i=1}^{k} B_i$ is normal.
(3) If all A_n are self-adjoint, then does it follow that $\sum_{n\in\mathbb{N}} A_n$ is self-adjoint?
(4) If all A_n are normal and pairwise commuting, then does it follow that $\sum_{n\in\mathbb{N}} A_n$ is normal?

Exercise 4.5.12. Let $A, B, C \in B(H)$ and let $T \in B(H \oplus H \oplus H)$ be defined by

$$T = \begin{pmatrix} A & 0 & 0 \\ 0 & 0 & C \\ 0 & B & 0 \end{pmatrix}.$$

Do we have

$$\|T\| = \max(\|A\|, \max(\|B\|, \|C\|))?$$

Exercise 4.5.13. Let $A \in B(H)$ be normal. Define T on $H \oplus H \oplus H$ by

$$T = \begin{pmatrix} 0 & 0 & A \\ A & 0 & 0 \\ 0 & A & 0 \end{pmatrix}.$$

Show that T is normal. When is T unitary?

Exercise 4.5.14. Let $T = A + iB \in B(H)$. When is T unitary?

Exercise 4.5.15. ([**50**]) Let $\varphi : [0,1] \to [0,1]$ be a C^1-diffeomorphism and define a composition operator $A : L^2[0,1] \to L^2[0,1]$ by

$$(Af)(x) = f(\varphi^{-1}(x)).$$

(1) Show that $A \in B(L^2[0,1])$.
(2) When is A self-adjoint?
(3) When is A normal?

Exercise 4.5.16. Give an example of an invariant non-reducing subspace for the bilateral shift operator on $\ell^2(\mathbb{Z})$.

CHAPTER 5

Positive Operators. Square Root

5.1. Basics

5.1.1. Positive Operators.

Definition 5.1.1. *Let H be a $\mathbb{C}$-Hilbert space and let $A \in B(H)$. We say that A is:*

(1) ***positive*** *(or* ***nonnegative****) if $< Ax, x > \geq 0$ for all $x \in H$. In symbols, $A \geq 0$.*
(2) ***strictly positive*** *if $< Ax, x >> 0$ for all $x \in H$ with $x \neq 0$. Symbolically, $A > 0$.*
(3) ***negative*** *if $< Ax, x > \leq 0$ for all $x \in H$. We then write $A \leq 0$.*

Remark. We would like to make a point clear concerning the positivity of an operator. First, it is easy to see that a positive operator $A \in B(H)$ is automatically self-adjoint because $< Ax, x > \in \mathbb{R}$ for all $x \in H$. The proof uses the fact that H is a $\mathbb{C}$-Hilbert.

However, this may pose a problem if H is an $\mathbb{R}$-Hilbert. Indeed, in that event, for all $x \in H$, then $< Ax, x >$ is in $\mathbb{R}$ anyway and regardless what A can be. So we adopt the following definition:

Definition 5.1.2. *If H is an $\mathbb{R}$-Hilbert, then $A \in B(H)$ is* ***positive*** *if $A = A^*$ and $< Ax, x > \geq 0$ for all $x \in H$.*

Remark. The reader should be wary of the terminology of positivity et al. For instance, what we call here positive may have a different signification elsewhere.

Remark. Since positive operators are self-adjoint, calling on Hellinger-Toeplitz Theorem, we could replace $A \in B(H)$ in the definition above by $A \in L(H)$.

Examples 5.1.3.

(1) *Let $A \in B(L^2[0,1])$ be defined by $Af(x) = xf(x)$. Then A is positive.*

(2) *The real matrix*

$$A = \begin{pmatrix} 1 & 2 \\ 0 & 2 \end{pmatrix}$$

is not positive as it is not self-adjoint.

(3) *Let $A \in B(H)$. Then AA^* and A^*A are both positive.*

Based on the definition of positive operators, we have

DEFINITION 5.1.4. *Let $A, B \in B(H)$. We say that $A \geq B$ if both A and B are* ***self-adjoint*** *and if $A - B \geq 0$.*

We give some basic properties concerning positive operators.

THEOREM 5.1.5. *Let $A, B, C \in B(H)$ be self-adjoint. Then*

(1) *$A = B$ iff $A \leq B$ and $A \geq B$.*
(2) *If $A, B \geq 0$ and $A + B = 0$, then $A = B = 0$.*
(3) *If $A \leq B$ and C is positive, and C commutes with both A and B, then $AC \leq BC$.*
(4) *In particular, if $A \leq B$, then:*
   (a) *For any $\alpha \geq 0$, we have $\alpha A \leq \alpha B$.*
   (b) *For any $\beta \leq 0$, we have $\beta A \geq \beta B$.*

PROPOSITION 5.1.6. *(a proof is provided in Exercise 5.3.8, cf. Proposition 5.1.11) Let $A, B, T \in B(H)$. Then*

(1)
$$A \geq 0 \Longrightarrow T^*AT \geq 0 \text{ and } TAT^* \geq 0.$$

(2) *Hence (if A and B are also self-adjoint)*
$$A \geq B \Longrightarrow T^*AT \geq T^*BT \text{ and } TAT^* \geq TBT^*.$$

COROLLARY 5.1.7. *(see the "True or False" Section) Assume that $A \in B(H)$ is positive and invertible. Then $A^{-1} \geq 0$.*

We already observed in the previous chapter that a skew symmetric operator has a self-adjoint square. In fact, we have more:

PROPOSITION 5.1.8. *Let $A \in B(H)$ be skew symmetric. Then $A^2 \leq 0$. In particular, if $B \in B(H)$, then $(B - B^*)^2 \leq 0$.*

Since orthogonal projections are self-adjoint, we may compare them. This has interesting and practical consequences:

PROPOSITION 5.1.9. *Let $P, Q \in B(H)$ be two orthogonal projections. The following assertions are pairwise equivalent:*

(1) *$P \leq Q$.*
(2) *$\|Px\| \leq \|Qx\|$, for all $x \in H$.*

(3) $QP = P$.
(4) $PQ = P$.
(5) $QPQ = P$.
(6) $Q - P$ *is an orthogonal projection.*

We have previously met the definition of a contraction and an equivalent statement in the "True or False" Section of Chapter 2. Here are some more equivalent properties:

PROPOSITION 5.1.10. *Let $A \in B(H)$. Then the following statements are equivalent:*

(1) A *is a contraction;*
(2) $A^*A \leq I$;
(3) $AA^* \leq I$;
(4) A^* *is a contraction;*
(5) A^*A *is a contraction.*

Related to Proposition 5.1.6, we have

PROPOSITION 5.1.11. *(shown in Exercise 5.3.29, cf. Proposition 5.1.17) If $A \in B(H)$ is positive and $K \in B(H)$ is a contraction such that $KA = AK$, then*

$$KAK^* \leq A.$$

REMARK. The reader may look for a counterexample showing that the commutativity condition of A and K may not just be dropped. Another natural question is whether $KAK^* \leq A$ implies $AK = KA$? The answer to this is also negative (find a simple counterexample using the shift operator on ℓ^2. If not found, then you may consult [**25**], Chapter VI, § 6, Exercise 14 for another example on $\ell^2(\mathbb{Z})$). The fact that we have taken infinite dimensional spaces is not random. See Test 5.4.2.

In another point, it is plain that $\|A - B\| \leq \|A\| + \|B\|$. What if both operators A and B are positive? Well, we can have a better estimate!

PROPOSITION 5.1.12. *(see Exercise 5.3.13) Let $A, B \in B(H)$ be both positive. Then*

$$\|A - B\| \leq \max(\|A\|, \|B\|).$$

The following result generalizes the Cauchy-Schwarz Inequality (a proof may be found in Exercise 5.3.10). See Exercises 6.5.7 & 11.5.17 for yet more general versions.

Theorem 5.1.13. *(**Generalized Cauchy-Schwarz Inequality**) Let $A \in B(H)$ be positive. Then*

$$|<Ax,y>|^2 \leq <Ax,x><Ay,y>$$

for all $x, y \in H$.

By the simple inequality $\sqrt{a}\sqrt{b} \leq \frac{1}{2}(a+b)$ holding for positive real numbers a, b, we have the following consequence of the preceding theorem:

Corollary 5.1.14. *Let $A \in B(H)$ be positive. Then*

$$|<Ax,y>| \leq \frac{1}{2}[<Ax,x> + <Ay,y>]$$

for all $x, y \in H$.

Another equally important inequality is the so-called Reid Inequality. Its first proof (to be found in Exercise 5.3.14) relies on the foregoing corollary. Another proof may be consulted in Exercise 6.3.1.

Theorem 5.1.15. *([**181**]) Let $A, K \in B(H)$ be such that A is positive and AK is self-adjoint. Then*

$$|<AKx,x>| \leq \|K\| <Ax,x> \text{ (\textbf{\textit{Reid Inequality}})}$$

for all $x \in H$. An equivalent statement is

$$\|K\|A \pm AK \geq 0.$$

In particular, if K is a contraction, then

$$|<AKx,x>| \leq <Ax,x>.$$

Remark. Theorem 6.1.8 below provides an improvement of Theorem 5.1.15. See Theorem 12.1.5 for yet another improvement.

See [**131**] for other types of generalizations of the Reid Inequality.

The natural question now is when the product of two positive operators is itself positive? We have:

Corollary 5.1.16. *(cf. Corollary 6.1.15) Let $A, B \in B(H)$ be positive and such that $AB = BA$. Then $AB \geq 0$.*

Remark. Corollary 5.1.16 will be shown in Exercise 5.3.16. Another proof may be found in Exercise 5.3.28. The proof in Exercise 5.3.28 is well spread among the mathematical community (see also Theorem 9.3.1 of [**123**]). It mainly uses positive square roots of positive operators (to be defined shortly afterwards). The proof given in Exercise 5.3.16 is square-root-free. It is a direct application of Theorem 5.1.15. This idea is due Reid himself in [**181**].

As a consequence of Theorem 5.1.15, we have

PROPOSITION 5.1.17. *(see Exercise 5.3.15) Let $A \in B(H)$ be positive and let $K \in B(H)$ be a contraction. If $AK^* = KA$, then*

$$K^2A = A(K^*)^2 = KAK^* \leq A.$$

It is clear that $[A, B] \in B(H)$ whenever $A, B \in B(H)$ and that

$$\|[A, B]\| = \|AB - BA\| \leq 2\|A\|\|B\|.$$

A natural question is: Can we improve the constant 2 by imposing a stronger condition on either A or B (or both!)? The answer turns out to be yes with a by Kittaneh in [**120**].

THEOREM 5.1.18. *(a proof is supplied in Exercise 5.3.38) Let A, $B \in B(H)$ be such that either A or B is positive. Then*

$$\|[A, B]\| \leq \|A\|\|B\|.$$

We finish with a word on the Invariant Subspace Problem. A result by Radjavi-Rosenthal [**172**] is:

THEOREM 5.1.19. *If $A \in B(H)$ is a non-scalar operator which is the product of a* ***positive*** *operator and a self-adjoint operator, then A has a non-trivial hyperinvariant subspace.*

An open problem was then posed in [**172**] (and to the best of my knowledge it is still unresolved): *Does a product of two self-adjoint operators have a non-trivial invariant subspace?*

5.1.2. Invertibility Revisited. We already know the definition of an invertible operator. In this subsection, we present more practical results which allow us to see when a given operator is (is not) invertible.

Recall that (cf. Theorem 2.1.64) $A \in B(H)$ is invertible iff $\operatorname{ran}(A)$ is dense in H and A is bounded below.

As a consequence of Theorem 3.1.66, we have

LEMMA 5.1.20. *$A \in B(H)$ is invertible iff A^* is injective and A is bounded below.*

EXAMPLE 5.1.21. *(already known) The shift operator S is not invertible for S^* is not injective.*

The previous lemma may be transformed into the following result

PROPOSITION 5.1.22. *Let $A \in B(H)$. Then A is invertible* ***iff*** *A^* is injective and A is bounded below* ***iff*** *$AA^* \geq \alpha I$ and $A^*A \geq \alpha I$ for some $\alpha > 0$.*

Remark. The condition $A^*A \geq \alpha I$ (or $AA^* \geq \alpha I$) alone is not sufficient in general to make A invertible. Indeed, if S is the shift operator, then $S^*S = I$ and yet S is not invertible.

Some easily verified properties of invertible operators are given next.

Proposition 5.1.23. *Let $A \in B(H)$. Then*

1. *If $A \geq \alpha I$ for some $\alpha > 0$, then A is invertible.*
2. *A is invertible iff A^* is bounded below and A is injective.*
3. *A is invertible iff A and A^* are both bounded below.*

The invertibility of normal operators (a fortiori self-adjoint and unitary ones) is thus easily characterized.

Corollary 5.1.24. *Let $A \in B(H)$ be normal. Then A is invertible iff A is bounded below.*

5.1.3. Square Root of Positive Operators.

First, we give a general definition.

Definition 5.1.25. *Let $A \in B(H)$. We say that $B \in B(H)$ is a* ***square root*** *of A if $B^2 = A$.*

Remark. We can similarly define a cube root. Indeed, we say that $B \in B(H)$ is a **cube root** of some $A \in B(H)$ if $B^3 = A$.

More generally, we say that $B \in B(H)$ is a n**th root** of some $A \in B(H)$ if $B^n = A$ (where $n \in \mathbb{N}$).

Let's go back to square roots:

Example 5.1.26. *A square root of*

$$\begin{pmatrix} 1 & 0 \\ 0 & 2 \end{pmatrix}$$

is:

$$\begin{pmatrix} 1 & 0 \\ 0 & \sqrt{2} \end{pmatrix}.$$

Another one is:

$$\begin{pmatrix} -1 & 0 \\ 0 & \sqrt{2} \end{pmatrix}.$$

Remark. Are there more? The answer is yes! In fact, it can be shown that if a 2×2 matrix has two non-zero distinct eigenvalues, then this matrix has four square roots. The "four" here corresponds to 2^2 as it can also be shown that an $n \times n$ matrix with n non-zero different eigenvalues has 2^n square roots. The fact that the non-zero eigenvalues must be different is essential as shown by the next example.

In fact, the authors in **[127]** and **[205]** gave, via the Cayley-Hamilton Theorem, explicit formulae for all square roots of 2×2 matrices. In particular, their method shows when a given matrix has square roots and how many of such roots there are.

An operator $A \in B(H)$ may have none (see the "True or False" Section) as it can have *infinitely* many square roots. For example,

EXAMPLE 5.1.27. *Let I be the identity 2×2 matrix, i.e.*

$$I = \begin{pmatrix} 1 & 0 \\ 0 & 1 \end{pmatrix}.$$

Then I has infinitely many square roots. Indeed,

$$A_x = \begin{pmatrix} x & 1 \\ 1 - x^2 & -x \end{pmatrix}$$

represents, for each $x \in \mathbb{R}$, a square root of I (as $A_x^2 = I$ whichever x).

REMARKS.

(1) We already know from an Advanced Linear Algebra Course (cf. **[10]**) that an invertible $A \in B(H)$ has a square root as long as $\dim H < \infty$. This, unfortunately, does not remain necessarily true if $\dim H = \infty$. Halmos et al. gave a counterexample in **[97]**. See also **[53]**.
(2) Readers may wish to consult **[107]** for more on square roots of matrices.
(3) Every self-adjoint operator has a cube root. See Chapter 11.
(4) In general, normal (a fortiori self-adjoint) operators have square roots. The proof in the finite dimensional setting, which is already known to readers, is based on the Complex Spectral Theorem. The proof for an arbitrary Hilbert space relies also on the Spectral Theorem for normal operators. See Chapter 11.
(5) Of course, an equation like $M^2 = N$, for normal $N \in B(H)$, may have other non normal square roots. See Exercise 6.3.9 for conditions guaranteeing that M be normal.

However, ***a positive operator has one, and only one, positive square root***. We have

THEOREM 5.1.28. *Let $A \in B(H)$ be positive. Then A possesses a **unique positive** square root denoted by $\sqrt{A}$ (or $A^{\frac{1}{2}}$). Moreover, if $B \in B(H)$ is such that $AB = BA$, then $\sqrt{A}B = B\sqrt{A}$.*

Remark. Exercise 5.3.26 provides a fairly short proof of the previous result. Using the Spectral Theorem for self-adjoint operators, the proof of existence of the square root is only two line-long. See Example 11.1.10. Another proof involving series of operators is prescribed in Exercise 5.5.7.

As for uniqueness, Exercise 11.3.3 gives a very simple proof (compared to the one in Exercise 5.3.26) of this property. See Exercise 5.3.27 for another proof.

Remark. Notice that we exclusively reserve the notation $\sqrt{A}$ to the positive square of A *and not to other square roots.* Indeed, the previous theorem does not exclude that a positive operator may have other square roots. It says that a positive operator has a unique positive square root. For instance, we have already observed that the identity 2×2 matrix has infinitely many square roots but, as far as positive operators are concerned, only

$$\sqrt{I} = I.$$

Remark. One of the important applications of Theorem 5.1.28, is that sometimes we may just guess ***a*** square root B of some $A \in B(H)$, and if $A \geq 0$ and $B \geq 0$, then B becomes ***the*** positive square root of A. For instance, define A on $L^2[0,1]$ by $Af(x) = xf(x)$. Then $Bf(x) = \sqrt{x}f(x)$ is clearly *a* square root of A for $B^2 = A$. Since A and B are *both positive*, B *is the square root of* A.

Remark. Obviously, there are self-adjoint operators without any positive (or even self-adjoint) square root. Indeed, if A is a self-adjoint operator and if B is such that $B^2 = A$, then this equation has no solution if A is not positive and if B is positive (or even self-adjoint) as B^2 would then be positive.

Now, we give some consequences of Theorem 5.1.28:

Corollary 5.1.29. *Let $A, B \in B(H)$ be positive. If $AB = BA$, then*

$$\sqrt{A}\sqrt{B} = \sqrt{B}\sqrt{A}.$$

Corollary 5.1.30. *(see the "True or False" Section) Let $A \in B(H)$ be positive and invertible. Then $A^{\frac{1}{2}}$ too is invertible and*

$$(A^{\frac{1}{2}})^{-1} = (A^{-1})^{\frac{1}{2}}.$$

The next consequence is shown in Exercise 5.3.37

COROLLARY 5.1.31. *Let $A, B, C \in B(H)$ be such that B and C are positive. If $BA = AC$, then*

$$\sqrt{B}A = A\sqrt{C}.$$

REMARK. It is worth noticing that if $A, B \in B(H)$ with $B \geq 0$, then in general

$$BA = A^*B \not\Longrightarrow \sqrt{B}A = A^*\sqrt{B}$$

(try to find a counterexample in the setting of 2×2 matrices).

Related to this, notice an interesting result by Sebestyén who proved in [**197**] that if $A, B \in B(H)$ with $B \geq 0$, then

$$BA = A^*B \Longrightarrow \sqrt{B}A = C\sqrt{B}$$

for some self-adjoint $C \in B(H)$. See also [**102**] for more related results.

Nonetheless, we have

COROLLARY 5.1.32. *(See Test 5.4.7) Let $A, B \in B(H)$ be such that $B \geq 0$. Then*

$$BA = A^*B = 0 \Longrightarrow \sqrt{B}A = A^*\sqrt{B}\ (= 0)$$

or equivalently

$$BA = 0 \Longleftrightarrow \sqrt{B}A = 0.$$

COROLLARY 5.1.33. *(For a proof, see Exercise 5.3.28) If $A, B \in B(H)$ are two commuting and positive operators, then AB is positive.*

COROLLARY 5.1.34. *(See Exercise 5.3.28) If $A, B \in B(H)$ are two commuting and positive, then*

$$\sqrt{AB} = \sqrt{A}\sqrt{B}.$$

Theorem 5.1.28 has an analogue for nth roots (see Exercise 5.5.10):

THEOREM 5.1.35. *Let $A \in B(H)$ be positive. Then A has a* ***unique positive*** *"nth" root denoted by $A^{\frac{1}{n}}$, where $n \in \mathbb{N}$. Moreover, if $B \in B(H)$ is such that $AB = BA$, then $A^{\frac{1}{n}}B = BA^{\frac{1}{n}}$.*

COROLLARY 5.1.36. *(see Test 5.4.5, cf. Example 11.1.12) If $A, B \in B(H)$ are two commuting and positive operators, then*

$$(AB)^{\frac{1}{n}} = A^{\frac{1}{n}}B^{\frac{1}{n}}.$$

As observed above, there are different proofs of Theorem 5.1.28. One proof utilizes bounded monotone sequences of self-adjoint operators (given in Exercise 5.3.26 as alluded to above). So, it seems appropriate to give the following definition and theorem here.

DEFINITION 5.1.37. *Let (A_n) be a sequence of self-adjoint operators on a Hilbert space H.*

(1) *We say that (A_n) is* ***bounded monotone increasing*** *if there is a self-adjoint $A \in B(H)$ such that*

$$A_1 \leq A_2 \leq \cdots \leq A_n \leq \cdots \leq A.$$

(2) *We say that (A_n) is* ***bounded monotone decreasing*** *if there is a self-adjoint $A \in B(H)$ such that*

$$A_1 \geq A_2 \geq \cdots \geq A_n \geq \cdots \geq A.$$

As in the case of sequences of real numbers, we have the following theorem.

THEOREM 5.1.38. *(a proof is provided in Exercise 5.3.25) Let (A_n) be a sequence of self-adjoint operators in $B(H)$. If (A_n) is bounded monotone increasing, then it is strongly convergent to a self-adjoint operator in $B(H)$.*

REMARK. The same conclusion holds in the previous theorem if we replace "increasing" by "decreasing".

A natural question which pops up is how positive operators behave in terms of monotony with respect to square roots or powers? More precisely, if $0 \leq A \leq B$, then do we have $\sqrt{A} \leq \sqrt{B}$? What about $A^2 \leq B^2$? How about the inverse operation?

A different approach from what is known in most textbooks (as e.g. in Exercise 19 on Page 310 of [**50**] or in Proposition 4.2.8 of [**112**]) is via Theorem 5.1.15:

THEOREM 5.1.39. *(see Exercise 5.3.30 for this approach) Let $A, B \in B(H)$ be such that $0 \leq A \leq B$. Then:*

(1) *$\sqrt{A} \leq \sqrt{B}$ (cf. Theorem 11.1.15).*
(2) *If further A is taken to be invertible, then B too is invertible and $0 \leq B^{-1} \leq A^{-1}$.*

REMARK. In [**131**], it is shown that in fact ***Reid Inequality is "equivalent" to the statement*** $0 \leq S \leq T \Rightarrow \sqrt{S} \leq \sqrt{T}$, $S, T \in B(H)$. See Exercise 6.3.1.

Now, if $A, B \in B(H)$ are two self-adjoint operators such that $A \geq B \geq 0$, then it is not necessary that $A^2 \geq B^2$. See Exercise 5.3.18. The conclusion holds if we assume further that $AB = BA$ (see again Exercise 5.3.18. See also Exercise 11.5.7). Another proof is based on the following:

PROPOSITION 5.1.40. *Let H be a complex Hilbert space. If $A, B \in B(H)$ are self-adjoint and $AB \geq 0$, then*

$$\forall x \in H : \|Ax\| \leq \|Bx\| \Longleftrightarrow \exists K \in B(H) \text{ **positive** contraction} : A = KB.$$

REMARK. By scrutinizing the proof of the previous result (which may be found in Exercise 5.3.34), we see that we can replace "positive" by "self-adjoint" in the statement and in the proof. Hence we also have:

PROPOSITION 5.1.41. *Let $A, B \in B(H)$ be self-adjoint and such that AB is self-adjoint (that is, iff $AB = BA$). Then*

$$\forall x \in H : \|Ax\| \leq \|Bx\| \Longleftrightarrow \exists K \in B(H) \text{ **self-adjoint** contraction} : A = KB.$$

COROLLARY 5.1.42. *(see Exercise 5.3.35, cf. Proposition 11.1.14) Let $A, B \in B(H)$ be commuting. Then*

$$0 \leq A \leq B \Longrightarrow A^2 \leq B^2.$$

The following consequence of Proposition 5.1.40 is obviously *weaker* than Theorem 5.1.39. It is, however, a good exercise. Besides, available proofs (in the commutativity case!) may be fairly long sometimes (as in Exercise 17 on Page 310 of [**50**]).

COROLLARY 5.1.43. *Let $A, B \in B(H)$ be positive and commuting. Then:*

(1) $0 \leq A \leq B \Rightarrow \sqrt{A} \leq \sqrt{B}$.
(2) $0 \leq A \leq B \Rightarrow B^{-1} \leq A^{-1}$, *whenever A is invertible.*

COROLLARY 5.1.44. *(see Exercise 5.3.31) Let $A, B \in B(H)$ be such that $AB = BA$ and $A, B \geq 0$. Then*

$$\sqrt{A+B} \leq \sqrt{A} + \sqrt{B} \leq \sqrt{2(A+B)}.$$

REMARK. It is easy to see that if only $A, B \geq 0$ (i.e. without $AB = BA$!), then

$$\sqrt{A} + \sqrt{B} \leq 2\sqrt{A+B}.$$

This is in fact a simple application of Theorem 5.1.39 (start with $A \leq A + B$...).

5.2. True or False: Questions

QUESTIONS. Comment on the following questions/statements and indicate those which are false and those which are true when this applies. Justify your answers.

(1) Let $A, B \in B(H)$. To compare A and B with respect to "$\leq$", we have to assume that A and B are self-adjoint. Can't we just say that $A \geq B$ if $A - B \geq 0$ without bothering about the self-adjointness of each of A and B. After all, $A - B \geq 0$ already means that $A - B$ is self-adjoint.
(2) If an operator is not positive, then it is negative.
(3) Let $A \in B(H)$ where H is a Hilbert space. Then A^2 is positive.
(4) Let $A \in B(H)$ be positive and invertible. Then $A^{-1} \geq 0$.
(5) Let (A_n) be a sequence of positive and bounded operators on H. If (A_n) converges in norm to some $A \in B(H)$, then $A \geq 0$.
(6) Any bounded operator on a Hilbert space admits a square root.
(7) (cf. Example 5.1.27) Let $A \in B(H)$ be ***self-adjoint***. Then

$$A^2 = I \Longrightarrow A = I \text{ or } A = -I.$$

(8) Let $A \in B(H)$ be self-adjoint and such that $A \geq I$. Then $A^{-1} \leq I$.
(9) If $A \in B(H)$ is an integral operator which is self-adjoint and with a positive kernel ($K(x, y) \geq 0$ for all x, y), then $A \geq 0$.
(10) Let $A \in B(H)$ be positive and invertible. Then $A^{\frac{1}{2}}$ too is invertible and

$$(A^{\frac{1}{2}})^{-1} = (A^{-1})^{\frac{1}{2}}.$$

(11) Let $A \in B(H)$ be positive. Then

$$\|\sqrt{A}\| = \sqrt{\|A\|}.$$

5.3. Exercises with Solutions

Exercise 5.3.1. Are the matrices

$$A = \begin{pmatrix} 1 & 1 \\ 1 & 1 \end{pmatrix}, \ B = \begin{pmatrix} 1 & 1 \\ 0 & 2 \end{pmatrix} \text{ and } C = \begin{pmatrix} 1 & 2 \\ 2 & 2 \end{pmatrix}.$$

positive?

Exercise 5.3.2. Let S be the shift operator on $\ell^2(\mathbb{N})$. Is $I - SS^*$ positive?

Exercise 5.3.3. (cf. Exercise 10.3.9) Let $A \in B(\ell^2)$ be the multiplication operator (see Exercise 3.3.43) defined by:

$$A(x_1, x_2, \cdots, x_n, \cdots) = (\alpha_1 x_1, \alpha_2 x_2, \cdots, \alpha_n x_n, \cdots)$$

where $(\alpha_n)_n \in \ell^\infty$. Show that

$$A \geq 0 \Longleftrightarrow \alpha_n \geq 0, \ \forall n \in \mathbb{N}.$$

Exercise 5.3.4. (cf. Exercise 4.3.33) Let $A \in B(H)$ be self-adjoint. Show that e^A is positive.

Exercise 5.3.5. (cf. Exercise 11.5.10) Let $A, B \in B(H)$ be both positive. Does it follow that $AB + BA \geq 0$?

Exercise 5.3.6. Let A and B be two bounded and positive operators on a *complex* Hilbert space H. Show that if $A + B = 0$, then $A = B = 0$.

Exercise 5.3.7. Let A be a matrix on a finite dimensional space such that $A \geq 0$ and tr$A$=0. Show that $A = 0$.

Exercise 5.3.8. Let $A, B, T \in B(H)$ where A and B are self-adjoint.

(1) Show that:

$$A \geq 0 \Longrightarrow T^*AT \geq 0 \text{ and } TAT^* \geq 0.$$

(2) Show that:

$$A \geq B \Longrightarrow T^*AT \geq T^*BT \text{ and } TAT^* \geq TBT^*.$$

Exercise 5.3.9. Let $P, Q \in B(H)$ be two orthogonal projections. Show that $P - Q$ is an orthogonal projection iff $P \geq Q$.

Exercise 5.3.10. Let $A \in B(H)$ be positive.

(1) Show that

$$| < Ax, y > |^2 \leq < Ax, x > < Ay, y >$$

for all $x, y \in H$.

(2) Infer that for every $x \in H$,

$$\|Ax\|^2 \leq \|A\| < Ax, x > .$$

Exercise 5.3.11. Let $A \in B(H)$ be self-adjoint.

(1) Show that

$$-I \leq A \leq I \Longleftrightarrow \|A\| \leq 1.$$

(2) Let $\alpha \geq 0$. Show that

$$-\alpha I \leq A \leq \alpha I \Longleftrightarrow \|A\| \leq \alpha.$$

Exercise 5.3.12. Let $A, B \in B(H)$ be self-adjoint where $A \geq 0$. Show that

$$-A \leq B \leq A \Longrightarrow \|B\| \leq \|A\|.$$

Exercise 5.3.13. Let $A, B \in B(H)$ be both positive. Show that

$$\|A - B\| \leq \max(\|A\|, \|B\|).$$

Exercise 5.3.14. Let $A, K \in B(H)$ be such that A is positive and AK is self-adjoint. Prove that

$$|<AKx, x>| \leq ||K|| <Ax, x>$$

for all $x \in H$.

Exercise 5.3.15. (cf. Exercise 5.3.29) Let $A \in B(H)$ be positive and let $K \in B(H)$ be a contraction. Show that if $AK^* = KA$, then

$$K^2A = A(K^*)^2 = KAK^* \leq A.$$

Exercise 5.3.16. Let $A, B \in B(H)$ be commuting and positive. Using the Reid Inequality, show that $AB \geq 0$.

Exercise 5.3.17. Let $A \in B(H)$ be positive. Show that A^n is also positive for each $n \in \mathbb{N}$.

Exercise 5.3.18. (cf. Exercise 5.3.19) Let $A, B \in B(H)$ be such that $A \geq B \geq 0$.

(1) Does it follow that $A^2 \geq B^2$?
(2) Show that $A^2 \geq B^2$ whenever $AB = BA$.

Exercise 5.3.19. (cf. Proposition 11.1.14) Let $A, B \in B(H)$ be such that $0 \leq A \leq B$ and $AB = BA$. Show that $0 \leq A^n \leq B^n$ for all $n \in \mathbb{N}$.

Exercise 5.3.20. Let A be a bounded self-adjoint operator on an $\mathbb{R}$-Hilbert space H such that

$$\exists c > 0, \forall x \in H : <Ax, x> \geq c||x||^2.$$

(1) Show that A is invertible.
(2) Let $p(t) = t^2 + at + b$ be a real polynomial having a *strictly negative* discriminant. Show that $p(A)$ is invertible.
(3) *Application*: Check that $A^2 + A + I$ is invertible whenever A is self-adjoint.
(4) Show that the hypothesis A being self-adjoint cannot be simply dropped.

Exercise 5.3.21. Let $A \in B(H)$ be self-adjoint. Let

$$U = (A - iI)(A + iI)^{-1}$$

(U is called the **Cayley Transform** of A).

(1) Explain why $A + iI$ is invertible (so that $(A + iI)^{-1}$ makes sense!).
(2) Show that U is unitary.

Exercise 5.3.22. ([**29**]) Let $U, V \in B(H)$ be both unitary. Show that the following assertions are equivalent:

(1) $\|U - V\| < 2$;
(2) $U + V$ is invertible.

Exercise 5.3.23. Let $A \in B(H)$. Show that

$$\text{Re}A \geq 0 \iff (A - \alpha I)^*(A - \alpha I) \geq \alpha^2 I, \ \forall \alpha < 0.$$

Exercise 5.3.24. Find the square root (if it exists) of the following operators:

(1) $A : \ell^2 \to \ell^2$ defined by

$$A(x_1, x_2, \cdots) = (0, 0, x_3, x_4, \cdots).$$

(2) S is the shift operator on ℓ^2. What about S^*?

Exercise 5.3.25. Let (A_n) be a sequence of self-adjoint operators in $B(H)$. Prove that if (A_n) is bounded monotone increasing, then it is strongly convergent to a self-adjoint operator in $B(H)$.

Exercise 5.3.26. Let $A \in B(H)$ be positive.

(1) Suppose that $\|A\| \leq 1$. Define a sequence (B_n) of operators in $B(H)$ by

$$\begin{cases} B_0 = 0, \\ B_{n+1} = B_n + \frac{1}{2}(A - B_n^2). \end{cases}$$

Show that (B_n) is a sequence of positive self-adjoint operators which is also bounded monotone increasing.

(2) Deduce that (B_n) strongly converges to a positive $B \in B(H)$ such that $B^2 = A$. Infer also that any operator which commutes with A commutes with B.

(3) Obtain the same conclusion by making no assumption this time on the norm $\|A\|$.

(4) Show that if B and C are positive and such that $B^2 = A$ and $C^2 = A$, then $B = C$.

Exercise 5.3.27. Give another proof of the uniqueness of the positive square root of positive operators (**hint**: if $T \in B(H)$ is self-adjoint, what is $\|T^4\|$?).

Exercise 5.3.28. Let A and B be two *positive* operators on a complex Hilbert space H.

(1) Show that if A and B commute, then AB (and hence BA) is positive. Infer that

$$(AB)^{\frac{1}{2}} = A^{\frac{1}{2}} B^{\frac{1}{2}}.$$

(2) Give an example showing the importance of the commutativity of A and B for the result to hold.
(3) Prove the converse of the result in Question 1, that is, prove that if A, B and AB are all positive operators, then A and B must commute.

Exercise 5.3.29. (cf. Exercise 5.3.15) Let $A, K \in B(H)$ where A is positive and K is a contraction. Show that if $AK = KA$, then $K^*AK \leq A$.

Exercise 5.3.30. Let $A, B \in B(H)$ be such that $0 \leq A \leq B$.
(1) Show that $\sqrt{A} \leq \sqrt{B}$ (cf. Exercise 6.3.1).
(2) If further A is taken to be invertible, then show that B too is invertible and that $B^{-1} \leq A^{-1}$.

Exercise 5.3.31. Let $A, B \in B(H)$ be such that $AB = BA$ and $A, B \geq 0$. Show that

$$\sqrt{A+B} \leq \sqrt{A} + \sqrt{B} \leq \sqrt{2(A+B)}.$$

Exercise 5.3.32. Let A be a self-adjoint operator on a complex Hilbert space H such that $\|A\| \leq 1$. Let I be the identity operator on H.
(1) Justify the existence of $(I - A^2)^{\frac{1}{2}}$.
(2) Set $U_{\pm} = A \pm i(I - A^2)^{\frac{1}{2}}$. Show that $U_{\pm}$ are unitary operators on H.

Exercise 5.3.33. Show that any $A \in B(H)$ may be written as a linear combination of four *unitary* operators.

Exercise 5.3.34. Let H be a complex Hilbert space. If $A, B \in B(H)$ are self-adjoint and $BA \geq 0$, then show that

$$\forall x \in H: \ \|Ax\| \leq \|Bx\| \Longleftrightarrow \exists K \in B(H) \textbf{ positive } \text{contraction}: \ A = KB.$$

Exercise 5.3.35. Let $A, B \in B(H)$ be positive and commuting. Using Proposition 5.1.40, show that

$$0 \leq A \leq B \Longrightarrow A^2 \leq B^2.$$

Exercise 5.3.36. ([**7**]) Let $A, B, C \in B(H)$ be such that $A, B \geq 0$. Define an operator T on $B(H \oplus H)$ by

$$T = \begin{pmatrix} A & C^* \\ C & B \end{pmatrix}.$$

Show that

$$T \geq 0 \Longleftrightarrow |< Cx, y >|^2 \leq < Ax, x >< By, y >, \ \forall x, y \in H.$$

Exercise 5.3.37. (cf. Exercise 11.3.10) Let $A, B, C \in B(H)$ be such that B and C are positive. Show that if $BA = AC$, then

$$\sqrt{B}A = A\sqrt{C}.$$

Exercise 5.3.38. ([**120**]). Let $A, B \in B(H)$ be such that either A *or* B is positive. We want to show that

$$\|[A, B]\| \leq \|A\|\|B\|...(1)$$

WLOG, we choose $A \geq 0$.

(1) If B is a self-adjoint contraction, show that

$$\|[A, B]\| \leq \|A\|.$$

(2) Deduce that if B is self-adjoint but not necessarily a contraction this time, then Inequality (1) still holds.

(3) Show, via an operator matrix trick, that Inequality (1) holds for any $B \in B(H)$.

Remark. The reader is asked to generalize this result to three operators in Exercise 5.5.9.

Exercise 5.3.39. ([**157**]) Let $T \in B(H)$ be such that $T^2 = 0$ and $\operatorname{Re} T \geq 0$ (or $\operatorname{Im} T \geq 0$). Show that T is normal and so $T = 0$.

5.4. Tests

Test 5.4.1. For any $a > 0$, set

$$A = \begin{pmatrix} a & 1 \\ 1 & a^{-1} \end{pmatrix}.$$

Show that

$$\sqrt{A} = \frac{1}{\sqrt{a^2+1}} \begin{pmatrix} a\sqrt{a} & \sqrt{a} \\ \sqrt{a} & \frac{1}{\sqrt{a}} \end{pmatrix}.$$

Test 5.4.2. Let H be *finite dimensional*. Let $A, B \in B(H)$ be self-adjoint. Show that if A is invertible, then

$$ABA^{-1} \leq B \Longrightarrow AB = BA.$$

Test 5.4.3. Apart from the identity operator, is there *another* operator $A \in B(H)$ which is self-adjoint, positive and an isometry at the same time?

Test 5.4.4. Let $A, B \in B(H)$ be such that $A + B \geq 0$. Do we have

$$\|AB - BA\| \leq \frac{1}{2}\|A + B\|\|A - B\|?$$

Test 5.4.5. Let $A, B \in B(H)$ be positive and commuting. Let $n \in \mathbb{N}$. Show that

$$(AB)^{\frac{1}{n}} = A^{\frac{1}{n}} B^{\frac{1}{n}}.$$

Test 5.4.6. (cf. [**102**]) Let $A, B \in B(H)$ be such that

$$A \geq 0, \ AB + BA = 0.$$

Show that $AB = 0$.

Test 5.4.7. Let $A, B \in B(H)$ with $B \geq 0$. Show that

$$BA = A^*B = 0 \Longrightarrow \sqrt{B}A = A^*\sqrt{B} \ (= 0).$$

5.5. More Exercises

Exercise 5.5.1. Let $T = A + iB \in B(H)$. When is T positive?

Exercise 5.5.2. Let $A \in B(H)$ be self-adjoint and such that $A \geq I$ and $\|A\| = 1$. Show that necessarily $A = I$.

Exercise 5.5.3. Let $S \in B(\ell^2)$ be the shift operator. Show that

$$SS^* \geq S^2 S^{*^2}.$$

Exercise 5.5.4. ([**216**]) Let $A \in B(H)$ be invertible. Find the square root of $B \in B(H \oplus H)$ given by

$$B = \begin{pmatrix} A & I \\ I & A^{-1} \end{pmatrix}$$

where I is the identity operator on H.

Hint: Have a wee look at Test 5.4.1.

Exercise 5.5.5. (cf. [**10**]) Let H be finite-dimensional and let $A \in B(H)$ be nilpotent and let I be the identity operator on H. Show that $I + A$ has a square root.

Application: Find a square root of $I+A$ where $A \in B(\mathbb{R}^5)$ defined by

$$A(x, y, z, t, u) = (2y, 3z, -t, 4u, 0).$$

Exercise 5.5.6. ([**61**]) Let $A, B \in B(H)$. Show that the following statements are equivalent:

(1) $\operatorname{ran}(A) \subset \operatorname{ran}(B)$,
(2) $AA^* \leq \alpha^2 BB^*$ for some $\alpha \geq 0$,
(3) there exists $C \in B(H)$ so that $A = BC$.

Exercise 5.5.7. ([**52**]) Let $A \in B(H)$ be positive. WLOG, we may assume that $\|A\| \leq 1$. First, recall that: *The power series*

$$\sum_{n=1}^{\infty} a_n z^n = 1 - (1-z)^{\frac{1}{2}}$$

converges for any complex z such that $|z| < 1$ where all a_n are positive. If $z \in (0,1)$, then taking the limit as $z \to 1$, we get

$$\sum_{n=1}^{\infty} a_n = 1.$$

Inspired by this power series, let

$$B = I - \sum_{n=1}^{\infty} a_n (I-A)^n.$$

(1) Show that the series which appears in the definition of B is absolutely convergent.
(2) Show that $B^2 = A$ and check that $B \geq 0$.
(3) Verify that B commutes with every bounded operator which commutes with A.

Exercise 5.5.8. Let (A_n) be a sequence of positive and bounded operators on H. Show that if (A_n) converges in norm to some $A \in B(H)$, then $\sqrt{A_n} \to \sqrt{A}$ in norm.

Exercise 5.5.9. ([**120**], cf. Exercise 5.3.38) Let $A, B, C \in B(H)$ such that A and C are positive. Show that

$$\|AB - BC\| \leq \max(\|A\|, \|C\|)\|B\|.$$

Exercise 5.5.10. ([**182**]) Let $A, B, C \in B(H)$ and let $n \in \mathbb{N}$.

(1) Show that if $BC = CB$ and $0 \leq B \leq C \leq I$, then

$$\frac{1}{n}(C^n - B^n) \leq C - B.$$

(2) By defining a sequence (R_k) in $B(H)$ by

$$R_0 = 0,\ R_{k+1} = R_k + \frac{1}{n}(A - R_k^n),$$

show that R is an nth root of A, i.e. $R^n = A$ (**hint**: show that R is the strong limit of (R_k)).

Exercise 5.5.11. Let (A_n) and (B_n) be two sequences in $B(H)$ such that $0 \leq B_n \leq A_n$ for all n. Show that if $\sum_{n=1}^{\infty} \sqrt{A_n}$ converges (in SOT), then so does $\sum_{n=1}^{\infty} \sqrt{B_n}$ (in SOT). Assuming we have the convergence of the two series, and if $\sum_{n=1}^{\infty}\sqrt{B_n}$ is invertible, then does it follow that $\sum_{n=1}^{\infty}\sqrt{A_n}$ is invertible?

Exercise 5.5.12. ([**157**]) Let $S = C + iD \in B(H)$ be such that $T^2 = S$ where $T = A + iB \in B(H)$.

(1) If $[\cdot,\cdot]$ denotes the usual commutator, then show that
$$[B,C] = [A,D]$$
and
$$[A,C] = [B,D].$$
(Consequently, $BC = CB \Leftrightarrow AD = DA$ and $AC = CA \Leftrightarrow BD = DB$).

(2) Show that if $A \geq 0$ (or $A \leq 0$), then
$$T \text{ is normal } \Longleftrightarrow AD = DA.$$

(3) Show that if $B \geq 0$ (or $B \leq 0$), then
$$T \text{ is normal } \Longleftrightarrow BD = DB.$$

Remark. In particular, if S is self-adjoint (i.e. $D = 0$) and $A \geq 0$ or $A \leq 0$ or $B \geq 0$ or $B \leq 0$, then T is always normal.

CHAPTER 6

Absolute Value. Polar Decomposition of an Operator

6.1. Basics

6.1.1. Definitions and Properties. It is clear that if $A \in B(H)$, then A^*A is positive. Thereupon, it is legitimate to define its positive square root. This gives rise to an important notion in Operator Theory.

DEFINITION 6.1.1. *Let $A \in B(H)$. The (unique) positive square root of A^*A is called the* ***absolute value*** *(or* ***modulus****) of A, and is denoted by $|A|$. In symbols,*

$$|A| = \sqrt{A^*A}$$

(if necessary we use the notation $|A|_{B(H)}$).

REMARK. The reader should bear in mind that this absolute value is the positive square root of A^*A and not that of AA^*. It is clear that $\sqrt{AA^*} = |A^*|$. In general, $|A| \neq |A^*|$, but plainly

$$A \text{ is normal} \iff |A| = |A^*|.$$

EXAMPLES 6.1.2.

(1) *If S is the shift, then*

$$|S| = \sqrt{S^*S} = \sqrt{I} = I$$

(by uniqueness of the positive square root).

(2) *If $A = \begin{pmatrix} 0 & 2 \\ 0 & 0 \end{pmatrix}$, then*

$$|A| = \sqrt{A^*A} = \begin{pmatrix} 0 & 0 \\ 0 & 2 \end{pmatrix}$$

and

$$|\operatorname{Re} A| = |\operatorname{Im} A| = I.$$

The reader should be very careful with the symbol $|A|$ and must always remember that it is an operator! For instance, while it is true that (see Exercise 6.3.3)

$$|\alpha A| = |\alpha|\,|A| \text{ and } |B + \alpha I| \leq |B| + |\alpha| I$$

for all $\alpha \in \mathbb{C}$ (where here B is self-adjoint), it is *not always true* that

$$|A+B| \leq |A| + |B| \text{ or } |AB| = |A||B|$$

if $A, B \in B(H)$ are arbitrary. For counterexamples see, Exercise 6.3.2.

Other nostalgic relations such as $|\,|A| - |B|\,| \leq |A+B|$ or $\|\,|A| - |B|\,\| \leq \|A+B\|$ are (in general) false either. Worse, while a priori if $A, B \in B(H)$ are arbitrary, then there is no reason why we should have $|A||B| = |B||A|$, even when $AB = BA$, the equality $|A||B| = |B||A|$ need not hold (the counterexample should avoid normality, cf. Proposition 6.1.13).

Nevertheless, if we take commuting normal operators, then things are much better. Notice also that commutativity is not so unnatural as we already have it in $(\mathbb{C}, \times)$. More details are given in the next two subsections.

Now, we summarize some of the straightforward properties of the absolute value of an operator related to norms.

Proposition 6.1.3. *Let $A \in B(H)$. Then*

(1) $\|A\| = \||A|^2\|^{\frac{1}{2}} = \||A|\| = \||A|^{\frac{1}{2}}\|^2$,

(2) *For all $x \in H$, we have*

$$< |A|x, x >= \||A|^{\frac{1}{2}}x\|^2 \leq \||A|x\|\|x\|.$$

(3) *For all $x \in H$, we have*

$$\|Ax\|^2 = \||A|x\|^2 \leq \|A\| < |A|x, x > .$$

The proof of the following practical result uses the polar decomposition which is presented in the next section.

Proposition 6.1.4. *(see Exercise 6.3.7) Let $A \in B(H)$ be self-adjoint. Then*

$$| < Ax, x > | \leq < |A|x, x > .$$

Corollary 6.1.5. *(see also Exercise 6.3.7) Let $A \in B(H)$ be self-adjoint. Then*

$$-|A| \leq A \leq |A|.$$

Remark. Let $A, B \in B(H)$ be self-adjoint with $B \geq 0$. Then, in general

$$-B \leq A \leq B \not\Longrightarrow |A| \leq B.$$

However, we have:

Proposition 6.1.6. *(a proof can be found in Exercise 6.3.8) Let $A, B \in B(H)$ be self-adjoint. Then*

$$|A| \leq B \Longrightarrow -B \leq A \leq B.$$

When looking at the proof of Proposition 6.1.4 which is given in Exercise 6.3.7, we see that it works perfectly well for a normal operator and with the same token. So, an improvement of Proposition 6.1.4 is (see also Proposition 12.1.6):

PROPOSITION 6.1.7. *Let $A \in B(H)$ be normal. Then*

$$|<Ax,x>| \leq <|A|x,x> .$$

Thanks to Proposition 6.1.7, we have an improved version of the Reid Inequality (Theorem 5.1.15). A proof may be found in Exercise 6.3.16, and a generalization may be found in Exercise 12.3.14.

THEOREM 6.1.8. *([**56**]) Let $A, K \in B(H)$ be such that A is positive and AK is normal. Then*

$$|<AKx,x>| \leq \|K\| <Ax,x>$$

for all $x \in H$.

Corollary 6.1.5 yields the following interesting definition:

DEFINITION 6.1.9. *Let $A \in B(H)$ be self-adjoint. Set*

$$A^+ = \frac{|A|+A}{2} \text{ and } A^- = \frac{|A|-A}{2}.$$

Then A^+ is called the ***positive part*** *of A whereas A^- is called the* ***negative part*** *of A.*

REMARK. Observe that thanks to Corollary 6.1.5, the positive part and the *negative* part are *both positive* operators.

The next result is clear now.

PROPOSITION 6.1.10. *Let $A \in B(H)$ be self-adjoint. Then the two positive operators A^+ et A^- satisfy:*

$$A = A^+ - A^- \text{ et } A^+A^- = A^-A^+ = 0.$$

We finish with an example of a matrix of operators.

DEFINITION 6.1.11. *Let A be a contraction defined on a Hilbert space H. On $H \oplus H$, set*

$$J(A) = \begin{pmatrix} (I-|A^*|^2)^{\frac{1}{2}} & A \\ -A^* & (I-|A|^2)^{\frac{1}{2}} \end{pmatrix}.$$

The operator matrix J is called the ***Julia Operator*** *.*

THEOREM 6.1.12. *(see Exercise 6.3.18 for a proof) The Julia operator is unitary.*

6.1.2. The Absolute Value and the Product. Let us state all the interesting properties of the absolute value as regards products. As we will move on, the reader will notice an amazing analogy to complex numbers. Notice that all of the results in this subsection as well as the next one appeared in [**155**] (note also that some of these results can be proved once we know the Spectral Theorem). So, the interesting fact is that these two subsections can be taught in courses which do not go as far as the Spectral Theorem because the proofs here are "spectral-theorem-free".

PROPOSITION 6.1.13. *(see Exercise 6.3.12) Let $A, B \in B(H)$ be such that $AB = BA$. If A (or B!) is normal, then $|A||B| = |B||A|$.*

REMARK. The preceding result was proved in [**217**] by assuming that *both* A and B are normal.

We have already observed above that in general $|AB| \neq |A||B|$. The following result is inspired by one in [**87**].

THEOREM 6.1.14. *(a proof is supplied in Exercise 11.3.33) Let $A, B \in B(H)$ be self-adjoint such that AB is normal. Then*

$$|AB| = |A||B|.$$

Since $|AB|$ is self-adjoint, we have:

COROLLARY 6.1.15. *Let $A, B \in B(H)$ be self-adjoint such that AB is normal. Then $|A||B|$ is self-adjoint, i.e. $|A||B| = |B||A|$.*

An akin result to Theorem 6.1.14 is:

THEOREM 6.1.16. *(see Exercise 6.3.12) Let $A, B \in B(H)$ be such that $AB = BA$. If A is normal, then*

$$|AB| = |A||B|.$$

Before generalizing the previous result, we give some direct consequences. The first one is a funny application.

COROLLARY 6.1.17. *(see Exercise 6.3.12) Let $A, B \in B(H)$ be such that $AB = BA$. If A and B are normal, then*

$$|AB| = |A^*B| = |AB^*| = |A^*B^*| = |B^*A^*| = |B^*A| = |BA^*| = |BA|.$$

COROLLARY 6.1.18. *(see Exercise 6.3.12) Let $A, B \in B(H)$ be such that $AB = BA$. If A is normal and B is invertible, then*

$$|AB^{-1}| = |A||B^{-1}|.$$

Corollary 6.1.19. *(see Exercise 6.3.12 and Test 11.4.11 for proofs) Let $A \in B(H)$ be normal and invertible. Then*

$$|A^{-1}| = |A|^{-1}.$$

Theorem 6.1.16 may be generalized as follows:

Proposition 6.1.20. *(see Test 6.4.2) Let $(A_i)_{i=1,\cdots,n}$ be a family of pairwise commuting elements of $B(H)$. If all $(A_i)_{i=1,\cdots,n}$* ***but one*** *are normal, then*

$$|A_1A_2\cdots A_{n-1}A_n| = |A_1||A_2|\cdots|A_{n-1}||A_n|.$$

It is simple to see that $|A^2| = |A|^2$ does not hold in general (find a counterexample). But for normal A (yet again), things are better.

Corollary 6.1.21. *(see Test 6.4.2) Let $A \in B(H)$ be normal and invertible. Let $n \in \mathbb{Z}$. Then*

$$|A^n| = |A|^n.$$

6.1.3. The Absolute Value and the Sum. The primary purpose of this subsection is to see when the triangle inequality with respect to $|\cdot|_{B(H)}$ holds.

Theorem 6.1.22. *(see Exercise 6.3.14) Let $A, B \in B(H)$ be such that $AB = BA$. If A and B are normal, then the following triangle inequality holds:*

$$|A+B| \leq |A| + |B|.$$

Remark. In Test 12.4.2, we show a stronger result than Theorem 6.1.22.

Corollary 6.1.23. *Let $A \in B(H)$ be normal. Then*

$$|A| \leq |\mathrm{Re}A| + |\mathrm{Im}A|.$$

We have another simple consequence of Theorem 6.1.22.

Corollary 6.1.24. *Let $A, B \in B(H)$ be such that $AB = BA$. If A and B are normal, then the following triangle inequality holds:*

$$|A-B| \leq |A| + |B|.$$

Theorem 6.1.22 may be generalized to a finite sum of operators. Before that, recall that the sum of a finite number of commuting normal operators remains normal.

Corollary 6.1.25. *Let $(A_i)_{i=1,\cdots,n}$ be a family of pairwise commuting elements of $B(H)$. If all $(A_i)_{i=1,\cdots,n}$ are normal, then*

$$|A_1 + A_2 + \cdots + A_n| \leq |A_1| + |A_2| + \cdots + |A_n|.$$

We have already observed above that $\||A|-|B|\| \le \|A \pm B\|$ is not true in general. However, we have

PROPOSITION 6.1.26. *(a proof is provided in Exercise 6.3.15, cf. Exercise 12.5.4) Let $A, B \in B(H)$ be such that $AB = BA$. If A and B are normal, then the following inequalities hold:*

$$\||A|-|B|\| \le \|A \pm B\|.$$

Proposition 6.1.26 can be regarded as a particular case of the following remarkable result:

PROPOSITION 6.1.27. *(a proof is given in Exercise 6.3.15, cf. Exercise 12.5.4) Let $A, B \in B(H)$ be such that $AB = BA$. If A and B are normal, then the following inequalities hold:*

$$||A|-|B|| \le |A \pm B|.$$

6.1.4. The Absolute Value and Other Inequalities. We have already observed a certain analogy between operators and complex numbers. The following result (whose proof is outlined in Exercise 6.5.10) corroborates the resemblance to complex numbers.

THEOREM 6.1.28. *([**76**]) If $A \in B(H)$ is such that $|A| \le \operatorname{Re} A$, then A is positive.*

We also have

THEOREM 6.1.29. *([**75**]) If $A \in B(H)$ satisfies $|A|^2 \le (\operatorname{Re}A)^2$, then A is self-adjoint.*

PROPOSITION 6.1.30. *(a proof is supplied in Exercise 6.3.17) Let A be a bounded normal operator. Then A is self-adjoint whenever $|A| \le |\operatorname{Re}A|$.*

REMARK. The result of the previous proposition is a partial answer to a conjecture which appeared in [**76**]. The authors then asked whether $|A| \le |\operatorname{Re}A|$ implies that A is self-adjoint? They discussed cases when the conjecture holds (see Exercise 6.5.10), but the conjecture as it stands is still open.

6.1.5. Polar Decomposition. It is known that any complex number z may be written as $z = re^{i\theta}$, where $r = |z|$ and $\theta = \arg z$. By analogy to complex numbers, we have:

DEFINITION 6.1.31. *Let $A \in B(H)$. If we can write $A = UP$, where U is unitary and P is positive, then UP is called the* ***Polar Decomposition*** *of A.*

Remark. The fact that we call UP the polar decomposition of A is purely conventional. There is no harm in calling PU a polar decomposition, but we have made our choice and we shall stick to it.

Remark. Not all bounded operators have a polar decomposition UP with U unitary and P positive! For a counterexample, see Exercise 6.3.19.

The first result is the following (a proof is given in Exercise 6.3.20).

Theorem 6.1.32. *Let $A \in B(H)$ be invertible. Then A has a unique polar decomposition $A = U|A|$.*

Normal operators too have a polar decomposition. We state the result here and the major part of the proof may be found in Exercises 6.3.24 & 6.3.25. A much simpler and a full proof may be found in Exercise 11.3.26.

Theorem 6.1.33. *Let $A \in B(H)$ be normal. Then A has a polar decomposition*

$$A = U|A| = |A|U$$

where $U \in B(H)$ is unitary.

In particular, if A is self-adjoint, then U may be taken to be a ***fundamental symmetry****, i.e. $U = U^*$ and $U^2 = I$.*

We have just seen that some particular classes of operators always have polar decompositions, and that not all of them may have one. The question is what if we write A as UP where U is something weaker than unitary? It turns out that the answer is yes. It is also the right place to introduce the following concept.

Definition 6.1.34. *Let $V \in B(H)$. We say that V is a* ***partial isometry*** *if V restricted to $(\ker V)^\perp$ is an isometry, that is, for all $x \in (\ker V)^\perp$: $\|Vx\| = \|x\|$.*

Remark. Trivially, an injective partial isometry is an isometry.

Examples 6.1.35.

(1) *Every orthogonal projection is a partial isometry.*
(2) *The shift operator, and an isometry in general, are partial isometries.*

Proposition 6.1.36. *A partial isometry is a contraction.*

Proposition 6.1.37. *(a proof is supplied in Exercise 6.3.23) The adjoint of an isometry is a surjective partial isometry.*

Before going further, some basic properties of partial isometries (see also Exercise 6.3.22) are gathered in the next result.

PROPOSITION 6.1.38. *Let $A \in B(H)$. Then the following statements are equivalent:*

1. *A is a partial isometry.*
2. *A^* is a partial isometry.*
3. *$AA^*A = A$.*
4. *$A^*AA^* = A^*$.*
5. *AA^* is an orthogonal projection.*
6. *A^*A is an orthogonal projection.*

Here is the promised factorization of an arbitrary $A \in B(H)$:

THEOREM 6.1.39. *(see Exercise 6.3.24) Every $A \in B(H)$ may be decomposed as*

$$A = V|A| \text{ and } |A| = V^*A$$

where $V \in B(H)$ is a partial isometry. Moreover, $\ker V = \ker A$.

PROPOSITION 6.1.40. *In the previous theorem, V is uniquely determined by the kernel condition* $\ker V = \ker A$. *Besides,* $\operatorname{ran} V = \overline{\operatorname{ran} A}$.

PROPOSITION 6.1.41. *(see e.g.* **[81]** *or* **[125]***) Let $A \in B(H)$ be normal. Then the partial isometry V which appeared in Theorem 6.1.39 may be taken to be a unitary operator.*

Theorem 6.1.33 becomes a consequence of the preceding proposition:

COROLLARY 6.1.42. *(see Exercise 6.3.25) Let $A \in B(H)$ be normal. Then*

$$A = U|A| = |A|U$$

for some unitary operator U.

REMARK. The decomposition of Theorem 6.1.39 may also be called a **polar decomposition**. Many authors call it such, however, we choose not to do so in this book.

6.2. True or False: Questions

QUESTIONS. Comment on the following questions/statements and indicate those which are false and those which are true when this applies. Justify your answers.

(1) If $A \in B(H)$, then

$$|A| = 0 \Longleftrightarrow A = 0.$$

(2) Let $A \in B(H)$. Then
$$A \geq 0 \Longleftrightarrow |A| = A.$$
(3) Let $A \in B(H)$. Then
$$|\mathrm{Re} A| \leq |A| \text{ and } |\operatorname{Im} A| \leq |A|.$$
(4) Let $A \in B(H)$. Then
$$|A| \text{ is invertible} \Longleftrightarrow A \text{ is invertible.}$$
(5) Let $A, B \in B(H)$. Then
$$|A| + |B| \text{ is invertible} \Longleftrightarrow A + B \text{ is invertible.}$$
(6) Proposition 6.1.7 is "equivalent" to Theorem 6.1.8.
(7) In the polar decomposition of an invertible operator $A = U|A|$ (see Theorem 6.1.32), we have $U|A| = |A|U$. Perhaps we have $UA = AU$ or $A|A| = |A|A$.
(8) Let $A \in B(H)$ be normal. Then just as in the case of invertibility, the polar decomposition of a normal operator is unique.

6.3. Exercises with Solutions

Exercise 6.3.1. Show that the Reid Inequality is "equivalent" to the operator monotony of the positive square root of positive operators.

Exercise 6.3.2. Let $A, B \in B(H)$. Show that the following properties need *not* hold:
(1) $|AB| = |A|\,|B|$,
(2) $|A + B| \leq |A| + |B|$,
where $|\cdot|$ is the usual absolute value of an operator.

Exercise 6.3.3. Let $A \in B(H)$ and let $\alpha \in \mathbb{C}$.
(1) Show that
$$|\alpha A| = |\alpha|\,|A|.$$
(2) Assume further here that A is self-adjoint. Show that
$$|A + \alpha I| \leq |A| + |\alpha| I.$$

Exercise 6.3.4. (cf. Test 6.4.8) Let $A, B \in B(H)$. Show that the following inequality *does hold*:
$$|A + B|^2 \leq 2(|A|^2 + |B|^2)$$
where $|\cdot|$ is the usual absolute value of an operator.

Exercise 6.3.5. Let $A, B \in B(H)$.
(1) Show that if $A + B$ is invertible, then so is $|A|^2 + |B|^2$.

(2) Deduce that if $A+B$ is invertible, then so are $|A|+|B|$ and $|A|^{2^n}+|B|^{2^n}$ ($n \in \mathbb{N}$) as well (see also Test 11.4.8).
(3) Assume here that $A, B \geq 0$ and let $\alpha, \beta \in \mathbb{C}$. Show that

$$\alpha A + \beta B \text{ invertible } \Longrightarrow A + B \text{ invertible.}$$

(4) Finally, deduce that if $C \in B(H)$ is invertible, then so is $|C-A|+|A|$ whichever $A \in B(H)$. In particular, $|I-A|+|A|$ is always invertible (for any $A \in B(H)$).

Remark. Most of the previous results appeared in **[29]**, but our proof is much simpler. Notice also that these results do hold for a finite sum as in **[29]** or by the same approach (by using the solution of Test 6.4.8 in case $a_k = 1$ for all k).

Exercise 6.3.6. (cf. Exercise 12.3.11) Let $A \in B(H)$ be normal. Show that

$$|\text{Re}A| \leq |A| \text{ and } |\operatorname{Im} A| \leq |A|.$$

Exercise 6.3.7. Let $A \in B(H)$ be self-adjoint.

(1) Show that:

$$|< Ax, x >| \leq < |A|x, x >$$

for all $x \in H$.
(2) Infer that

$$-|A| \leq A \leq |A|.$$

Exercise 6.3.8. Let $A, B \in B(H)$ be self-adjoint. Show that

$$|A| \leq B \Longrightarrow -B \leq A \leq B.$$

Exercise 6.3.9. (A result inspired by Exercise 10 on Page 22 of **[188]** about complex numbers). Let $N = C + iD \in B(H)$ be normal with either $D \geq 0$ or $D \leq 0$. Let $T \in B(H)$ be a solution of $T^2 = N$. Then T is normal.

Exercise 6.3.10. (**[67]**) Let $A \in B(H)$. Prove that the following are equivalent:

(1) A is normal.
(2) Each of $|A^*|$ and $|A|$ commutes with ReA.
(3) Give an example showing the importance of both hypotheses in the foregoing question.

Exercise 6.3.11. (**[118]**) Let $A \in B(H)$. Show that

$$T = \begin{pmatrix} |A| & A^* \\ A & |A^*| \end{pmatrix}$$

is a positive operator on $B(H \oplus H)$.

Exercise 6.3.12. Let $A, B \in B(H)$ be such that $AB = BA$ and where A is normal.

(1) Prove that
$$|A||B| = |B||A|.$$

(2) Show that
$$|AB| = |A||B|.$$

(3) Only in this question assume that B is normal. Show that
$$|AB| = |A^*B| = |AB^*| = |A^*B^*| = |B^*A^*| = |B^*A| = |BA^*| = |BA|.$$

(4) Assume here that B is invertible. Show that
$$|AB^{-1}| = |A||B^{-1}|.$$

(5) Deduce that if A is invertible, then
$$|A^{-1}| = |A|^{-1}.$$

Exercise 6.3.13. Let $A, B \in B(H)$ be such that $AB = BA$. If A is normal and $A^*B + B^*A \leq 0$, then
$$|A + B| \leq |A| + |B|.$$

Exercise 6.3.14. Let $A, B \in B(H)$ be such that $AB = BA$. Show that if A and B are normal, then the following triangle inequality holds:
$$|A + B| \leq |A| + |B|.$$

Exercise 6.3.15. Let $A, B \in B(H)$ be both normal and such that $AB = BA$.

(1) Prove that
$$\||A| - |B|\| \leq \|A + B\|.$$

(2) Infer that
$$\||A| - |B|\| \leq \|A - B\|.$$

(3) Establish that
$$||A| - |B|| \leq |A \pm B|.$$

Exercise 6.3.16. Prove Theorem 6.1.8.

Exercise 6.3.17. Let A be a bounded normal operator. Show that A is self-adjoint whenever $|A| \leq |\mathrm{Re}A|$.

Exercise 6.3.18. Let $A \in B(H)$ be a contraction. Show that the Julia Operator $J(A)$ is unitary on $H \oplus H$.

Exercise 6.3.19. Give an example of an operator which does not have a polar decomposition UP, where U is unitary and P is positive.

Exercise 6.3.20. Let $A \in B(H)$ be invertible. Show that A has a *unique* polar decomposition $A = U|A|$.

Exercise 6.3.21. Write the matrix

$$A = \begin{pmatrix} -1 & -2 \\ 2 & 1 \end{pmatrix}$$

as a product of a rotation and a positive matrix.

Exercise 6.3.22. First, we give a definition: *Let $A \in B(H)$. We say that A has a* ***generalized inverse*** *if there exists a $B \in B(H)$ such that $ABA = A$ and $BAB = B$.*

(1) We already know that the shift S does not possess an inverse. Show that S has a generalized inverse. Can this be generalized to some class of operators?
(2) Show that if B is an (ordinary!) inverse of A, then it is a generalized inverse of A.
(3) ([**137**]) Show that the following statements are equivalent:
    (a) A is a partial isometry.
    (b) A is a contraction, has a generalized inverse which is again a contraction.

Exercise 6.3.23. Let $V \in B(H)$ be an isometry. Show that V^* is a surjective partial isometry.

Exercise 6.3.24. Let $A \in B(H)$. Show that

$$A = V|A| \text{ and } |A| = V^*A,$$

where $V \in B(H)$ is a partial isometry. Check also that $\ker V = \ker A$.

Exercise 6.3.25. Let $A \in B(H)$ be normal. Show that

$$A = U|A| = |A|U$$

where $U \in B(H)$ is unitary.

Exercise 6.3.26. Let $A \in B(H)$ be a right (or left) invertible normal operator. Show that A is invertible.

Remarks. In fact a more general result holds (and for non necessarily bounded operators). It was shown in [**55**] that:

(1) If $A \in B(H)$ is right invertible and $\ker A \subset \ker A^*$, then A is invertible.
(2) If $A \in B(H)$ is left invertible and $\ker A \supset \ker A^*$, then A is invertible.
(3) Therefore, if $\ker A = \ker A^*$, then A is invertible iff it is right invertible iff it is left invertible.

6.4. Tests

Test 6.4.1. Let $A \in B(H)$ be normal. Show that if $A + A^*$ is invertible, then so is A.

Test 6.4.2. Let $(A_i)_{i=1,\cdots,n}$ be a family of pairwise commuting elements of $B(H)$. Suppose that all $(A_i)_{i=1,\cdots,n}$ **but one** are normal. Establish the following

$$|A_1A_2\cdots A_{n-1}A_n| = |A_1||A_2|\cdots|A_{n-1}||A_n|.$$

Deduce that if $A \in B(H)$ is normal and invertible, then

$$|A^n| = |A|^n$$

whichever $n \in \mathbb{Z}$.

Test 6.4.3. Does there exist a non-self-adjoint $A \in B(H)$ such that $A^2 = 0$ and $AA^* \geq 3A^*A$?

Test 6.4.4. Let $A \in B(H)$ be normal. Using the polar decomposition of A, show that

$$|\mathrm{Re}A| \leq |A| \text{ and } |\operatorname{Im} A| \leq |A|.$$

Test 6.4.5. Let $A \in B(H)$. Using Proposition 6.1.3, show that

$$\|A\| = \sup_{\|x\|=1} < |A|x, x > .$$

In particular, if $A \geq 0$, then

$$\|A\| = \sup_{\|x\|=1} < Ax, x > .$$

Test 6.4.6. Let $A, B \in B(H)$ be such that $A + B = AB$. Assume that A (or B) is normal. Show that $AB = BA$.

Test 6.4.7. Let $A, B \in B(H)$ be such that $AB \geq 0$ and AB is invertible. Show that $|A^*|^2 + |B|^2$ is invertible.

Remark. The previous result and many more related results may be found in in **[156]**. The reader may find in it a counterexample showing that the assumption AB being invertible does not mean that we have run out of ideas (this assumption is really necessary).

Test 6.4.8. Let $(A_k)_{k=1,\cdots,n}$ be in $B(H)$ and let $(a_k)_{k=1,\cdots,n}$ be in $\mathbb{C}$. Show that

$$\left|\sum_{k=1}^{n} a_k A_k\right|_{B(H)} \leq \sqrt{\sum_{k=1}^{n} |a_k|^2_{\mathbb{C}}}\sqrt{\sum_{k=1}^{n} |A_k|^2_{B(H)}}.$$

6.5. More Exercises

Exercise 6.5.1. Let (A_n) be a sequence in $B(H)$ which converges in norm to some $A \in B(H)$. Show that

$$\lim_{n\to\infty} \||A_n| - |A|\| = 0.$$

Exercise 6.5.2. Let (A_n) be a sequence in $B(H)$. Assume that (A_n) converges strongly to A and that (A_n^*) converges strongly to A^*. Show that

$$\lim_{n\to\infty} \||A_n|x - |A|x\| = 0$$

for all $x \in H$.

Exercise 6.5.3. Find $|A|$ in the case of Exercise 4.5.15.

Exercise 6.5.4. ([**50**]) Define a *linear and bounded* composition operator $A : L^2[0,1] \to L^2[0,1]$ by

$$(Af)(x) = f\left(\frac{2x}{1+x}\right).$$

Find $|A|$ and decompose A as in Theorem 6.1.39.

Exercise 6.5.5. Let $A, B \in B(H)$ be positive, invertible and commuting. Show that $A + B$ is invertible.

Exercise 6.5.6. ([**81**]) Let $A \in B(H)$ be decomposed as $A = V|A|$ where V is a partial isometry (Theorem 6.1.39).

(1) Show that
$$A^2 = 0 \Longleftrightarrow V^2 = 0.$$
(2) Can the previous be generalized to higher powers?

Exercise 6.5.7. (see also Exercise 11.5.17 for a more general form) Let $A \in B(H)$.

(1) Show that
$$|< Ax, y >|^2 \leq < |A|x, x >< |A^*|y, y >$$
for all $x, y \in H$.

REMARK. This is yet another **Generalized Cauchy-Schwarz Inequality**.

(2) Show that the equality holds in the previous inequality iff $|A|x$ and A^*y are linearly dependent iff Ax and $|A^*|y$ are linearly dependent.
(3) Based on the previous question, give an example when the inequality above is strict even when $|A^*|y$ and $|A|x$ are linearly dependent.

(4) Do we always have for all $x, y \in H$:

$$| < Ax, y > |^2 \leq < |A|x, x > < |A|y, y > ?$$

How about

$$| < Ax, y > |^2 \leq < |A^*|x, x > < |A^*|y, y > ?$$

And

$$| < Ax, y > |^2 \leq < |A^*|x, x > < |A|y, y > ?$$

Exercise 6.5.8. Can you use the result of Test 6.4.5 and the polar decomposition (or else) of a self-adjoint (or normal) $A \in B(H)$ to provide a simple proof of

$$\|A\| = \sup_{\|x\|=1} | < Ax, x > | ?$$

Exercise 6.5.9. Let $A \in B(H)$. Show that A^n converges to zero strongly iff $|A^n|$ converges to zero strongly iff $|A^n|$ converges to zero weakly.

Exercise 6.5.10. ([**76**]) We propose to prove that: *If* $A \in B(H)$ *is such that* $|A| \leq \text{Re}A$*, then* A *is positive.*

(1) Let $P, V \in B(H)$, where P is positive. Show that if $P \leq \text{Re}(VP)$, then $P \leq VPV^*$. Show also that if $P \leq \text{Re}(VP)$ and $P = VPV^*$, then $VP = P$.
(2) Set $A = VP$ with $P, V \in B(H)$ where P is positive and V is power bounded. Using the previous question, show that $A = P$.
(3) Using the "polar decomposition of Theorem 6.1.39" of an $A \in B(H)$, show that $|A| \leq \text{Re}A \Rightarrow A \geq 0$.

As alluded to in the "Cartesian Form" subsection, the authors in [**76**], proposed the following conjecture (which generalizes the result outlined in the previous questions): *If* $A \in B(H)$ *is such that* $|A| \leq |\text{Re}A|$*, then is* A *self-adjoint?* The conjecture holds if one of the following occurs:

(1) $\dim H < \infty$ ([**76**]),
(2) A belongs to the **Θ-class** (introduced in [**36**]), that is, the class of operators $A \in B(H)$ such that $A + A^*$ commutes with A^*A, which means that the conjecture is true for isometries and normal operators (see Exercise 6.3.17) among others ([**154**]).

Remark. The conjecture is true for the class of compact operators. See Exercise 9.5.4. It also holds for the class "hyponormal" operators as can be seen in Exercise 12.3.12.

CHAPTER 7

Spectrum of an Operator

7.1. Basics

7.1.1. Definitions and Properties. In this subsection, a very useful subset in $\mathbb{C}$ is introduced, from which a whole theory has been developed, which is Spectral Theory. The spectrum represents is Quantum Mechanics the possible values of the energy of a **quantum observable** (a self-adjoint operator!). More on this may be found in [**89**], [**90**] or [**207**].

DEFINITION 7.1.1. *Let H be a Hilbert space and let $A \in B(H)$. The set*

$$\sigma(A) = \{\lambda \in \mathbb{C} : \lambda I - A \text{ is not invertible}\}.$$

*is called the **spectrum** of A.*

*The **resolvent set** of A, denoted by $\rho(A)$, is the complement of $\sigma(A)$, i.e.*

$$\rho(A) = \mathbb{C} \setminus \sigma(A).$$

REMARKS.

(1) The spectrum of an operator may be defined in a similar manner on a Banach space. In fact, we can equally define the spectrum of elements of a Banach algebra with a lot of interesting consequences. See e.g. [**190**].
(2) In most cases, we will write $\lambda - A$ to mean $\lambda I - A$. We may also use $A - \lambda I$ in the definition.
(3) By the Banach Isomorphism Theorem, $\lambda - A$ is invertible, i.e. it has a bounded inverse, whenever it is bijective.
(4) In a finite dimensional setting, the spectrum of a bounded operator, i.e. of a matrix is reduced to the set of its eigenvalues. This is clear from the following well known observation: If $\dim H < \infty$ and $A : H \to H$ is linear (hence bounded!), then

$$A \text{ is injective} \Longleftrightarrow A \text{ is surjective} \Longleftrightarrow A \text{ is bijective}.$$

EXAMPLES 7.1.2. *Let H be a complex Hilbert space.*

(1) *If I is the identity on H, then $\sigma(I) = \{1\}$.*

(2) *If*

$$A = \begin{pmatrix} 1 & 2 \\ 0 & -3 \end{pmatrix},$$

then $\sigma(A) = \{-3, 1\}$.

(3) *If* $\mathcal{F}$ *is the Fourier Transform on* $L^2(\mathbb{R})$, *then*

$$\sigma(\mathcal{F}) = \{1, -1, i, -i\}$$

(this is not obvious and will be proved in Exercise 7.3.26).

(4) *(Exercise 7.3.14) Let* S *be the shift operator on* ℓ^2. *Then*

$$\sigma(S) = \{\lambda \in \mathbb{C} : |\lambda| \leq 1\}.$$

Let A be the multiplication operator by the continuous function φ on $[a, b]$, i.e. $Af(x) = \varphi(x)f(x)$ for all $f \in L^2[a, b]$. Then (cf. Exercise 2.3.26) A is invertible iff $\varphi(x) \neq 0$ for all $x \in [a, b]$. This tells us that $\lambda - A$ is invertible iff $\lambda \neq \varphi(x)$ for all $x \in [a, b]$. Therefore:

PROPOSITION 7.1.3. *(cf. Exercise 10.3.9) Let* A *be the multiplication operator on* $L^2[a, b]$ *by the continuous function* φ *on* $[a, b]$. *Then*

$$\sigma(A) = \{\varphi(x) : x \in [a, b]\} = \operatorname{ran}\varphi.$$

Some of the most fundamental properties of $\sigma(A)$ are given in the next theorem.

THEOREM 7.1.4. *Let* H *be a* ***complex*** *Hilbert space such that* $H \neq \{0\}$. *Let* $A \in B(H)$. *Then:*

(1) $\sigma(A)$ *is never empty (for a proof see Exercise 7.3.1).*

(2) $\sigma(A)$ *is a bounded and a closed subset of* $\mathbb{C}$, *i.e.* $\sigma(A)$ *is compact. Moreover,* $\sigma(A) \subset B(0, \|A\|)$ *(for a proof, see Exercise 7.3.2).*

REMARK. It is important here to insist on H being a *complex* Hilbert space. If the scalar field is $\mathbb{R}$, then the spectrum may well be empty even when $\dim H < \infty$. See Exercise 7.3.22 for a counterexample.

We have somehow a converse to the second property in the last theorem in some particular setting.

PROPOSITION 7.1.5. *(see Exercise 7.3.6 for a proof) Any compact set* K *in* $\mathbb{C}$ *can be regarded as the spectrum of some bounded operator* A *on* ℓ^2.

The next result pinpoints a relation between spectra of A and A^*, and also a similar result for the resolvent set.

PROPOSITION 7.1.6. *Let* $A \in B(H)$. *Then*

$$\sigma(A^*) = \{\overline{\lambda}: \lambda \in \sigma(A)\} \text{ and } \rho(A^*) = \{\overline{\lambda}: \lambda \in \rho(A)\}.$$

EXAMPLE 7.1.7. *(Exercise 7.3.14) Let* S *be the shift operator on* ℓ^2. *Then*

$$\sigma(S^*) = \{\lambda \in \mathbb{C}: |\lambda| \leq 1\}.$$

PROPOSITION 7.1.8. *Let* $A \in B(H)$ *be invertible. Then*

$$\sigma(A^{-1}) = \{\lambda^{-1}: \lambda \in \sigma(A)\}.$$

The spectrum of a diagonal matrix of operators is very easy to characterize using Proposition 4.1.33:

PROPOSITION 7.1.9. *Let* H *and* K *be two Hilbert spaces and consider the operator* $T = A \oplus B$ *on* $H \oplus K$. *Then*

$$\sigma(T) = \sigma(A) \cup \sigma(B).$$

That is,

$$\sigma\begin{pmatrix} A & 0 \\ 0 & B \end{pmatrix} = \sigma(A) \cup \sigma(B).$$

We can have a good information about spectra of some classes of operators.

THEOREM 7.1.10. *Let* $A \in B(H)$. *Then*

(1) $\sigma(A)$ *is real whenever* A *is self-adjoint (see Exercise 7.3.3).*
(2) $\sigma(A) \subset \{\lambda \in \mathbb{C}: |\lambda| = 1\}$ *if* A *is unitary (see Exercise 7.3.3).*
(3) $\sigma(A) = \{0, 1\}$ *if* A *is a non trivial orthogonal projection (a proof is given in Exercise 7.3.24).*

REMARK. Each backward implication in the previous theorem is false in general (this is quite easy to see via simple counterexamples in $\mathbb{R}^2$, say). However, these backward implications do hold if we further assume that A is normal. See Proposition 11.1.39.

In fact, we have a better estimate of the first assertion of the previous theorem:

PROPOSITION 7.1.11. *Let* $A \in B(H)$ *be self-adjoint. Set*

$$m = \inf_{\|x\|=1} < Ax, x >$$

and

$$M = \sup_{\|x\|=1} < Ax, x > .$$

Then $\sigma(A) \subset [m, M]$ *and* $m, M \in \sigma(A)$.

Corollary 7.1.12. *(see Exercise 7.3.11 for another proof) Let $A \in B(H)$ be self-adjoint. Then at least one of $\|A\|$ or $-\|A\|$ is in $\sigma(A)$.*

Corollary 7.1.13. *(cf. Proposition 11.1.39) Let $A \in B(H)$ be self-adjoint. Then*

$$\sigma(A) \subset \mathbb{R}^+ \Longleftrightarrow A \geq 0.$$

In such case, $\|A\| \in \sigma(A)$.

7.1.2. The Resolvent Function.

Definition 7.1.14. *Let $A \in B(H)$ and let $\lambda \in \rho(A)$. Then the function from $\rho(A)$ into $B(H)$ defined by $\lambda \mapsto (\lambda I - A)^{-1}$ is well defined and it is called the* ***resolvent function*** *(or just the* ***resolvent****). It is denoted by $R(\lambda, A)$.*

Remark. It is clear that unless $H = \{0\}$, then $R(\lambda, A) \neq 0$.

Proposition 7.1.15. *(cf. Theorem 2.1.88) Let $A \in B(H)$. If $|\lambda| > \|A\|$, then $\lambda \in \rho(A)$ and we have*

$$\|R(\lambda, A)\| \leq \frac{1}{|\lambda| - \|A\|}.$$

In particular,

$$\lim_{|\lambda| \to \infty} R(\lambda, A) = 0.$$

In the next proposition, we list some other easily verified properties.

Proposition 7.1.16. *Let $A \in B(H)$ and let $\lambda, \mu \in \rho(A)$. Then*

(1) $R(\lambda, A)R(\mu, A) = R(\mu, A)R(\lambda, A)$.
(2) *If $B \in B(H)$ commutes with A, then it commutes with $R(\lambda, A)$.*
(3)
$$R(\lambda, A) - R(\mu, A) = (\mu - \lambda)R(\lambda, A)R(\mu, A)$$
(more known as the ***resolvent equation****).*

Proposition 7.1.17. *Let $A, B \in B(H)$. Then for all $\lambda \in \rho(A) \cap \rho(B)$:*

$$AB = BA \Longleftrightarrow R(\lambda, A)R(\lambda, B) = R(\lambda, B)R(\lambda, A).$$

The next result tells us that the resolvent function is analytic.

Proposition 7.1.18. *The function $\lambda \mapsto R(\lambda, A)$ is analytic at each point in the open set $\rho(A)$. In particular, it is continuous. We also have*

$$\frac{d}{d\lambda}R(\lambda, A) = -(R(\lambda, A))^2.$$

Remark. The previous propositions are the building blocks to prove Theorem 7.1.4.

Proposition 7.1.19. *Let $A \in B(H)$. Then for all $\lambda \notin \sigma(A)$, we have*

$$\|(A-\lambda)^{-1}\| \geq \frac{1}{d(\lambda,\sigma(A))}.$$

Proposition 7.1.20. *Let $A \in B(H)$. Then*

$$[R(\lambda,A)]^* = R(\overline{\lambda},A^*).$$

7.1.3. Subsets of the Spectrum.

Definition 7.1.21. *Let H be a Hilbert space and let $A \in B(H)$.*

(1) *The set of **eigenvalues** of A, denoted by $\sigma_p(A)$, is called the **point spectrum** of A and is given by:*

$$\sigma_p(A) = \{\lambda \in \mathbb{C} : \exists x \in H, x \neq 0 : Ax = \lambda x\},$$

*that is, the set of λ for which $\lambda I - A$ is not injective. The non-zero vector x is called an **eigenvector**. The space* $\ker(\lambda I - A)$ *is called an **eigenspace**.*

(2) *The **continuous spectrum** is defined by*

$$\sigma_c(A) = \{\lambda \in \mathbb{C} : \lambda I - A \text{ is injective, has a dense range, but not surjective}\}.$$

(3) *The **residual spectrum** is defined by*

$$\sigma_r(A) = \{\lambda \in \mathbb{C} : \lambda I - A \text{ is injective but it does not have a dense range}\}.$$

Remark. In older literature, they used to say "**proper value**" as in e.g. **[11]** and **[25]** for an eigenvalue. The French are still using it by saying "valeur propre".

Remark. Remember that if H is finite dimensional, and $A \in L(H)$, then

$$\sigma_r(A) = \sigma_c(A) = \varnothing.$$

Remark. Other decompositions may be obtained. The interested reader may wish to consult **[178]**.

Proposition 7.1.22. *The three different subsets of $\sigma(A)$ defined above constitute a partition of $\sigma(A)$.*

Proposition 7.1.23. *(see Exercise 7.3.8). Let $A \in B(H)$ be normal. Then*

$$\sigma_r(A) = \varnothing.$$

Examples 7.1.24.

(1) *The shift operator has no eigenvalues. See Exercise 7.3.14.*

(2) *The shift's adjoint S^* has lots of eigenvalues. In fact,*

$$\sigma_p(S^*) = \{\lambda \in \mathbb{C} : \ |\lambda| < 1\}.$$

(3) *The multiplication operator on $L^2[a, b]$ by the function x, i.e. $Af(x) = xf(x)$ does not have any eigenvalue since $(\lambda - x)f(x) = 0$ gives $f(x) = 0$ almost everywhere, i.e. $f = 0$ on L^2 (cf. Exercise 10.3.9).*

(4) *Since the Fourier transform $\mathcal{F}$ on $L^2(\mathbb{R})$ is unitary, we have*

$$\sigma_r(\mathcal{F}) = \varnothing.$$

By Exercise 7.3.26

$$\sigma(\mathcal{F}) = \sigma_p(\mathcal{F}) = \{1, -1, i, -i\}.$$

Therefore,

$$\sigma_c(\mathcal{F}) = \varnothing.$$

Proposition 7.1.25. *Let $A \in B(H)$ be self-adjoint and let $B \in B(H)$ be normal. Then*

(1) *The eigenvalues of A (when they exist) are real.*
(2) *If λ and μ are two distinct eigenvalues for A, then the corresponding eigenvectors are orthogonal.*
(3) *If $\lambda \in \mathbb{C}$ is an eigenvalue for B, then so is $\overline{\lambda}$ for B^*.*
(4) *If λ and μ are two distinct eigenvalues for B, then the corresponding eigenvectors are orthogonal.*

The next results are the analogue of Proposition 7.1.6 for other types of (sub)spectra (a proof is given in Exercise 7.3.9).

Proposition 7.1.26. *Let $A \in B(H)$. Then*

$$\sigma_c(A^*) = \{\overline{\lambda} : \ \lambda \in \sigma_c(A)\}$$

and

$$\sigma_r(A) = \{\overline{\lambda} : \ \lambda \in \sigma_p(A^*)\} \setminus \sigma_p(A).$$

Corollary 7.1.27. *Let $A \in B(H)$. Then*

$$\sigma_c(A) = \sigma(A) \setminus (\sigma_p(A) \cup \sigma_r(A))$$

and

$$\sigma_c(A) = \sigma(A) \setminus (\sigma_p(A) \cup \{\overline{\lambda} : \ \lambda \in \sigma_p(A^*)\}).$$

Corollary 7.1.28. *Let H be a complex Hilbert space and let $A \in B(H)$. If $\lambda \in \sigma_r(A)$, then $\overline{\lambda} \in \sigma_p(A^*)$.*

Lastly, we introduce the useful notion of approximate point spectrum.

Definition 7.1.29. *Let $A \in B(H)$. A complex number λ is called an **approximate eigenvalue** of A is for any $\varepsilon > 0$, there exists $x \in H$, with $\|x\| = 1$, such that*

$$\|(\lambda - A)x\| < \varepsilon.$$

*The set of approximate eigenvalues is denoted by $\sigma_a(A)$. We shall call it the **approximate point spectrum**.*

Remark. It is clear (by considering $\frac{x}{\|x\|}$) that λ is an approximate eigenvalue if for any $\varepsilon > 0$, there exists $x \in H$ such that

$$\|(\lambda - A)x\| < \varepsilon\|x\|.$$

This also means that λ is *not* an approximate eigenvalue iff $\lambda I - A$ is bounded below.

The following proposition gives a practical characterization of the approximate point spectrum (a proof may be found in Exercise 7.3.10).

Proposition 7.1.30. *Let $A \in B(H)$. Then*

$$\lambda \in \sigma_a(A) \Longleftrightarrow \exists x_n \in H,\ \|x_n\| = 1 \text{ such that } \lim_{n\to\infty} \|(\lambda - A)x_n\|_H = 0.$$

One interesting question is: Can $\sigma_a(A)$ be empty? An answer and more properties are given in the next result (whose proof may also be found in Exercise 7.3.10).

Proposition 7.1.31. *Let $A \in B(H)$. Then $\sigma_a(A)$ is a nonempty and closed subset of $\mathbb{C}$. Moreover,*

$$\partial\sigma(A) \subset \sigma_a(A),$$

where $\partial\sigma(A)$ designates the boundary (or frontier) of the spectrum $\sigma(A)$.

Corollary 7.1.32. *Let $A \in B(H)$. Then $\sigma_a(A)$ is a compact subset of $\mathbb{C}$.*

7.1.4. Spectrum of the Product of Operators. If A and B are two (finite!) square matrices, then clearly

$$\sigma_p(AB) = \sigma_p(BA),$$

that is

$$\sigma(AB) = \sigma(BA).$$

Does the previous hold for any bounded operators A and B on an infinite dimensional space? The answer is no in general. A counterexample may be found in Exercise 7.3.17.

The only thing which we are sure of is that the non-zero elements of AB and BA are the same. This is the first important result (usually called the **Jacobson Lemma**), whose proof is given in Exercise 7.3.17.

Theorem 7.1.33. *Let $A, B \in B(H)$. Then*

$$\sigma(AB) \cup \{0\} = \sigma(BA) \cup \{0\}.$$

Remark. The conclusion of the previous theorem may be also interpreted as

$$\sigma(AB) - \{0\} = \sigma(BA) - \{0\}.$$

Corollary 7.1.34. *Let $A, B \in B(H)$. Then*

$$\sigma(AB) = \{0\} \Longleftrightarrow \sigma(BA) = \{0\}.$$

The rest of this subsection is devoted to when $\sigma(AB) = \sigma(BA)$ holds. One straightforward result is to assume that one of the operators is invertible. See Exercise 7.3.17.

Proposition 7.1.35. *Let $A, B \in B(H)$. If A (or B) is invertible, then*

$$\sigma(AB) = \sigma(BA).$$

Remark. The Jacobson Lemma also holds if one of the operators is "compact" (not yet defined). See Test 9.4.3 for a proof.

From now on, we list the results as they appeared "historically" in the literature. The first result is given in the following theorem (a proof is given in Exercise 7.3.19).

Theorem 7.1.36. *([**106**]) Let $A, B \in B(H)$. Assume that B is positive and let P be its unique positive square root. Then*

$$\sigma(AB) = \sigma(BA) = \sigma(PAP).$$

It is clear that if we further assume that A is self-adjoint, then PAP is always self-adjoint. This leads to the next interesting result too.

Corollary 7.1.37. *Let $A, B \in B(H)$ be both self-adjoint, one of them is positive. Then $\sigma(AB)$ (and also $\sigma(BA)$) is a subset of the real line.*

Remark. Of course, $\sigma(AB) \subset \mathbb{R}$ does not make AB self-adjoint! If for instance AB is normal with $B \geq 0$, say, then thanks to the previous corollary, AB will be self-adjoint. See Exercise 11.3.34, and Exercise 11.3.35 for a different method.

Theorem 7.1.36 has been generalized as follows (two proofs are supplied in Exercise 7.3.20):

Theorem 7.1.38. *([**15**]) Let $A, B \in B(H)$. If B is normal, then*

$$\sigma(AB) = \sigma(BA).$$

In fact, less than normality suffices. The ultimate generalization was obtained in [**55**] and it reads:.

Theorem 7.1.39. *Let $A, B \in B(H)$. If $\ker(A^*) = \ker(A)$, then*

$$\sigma(BA) = \sigma(AB).$$

Remark. It is worth noticing that the foregoing result holds for non-necessarily bounded operators (see again [**55**]).

In fine, we give interesting properties on the spectra of products and sums.

Proposition 7.1.40. *Let $A, B \in B(H)$ be commuting. Then*

$$\sigma(AB) \subset \sigma(A)\sigma(B) \text{ and } \sigma(A+B) \subset \sigma(A) + \sigma(B),$$

where

$$\sigma(A)\sigma(B) = \{\lambda\mu : \ \lambda \in \sigma(A), \mu \in \sigma(B)\}$$

and

$$\sigma(A) + \sigma(B) = \{\lambda + \mu : \ \lambda \in \sigma(A), \mu \in \sigma(B)\}.$$

Corollary 7.1.41. *Let $A, B \in B(H)$ be commuting and positive. Then AB is positive.*

Remark. The proof of the previous proposition is very simple if we know the so-called **Gelfand Transform** related to commutative Banach Algebras (a topic not included in this manuscript). See [**112**] or [**190**]. Can you provide a direct proof of Proposition 7.1.40, i.e. without calling on the Gelfand Transform?

Remark. Arendt et al. showed in [**8**] that $\sigma(A+B) \subset \sigma(A)+\sigma(B)$ holds even when one of the operators is unbounded.

The following corollary will be shown in Exercise 8.3.9 (the proof is from [**156**]).

Corollary 7.1.42. *(cf. Proposition 11.1.39) Let $T = A + iB$ be normal and such that $\sigma(T) \subset \mathbb{R}$. Then T is self-adjoint.*

As another consequence of Proposition 7.1.40, there is a result closely related to the next subsection besides its own interest. It is due to Lumer and Rosenblum (in [**133**]).

Theorem 7.1.43. *(see Exercise 7.3.30) Let $A, B \in B(H)$. We define an operator T on $B(H)$ (for all $X \in B(H)$) by*

$$T(X) = AX - BX$$

(T is an operator on the vector space of operators!). Then $T \in B(B(H))$ and

$$\sigma(T) \subset \sigma(A) - \sigma(B)$$

where

$$\sigma(A) - \sigma(B) = \{\lambda - \mu : \lambda \in \sigma(A), \mu \in \sigma(B)\}.$$

7.1.5. On The Operator Equation $AX - XB = C$. We have already dealt with operator equations (e.g. Exercises 2.3.37 & 2.3.38).

In this subsection, we consider the operator equation

$$AX - XB = C,$$

where $A, B, C \in B(H)$ and $X \in B(H)$ is the unknown. This equation is more commonly known as the **Sylvester Equation**. Indeed, Sylvester [**206**] was apparently the first one who investigated this sort of equations in a matrix setting (finite dimensional!).

In this subsection, we are mainly interested in this equation on infinite dimensional spaces. The foremost questions when solving an equation are: The *existence* of a solution and the *uniqueness* of such solution.

Separation of the spectra of A and B is sufficient (is it necessary? See the "True or False" Section):

Theorem 7.1.44. *(**Rosenblum**, [**183**]) Let $A, B \in B(H)$ be such that $\sigma(A) \cap \sigma(B) = \varnothing$. Then the equation $AX - XB = C$ has a unique solution X (in $B(H)$) for each $C \in B(H)$.*

Remark. Writing $T(X) = AX - BX$ as in Theorem 7.1.43, the existence and uniqueness of the solution of $T(X) = AX - BX = C$ for every $C \in B(H)$, is clearly reduced to showing that T is invertible. Theorem 7.1.43 then makes this statement very simple. See Exercise 7.3.31.

A simple observation is the following:

Corollary 7.1.45. *Let $A \in B(H)$ be such that $A > 0$. Then the equation $AX + XA = C$ has a unique solution (in $B(H)$) for each $C \in B(H)$. In particular, the unique solution of $AX + XA = 0$ is $X = 0$.*

A more general equation is one of the kind: $AXB - CXD = E$. To solve it, we may use a similar idea of the proof of Theorem 7.1.44. We have:

THEOREM 7.1.46. *(see Exercise 7.3.32) Let* $A, B, C, D \in B(H)$ *be such that*

$$AC = CA,\ \ BD = DB \text{ and } \sigma(A)\sigma(B) \cap \sigma(C)\sigma(D) = \varnothing.$$

Then the equation $AXB - CXD = E$ *has a unique solution* $X \in B(H)$ *for each* $E \in B(H)$.

REMARK. In fact, the method of the solution of Exercise 7.3.32 can be applied to prove the existence and uniqueness of a solution of a more general type of equations, namely:

$$T(X) := \sum_{k=1}^{n} A_k X B_k = C$$

where A_k, B_k, C, X are all in $B(H)$ (for each $k = 1, \cdots, n$). Then the equation

$$\sum_{k=1}^{n} A_k X B_k = C$$

has a unique solution X in $B(H)$ for each $C \in B(H)$ if

$$\forall j, k \in 1, \cdots, n :\ A_k A_j = A_j A_k,\ B_k B_j = B_j B_k$$

and

$$\sigma(T) \subset \sum_{k=1}^{n} \sigma(A_k)\sigma(B_k).$$

DEFINITION 7.1.47. *Let* $(A_k)_{1\leq k\leq n}$ *and* $(B_k)_{1\leq k\leq n}$ *be in* H. *The operator* $T : B(H) \to B(H)$ *defined by*

$$T(X) = \sum_{k=1}^{n} A_k X B_k$$

is called an ***Elementary Operator*** *(see* **[221]** *for more on elementary operators).*

Surely, readers long to see explicit solutions. Adopting a similar method to that of Exercise 2.3.37, say, we can find in some cases an explicit solution to the Sylvester Equation. The simple proof, which uses the "spectral radius", is postponed to next chapter (see Exercise 8.3.7). We have

THEOREM 7.1.48. *Let* $r > 0$ *and let* $A, B \in B(H)$ *be such that*

$$\sigma(A) \subset \{z \in \mathbb{C} :\ |z| > r\} \text{ and } \sigma(B) \subset \{z \in \mathbb{C} :\ |z| < r\}.$$

Then for every $C \in B(H)$, the solution of $AX - XB = C$ is given by:

$$X = \sum_{n=0}^{\infty} A^{-n-1}CB^n.$$

The same method of the proof of Theorem 7.1.48 (to be given in Exercise 8.3.7) mutatis mutandis may be applied to find an explicit solution of the generalized equation which appeared in Theorem 7.1.46.

THEOREM 7.1.49. *Let $r > 0$ and let $A, B, C, D \in B(H)$ be such that $AC = CA$, $BD = DB$,*

$$\sigma(A)\sigma(B) \subset \{z \in \mathbb{C} : |z| > r\} \text{ and } \sigma(C)\sigma(D) \subset \{z \in \mathbb{C} : |z| < r\}.$$

Then the equation $AXB - CXD = E$ has the unique solution X expressed as:

$$X = \sum_{n=0}^{\infty} C^n A^{-n-1} E B^{-n-1} D^n$$

for each $E \in B(H)$.

REMARK. See Exercise 11.5.19 for another explicit solution. See [**28**] for more constructions of solutions.

An interesting application of Theorem 7.1.44 is:

PROPOSITION 7.1.50. *([**69**], see Exercise 7.3.33) Let $A, B \in B(H)$ be such that $\sigma(A) \cap \sigma(B) = \varnothing$. Then any operator $C \in B(H)$ which commutes with $A + B$ and AB, will commute with A and B as well.*

REMARK. The condition $\sigma(A) \cap \sigma(B) = \varnothing$ in the preceding proposition may not just be dropped. We ask readers to find a (simple) counterexample.

COROLLARY 7.1.51. *Let $A, B \in B(H)$ satisfy $\sigma(A) \cap \sigma(B) = \varnothing$. Then A commutes with B if and only if $A + B$ commutes with AB.*

As another application of Theorem 7.1.44, we have a result on similarity:

THEOREM 7.1.52. *(see Exercise 7.3.34) Let $A, B \in B(H)$ be such that $\sigma(A) \cap \sigma(B) = \varnothing$. Then for any operator $C \in B(H)$, the matrix operator $\begin{pmatrix} A & C \\ 0 & B \end{pmatrix}$ is similar to $\begin{pmatrix} A & 0 \\ 0 & B \end{pmatrix}$.*

REMARK. Roth [**186**] showed in a *finite dimensional setting* that the similarity of the two matrices of operators above is equivalent to the existence of a solution X of $AX - XB = C$. A natural reflex is to ask whether the result of Roth is valid if $\dim H = \infty$? The answer is no even for unitary equivalence! The counterexample is due to Rosenblum in [**185**]. See Exercise 7.3.35. It is worth noticing that Roth's result does hold if $\dim H = \infty$ if we further assume that A and B are self-adjoint (possibly unbounded). This is due to Rosenblum again in [**185**]. Later, this was generalized by Schweinsberg to normal operators as well as finite rank operators (a class to be defined in the chapter of compact operators). See [**196**].

Readers interested in this topic may consult the good survey [**28**] for further reading and more applications.

7.1.6. Polynomial Calculus. We have come quickly across polynomials of operators. Here we treat this topic with some more detail.

DEFINITION 7.1.53. *Let $A \in B(H)$. Let $p = a_0 + a_1 x + \cdots + a_n x^n$, where $a_0, a_1, \cdots, a_n \in \mathbb{C}$, be a polynomial. Then we can associate with A an operator, denoted by $p(A)$, defined by*

$$p(A) = a_0 I + a_1 A + \cdots + a_n A^n,$$

where I is the usual identity operator on H.

REMARK. The previous definition makes sense for any $A \in B(H)$ and even if H is replaced by a Banach space.

EXAMPLES 7.1.54.

(1) *If $p(x) = a$, $a \in \mathbb{R}$, then $p(A) = aI$.*
(2) *If $p(x) = x^2 - x$, then $p(A) = A^2 - A$.*

PROPOSITION 7.1.55. *Let $A \in B(H)$, p, q be two polynomials and $\alpha, \beta \in \mathbb{C}$. Then*

$$1(A) = I, \ (\alpha p + \beta q)(A) = \alpha p(A) + \beta q(A) \text{ and } pq(A) = p(A)q(A).$$

REMARK. By the previous proposition, $p \mapsto p(A)$, defined from $\mathbb{C}[X]$ to $B(H)$, is a homomorphism of algebras with a unity.

One of the important results is the following theorem (for a proof, see Exercise 7.3.22)

THEOREM 7.1.56. ***(Spectral Mapping Theorem)*** *Let $p \in \mathbb{C}[X]$ be a polynomial and let $A \in B(H)$. Then*

$$\sigma(p(A)) = p(\sigma(A)).$$

REMARKS.

(1) If $p \in \mathbb{R}[X]$, then we are only sure that

$$\sigma(p(A)) \supset p(\sigma(A)).$$

For a counterexample to the reverse inclusion, see Exercise 7.3.22.

(2) The result in the previous theorem is the most basic form of Spectral Mapping Theorems. The reader will see other versions throughout this book.

If $p = a_0 + a_1x + \cdots + a_nx^n$, then set $\overline{p} = \overline{a_0} + \overline{a_1}x + \cdots + \overline{a_n}x^n$. Also, $|p|^2 = p\overline{p}$. We have the following useful result:

Proposition 7.1.57. *Let $p \in \mathbb{C}[X]$. If $A \in B(H)$ is self-adjoint, then*

$$[p(A)]^* = \overline{p}(A) \text{ and } \|p(A)\| = \max_{\lambda\in\sigma(A)} |p(\lambda)|.$$

Corollary 7.1.58. *Let $p, q \in \mathbb{C}[X]$ and let $A \in B(H)$ be self-adjoint. If p and q coincide on $\sigma(A)$, then $p(A)$ coincides with $q(A)$.*

In fact, a stronger version holds

Proposition 7.1.59. *(see Exercise 8.3.4) Let $p \in \mathbb{C}[X]$. If $A \in B(H)$ is normal, then $p(A)$ is normal and*

$$\|p(A)\| = \max_{\lambda\in\sigma(A)} |p(\lambda)|.$$

Remark. We do have $[p(A)]^* = \overline{p}(A)$ if A is normal (see e.g. the Functional Calculi Chapter for an even more general result), but $\overline{p}$ differs from the above notation.

Corollary 7.1.60. *Let $p, q \in \mathbb{C}[X]$ and let $A \in B(H)$ be normal. If p and q coincide on $\sigma(A)$, then $p(A)$ coincides with $q(A)$.*

In a coming chapter, we will give a sense $f(A)$ for a particular class of A and/or a particular class of f. But, based on the polynomial calculus we can define $f(A)$ if f is a rational function of the form $\frac{p(x)}{q(x)}$ where p and q are two polynomials of degrees n and m, say. Then a possible candidate for $f(A)$ will be $[q(A)]^{-1}p(A)$ (or $p(A)[q(A)]^{-1}$). But, we have to face a difficulty which is: Is $q(A)$ invertible? In case of invertibility of $q(A)$, then we set

$$f(A) = [q(A)]^{-1}p(A).$$

Having said this, when is $q(A)$ invertible anyway? The answer is given next.

Proposition 7.1.61. *Let $A \in B(H)$. Let $p \in \mathbb{C}(X)$ be such that $p(x) \neq 0$ for all $x \in \sigma(A)$. Then $p(A)$ is invertible.*

7.2. True or False: Questions

Questions. Comment on the following questions/statements and indicate those which are false and those which are true when this applies. Justify your answers.

(1) Let $A, B \in B(H)$. Then

$$\sigma(A+B) = \sigma(A) + \sigma(B).$$

(2) Let $A, B \in B(H)$. Then $\sigma(A) + \sigma(B)$ is a closed set.

(3) Let $A \in B(H)$. Then

$$A \text{ is a non trivial orthogonal projection} \Longleftrightarrow \sigma(A) = \{0, 1\}.$$

(4) Let $A \in B(H)$. Then

$$\sigma_p(A^*) = \{\overline{\lambda} : \lambda \in \sigma_p(A)\}.$$

(5) Let $A \in B(H)$. Then

$$A \text{ is normal } \Longleftrightarrow \sigma_r(A) = \varnothing.$$

(6) Let $P \in B(H)$ be an orthogonal projection, i.e. $P^2 = P$ and $P = P^*$. Then $\sigma(P^2) = (\sigma(P))^2 = \sigma(P)$. Then if λ is in $\sigma(P)$, then $\lambda^2 = \lambda$ and thus $\sigma(P) = \{0, 1\}$. Here is a nice proof or is it not?

(7) The condition $\sigma(A) \cap \sigma(B) = \varnothing$ in Theorem 7.1.44 is necessary for the existence of a solution to the Sylvester Equation.

7.3. Exercises with Solutions

Exercise 7.3.1. (cf. Exercises 10.3.23 & 10.3.24) Let H be a non-trivial $\mathbb{C}$-Hilbert space. Let $A \in B(H)$. Show that $\sigma(A)$ is never empty.

Exercise 7.3.2. Let H be a $\mathbb{C}$-Hilbert space. Let $A \in B(H)$.

(1) Show that if $|\lambda| > \|A\|$, then $\lambda \notin \sigma(A)$.

(2) Prove that $\sigma(A)$ is a closed set. Deduce that $\sigma(A)$ is a compact set.

Exercise 7.3.3. (cf. Exercise 7.3.24, and for a converse, see Exercise 11.3.28) Let $A, U \in B(H)$. Show that:

(1) If U is unitary, then

$$\sigma(U) \subset \{\lambda \in \mathbb{C} : |\lambda| = 1\}.$$

(2) If A is self-adjoint, then $\sigma(A)$ is a subset of the real line.

(3) If A is positive, then $\sigma(A) \subset \mathbb{R}^+$.

Exercise 7.3.4. Let H be a Hilbert space. Let (λ_n) be a sequence in $\rho(A)$ such that $\lambda_n \to \lambda$ ($\in \mathbb{C}$). Show that if the sequence $(R(\lambda_n, A))_n$ is bounded, then $\lambda \in \rho(A)$ (where $R(\cdot, A)$ is the usual resolvent function).

Exercise 7.3.5. Let $(\lambda_n)_n$ be a bounded sequence of complex numbers. Define on ℓ^2 the following multiplication operator:

$$Ax = A(x_1, x_2, \cdots) = (\lambda_1 x_1, \lambda_2 x_2, \cdots).$$

(1) Find $\sigma_p(A)$. Also, show that $\sigma(A) = \overline{\sigma_p(A)}$.
(2) What is $\sigma_r(A)$? What is $\sigma_c(A)$?

Exercise 7.3.6. Show that any compact set K in $\mathbb{C}$ can be the spectrum of some bounded operator A on $\ell^2(\mathbb{N})$.

Exercise 7.3.7. First, we give a definition: *Let $A \in B(H)$. We say that A satisfies the* ***Daugavet Equation*** *(in short we write D.E.) if*

$$\|A + I\| = \|A\| + 1$$

where I is the usual identity on H (a whole chapter is devoted to this topic in **[3]***).*

(1) Show that $A \in B(H)$ satisfies D.E. if and only if $\|A\| + 1 \leq \|A + I\|$.
(2) Give an example of $A \in B(H)$ satisfying D.E. and a $B \in B(H)$ not satisfying it.
(3) Show that A satisfies the D.E. iff A^* does.
(4) Show that if $\|A\| \in \sigma(A)$, then A satisfies D.E.
(5) Prove that if A satisfies D.E., then αA also satisfies D.E. for any $\alpha \geq 0$.

Exercise 7.3.8. Show that a normal operator A has always an empty residual spectrum.

Exercise 7.3.9. Let $A \in B(H)$. Show that

$$\sigma_c(A^*) = \{\overline{\lambda} : \ \lambda \in \sigma_c(A)\}$$

and

$$\sigma_r(A) = \{\overline{\lambda} : \ \lambda \in \sigma_p(A^*)\} \setminus \sigma_p(A).$$

Exercise 7.3.10. Let H be a Hilbert space and let $A \in B(H)$.

(1) Let $\lambda \in \mathbb{C}$ and define a function $f : \mathbb{C} \to \mathbb{R}^+$ by

$$f(\lambda) = \inf_{\|x\|=1} \|\lambda x - Ax\|.$$

(a) Prove that the following statements are equivalent:
(i) $\exists x_n \in H, \ \|x_n\| = 1, \ \lim_{n\to\infty} \|(\lambda - A)x_n\| = 0$;
(ii) $f(\lambda) = 0$;

(iii) $\lambda \in \sigma_a(A)$.

(b) Show that f is continuous.

(2) Show that if $\lambda \in \rho(A)$, then

$$f(\lambda) = \frac{1}{\|R(\lambda, A)\|}.$$

(3) (a) Prove that $\sigma_a(A)$ is closed.

(b) Show that

$$\overline{\sigma_p(A)} \subset \sigma_a(A) \subset \sigma(A)...(1)$$

(c) Deduce that $\sigma_a(A)$ is compact.

(4) Denote the boundary of $\sigma(A)$ by $\partial\sigma(A)$.

(a) Show that

$$\partial\sigma(A) \subset \sigma_a(A).$$

(b) Infer that $\sigma_a(A)$ is not empty.

(5) Show that

$$\sigma(A) = \sigma_a(A) \cup \sigma_r(A)$$

(where the union need not be disjoint).

(6) Deduce that if A is is normal (a fortiori self-adjoint!), then

$$\sigma(A) = \sigma_a(A).$$

Exercise 7.3.11. Let $A \in B(H)$ be self-adjoint. Show that at least $\|A\|$ or $-\|A\|$ is in $\sigma(A)$.

Exercise 7.3.12. Let $A, B \in B(H)$ be similar.

(1) Show that $\sigma(A) = \sigma(B)$.
(2) Do A and B have the same eigenvalues?
(3) Do we have $\sigma_r(A) = \sigma_r(B)$? $\sigma_c(A) = \sigma_c(B)$?

Exercise 7.3.13. Let $A, B \in B(H)$ be similar.

(1) Show that A is bounded below iff B is bounded below.
(2) Infer that $\sigma_a(A) = \sigma_a(B)$.

Exercise 7.3.14. Let S be the shift operator defined on ℓ^2 by

$$S(x_1, x_2, \cdots, x_n, \cdots) = (0, x_1, x_2, \cdots, x_n, \cdots).$$

(1) Show that S does not possess any eigenvalue.
(2) Find $\sigma_p(S^*)$.
(3) What is $\sigma(S^*)$?
(4) Show that $\sigma(S) = \{\lambda \in \mathbb{C} : |\lambda| \leq 1\}$ (cf. Exercise 7.5.1).
(5) Show that $\sigma_a(S) = \{\lambda \in \mathbb{C} : |\lambda| = 1\}$ (cf. Exercise 7.5.1).
(6) Find $\sigma_r(S)$, $\sigma_c(S)$, $\sigma_r(S^*)$ and $\sigma_c(S^*)$ (using two methods).
(7) What is $\sigma(S^*S)$?

(8) Is $0 \in \sigma_p(SS^*)$? What is then $\sigma(SS^*)$?

Exercise 7.3.15. (cf. Exercise 7.5.3) On $\ell^2(\mathbb{Z})$, let R the bilateral shift defined by

$$R(x_n)_{n\in\mathbb{Z}} = R(\cdots, x_{-1}, \boldsymbol{x_0}, x_1, \cdots) = (\cdots, x_{-2}, \boldsymbol{x_{-1}}, x_0, \cdots),$$

where the "bold" indicates the zeros positions of the sequence.

(1) Show that $\sigma_p(R) = \varnothing$.
(2) Prove that

$$\sigma(R) = \sigma_c(R) = \sigma_a(R) = \{\lambda \in \mathbb{C} : |\lambda| = 1\}.$$

(3) Using the operator $U \in B(\ell^2(\mathbb{Z}))$ defined below, or else, find $\sigma_p(R^*)$, $\sigma(R^*)$, $\sigma_a(R^*)$, $\sigma_r(R^*)$ and $\sigma_c(R^*)$:

$$U(\cdots, x_{-2}, x_{-1}, \boldsymbol{x_0}, x_1, x_2, \cdots) = (\cdots, x_2, x_1, \boldsymbol{x_0}, x_{-1}, x_{-2}, \cdots).$$

Exercise 7.3.16. Let H be a separable complex Hilbert space and let $A \in B(H)$ be normal.

(1) Show that $\sigma_p(A)$ is countable.
(2) Give an example which shows that normality cannot be dispensed with.

Exercise 7.3.17. Let A and B be two (finite) square matrices. Let S and T be two bounded operators (not necessarily matrices).

(1) Show that AB and BA have the same spectrum, i.e. in this case, the same eigenvalues.
(2) Show that

$$\sigma(ST) \cup \{0\} = \sigma(TS) \cup \{0\}.$$

(3) Give an example of S and T such that $\sigma(ST) \neq \sigma(TS)$.
(4) Deduce that $\sigma(ST) = \{0\}$ iff $\sigma(TS) = \{0\}$.
(5) Show that if S (or T) is invertible, then $\sigma(ST) = \sigma(TS)$.

Exercise 7.3.18. Let $A, B \in B(H)$.

(1) Assume that B is onto and that $\overline{\operatorname{ran}(B^*)} = H$. Show that B is invertible.
(2) Show that if BA is invertible and $0 \notin \sigma_r(B^*)$, then B is invertible.

Exercise 7.3.19 ([**106**]). Let $A, B \in B(H)$ be such that B is positive. Let P be the (unique) square root of B. Prove, using Exercise 7.3.18 (or else!) that

$$\sigma(AB) = \sigma(BA) = \sigma(PAP).$$

Exercise 7.3.20. Let $A, B \in B(H)$ be such that one of them is normal. Using two proofs, show that

$$\sigma(AB) = \sigma(BA).$$

Exercise 7.3.21. Using Theorem 7.1.33, show that the equation

$$AB - BA = I$$

does not hold in $B(H)$.

Exercise 7.3.22. Let $p \in \mathbb{C}[X]$ be a polynomial (of degree at least one) and let $A \in B(H)$.

(1) Show that

$$\sigma(p(A)) = p(\sigma(A)).$$

(2) Provide an example to show that the result is not true anymore if $p \in \mathbb{R}[X]$.

Exercise 7.3.23. Let

$$A = \begin{pmatrix} 1 & 2 \\ 2 & 1 \end{pmatrix}$$

By computing only the spectrum of A, find

(1) $\sigma(A^2)$,
(2) $\sigma(A^n)$, $n \in \mathbb{N}$,
(3) $\sigma(A^{-1})$,

Exercise 7.3.24. (cf. Exercise 8.3.6 & 11.3.28) Let P be an orthogonal projection on a Hilbert space. In order to avoid trivialities, assume further that $P \neq 0$ and $P \neq I$ (I being the identity operator on H). Finally, consider the operator $A = 2P - I$.

(1) By computing A^2, show that $\sigma(P) \subseteq \{0, 1\}$.
(2) What are the eigenvalues of P?
(3) What is then $\sigma(P)$?

Exercise 7.3.25. Let $A \in B(\ell^2)$ be defined by

$$A(x_1, x_2, x_3, x_4, \cdots) = (-x_1, x_2, -x_3, x_4, \cdots).$$

(1) Show that 1 and -1 are two eigenvalues for A.
(2) Verify that $A^2 = I_{B(\ell^2)}$.
(3) Deduce that $\sigma(A) = \{1, -1\}$.

Exercise 7.3.26. Let $\mathcal{F}$ be the $L^2(\mathbb{R})$-Fourier transform.

(1) Show that $\mathcal{F}^4 = I_{L^2(\mathbb{R})}$ but $\mathcal{F}^2 \neq I_{L^2(\mathbb{R})}$.
(2) Deduce that $\sigma(\mathcal{F}) \subset \{1, -1, i, -i\}$.

(3) By considering the following operators on $L^2(\mathbb{R})$

$$p = \frac{1}{4}(I_{L^2(\mathbb{R})} + \mathcal{F} + \mathcal{F}^2 + \mathcal{F}^3),\ q = \frac{1}{4}(I_{L^2(\mathbb{R})} - i\mathcal{F} - \mathcal{F}^2 + i\mathcal{F}^3),$$

$$r = \frac{1}{4}(I_{L^2(\mathbb{R})} - \mathcal{F} + \mathcal{F}^2 - \mathcal{F}^3),\ s = \frac{1}{4}(I_{L^2(\mathbb{R})} + i\mathcal{F} - \mathcal{F}^2 - i\mathcal{F}^3),$$

show that $1, -1, i$ and $-i$ are eigenvalues for $\mathcal{F}$. What is then $\sigma(\mathcal{F})$?

Exercise 7.3.27. Give an example of a bounded operator $A \neq 0$ such that $\sigma(A) = \{0\}$.

Exercise 7.3.28.

(1) Find the spectrum of a nilpotent operator on some complex Hilbert space H.
(2) What can you deduce from the preceding answer?

Exercise 7.3.29. Let $A, N \in B(H)$ be commuting. Show that if N is nilpotent, then

$$\sigma(A + N) = \sigma(A).$$

Exercise 7.3.30. Let $A, B \in B(H)$. We define an operator T on $B(H)$ by

$$T(X) = AX - XB$$

for all $X \in B(H)$.

(1) Show that $T \in B(B(H))$.
(2) Write $T = L_A - R_B$ where $L_A, R_B : B(H) \to B(H)$ are defined for any $X \in B(H)$ by

$$L_A(X) = AX \text{ and } R_B(X) = XB$$

respectively. Show that

$$\sigma(L_A) \subset \sigma(A) \text{ and } \sigma(R_B) \subset \sigma(B).$$

(3) Infer that

$$\sigma(T) \subset \sigma(A) - \sigma(B).$$

Exercise 7.3.31. Let $A, B \in B(H)$ be such that $\sigma(A) \cap \sigma(B) = \varnothing$. Show that the equation

$$AX - XB = C$$

has a unique solution (in $B(H)$) for each $C \in B(H)$.

Exercise 7.3.32. (cf. Exercises 7.3.30 & 7.3.31) Let $A, B, C, D \in B(H)$ be such that

$$AC = CA,\ BD = DB \text{ and } \sigma(A)\sigma(B) \cap \sigma(C)\sigma(D) = \varnothing.$$

Show that the equation

$$AXB - CXD = E$$

has a unique solution $X \in B(H)$ for each $E \in B(H)$.

Exercise 7.3.33. Let $A, B \in B(H)$ be such that $\sigma(A) \cap \sigma(B) = \varnothing$. Assume that a $C \in B(H)$ is such that $C(A+B) = (A+B)C$ and $C(AB) = (AB)C$. Show that

$$AC = CA \text{ and } BC = CB.$$

Exercise 7.3.34. Let $A, B \in B(H)$ be such that $\sigma(A) \cap \sigma(B) = \varnothing$. Show that for any operator $C \in B(H)$, the matrix operator $\begin{pmatrix} A & C \\ 0 & B \end{pmatrix}$ is similar to $\begin{pmatrix} A & 0 \\ 0 & B \end{pmatrix}$.

Exercise 7.3.35. Let $A, B, C, X \in B(H)$ where H is a Hilbert space such that $\dim H = \infty$. Show that the similarity of the matrices $\begin{pmatrix} A & C \\ 0 & B \end{pmatrix}$ and $\begin{pmatrix} A & 0 \\ 0 & B \end{pmatrix}$ need not imply the existence of a solution to $AX - XB = C$.

Exercise 7.3.36. Let H be a complex Hilbert space.

(1) Show that if $A \in B(H)$ is non zero nilpotent, then A has a nontrivial hyperinvariant subspace (here H may be over $\mathbb{R}$ as well).
(2) Show that any non scalar algebraic operator $A \in B(H)$ has a nontrivial hyperinvariant subspace.

7.4. Tests

Test 7.4.1. Does there exist a self-adjoint $A \in B(\mathbb{R}^3)$ such that $A(1,1,1) = (0,0,0)$ and $A(1,2,2) = (2,4,4)$?

Test 7.4.2. Give an example of a bounded multiplication operator M_φ (defined on $L^2(0,1)$) such that

$$\sigma_p(M_\varphi) = \{0,1\},\ \sigma(M_\varphi) = \sigma_c(M_\varphi) = (0,1).$$

Test 7.4.3. Give an example showing that the inclusions (1) in Exercise 7.3.10 may be strict.

Test 7.4.4. Does the shift operator satisfy the Daugavet equation (cf. Exercise 7.3.7)?

Test 7.4.5. Consider the **Lyapunov Equation**

$$AX + XA^* = -I$$

where $A, X \in B(H)$ and I is the identity operator on H. Show that if $\operatorname{Re}\sigma(A) < 0$, then the solution X of the Lyapunov Equation in this case is self-adjoint.

Test 7.4.6. Let $A \in B(H)$. Prove that for any polynomial p (in $\mathbb{C}$),

$$\sigma_a(p(A)) = p(\sigma_a(A)).$$

Test 7.4.7. Let $A \in B(H)$ and let $\lambda \in \mathbb{C}$.

(1) Show that $\ker(\lambda I - A)$ is hyperinvariant for A.
(2) Deduce that if a non scalar operator A has an eigenvalue λ, then $\ker(\lambda I - A)$ is a nontrivial hyperinvariant subspace for A.

7.5. More Exercises

Exercise 7.5.1. ([**48**]) Let (α_n) be an increasing real sequence which is convergent to some r where also $\alpha_1 > 0$. Define $A : \ell^2 \to \ell^2$ by

$$A(x_1, x_2, \cdots) = (0, \alpha_1 x_1, \alpha_2 x_2, \cdots).$$

(1) Show that $\sigma(A) = \{\lambda \in \mathbb{C} : \ |\lambda| \leq r\}$.
(2) What is $\sigma_a(A)$?
(3) What is $\sigma_p(A)$?

Exercise 7.5.2. Let $A \in B(H)$ be an isometry.

(1) Prove that $\sigma(A)$ is either included in $\{\lambda \in \mathbb{C} : \ |\lambda| = 1\}$ or equal to $\{\lambda \in \mathbb{C} : \ |\lambda| \leq 1\}$.
(2) Show that

$$\sigma_a(A) = \sigma(A) \cap \{\lambda \in \mathbb{C} : \ |\lambda| = 1\}.$$

Exercise 7.5.3. Let

$$L^2(\mathbb{T}) = \{f : \mathbb{T} \to \mathbb{C} : \int_{\mathbb{T}} |f(z)|^2 dz < \infty\},$$

where $\mathbb{T}$ is the unit circle. Assume that the Lebesgue measure is normalized so that $\int_{\mathbb{T}} dz = 1$. Consider the *orthonormal basis* in $L^2(\mathbb{T})$ constituted of $\{z^n : \ n \in \mathbb{Z}\}$ (call its elements e_n). Finally, let M_z be defined on $D(M_z) = \{f \in L^2(\mathbb{T}) : \ zf \in L^2(\mathbb{Z})\}$ by

$$(M_z f)(z) = zf(z).$$

(1) Show that $M_z \in B(L^2(\mathbb{T}))$. In fact, show that M_z is unitary. What is M_z^{-1}?.
(2) Find $\sigma(M_z)$, $\sigma_p(M_z)$, $\sigma_r(M_z)$, $\sigma_a(M_z)$ and $\sigma_c(M_z)$.
(3) Show that for each $z \in L^2(\mathbb{T})$, we have

$$(M_z e_n)(z) = e_{n+1}(z),$$

i.e. M_z may be seen as the bilateral shift R.
(4) Deduce $\sigma(R)$, $\sigma_p(R)$, $\sigma_r(R)$, $\sigma_a(R)$ and $\sigma_c(R)$.

Exercise 7.5.4. Let R be the bilateral (forward) shift on $\ell^2(\mathbb{Z})$. Define $\Delta : \ell^2(\mathbb{Z}) \to \ell^2(\mathbb{Z})$ by

$$\Delta = R + R^* - 2I,$$

where I is the usual identity. Then Δ is called the **discrete Laplacian**.
(1) Check that $\Delta \in B(\ell^2(\mathbb{Z}))$ and that it is self-adjoint. Is $\Delta \leq 0$?
(2) What is $\sigma_p(\Delta)$?
(3) Find $\sigma(\Delta)$.

Exercise 7.5.5. Let $C \in B(H)$ be self-adjoint and let $T = A+iB \in B(H)$ be such that $T^2 = C$.
(1) If $\sigma(A) \cap \sigma(-A) = \varnothing$, then show that T is self-adjoint and invertible.
(2) If $\sigma(B) \cap \sigma(-B) = \varnothing$, then show that T is skew symmetric (that is, $T^* = -T$) and invertible.

CHAPTER 8

Spectral Radius. Numerical Range

8.1. Basics

8.1.1. Spectral Radius.

Definition 8.1.1. *Let A be in $B(H)$. The* ***spectral radius*** *of A is defined as*

$$r(A) = \sup\{|\lambda| : \ \lambda \in \sigma(A)\}.$$

Remark. It is clear that $r(A)$ is a positive *finite* number.

There is a nice relation between the spectral radius of an operator and the norm of its powers.

Proposition 8.1.2 ([**82**], **Gelfand-Beurling Formula**). *Suppose $A \in B(H)$. Then the limit $\lim_{n\to\infty} \|A^n\|^{\frac{1}{n}}$ exists, and we have*

$$\lim_{n\to\infty} \|A^n\|^{\frac{1}{n}} = \inf_{n\in\mathbb{N}} \|A^n\|^{\frac{1}{n}} = r(A).$$

Remark. We already proved in Exercise 2.3.9 that $\lim_{n\to\infty} \|A^n\|^{\frac{1}{n}}$ exists and that it equals $\inf_{n\in\mathbb{N}} \|A^n\|^{\frac{1}{n}}$. The hardest part in the proof is to show that $\lim_{n\to\infty} \|A^n\|^{\frac{1}{n}} = r(A)$ (see e.g. [**105**]).

Here is the main theorem of this subsection.

Theorem 8.1.3. *(see Exercise 8.3.2 and Exercise 8.3.3, see also Test 8.4.2) Let $A \in B(H)$ be self-adjoint (or normal). Then*

$$r(A) = \|A\|.$$

Remark. The previous theorem and its variants will be referred to as "**The Spectral Radius Theorem**".

Since A^*A (and AA^*) is self-adjoint, finding the norm of A for matrices, say, becomes very easy. Indeed, we may find the eigenvalues of A^*A, then we use the next result:

Corollary 8.1.4. *Let $A \in B(H)$. Then*

$$\|A\| = \sqrt{r(A^*A)} = \sqrt{r(AA^*)}.$$

The conclusion in the previous theorem is not only satisfied by self-adjoint or normal operators. We will see that it holds, for instance, for "hyponormal" operators (see Exercise 12.3.6). So giving this property a name seems to be appropriate.

Definition 8.1.5. *An operator $A \in B(H)$ satisfying*

$$r(A) = \|A\|$$

is called ***normaloid***.

Examples 8.1.6.

(1) *We have just observed that normal (hence unitary and self-adjoint) operators are normaloid.*
(2) *The shift operator is normaloid (see Exercise 8.3.1).*
(3) *Let $A : L^2(\mathbb{R}) \to L^2(\mathbb{R})$ be the bounded operator defined by*

$$(Af)(x) = e^{-x^2} f(x-1)$$

where $f \in L^2(\mathbb{R})$. Then A is normaloid (see Exercise 10.3.23).

In relation with the limits appearing in Proposition 8.1.2 above, we have the following definition.

Definition 8.1.7. *Let $A \in B(H)$. We say that A is* ***quasinilpotent*** *if*

$$\lim_{n\to\infty} \|A^n\|^{\frac{1}{n}} = 0.$$

Remark. In some textbooks, e.g. in [**169**], they use instead the term **generalized nilpotent**.

Remark. Related to the Invariant Subspace Problem, Read [**177**] constructed quasinilpotent operators without non-trivial invariant subspaces.

Examples 8.1.8.

(1) *Every nilpotent operator is clearly quasinilpotent.*
(2) *The identity operator is not quasinilpotent.*
(3) *Define the Volterra Operator on $L^2[0,1]$. Then it is quasinilpotent (see Exercise 9.3.21).*

We have the following equivalent definition of quasinilpotence.

Proposition 8.1.9. *Let $A \in B(H)$. Then:*

$$A \textit{ is quasinilpotent iff } \sigma(A) = \{0\}.$$

The following result, its proof and related brief historical notes may be found in [**169**].

Theorem 8.1.10. *Let $A, B \in B(H)$ and set $C = [A, B]$ (the commutator of A and B). If $AC = CA$, then C is quasinilpotent.*

Finally, we give properties relating the spectral radius of products and sums of linear operators.

Proposition 8.1.11 (for a proof, see Exercise 8.3.5). *Let A and B be two operators on a Hilbert space. Then*

$$r(AB) = r(BA).$$

Proposition 8.1.12. *Let $A, B \in B(H)$ be commuting. Then*

$$r(A + B) \leq r(A) + r(B).$$

Proposition 8.1.13. *Let $A, B \in B(H)$ be commuting. Then*

$$r(AB) \leq r(A)r(B).$$

8.1.2. Numerical Range.

We introduce another important subset of $\mathbb{C}$.

Definition 8.1.14. *Let $A \in B(H)$. The* ***numerical range*** *of A is defined by*

$$W(A) = \{< Ax, x >: \ x \in H, \ \|x\| = 1\}.$$

Remark. Different names for the numerical range exist such as: **Hausdorff domain** or the **fields of values** (especially for finite dimensional matrices).

Remark. It is clear that $W(A)$ is just the image of the Euclidean unit sphere under the bounded map $x \mapsto < Ax, x >$. This justifies the appellation above.

Examples 8.1.15.

(1) *Let*

$$A = \begin{pmatrix} 1 & 0 \\ 0 & 0 \end{pmatrix}.$$

Then $W(A) = [0, 1]$ (see Exercise 8.3.10).

(2) *Let*

$$B = \begin{pmatrix} 0 & 1 \\ 0 & 0 \end{pmatrix}.$$

Then $W(A) = \{z \in \mathbb{C} : \ |z| \leq \frac{1}{2}\}$ (see Exercise 8.3.10).

(3) *The numerical range of the shift S is given by:*

$$W(S) = B(0, 1), \text{ i.e. the open unit disk}$$

(a proof is given in Exercise 8.3.14).

Remark. ($W(A)$ Vs. $\sigma(A)$) One of the great facts about the numerical range is that it contains much more information about the operator than the spectrum does. For instance, many operators have a spectrum reduced to a singleton $\{a\}$ ($a \in \mathbb{C}$), but as far as the numerical range is concerned, only a multiple of the identity (namely aI) has this property. Also, if the spectrum of a given operator is a subset of the real line, then not much can be said about the operator itself. However, the numerical range is a subset of the real line iff the operator in question is self-adjoint. The last properties and others are gathered in the next proposition (and as quoted by Shapiro in [**198**]: "Very little about the numerical range is obvious-here is a more-or-less complete list of what is"):

Proposition 8.1.16. *(see Exercise 8.3.10) Let $A, B \in B(H)$ and $\alpha, \beta \in \mathbb{C}$. Let $U \in B(H)$ be unitary. Then*

(1) $\sigma_p(A) \subset W(A)$.
(2) $W(A) \subset B_c(0, \|A\|)$.
(3) $W(A) = \{\alpha\}$ *iff* $A = \alpha I$, *where I is the identity operator on H. More generally, we have* $W(\alpha A + \beta I) = \alpha W(A) + \beta$.
(4) $W(A + B) \subset W(A) + W(B)$.
(5) $W(A^*) = \{\overline{\lambda} : \lambda \in W(A)\}$.
(6) $W(A) \subset \mathbb{R}$ *iff A is self-adjoint.*
(7) $W(A) = W(UAU^*)$.

The next result gives us a characterization in a bi-dimensional setting (usually called the "**Ellipse Lemma**").

Theorem 8.1.17. *([**88**]) Let $A \in B(H)$ where* $\dim H = 2$. *Then $W(A)$ is an ellipse whose foci are the eigenvalues of A.*

Proposition 8.1.18. *(for a proof, see Exercise 8.3.15) Let H be a finite dimensional Hilbert space and let $A \in B(H)$. Then $W(A)$ is compact.*

Now, we give the numerical range of a multiplication operator.

Proposition 8.1.19. *Let H be a separable Hilbert space. Let $A \in B(H)$ be the diagonal operator defined for all n by*

$$Ae_n = \lambda_n e_n,$$

where $(\lambda_n)_n$ is a bounded sequence of complex numbers and (e_n) is an orthonormal basis in H. Then $W(A) = \operatorname{conv}(\{\lambda_n\}_n)$.

Remark. The previous result is important in the sense that it may be used to find interesting counterexamples. Let A be self-adjoint on $B(\ell^2)$ such that $\sigma(A) = [0,1]$ (cf. Exercise 7.3.6). Then depending on whether 0 or 1 or both or none of them appear on the diagonal, it is clear that the numerical range of A can be any of the following intervals: $(0,1]$ or $[0,1)$ or $(0,1)$ or $[0,1]$.

We have seen (and we will see) more or less "different shapes" of numerical ranges. But, they all belong to *one* class, namely:

THEOREM 8.1.20. ***(Toeplitz-Hausdorff)*** *Let $A \in B(H)$. Then $W(A)$ is a convex set in the complex plane.*

The following result tells us the relationship between the spectrum and the numerical range of a bounded operator.

THEOREM 8.1.21. *Let $A \in B(H)$. Then $\sigma(A) \subset \overline{W(A)}$.*

COROLLARY 8.1.22. *Let $A \in B(H)$ be self-adjoint. Then $\sigma(A) \subset \mathbb{R}$.*

The class of normal operators has interesting properties as regards the numerical range. In fact, the closure of the numerical range of a normal operator is completely determined by its spectrum.

THEOREM 8.1.23. *Let $A \in B(H)$ be normal. Then*

$$\overline{W(A)} = \operatorname{conv}(\sigma(A)).$$

REMARK. The spectral radius is associated with the spectrum. We may also introduce another "radius" related to the numerical range. This is the so-called **numerical radius**, denoted by $w(A)$, and defined as

$$w(A) = \sup\{|\lambda| : \lambda \in W(A)\}.$$

For more on this topic, see e.g. **[81]** or **[88]**. One of the interesting results (see **[125]**) is: $A \in B(H)$ *is normaloid iff*

$$\|A\| = \sup_{\|x\|=1} |< Ax, x >|.$$

8.2. True or False: Questions

QUESTIONS. Comment on the following questions/statements and indicate those which are false and those which are true when this applies. Justify your answers.

(1) Let $A \in B(H)$. Then

$$r(A) = r(A^*).$$

(2) Let $A, B \in B(H)$ be both self-adjoint. Then

$$\|A\| \leq \|B\| \Longleftrightarrow \sigma(A) \subset \sigma(B).$$

(3) Let $A, B \in B(H)$ be unitarily equivalent. Then
$$r(A) = r(B).$$
(4) Let $A, B \in B(H)$. Give an example which shows that it may well happen that
$$r(A + B) > r(A) + r(B).$$
(5) Let $A \in B(H)$ be a non-scalar normal operator. Then $\sigma(A)$ contains at least two distinct points.

(6) We know from Proposition 8.1.16 that if $A \in B(H)$, then $W(A) = \{\alpha\}$ iff $A = \alpha I$ for some α. But if we consider
$$A = \begin{pmatrix} 0 & 1 \\ -1 & 0 \end{pmatrix},$$
then $< Ax, x >= 0$ for any $x \in \mathbb{R}^2$ so that $W(A) = \{0\}$ and yet $A \neq 0$! Why does this example not violate the result of Proposition 8.1.16?

(7) Let H be a complex Hilbert space and let $A \in B(H)$. Then $W(A)$ can be empty.

(8) Let $A, B \in B(H)$ be both self-adjoint with equal spectra. Then $W(A) = W(B)$.

(9) The numerical range of a bounded operator is always closed.

(10) Let $A \in B(H)$ be self-adjoint. Then *either* $\sigma(A) \subset W(A)$ *or* $W(A) \subset \sigma(A)$.

8.3. Exercises with Solutions

Exercise 8.3.1. Let S be the shift operator and let $n \in \mathbb{N}$.

(1) Find $S^n = \underbrace{SS \cdots S}_{n \text{ times}}$.

(2) Deduce that S is normaloid.

Exercise 8.3.2. Let $A \in B(H)$ be self-adjoint.

(1) (a) Verify that $\|A^2\| = \|A\|^2$.

(b) Show that
$$\|A^{2^n}\| = \|A\|^{2^n} \text{ for all } n \in \mathbb{N}.$$
(c) Infer that
$$r(A) = \|A\|,$$
where $r(A)$ denotes the spectral radius of A.

(2) Deduce that $\|A^n\| = \|A\|^n$ for all $n \in \mathbb{N}$.

Exercise 8.3.3. Let $A \in B(H)$ be normal.

(1) Show that $\|A^2\| = \|A\|^2$.

(2) Show that $\|A^{2^n}\| = \|A\|^{2^n}$ for all $n \in \mathbb{N}$.
(3) Deduce that $\lim\limits_{n\to\infty} \|A^n\|^{\frac{1}{n}} = \|A\|$.
(4) Check that if $\sigma(A)$ is a singleton $\{\lambda\}$, then $A = \lambda I$ where I is the identity operator on H.
(5) Infer that if A is further assumed to be non-scalar, then $\sigma(A)$ contains at least two distinct points.

Exercise 8.3.4. Let $p \in \mathbb{C}[X]$ and let $A \in B(H)$ be normal. Show that $p(A)$ is normal and that

$$\|p(A)\| = \max_{\lambda\in\sigma(A)} |p(\lambda)|.$$

Exercise 8.3.5. Let A and B be two bounded operators on a Hilbert space. Show that

$$r(AB) = r(BA),$$

where $r(\cdot)$ stands for the spectral radius.

Exercise 8.3.6. Let A be a self-adjoint operator on a Hilbert space H. Suppose that $\sigma(A) = \{0, 1\}$. Show that A is an orthogonal projection.

Exercise 8.3.7. ([**28**]) Let $r > 0$ and let $A, B \in B(H)$ be such that

$$\sigma(A) \subset \{z \in \mathbb{C} : \ |z| > r\} \text{ and } \sigma(B) \subset \{z \in \mathbb{C} : \ |z| < r\}.$$

Show that for every $C \in B(H)$, **the** solution of $AX - XB = C$ is given by:

$$X = \sum_{n=0}^{\infty} A^{-n-1}CB^n.$$

Exercise 8.3.8. ([**21**]) Let $A \in B(H)$ be self-adjoint. Let B be a positive operator. Assume further that $BAB + A = 0$.
(1) Show that $\sigma[(BA)^2] \subset \mathbb{R}^-$.
(2) Deduce that the spectrum of BA is purely imaginary.
(3) Show that $\sigma(A) = \{0\}$.
(4) Infer that $A = 0$.

Exercise 8.3.9. Let $T = A+iB$ be normal and such that $\sigma(T) \subset \mathbb{R}$. Show that T is self-adjoint.

Exercise 8.3.10. Let $A, B \in B(H)$ and let $\alpha, \beta \in \mathbb{C}$. Show that
(1) $\sigma_p(A) \subset W(A)$.
(2) $W(A) \subset B_c(0, \|A\|)$.
(3) $W(A) = \{\alpha\}$ iff $A = \alpha I$, where I is the identity operator on H. More generally, we have $W(\alpha A + \beta I) = \alpha W(A) + \beta$.

(4) $W(A+B) \subset W(A)+W(B)$.
(5) $W(A^*) = \{\overline{\lambda} : \lambda \in W(A)\}$.
(6) $W(A) \subset \mathbb{R}$ iff A is self-adjoint.
(7) $W(A) = W(UAU^*)$ whenever U is unitary.

Exercise 8.3.11. Without using the Ellipse Lemma, find the numerical range of the following operators:

(1)
$$A = \begin{pmatrix} 1 & 0 \\ 0 & 0 \end{pmatrix}.$$

(2)
$$B = \begin{pmatrix} 0 & 1 \\ 0 & 0 \end{pmatrix}.$$

(3)
$$C = \begin{pmatrix} 3 & 0 \\ 0 & 1 \end{pmatrix}.$$

(4)
$$D = \begin{pmatrix} 0 & 0 \\ 1 & 0 \end{pmatrix}.$$

Exercise 8.3.12.

(1) Let P be a non-trivial orthogonal projection. Find $W(P)$.
(2) Apply this result to another example already treated above.

Exercise 8.3.13. Let $A, B \in B(H)$. Do we have

$$W(A+B) \supset W(A)+W(B)?$$

Exercise 8.3.14. Find the numerical range of the shift operator.

Exercise 8.3.15. (cf. Exercise 8.5.1) Let H be a *finite* dimensional Hilbert space and let $A \in B(H)$.

(1) Show that $W(A)$ is compact.
(2) Give an example which shows that the hypothesis "finite dimensional" may not just be omitted.

8.4. Tests

Test 8.4.1. Let

$$A = \begin{pmatrix} 2 & 2i \\ -2i & 2 \end{pmatrix} \text{ et } B = \begin{pmatrix} i & 1 \\ 1 & -i \end{pmatrix}.$$

(1) Find $\|A\|$.
(2) Deduce $\|B\|$.

Test 8.4.2. Using Exercise 7.3.11, show that self-adjoint operators are normaloid.

Test 8.4.3. Let $P \neq 0$ be an orthogonal projection. Using the Spectral Radius Theorem, show that $\|P\| = 1$.

Test 8.4.4. What is the numerical range of the operator A, defined on $\ell^2(\mathbb{N})$ by

$$A(x_1, x_2, x_3, x_4, x_5, \cdots) = (x_1, 0, x_3, 0, x_5, \cdots)?$$

Test 8.4.5. Let $A \in B(H)$ be such that $0 \notin \overline{W(A)}$. Justify the invertibility of A and show that $0 \notin \overline{W(A^{-1})}$.

8.5. More Exercises

Exercise 8.5.1. ([**14**], cf. Exercise 8.3.15) Let A be a *compact* operator on an *infinite* dimensional Hilbert space H.

(1) Prove that if $0 \in W(A)$, then $W(A)$ is closed.
(2) Provide an example of a compact operator A such that $W(A)$ is not closed to show that the hypothesis "$0 \in W(A)$" may not just be dropped.

Exercise 8.5.2. Let $A, B \in B(H)$ where H is a Hilbert space. Let $r(\cdot)$ denote the spectral radius. Prove that if A and B *commute*, then

$$r(A+B) \leq r(A) + r(B).$$

Exercise 8.5.3. Find the numerical range of the following matrix:

$$A = \begin{pmatrix} 0 & 0 & 0 \\ 1 & 0 & 0 \\ 0 & 0 & 1 \end{pmatrix}.$$

Exercise 8.5.4. Let A be a linear operator on a complex pre-Hilbert space H

(1) Show that A is bounded iff $W(A)$ is a bounded set of the complex plane (you may wish to use the general polarization formula).
(2) Does the previous result remain valid if we work with a real pre-Hilbert space?

CHAPTER 9

Compact Operators

9.1. Basics

9.1.1. Compact Operators. We introduce here an important class of bounded operators which is closely related to the study of Integral Equations.

DEFINITION 9.1.1. *Let H and H' be two Hilbert spaces and let $A \in B(H, H')$. We say that A is* ***compact*** *if for any sequence (x_n) in H with $\|x_n\| = 1$, (Ax_n) has a convergent subsequence in H'.*

The set of compact operators from H into H' will be denoted by $K(H, H')$.

REMARK. The definition of a compact operator is valid on Banach spaces as well.

REMARK. We will assume in the sequel that $H = H'$.

REMARK. In the previous definition, it is not compulsory to have $\|x_n\| = 1$. In fact, any bounded (x_n) will do (why?).

From topological considerations, there are equivalent definitions of compactness.

PROPOSITION 9.1.2. *Let $A \in B(H)$. Then $A \in K(H)$ iff one of the following occurs:*

1. *The image of the closed unit ball of H under A is a relatively compact subset of H.*
2. *$A(B)$ is totally bounded, where B is the open unit ball in H.*

Most of the following examples will be treated in detail in Exercise 9.3.1.

EXAMPLES 9.1.3.

1. *Let $A \in B(H)$ be such that $Ax = 0$ for all $x \in H$. Obviously, $A \in K(H)$.*
2. *An $A \in B(H)$ is in $K(H)$ whenever $\dim H < \infty$.*
3. *The identity operator on H is not compact if $\dim H = \infty$.*

Proposition 9.1.4. *(for a proof, see Exercise 9.3.2) Let H be a Hilbert space and let $A \in K(H)$. Then $A \in B(H)$. That is,*

$$K(H) \subset B(H).$$

Remark. Contra positively, by the last result, unbounded operators are not compact.

One may ask how stable are compact operators as regards multiplications and products? The answer is given in the following result and a proof is supplied in Exercise 9.3.8.

Proposition 9.1.5.

(1) *The sum of two compact operators is compact.*
(2) *The products AB and BA are compact if A is compact and B is (only) bounded.*

The fact that the identity is not compact on an infinite dimensional space combined with the second statement of the foregoing proposition give us an important property of compact operators. Indeed, a compact operator A on an infinite dimensional space is never invertible (for if it were, then $AA^{-1} = I$ would be compact!).

Corollary 9.1.6. *Let $A \in K(H)$ where H is infinite dimensional. Then A is not invertible.*

The next two results furnish a little more the set $K(H)$.

Proposition 9.1.7. *An operator $A \in B(H)$ is compact iff its adjoint A^* is compact.*

Theorem 9.1.8. *The set $K(H)$ is closed in $B(H)$ (with respect to the topology of the operator norm). That is, if (A_n) is a sequence of operators in $K(H)$ such that $\lim_{n\to\infty} \|A_n - A\| = 0$ for some $A \in B(H)$, then $A \in K(H)$.*

Before giving a practical consequence, we recall here the following definition.

Definition 9.1.9. *Let A be a linear operator. We say that A has **finite rank** if $\dim[\operatorname{ran}(A)]$ is finite.*

Proposition 9.1.10. *If $A \in B(H)$ is of a finite rank, then A is compact.*

Corollary 9.1.11. *If (A_n) is a sequence of bounded and finite rank operators such that $\|A_n - A\| \to 0$ as n tends to ∞, where $A \in B(H)$, then $A \in K(H)$.*

Conversely, if $A \in K(H)$, then there exists a sequence of finite rank operators (A_n) which converges to $A \in B(H)$.

The next result prescribes a necessary condition of compactness.

Proposition 9.1.12. *Let H be a Hilbert space with* $\dim H = \infty$ *and let (e_n) be an orthonormal basis in H. If $A \in K(H)$, then*

$$\lim_{n\to\infty} \|Ae_n\| = 0.$$

One may wonder when an operator matrix is compact? The following result answers this question.

Proposition 9.1.13. *Let $A_{ij} \in B(H)$, where $i, j = 1, 2$. Let*

$$A = \begin{pmatrix} A_{11} & A_{12} \\ A_{21} & A_{22} \end{pmatrix}.$$

Then A is compact on $B(H \oplus H)$ iff each of the four operators A_{ij} is compact.

The class of compact operators seems to be "nice". Maybe the invariant subspace problem holds for it? The first result is very simple:

Proposition 9.1.14. *(see Test 9.4.6) Every compact self-adjoint operator on an infinite dimensional space has a non-trivial invariant subspace.*

More generally, we have the following tremendous result (holding even in a *Banach* space). This is more known as the **Lomonosov Theorem** ([**132**]):

Theorem 9.1.15. *Let H be an infinite-dimensional complex Hilbert (in fact Banach!) space. Let $A \in B(H)$ be a not multiple of the identity. If there exists a non-zero compact operator B which commutes with A, then A has a non trivial hyperinvariant subspace.*

9.1.2. Hilbert-Schmidt Operators. Sometimes it can be a little hard to prove that an operator is compact even with all the practical results that we have seen so far. However, there is another class of operators which is stronger than compactness but quite easier to verify.

Definition 9.1.16. *Let H be an infinite-dimensional Hilbert space and let (e_n) be an orthonormal basis in H. If $A \in B(H)$ is such that*

$$\sum_{n=1}^{\infty} \|Ae_n\|^2 < \infty,$$

then A is called a ***Hilbert-Schmidt*** *operator.*

The family of all Hilbert-Schmidt operators is denoted by $B_2(H)$. *If* $A \in B_2(H)$, *then*

$$\|A\|_2 = \left(\sum_{n=1}^{\infty} \|Ae_n\|^2 \right)^{\frac{1}{2}}$$

defines a norm on $B_2(H)$.

Remark. The whys and wherefores of the notation $B_2(H)$ will be known in Exercise 9.5.6.

Before giving the link to compact operators, we have some elementary results.

Proposition 9.1.17.

(1) *If* $A \in B_2(H)$, *then* $\|A\| \leq \|A\|_2$ *and* $\|A\|_2 = \|A^*\|_2$. *Moreover,* $(B_2(H), \|\cdot\|_2)$ *is a Banach space, but* $(B_2(H), \|\cdot\|)$ *is not closed.*

(2) *If* $A \in B(H)$ *and* $B \in B_2(H)$, *then* $AB, BA \in B_2(H)$ *and*

$$\|AB\|_2 \leq \|A\|\|B\|_2 \text{ and } \|BA\|_2 \leq \|A\|\|B\|_2.$$

Theorem 9.1.18. *(Exercise 9.3.10) Hilbert-Schmidt operators are compact.*

Corollary 9.1.19. *If* H *is infinite dimensional, then the identity* $I \in B(H)$ *is not Hilbert-Schmidt.*

Remark. There exist compact operators which are not Hilbert-Schmidt. One example is provided just after the next theorem.

A priori, the definition of a Hilbert-Schmidt operator seems to be a little bit non-practical when using to prove that an operator is *not* Hilbert-Schmidt since we have to do it for all bases! Thankfully, the following result tells us that this is not the case, i.e. the definition does not depend on the choice of the orthonormal basis.

Theorem 9.1.20. *Let* H *be an infinite-dimensional Hilbert space and let* (e_n) *and* (f_n) *be two orthonormal bases in* H. *Let* $A \in B(H)$. *Then*

$$\sum_{n=1}^{\infty} \|Af_n\|^2 = \sum_{n=1}^{\infty} \|Ae_n\|^2 = \sum_{n=1}^{\infty} \|A^*e_n\|^2$$

(in the sense that all three can be infinite simultaneously or finite simultaneously).

EXAMPLE 9.1.21. *(see Exercise 9.3.12) Let $A \in B(\ell^2)$ be defined by*

$$A(x_1, x_2, x_3, \cdots) = \left(x_1, \frac{1}{\sqrt{2}}x_2, \frac{1}{\sqrt{3}}x_3, \cdots\right).$$

Then A is compact without being Hilbert-Schmidt.

PROPOSITION 9.1.22. *Let $A \in B(H)$. Then A is Hilbert-Schmidt iff A^* is Hilbert Schmidt.*

The primary application of the Hilbert-Schmidt theory is that an integral operator on an adequate setting is Hilbert-Schmidt.

THEOREM 9.1.23. *(see Exercise 9.3.11) Let $K \in L^2((a,b) \times (c,d))$ and define $A : L^2(a,b) \to L^2(c,d)$ by*

$$(Af)(x) = \int_c^d K(x,y)f(y)dy, \ a < x < b.$$

Then A is Hilbert-Schmidt (and so compact).

REMARK. The intervals (a,b) and (c,d) are allowed to be unbounded (even equal to $\mathbb{R}$).

REMARK. Surprisingly, the converse of the previous theorem also holds, i.e. a Hilbert-Schmidt operator may be expressed as an integral operator on some $L^2(X, d\mu)$ space. This result may be found in e.g. ([**178**], Theorem VI.23).

9.1.3. Spectrum and Compact Operators. Using the approximate spectrum, we can prove the following interesting result (see Exercise 9.3.20):

THEOREM 9.1.24. *Let $A \in B(H)$ be self-adjoint. Then at least $\|A\|$ or $-\|A\|$ is in $\sigma(A)$ (this is already known to us from before). If the self-adjoint A is also taken to be compact, then $\|A\|$ or $-\|A\|$ is actually in $\sigma_p(A)$.*

The next theorem prescribes more spectral properties of compact operators.

THEOREM 9.1.25. *Let $A \in B(H)$ be compact. Then*

(1) *If $\dim H = \infty$, then $0 \in \sigma(A)$.*
(2) *Also,*
$$\sigma(A) = \sigma_p(A) \cup \{0\}.$$
(3) *$\sigma(A)$ is countable. In case $\sigma(A)$ is infinite, then the eigenvalues (λ_n) may be arranged so that for all n*
$$|\lambda_{n+1}| \leq |\lambda_n| \text{ and } \lim_{n\to\infty} \lambda_n = 0.$$

We close this subsection with a fundamental theorem and a nice result:

THEOREM 9.1.26. *(**The Spectral Theorem for Compact Self-adjoint Operators**, see e.g.* **[48]***) Let A be a compact and self-adjoint operator on a Hilbert space H. Then there exists an orthonormal sequence (finite or infinite) $\{\varphi_n\}$ of eigenvectors of A, with corresponding real eigenvalues (λ_n), such that for every $x \in H$, we have*

$$Ax = \sum_{n=1}^{\infty} \lambda_n < x, \varphi_n > \varphi_n.$$

PROPOSITION 9.1.27. *Let A be the compact self-adjoint integral operator defined on $L^2(X, \mu)$ (where μ is a σ-finite measure on a measure space X) by*

$$Af(x) = \int_X K(x,y)f(y)d\mu(y)$$

(where $K \in L^2(X^2, \mu \times \mu)$ and $K(x,y) = \overline{K(y,x)}$). If (λ_n) is the sequence of (non-zero) eigenvalues, then

$$\iint_{X\times X} |K(x,y)|^2 d\mu(x)d\mu(y) = \sum_{n=0}^{\infty} \lambda_n^2.$$

9.2. True or False: Questions

QUESTIONS. Comment on the following questions/statements and indicate those which are false and those which are true when this applies. Justify your answers.

(1) Let H be a Hilbert space. Then $(K(H), \|\cdot\|_{B(H)})$ is a Banach space.
(2) Let $A \in B(H)$. Then
$$A \text{ is compact} \Longleftrightarrow A^2 \text{ is compact}.$$
(3) Let $A \in B(H)$ be such that $A^n = I$ for some n in $\mathbb{N}$. Then surely A is compact.
(4) Compact operators are normaloid.
(5) Let $A, B \in B(H)$ be two unitarily equivalent operators. Then
$$A \text{ is compact} \Longleftrightarrow B \text{ is compact}.$$
(6) Let $A \in B(H)$. Then
$$A \text{ is compact} \Longleftrightarrow |A| \text{ is compact}.$$

(7) Let $A, B \in B(H)$ be such that $0 \le A \le B$. Then

$$A \in K(H) \text{ iff } B \in K(H).$$

9.3. Exercises with Solutions

Exercise 9.3.1. Let $A \in B(H)$.

(1) Show that $A \in K(H)$ if A is of finite rank.
(2) (We may assume $A \in L(H)$ only!) Show that $A \in K(H)$ whenever $\dim H < \infty$.
(3) Prove that $I \notin K(H)$ if $\dim H = \infty$.

Exercise 9.3.2. Show that a compact operator is necessarily bounded.

Exercise 9.3.3. We already know from Exercise 2.5.7 that the Hardy operator $Hf(x) = \frac{1}{x}\int_0^x f(y)dy$ is bounded on $L^2(0,\infty)$. Is it compact?

Exercise 9.3.4. Let H be a Hilbert space and let $x, t \in H$. Define an operator A on H by

$$Ax = <x, t> t.$$

(1) Show that $A \in B(H)$. Find $\|A\|$.
(2) Show that A is self-adjoint and positive.
(3) Prove that for some $\lambda \ge 0$, $A^2 = \lambda A$. Deduce the expression of $\sqrt{A}$.
(4) Find $r(A)$ and find again $\|A\|$.
(5) Show that A is compact.

Exercise 9.3.5. Let $A \in B(\ell^2)$ be the multiplication operator by the bounded (λ_n), that is,

$$A(x_1, x_2, \cdots, x_n, \cdots) = (\lambda_1 x_1, \lambda_2 x_2, \cdots, \lambda_n x_n, \cdots).$$

(1) Prove that A is compact iff $\lim_{n\to\infty} \lambda_n = 0$.
(2) Deduce another proof why the identity is not compact on ℓ^2.
(3) Show that A is Hilbert-Schmidt iff

$$\sum_{n=1}^{\infty} |\lambda_n|^2 < \infty.$$

Exercise 9.3.6. Let $A : \ell^2 \to \ell^2$ be the bounded operator defined by

$$A(x_1, x_2, \cdots, x_n, \cdots) = \left(x_1, \frac{x_2}{2}, \cdots, \frac{x_n}{n}, \cdots\right).$$

Show that A is strictly positive but A is not invertible.

Exercise 9.3.7. Let $A \in B(\ell^2)$ be defined by

$$A(x_1, x_2, x_3, \cdots) = (0, x_1, 0, x_3, \cdots).$$

Show that A is not compact whereas A^2 is compact.

Remark. If A^2 is compact and A is normal, then A is automatically compact. In fact, something weaker than normality (namely hyponormality) leads to the desired result. The proof may be found in Exercise 12.3.10.

Exercise 9.3.8. Let $A, B \in B(H)$ and let $\alpha, \beta \in \mathbb{C}$.

(1) Let $A, B \in K(H)$. Show that $\alpha A + \beta B \in K(H)$.
(2) Show that if $A \in K(H)$, then AB and BA are both compact.
(3) Show that A^*A (or AA^*) is compact iff A is compact.
(4) Deduce that A is compact iff A^* is compact.

Exercise 9.3.9. (see Test 9.4.4 for another proof) Assume that $\dim H = \infty$ and let $A \in B(H)$ be bounded below. Can A be compact?

Exercise 9.3.10. Prove that Hilbert-Schmidt operators are compact.

Exercise 9.3.11. Let $-\infty \leq a < b \leq \infty$ and $-\infty \leq c < d \leq \infty$. Let $K \in L^2((a,b) \times (c,d))$. Define the integral operator $A : L^2(a,b) \to L^2(c,d)$ by

$$Af(x) = \int_c^d K(x,y)f(y)dy.$$

Show that A is Hilbert-Schmidt (hence compact!).

Exercise 9.3.12. Let $A \in B(\ell^2)$ be defined by

$$A(x_1, x_2, \cdots, x_n, \cdots) = (x_1, \frac{1}{\sqrt{2}}x_2, \cdots, \frac{1}{\sqrt{n}}x_n, \cdots).$$

Show that A is compact but it is not Hilbert-Schmidt.

Exercise 9.3.13. Using different methods, show that the shift operator is not compact.

Exercise 9.3.14. Let $A \in K(H)$ where H is infinite dimensional.

(1) Show that $0 \in \sigma(A)$.
(2) Show that $\sigma(A)$ has empty interior.
(3) Infer that $0 \in \partial\sigma(A)$. Does it follow that $0 \in \sigma_a(A)$?
(4) Show that

$$\sigma_p(A) - \{0\} = \sigma_a(A) - \{0\}.$$

(5) Deduce that
$$\sigma(A) = \sigma_a(A).$$

Exercise 9.3.15. Let A be the multiplication operator on ℓ^2 by a bounded sequence (α_n) which also tends to zero (hereinafter $A \in K(\ell^2)$). Let S be the usual shift operator. Find $\sigma_p(AS)$ and $\sigma(AS)$.

Exercise 9.3.16. Let $A \in B(H)$.

(1) Show that if A is *positive*, then A is compact iff $\sqrt{A}$ is compact.
(2) Deduce that (any $A \in B(H)$) A is compact iff $|A|$ is compact.
(3) Let $B \in K(H)$ be such that $0 \leq A \leq B$. Prove that $A \in K(H)$.

Exercise 9.3.17. ([**40**]) Let $A \in B(L^2[0,1])$ be defined for all $f \in L^2[0,1]$ and all $x \in [0,1]$ by
$$Af(x) = \int_0^1 K(x,y)f(y)dy$$
where
$$K(x,y) = \begin{cases} y(x+1), & 0 \leq x \leq y \leq 1, \\ x(y+1), & 0 \leq y \leq x \leq 1. \end{cases}$$

(1) Show that A is self-adjoint and compact.
(2) Show that any eigenvector associated with a non zero eigenvalue must be continuous.
(3) Is $0 \in \sigma_p(A)$? Find the non-zero eigenvalues of A.
(4) Is A positive?

Exercise 9.3.18. Change the kernel of A in the previous exercise by
$$K(x,y) = \begin{cases} x(1-y), & 0 \leq x \leq y \leq 1, \\ y(1-x), & 0 \leq y \leq x \leq 1. \end{cases}$$

(1) Find $\sigma_p(A)$ and $\sigma(A)$.
(2) Is A positive?
(3) Give $\|A\|$.

Exercise 9.3.19. Show that a finite rank operator has a finite spectrum.

Exercise 9.3.20. Let $A \in B(H)$ be self-adjoint and compact. Check that $\|A\|$ or $-\|A\|$ is in $\sigma_p(A)$.

Exercise 9.3.21. We define on $L^2[0,1]$ the operator V (called **Volterra Operator**)
$$(Vf)(x) = \int_0^x f(t)dt$$
(with $0 < x < 1$).

(1) Show that V is bounded and that $\|V\| \leq \frac{1}{\sqrt{2}}$.
(2) Prove that, in fact, we have

$$\|V\| = \frac{2}{\pi}.$$

(3) Let $n \in \mathbb{N}$. Find V^n, the expression of powers (or iterates) of V.
(4) Show that for all x and all $n \geq 1$, we have:

$$|(V^{n+1}f)(x)| \leq \frac{x^n}{n!} \text{ for all } f \text{ with } \|f\|_2 \leq 1.$$

(5) Deduce that $\sigma(V) = \{0\}$. Can V be self-adjoint?
(6) Find V^*. Is V normal?
(7) Prove that $V + V^*$ is a positive operator.
(8) What is $\sigma_r(V)$? $\sigma_p(V)$? $\sigma_c(V)$?
(9) Why is V compact?
(10) Find the eigenvalues of V^*V and the corresponding eigenvectors.
(11) Diagonalize V^*V.
(12) Compute $\|V^*V\|$ and find (again!) $\|V\|$.
(13) Explain why $I + V$ and $I - V$ are invertible. Find $(I - V)^{-1}$ and $(I + V)^{-1}$ explicitly.
(14) Prove that $(I + V)^{-1}$ is similar to $I - V$.
(15) (A funny application!) Find

$$\sum_{n=0}^{\infty} \frac{1}{(2n+1)^4} \text{ and } \sum_{n=1}^{\infty} \frac{1}{n^4}.$$

Remark. For more properties of the Volterra Operator, see Exercises 9.5.8 & 9.5.9 and Test 12.4.3.

Exercise 9.3.22. (cf. [**94**]) Using the Volterra Operator or else, find an example of a bounded operator A such that $A \neq I$, $\sigma(A) = \{1\}$ and $\|A\| = 1$.

9.4. Tests

Test 9.4.1. Let $A, B \in B(H)$ be such that AB is compact. Can we say that both A and B are compact?

Test 9.4.2. Let $A, B, C \in B(H)$ be such that one of them is compact. Which of the following is compact

$$ABC, \ BAC \text{ and } ACB?$$

Test 9.4.3. Let $A, B \in B(H)$. Show that if A or B is compact, then

$$\sigma(AB) = \sigma(BA).$$

Test 9.4.4. Assume that $\dim H = \infty$ and let $A \in B(H)$ be bounded below. Show that A is not compact with a different proof from the one of Exercise 9.3.9 .

Test 9.4.5. (cf. Exercise 11.5.13) Let H be an infinite dimensional Hilbert space. Let $A \in K(H)$ and let p be a complex polynomial of degree n. Show that $p(A)$ is compact iff $p(0) = 0$.

Test 9.4.6. Let $A \in B(H)$ be compact and self-adjoint. Assume further that $\dim H = \infty$. Show that A has always a non-trivial invariant subspace.

9.5. More Exercises

Exercise 9.5.1. Let $A \in B(H)$ be a compact projection (*not necessarily orthogonal*). Show that A is a finite rank operator (**Hint:** Show that $B : \operatorname{ran}(A) \to \operatorname{ran}(A)$ is the identity operator, where B is the restriction of A to $\operatorname{ran}(A)$).

Exercise 9.5.2. Let $A \in B(H)$ be such that $I - A \in K(H)$. Show that A is right invertible iff A left invertible.

Exercise 9.5.3. Let $A \in B(L^2[0,1])$ be defined for all $f \in L^2[0,1]$ and all $x \in [0,1]$ by

$$Af(x) = \int_0^1 e^{-|x-y|} f(y)dy.$$

(1) Show that A is self-adjoint, positive, compact and a contraction.
(2) Show that $0 \notin \sigma_p(A)$.

Exercise 9.5.4. ([**76**])(cf. Exercise 6.5.10) Show that if $A \in K(H)$ is such that $|A| \leq |\mathrm{Re} A|$, then A self-adjoint.

Exercise 9.5.5. Let H be a *separable* Hilbert space and let (x_n) be an orthonormal basis. Let $A \in B(H)$ be positive. Define

$$\mathrm{tr} A = \sum_{n=1}^{\infty} < Ax_n, x_n > .$$

Then $\mathrm{tr}A$ is called the **trace** of A (we notice that $\mathrm{tr}A \in [0, \infty]$).

(1) Show that $\mathrm{tr}A$ is independent of the chosen basis.
(2) Show also that:

(a) Verify that $A \in B(H)$ is Hilbert-Schmidt iff $\text{tr}A^*A < \infty$.
(b) $\text{tr}(\alpha A + B) = \alpha \text{tr}A + \text{tr}B$ for all $\alpha \geq 0$ and all positive $A, B \in B(H)$.
(c) $\text{tr}(UAU^*) = \text{tr}A$ for any unitary $U \in B(H)$.
(d) $\text{tr}A \leq \text{tr}B$ whenever $0 \leq A \leq B$.

Exercise 9.5.6. (more details may be found in [**178**]) Let $A \in B(H)$. We say that A is **trace class** if $\text{tr}|A| < \infty$. The set of all trace class operators from H into H is denoted by $B_1(H)$.

(1) Show that $B_1(H)$ is a linear subspace of $B(H)$.
(2) Show that if $A \in B(H)$ and $B \in B_1(H)$, then both AB and BA belong to $B_1(H)$.
(3) Show that $A \in B_1(H)$ iff $A^* \in B_1(H)$.
(4) Show that $(B_1(H), \|\cdot\|_1)$ is a normed vector space where

$$\|A\|_1 = \text{tr}|A|$$

and that $\|A\| \leq \|A\|_2 \leq \|A\|_1$ (where $\|\cdot\|_2$ is the norm which we defined on $B_2(H)$). Show also the "*Cauchy-Schwarz like*" inequality

$$\|AB\|_1 \leq \|A\|_2 \|B\|_2.$$

(5) Prove that $(B_1(H), \|\cdot\|_1)$ is a Banach space. Is $(B_1(H), \|\cdot\|)$ complete?
(6) Show that $B_1(H) \subset K(H)$. Conversely, show that a compact A is in $B_1(H)$ iff $\sum_{n=1}^{\infty} \lambda_n < \infty$, where (λ_n) are the eigenvalues of $|A|$.

Exercise 9.5.7. First, recall the following definition:
Let H, K be two Hilbert spaces and let $A \in B(H, K)$. Then A is called a ***Fredholm Operator*** *if the dimension of* $\ker A$ *and the codimension of* $\text{ran}A$ *are finite. The* ***index*** *of A is defined to be the number*

$$\text{ind}A = \dim \ker A - \text{codim ran}A.$$

(1) If $\dim H, \dim K < \infty$, show that A is always Fredholm and find ind A.
(2) Show that the shift operator, its powers and its adjoint are Fredholm operators. Find the corresponding index in each case.
(3) Let A, B be two Fredholm operators.
   (a) Show that $\text{ran}A$ is closed.
   (b) Show that A is Fredholm iff there is a $C \in B(K, H)$ such that $I - CA$ and $I - AC$ are compact operators.
   (c) Demonstrate that AB is a Fredholm operator with

$$\text{ind}AB = \text{ind}A + \text{ind}B.$$

Exercise 9.5.8. ([**73**]) Let V be the Volterra Operator on $L^2(0,1)$. Prove that V is not similar to λV unless $\lambda = 1$.

Exercise 9.5.9. ([**94**]) Let V be the Volterra Operator (Exercise 9.3.21). Show that $W(V)$ is the set lying between the curves

$$x \mapsto \frac{1-\cos x}{x^2} \pm i\frac{x-\sin x}{x^2},\ 0 \le x \le 2\pi,$$

where the value at zero is taken to be the limit from the right.

Exercise 9.5.10. (**Sturm-Liouville Problem**) Let $H = L^2([a,b])$ and let

$$X = \{f \in C^2([a,b]) : \ f(a) = f(b) = 0\}$$

be considered as a subset of H. If g is continuous on $[a,b]$, let

$$-f'' + pf = g, ...(1)$$

where p is a real-valued continuous function on $[a,b]$. Assume that the differential equation

$$-f'' + pf = 0$$

has two non zero solutions f_1 and f_2 verifying

$$f_1(a) = f_2(b) \text{ and } f_1(b) \neq 0,\ f_2(a) \neq 0.$$

Finally, let W be the **Wronskian** of f_1 and f_2, that is,

$$W = \begin{vmatrix} f_1 & f_2 \\ f_1' & f_2' \end{vmatrix}.$$

(1) Show that the unique solution of (1) is given by

$$f(x) = \int_a^b K(x,y)g(y)dy$$

where

$$K(x,y) = \begin{cases} -W^{-1}f_1(x)f_2(y), & a \le x \le y \le b, \\ -W^{-1}f_1(y)f_2(x), & a \le y \le x \le b. \end{cases}$$

(2) Define $A : H \to H$ to be an integral operator with K as a kernel (then A is Hilbert-Schmidt) and $B : X \to H$ by $Bf = -f'' + pf$. Show that

$$\sigma_p(B) = \left\{\frac{1}{\lambda} :\ \lambda \neq 0,\ \lambda \in \sigma_p(A)\right\}.$$

CHAPTER 10

Closed Operators

10.1. Basics

Before talking about closed operators, we give a quick preview of basic properties of unbounded operators (cf. Definition 2.1.2). In many applications, e.g. in Quantum Mechanics, operators are not necessarily bounded. We take this opportunity to recommend [**90**], which is intended as an introduction to Quantum Mechanics for mathematicians (so we wouldn't find what we see as weird symbols!).

Two major examples of unbounded operators in Quantum Mechanics are the **position operator** which is the multiplication operator by an unbounded function, and the **momentum operator** which is $-i\frac{d}{dx}$. We can add differential operators in general.

Notice that in the majority of cases, unbounded operators on a Hilbert space H have a **domain** which is a *proper* subspace of H. We can, however, have an unbounded operator which is defined everywhere and a bounded operator which is not defined everywhere. We may also remind readers that the "same symbol" may be defined on different domains. For example, d/dx may be defined on $C_0^\infty(\mathbb{R})$. It can also be defined on $\mathcal{S}(\mathbb{R})$ or on

$$\left\{f \in L^2(\mathbb{R}) : \frac{d}{dx}f \in L^2(\mathbb{R})\right\}$$

(where the derivative is in the sense of distributions, see the last subsection of this chapter). This gives rise to three different operators: d/dx on each of the three previous domains.

Let A, B, C be three non necessarily bounded operators with domains $D(A)$, $D(B)$ and $D(C)$ respectively (remember that $D(A)$, $D(B)$ and $D(C)$ must be linear subspaces of H).

Definition 10.1.1. *We say that B is an **extension** of A, and we write $A \subset B$, if*

$$D(A) \subset D(B) \textit{ and } \forall x \in D(A) : \ Ax = Bx.$$

Remark. If A and B are two operators, then

$$A = B \Longleftrightarrow A \subset B \text{ and } B \subset A.$$

Now, we define the natural domains of AB and $A+B$.

Definition 10.1.2. *If A and B are two operators with domains $D(A)$ and $D(B)$ respectively, then*

$$D(AB) = \{x \in D(B) : \ Bx \in D(A)\} = B^{-1}[D(A)]$$

and

$$D(A+B) = D(A) \cap D(B).$$

Remark. It is conceivable that the domain $D(AB)$ and $D(A+B)$ may be trivial, i.e. reduced to the singleton $\{0\}$. In fact $D(A^2)$ alone may be trivial. For example:

Example 10.1.3. *(see also the remark below Proposition 10.1.11) Consider $Af = \mathcal{F}f = \hat{f}$ where $\mathcal{F}$ is the Fourier transform which we define on $D(A) = C_0^\infty(\mathbb{R}) \subset L^2(\mathbb{R})$. Then*

$$D(A^2) = \{f \in D(A) : \ Af \in D(A)\} = \{f \in C_0^\infty(\mathbb{R}) : \ \hat{f} \in C_0^\infty(\mathbb{R})\} = \{0\}$$

by Exercise 2.5.6.

With the idea of the previous example, we can find examples for the triviality of the domains $D(AB)$ and $D(A+B)$. More sophisticated counterexamples related to the sum can be found in [**58**], [**59**] and [**121**].

Remark. There are other cantankerous facts about unbounded operators. The reader may wish to consult any textbook on unbounded operators (as e.g. [**178**] or [**195**]) to see more pathologies and for further reading.

Remark. It is clear that if $D(A) = H$ (for instance if A is bounded and everywhere defined), then

$$D(AB) = \{x \in D(B) : \ Bx \in D(A)\} = D(B).$$

Proposition 10.1.4. *Let A, B, C be three operators with domains $D(A)$, $D(B)$ and $D(C)$ respectively. Then*

$$(A+B)+C = A+(B+C) \text{ and } (AB)C = A(BC)$$

(with equalities holding on e.g. the domains $D(A) \cap D(B) \cap D(C)$ and $D(ABC)$ respectively).

Remark. As regards the distributive laws, one has to be a little more careful. Indeed, while

$$(A+B)C = AC + BC$$

holds, one only has (see Exercise 10.3.1)

$$AB + AC \subset A(B + C).$$

It is easy to see that the equality holds if for instance $A \in B(H)$.

Proposition 10.1.5. *(see Exercise 10.3.3) Let A, B, C be three operators. Then*

$$A \subset B \Longrightarrow CA \subset CB \text{ and } AC \subset BC.$$

Before turning to closed operators, we give a definition of commutativity.

Definition 10.1.6. *Let A be an unbounded operator and let B be a bounded operator. We say that B **commutes** with A if*

$$BA \subset AB.$$

Now, we introduce closed operators.

Definition 10.1.7. *Let H be a Hilbert space. Let A be a linear but non necessarily bounded operator. Assume that A has a domain $D(A)$. We say that A is **closed** if its graph $G(A)$ is closed in $H \times H$. Since*

$$G(A) = \{(x, Ax) : x \in D(A)\},$$

we may write

$$A \text{ is closed} \Longleftrightarrow \left(\forall x_n \in D(A) : \begin{cases} x_n \longrightarrow x \\ Ax_n \longrightarrow y \end{cases} \Longrightarrow \begin{cases} x \in D(A), \\ y = Ax. \end{cases}\right).$$

Remark. If $D(A) = H$, by the Closed Graph Theorem we have

$$A \text{ is closed } \Longleftrightarrow A \text{ is bounded.}$$

Remark. It is clear that the same "symbol" may be closed with respect to a domain D_1 but not closed with respect to some other domain D_2. For example, see Exercise 10.3.8 (or Exercise 10.3.9 with Test 10.4.1).

Here are examples showing that the none of the notions "closed" and "bounded" needs to imply the other.

Examples 10.1.8.

(1) *The differential operator $Af = f'$ on the domain*

$$D(A) = \{f \in AC[0,1],\ f' \in L^2(0,1),\ f(0) = 0\} \subset L^2(0,1)$$

is closed but unbounded (the space $AC[0,1]$ is recalled below).

(2) *The identity operator I defined on a (non closed) subspace of $L^2(\mathbb{R})$ (e.g. on $C_0^\infty(\mathbb{R})$) is bounded but not closed.*

Remark. The reader should be aware that the notions of a "closed operator" and a "closed map" *are totally different.*

Remark. The concepts of *unbounded* self-adjoint or normal operators do exist (not unitary or compact ones, why?), but we content ourselves in the present manuscript with closed operators.

We may also keep the definition of the graph norm in the case of unbounded operators. We have

Definition 10.1.9. *Let H be a Hilbert space and A a given linear operator with domain $D(A) \subset H$. The* **graph norm** *of A is defined by*

$$\|x\|_{G(A)} = \sqrt{\|x\|^2 + \|Ax\|^2}.$$

There is a practical connection between closed operators and the corresponding graph norms.

Theorem 10.1.10. *(see Exercise 10.3.6) Let H be a Hilbert space and A a given linear operator with domain $D(A) \subset H$. Then*

$$A \text{ is closed} \Longleftrightarrow (D(A), \|\cdot\|_{G(A)}) \text{ is a Banach space.}$$

Remark. An easy interpretation of the previous result is: *A closed unbounded operator is bounded with respect to its graph norm.* This observation, however, does not make closed unbounded operators easier to handle.

The second example of Examples 10.1.8 is not coincidental. It belongs to a larger class of examples. We have:

Proposition 10.1.11. *(see Exercise 10.3.7) Let A be a bounded linear operator with domain $D(A)$. Then*

$$A \text{ is closed} \Longleftrightarrow D(A) \text{ is closed.}$$

Remark. The operator $\mathcal{F}$ in Example 10.1.3 is not closed (by Proposition 10.1.11). So one may wonder whether we can find a closed operator A such that $D(A)$ is dense and $D(A^2) = \{0\}$? Such operators do exist! A construction was first given by Naimark in **[159]** where A is also "symmetric". Then, Chernoff gave a much simpler counterexample in **[43]**. The example of Chernoff is important in the sense that the operator he considered has a dense domain and it was also "closed, symmetric and semi-bounded", yet the domain of its square was trivial.

Schmüdgen in **[194]** investigated among other questions when $D(A^n)$ is dense (cf. Proposition 10.1.18).

10.1.1. Spectrum. First, we say a few words on invertibility of unbounded operators.

Definition 10.1.12. *Let A be an injective operator (not necessarily bounded) from H into H. Then $A^{-1}: \operatorname{ran}(A) \to H$ is called the* ***inverse*** *of A with domain $D(A^{-1}) = \operatorname{ran}(A)$.*

Remark. We may interpret the previous definition as A being invertible if $AA^{-1} \subset I$ and $A^{-1}A \subset I$.

The inverse of an unbounded operator may be bounded and defined everywhere if e.g. $\operatorname{ran}(A) = H$. In such case, an unbounded A is invertible (we may also employ the terminology **boundedly invertible**) if there is a $B \in B(H)$ such that

$$AB = I \text{ and } BA \subset I$$

(hence $B = A^{-1}$).

Proposition 10.1.13. *If A is invertible, then it is closed.*

The spectrum of unbounded operators is defined as its analogue of bounded operators.

Definition 10.1.14. *Let A be an operator on H. The* ***resolvent set*** *of A, denoted by $\rho(A)$, is defined by*

$$\rho(A) = \{\lambda \in \mathbb{C} : \lambda I - A \text{ is bijective and } (\lambda I - A)^{-1} \in B(H)\}.$$

The complement of $\rho(A)$, denoted by $\sigma(A)$,

$$\sigma(A) = \mathbb{C} \setminus \rho(A)$$

is called the ***spectrum*** *of A.*

Also, $(\lambda I - A)^{-1}$ is called the ***resolvent*** *of A in λ and it is denoted by $R(\lambda, A)$ as in the bounded case.*

Remark. It is clear that $\lambda \in \rho(A)$ iff there exists a $B \in B(H)$ such that

$$(\lambda I - A)B = I \text{ and } B(\lambda I - A) \subset I.$$

The next result is important (a proof is supplied in Exercise 10.3.20).

Proposition 10.1.15. *Let A be a closed operator and let $\lambda \in \mathbb{C}$. If $\lambda I - A$ is bijective, then $(\lambda I - A)^{-1} \in B(H)$.*

Remark. We notice that we may also divide $\sigma(A)$ into three parts $\sigma_p(A)$, $\sigma_r(A)$ and $\sigma_c(A)$ as done for bounded A. This is not considered in this book.

The next simple result is interesting.

Proposition 10.1.16. *(see Exercise 10.3.21) Let A be a linear operator with $\rho(A) \neq \varnothing$. Then A is closed.*

Corollary 10.1.17. *Let A be a linear operator. If A is not closed, then $\sigma(A) = \mathbb{C}$.*

As observed in the remark just below Definition 10.1.2, it is quite likely that $D(A^2) = \{0\}$. The next result tells us when $D(A^2)$ is dense.

Proposition 10.1.18. *([**122**]) Let A be an unbounded operator with a domain $D(A) \subset H$, where H is a Hilbert space. If $\rho(A) \neq \varnothing$, then*

$$\overline{D(A)} = H \Longrightarrow \overline{D(A^2)} = H.$$

Remark. The previous result also holds for the density of $D(A^n)$ for $n \in \mathbb{N}$.

10.1.2. Closedness of Products and Sums. We start with a relatively simple result on the closedness of the sum and the product of operators.

Theorem 10.1.19. *(see Exercise 10.3.13) Let A and B be two linear operators on a Hilbert space. Then*

(1) *AB is closed if **either** A is closed and B is bounded **or** A is invertible and B is closed.*
(2) *$A + B$ is closed if B is bounded and A is closed.*

Before giving an interesting perturbation result on the closedness of the sum, we introduce the following.

Definition 10.1.20. *Let A and B be two unbounded operators with domains $D(A)$ and $D(B)$ respectively (on the same Hilbert space). Assume that $D(B) \subset D(A)$ and*

$$\|Ax\| \leq a\|Bx\| + b\|x\|, \ \forall x \in D(B)$$

*where "a, b" are positive reals. Then A is said to be **relatively bounded** with respect to B (or that A is B-**bounded**). The greatest lower bound of all "a" is called the **relative bound** of A.*

Example 10.1.21. *By Exercise 10.3.19, we may show that a multiplication operator by a function in $L^2(\mathbb{R}^3) + L^\infty(\mathbb{R}^3)$ is Δ-bounded, where Δ is the **Laplacian**, i.e.*

$$\Delta = \frac{\partial^2}{\partial x^2} + \frac{\partial^2}{\partial y^2} + \frac{\partial^2}{\partial z^2}.$$

Remark. For other classical examples of relatively bounded operators, see [**114**] or [**179**]. For new and different examples, see [**37**], [**143**] and [**144**].

THEOREM 10.1.22. *([**158**], see Exercise 10.3.17) Let B be a closed operator on $D(B)$. If A is B-bounded with relative bound "$a < 1$", then $A + B$ is closed on $D(B)$.*

As a direct application of the previous theorem, we have:

COROLLARY 10.1.23. *Let B be a closed operator on $D(B)$ and let $(A_k)_{k=1,\cdots,n}$ be a collection of operators. Assume that each A_k is B-bounded with a relative bound a_k. If $\sum_{k=1}^{n} a_k < 1$, then $B+A_1+\cdots+A_n$ is closed on $D(B)$.*

10.1.3. A Word on Distributional Derivatives. Distribution Theory is such a vast domain in Modern Mathematical Analysis. It is treated in many textbooks. We could cite [**77**], [**129**] and [**178**]. In this subsection, we recall very briefly a few facts needed from this theory. A good portion of the material here is borrowed from [**195**].

As it is customary in Distribution Theory, we shall denote $C_0^\infty(\mathbb{R}^n)$ by $\mathcal{D}(\mathbb{R}^n)$.

We also introduce the following notations: $\partial_k = \frac{\partial}{\partial x_k}$,

$$\partial^\alpha = \left(\frac{\partial}{\partial x_1}\right)^{\alpha_1} \cdots \left(\frac{\partial}{\partial x_n}\right)^{\alpha_n}$$

and $|\alpha| = \alpha_1 + \cdots + \alpha_n$ for $\alpha = (\alpha_1, \cdots, \alpha_n)$, and each α_k is in $\mathbb{Z}^+$.

DEFINITION 10.1.24. *A linear functional T on $\mathcal{D}(\mathbb{R}^n)$ is called a **distribution** on $\mathbb{R}^n$ if for each compact set K of $\mathbb{R}^n$, there are $n_K \in \mathbb{Z}^+$ and $C_K > 0$ such that*

$$|T(\varphi)| \leq C_K \sup\{|\partial^\alpha \varphi(x)| : \ x \in K, \ |\alpha| \leq n_K\}$$

for all $\varphi \in \mathcal{D}(\mathbb{R}^n)$ with $\operatorname{supp} \varphi \subset K$.

The vector space of distributions on $\mathbb{R}^n$ is designated by $\mathcal{D}'(\mathbb{R}^n)$.

REMARK. Obviously, the continuity of a distribution T may be interpreted as: Whenever $\varphi_m \in \mathcal{D}(\mathbb{R}^n)$ with $\varphi_m \to \varphi$ in $\mathcal{D}(\mathbb{R}^n)$, then (in $\mathbb{C}$)

$$T(\varphi_m) \to T(\varphi),$$

where $\varphi_m \to \varphi$ in $\mathcal{D}(\mathbb{R}^n)$ means that there is some fixed compact set $K \subset \mathbb{R}^n$ such that $\operatorname{supp}(\varphi_m - \varphi) \subset K$ and for each α (as defined above)

$$\partial^\alpha \varphi_m \longrightarrow \partial^\alpha \varphi$$

as $m \to \infty$ uniformly on K.

Therefore, a distribution is a continuous linear functional on $\mathcal{D}(\mathbb{R}^n)$.

Remark. We can also define distributions on another equally "nice" space, namely the Schwartz space $\mathcal{S}(\mathbb{R}^n)$ (defined in Chapter 1). Continuous linear functionals defined on $\mathcal{S}(\mathbb{R}^n)$ are called **tempered distributions** (see e.g. [**178**]).

All $L^p(\mathbb{R}^n)$-functions can be regarded as distributions. To see this, we introduce the following space:

$$L^p_{\text{loc}}(\mathbb{R}^n) = \{f : \mathbb{R}^n \to \mathbb{C} \text{ measurable}, \int_K |f|^p < \infty \text{ for each compact } K \subset \mathbb{R}^n\}$$

where $p \geq 1$.

Now if $f \in L^1_{\text{loc}}(\mathbb{R}^n)$, then for any $\varphi \in \mathcal{D}(\mathbb{R}^n)$, it can be shown that

$$T_f(\varphi) = \int_{\mathbb{R}^n} f\varphi dx$$

defines a continuous linear functional on $\mathcal{D}(\mathbb{R}^n)$, i.e. $T_f \in \mathcal{D}'(\mathbb{R}^n)$.

We have already observed in Chapter 1 that $L^p(\mathbb{R}^n)$ and $L^q(\mathbb{R}^n)$ were not comparable. Nonetheless, $L^p_{\text{loc}}(\mathbb{R}^n)$ and $L^q_{\text{loc}}(\mathbb{R}^n)$ are always comparable. More precisely,

$$q \geq p \Longrightarrow L^q_{\text{loc}}(\mathbb{R}^n) \subset L^p_{\text{loc}}(\mathbb{R}^n)$$

by a simple application of the Hölder Inequality. Hence $L^p_{\text{loc}}(\mathbb{R}^n) \subset L^1_{\text{loc}}(\mathbb{R}^n)$. Finally, since clearly $L^p(\mathbb{R}^n) \subset L^p_{\text{loc}}(\mathbb{R}^n)$, we have that any $L^p(\mathbb{R}^n)$-function yields a distribution.

One of the main virtues of distributions is that we can always differentiate them in the following sense:

Definition 10.1.25. *Let $T \in \mathcal{D}'(\mathbb{R}^n)$. We define $\partial^\alpha T$ by its action on every $\varphi \in \mathcal{D}(\mathbb{R}^n)$ as*

$$(\partial^\alpha T)(\varphi) = (-1)^{|\alpha|} T(\partial^\alpha \varphi).$$

*We call $\partial^\alpha T$ the **distributional derivative** of T.*

Remark. In terms of $L^1_{\text{loc}}(\mathbb{R}^n)$-functions, we have:

We say that a function $g \in L^1_{\text{loc}}(\mathbb{R}^n)$ is the **weak derivative** of $f \in L^1_{\text{loc}}(\mathbb{R}^n)$ if the distributional derivative $\partial^\alpha f$ is given by the function g, i.e.

$$(-1)^{|\alpha|} \int_{\mathbb{R}^n} f \partial^\alpha(\varphi) dx = \int_{\mathbb{R}^n} g\varphi dx$$

for all $\varphi \in \mathcal{D}(\mathbb{R}^n)$.

Remark. All definitions and results make sense if $\mathbb{R}^n$ is replaced by an open set $\Omega \subset \mathbb{R}^n$.

All that to define Sobolev spaces (readers may wish to consult [**105**], [**129**] or [**179**] for further reading).

Definition 10.1.26. *Let* $m \in \mathbb{Z}_+$. *Set*

$$H^m(\mathbb{R}^n) = \{f \in L^2(\mathbb{R}^n) : \; \partial^\alpha f \in L^2(\mathbb{R}^n) \text{ for all } \alpha \in \mathbb{Z}_+^n, \; |\alpha| \leq m\},$$

where derivatives are taken in the distributional sense. Then $H^m(\mathbb{R})$ *is a called a* ***Sobolev space****.*

Remark. The vector space $H^m(\mathbb{R}^n)$ is a Hilbert space with respect to the inner product:

$$< f, g >_{H^m(\mathbb{R}^n)} = \sum_{|\alpha| \leq m} < \partial^\alpha f, \partial^\alpha g >_{L^2(\mathbb{R}^n)} .$$

Using the Fourier transform, we may describe Sobolev spaces as follows:

Theorem 10.1.27.

$$H^m(\mathbb{R}^n) = \{f \in L^2(\mathbb{R}^n) : \; |x|^{\frac{m}{2}} \hat{f}(t) \in L^2(\mathbb{R}^n)\}$$

(where $|x|^2 = x_1^2 + \cdots + x_n^2$*). Besides, if* $p \in \mathbb{Z}_+$ *and* $m > p + \frac{n}{2}$ *, then* $H^m(\mathbb{R}^n) \subset C^p(\mathbb{R}^n)$.

The last stop here is at absolutely continuous functions (more on this topic may be found in [**211**]).

Definition 10.1.28. *Let* $a, b \in \mathbb{R}$ *with* $a < b$. *A function* f *on* $[a, b]$ *is said to be* ***absolutely continuous*** *if for each* $\varepsilon > 0$*, there exists* $\delta > 0$ *such that for any collection of pairwise disjoint subintervals* (a_k, b_k)

$$\sum_{k=1}^n |f(b_k) - f(a_k)| < \varepsilon \text{ if } \sum_{k=1}^n (b_k - a_k) < \delta.$$

We denote the set of all absolutely continuous functions on $[a, b]$ *by* $AC[a, b]$.

Remark. Absolute continuous functions are continuous. Are they differentiable? The answer is "almost yes":

Theorem 10.1.29. *A function* f *is absolutely continuous on* $[a, b]$ *if and only if* f' *exists almost everywhere on* $[a, b]$, $f' \in L^1[a, b]$ *and*

$$f(x) - f(a) = \int_a^x f'(t) dt$$

for all $x \in [a, b]$.

10.2. True or False: Questions

Questions. Comment on the following questions/statements and indicate those which are false and those which are true when this applies. Justify your answers.

(1) Let A be a linear operator with a dense domain $D(A) \subset H$. Then

$$\forall x \in D(A) : < Ax, x >= 0 \Longrightarrow A = 0.$$

(2) We know that if A is bounded, then A^2 is bounded. But if A is an unbounded operator, then is A^2 also unbounded?

(3) Let A be a linear operator. Then

$$\sigma(A) \text{ is not bounded } \Longleftrightarrow A \text{ is an unbounded operator.}$$

(4) Theorem 7.1.33 holds for unbounded operators too.

(5) Let A, B be two linear operators on a Hilbert space H. Assume also that $B \in B(H)$. Assume further that A has a domain $D(A)$. If $\overline{D(A)} = H$, then

$$A \subset B \Longrightarrow A = B.$$

(6) Let A, B be two linear operators on a Hilbert space H. Assume also that $B \in B(H)$. Assume further that A has a domain $D(A)$. If A is closed, then

$$A \subset B \Longrightarrow A = B.$$

10.3. Exercises with Solutions

Exercise 10.3.1. Let A, B, C be three operators with domains $D(A)$, $D(B)$ and $D(C)$ respectively. Show that in general

$$A(B + C) \not\subset AB + AC.$$

Exercise 10.3.2. Let A and B be two operators on a Hilbert space H such that $A \subset B$. Show that if A is surjective and B is injective, then $A = B$.

Remark. The previous result is a maximality result. For other results on maximality (as is the case with self-adjoint and normal *unbounded* operators), see **[190]**. See also, **[57]** and **[152]**.

Exercise 10.3.3. Let A, B, C be three operators with domains $D(A)$, $D(B)$ and $D(C)$ respectively. Show that

$$A \subset B \Longrightarrow CA \subset CB \text{ and } AC \subset BC.$$

Exercise 10.3.4. (cf. Exercise 11.3.11) Let H be a Hilbert space. Let A be a non necessarily bounded operator with domain $D(A) \subset H$ and let $B \in B(H)$. Assume further that $BA \subset AB$. Show that every complex polynomial in B commutes with A.

Exercise 10.3.5. (cf. Exercise 2.3.12) Let A be a linear operator in a Hilbert space H, with domain $D(A)$. Assume that A is closed. Assume further that for some $\alpha > 0$,

$$\|Ax\| \geq \alpha\|x\| \text{ for all } x \in D(A).$$

Show that $\text{ran}(A)$ is closed.

Exercise 10.3.6. Let A be a linear operator on a Hilbert space H with domain $D(A)$. Denote its graph by graph $G(A)$ and consider the graph norm

$$\|x\|_{G(A)} = (\|x\|_H^2 + \|Ax\|_H^2)^{\frac{1}{2}}.$$

(1) Show that: A is closed iff $(D(A), \|\cdot\|_{G(A)})$ is a Banach space.
(2) **Application:** Let

$$H^1(\mathbb{R}) = \{f \in L^2(\mathbb{R}) : \ f' \in L^2(\mathbb{R})\},$$

be the Sobolev space. Consider now an operator

$$A : H^1(\mathbb{R}) \to L^2(\mathbb{R}) \text{ defined by } Af = f'.$$

Show that A is an unbounded closed operator.

Exercise 10.3.7. Let A be a *bounded* linear operator with *domain* $D(A)$. Show that

$$A \text{ is closed} \Longleftrightarrow D(A) \text{ is closed}.$$

Exercise 10.3.8. Let $A : D(A) \subset \ell^2 \to \ell^2$ be defined by

$$Ax = A(x_n) = (nx_n) = (x_1, 2x_2, \cdots, nx_n, \cdots).$$

Investigate the closedness of A in the following two cases:
(1) $D_1 = D(A) = \{x = (x_n) \in \ell^2 : \ (nx_n) \in \ell^2\}$.
(2) $D_2 = D(A) = c_{00}$.
Is D_1 dense in ℓ^2? What about D_2?

Exercise 10.3.9. Let $\varphi : \mathbb{R} \to \mathbb{C}$ be a measurable function. Define *formally* $M_\varphi : L^2(\mathbb{R}) \to L^2(\mathbb{R})$ by

$$M_\varphi f(x) = \varphi(x) f(x)$$

with $x \in \mathbb{R}$. Then M_φ is (also!) called a **multiplication operator**, whereas the function φ is said to be its **symbol** or its **multiplier**.
(1) Is M_φ well-defined?

(2) Set
$$D(M_\varphi) = \{f \in L^2(\mathbb{R}) : \varphi f \in L^2(\mathbb{R})\}.$$
Show that $D(M_\varphi)$ is not empty. In fact, show that $D(M_\varphi)$ is dense in $L^2(\mathbb{R})$.

(3) Prove that M_φ is closed.

(4) Show that M_φ is bounded if and only if φ is essentially bounded. Show that in this case
$$\|M_\varphi\|_{B(L^2(\mathbb{R}))} = \|\varphi\|_\infty.$$

(5) Show that $D(M_\varphi) = L^2(\mathbb{R})$ iff φ essentially bounded.

In the remaining questions of this exercise, assume that φ is essentially bounded (and so $D(M_\varphi) = L^2(\mathbb{R})$!).

(6) Find M_φ^*. When is M_φ self-adjoint? Positive (find its positive square root in this case)? Normal? Unitary?

(7) Prove that the following assertions are equivalent:
 (a) $\operatorname{ran} M_\varphi$ is dense.
 (b) $\varphi(x) \neq 0$ for almost every $x \in \mathbb{R}$.
 (c) M_φ is injective.

(8) When is M_φ invertible? Give M_φ^{-1} in this case.

(9) Find $\sigma_p(M_\varphi)$ and $\sigma(M_\varphi)$.

(10) Prove that M_φ is compact iff $\varphi = 0$ a.e. in $\mathbb{R}$.

Remark. The setting of the previous exercise is not the most general. One can consider a σ-finite measure space instead of $\mathbb{R}$. See e.g. [**48**] (there is still a more general setting, see [**115**]).

Exercise 10.3.10. Let $a \in \mathbb{R}$. What is the norm of the linear operator A defined on $L^2(\mathbb{R})$ by
$$(Af)(x) = \varphi(x)f(x-a),$$
where φ is essentially bounded on $\mathbb{R}$?

Exercise 10.3.11. First, we give a definition: *If A is a linear but not necessarily bounded operator with domain $D(A)$, then we say that A is* ***symmetric*** *if*
$$< Ax, y >=< x, Ay >, \ \forall x, y \in D(A).$$

Remark. It is important to keep in mind that an "unbounded symmetric operator" is different from an "unbounded self-adjoint operator". In fact, self-adjoint implies symmetric but not vice versa.

Let A be symmetric. Prove the following:

(1) $\|Ax + ix\|^2 = \|x\|^2 + \|Ax\|^2$ (for each $x \in D(A)$).

(2) A is closed iff $\operatorname{ran}(A + iI)$ is closed.

Exercise 10.3.12. Consider the unbounded operator A

$$Af(x) = (x^2 + 1)f(x)$$

defined on the domain

$$D(A) = \{f \in L^2(\mathbb{R}) : \ (1 + x^2)f \in L^2(\mathbb{R})\}.$$

Show that A is invertible.

Exercise 10.3.13. Let A and B be two operators on a Hilbert space. Assume that A is closed and that B is bounded on all of H. Let C be boundedly invertible.

(1) Is BA closed?
(2) Show that AB (in this order!) is closed on $D(AB)$.
(3) Prove that CA is closed on $D(CA)$.
(4) Is the sum of two closed operators always closed?
(5) Show that $A + B$ is closed on $D(A)$.

Exercise 10.3.14. Let (α_n) be a complex sequence and let A be an unbounded operator on ℓ^2 defined by

$$A(x_1, x_2, \cdots) = (\alpha_1 x_2, \alpha_2 x_3, \cdots)$$

on the domain

$$D(A) = \{(x_n) \in \ell^2 : \ A(x_n) \in \ell^2\}.$$

Show that A is closed on $D(A)$.

Exercise 10.3.15. Let A, B, C be operators on a Hilbert space H. Assume that B is bounded and that B^2A is closed.

(1) If CA is closed, then does it follow that C is closed? Does it follow that A is closed?
(2) Show that BA is closed if B is injective.
(3) Show that BA is closed if B is either self-adjoint or normal.

Exercise 10.3.16. Let A be a closed but non necessarily bounded operator and $B \in B(H)$ such that BA is invertible with a bounded inverse defined everywhere. Show that B is right invertible (so that B becomes invertible if e.g. B is normal. See Exercise 6.3.26).

Exercise 10.3.17. Let A and B be two unbounded closed operators with domains $D(A)$ and $D(B)$ respectively.

(1) Show that if A is closed and B is A-bounded with relative bound $a < 1$, then $A + B$ is closed.
(2) Does the result remain valid if $a = 1$?

Exercise 10.3.18. ([**143**])Let

$$A\varphi(x,y) = \frac{\partial^2 \varphi(x,y)}{\partial x \partial y}$$

be defined on

$$D(A) = \{\varphi \in L^2(\mathbb{R}^2) : \ A\varphi \in L^2(\mathbb{R}^2)\}$$

(by a change of variables, A may be regarded as the the usual **wave** operator $\Box$).

(1) Show that A is an unbounded closed operator on $D(A)$.
(2) Show that $D(A) \not\subset L^\infty(\mathbb{R}^2)$.

> **Remark**. In fact, it was shown in [**143**] that any function in $D(A)$ is a BMO function (**B.M.O.: Bounded Mean Oscillation**, see [**65**] for the definition of BMO, and some of its properties).

Exercise 10.3.19. Let $A = \sum_{k=1}^n \frac{\partial^2}{\partial x_k^2}$ be defined on $H^2(\mathbb{R}^n)$, where

$$H^2(\mathbb{R}^n) = \{f \in L^2(\mathbb{R}^n) : \ Af \in L^2(\mathbb{R}^n) \text{ in the distributional sense}\}.$$

(1) Show that A is unbounded and closed on $H^2(\mathbb{R}^n)$.
(2) Assume that $n \leq 3$. Show that for all $a > 0$, there is a $b > 0$ such that

$$\|f\|_\infty \leq a\|Af\|_2 + b\|f\|_2,$$

i.e. $H^2(\mathbb{R}^n) \subset L^\infty(\mathbb{R}^n)$.
(3) Show that if $n \geq 4$, then $H^2(\mathbb{R}^n) \not\subset L^\infty(\mathbb{R}^n)$.
(4) Let $V \in L^2(\mathbb{R}^n) + L^\infty(\mathbb{R}^n)$, $n \leq 3$. Show that $A + M$ is closed, where M represents a multiplication operator by the function V.

Exercise 10.3.20. (cf. [**138**]). Let A be a closed operator defined on $D(A) \subset H$, where H is a Hilbert space.

(1) Assume that A is injective. Show that $A^{-1} : D(A^{-1}) \to D(A)$ is closed (where $D(A^{-1}) = \text{ran}(A)$).
(2) Show that for all complex λ, such that $\lambda I - A$ is bijective, then $R(\lambda, A) \in B(H)$.
(3) Show that $\sigma(A)$ is closed.

Exercise 10.3.21. Let A be a linear operator such that $\rho(A) \neq \varnothing$. Show that A is closed.

Exercise 10.3.22. Let A be a linear operator in a Hilbert space H with domain $D(A) \subset H$ and $\text{ran}(A)$ as its range. Assume that $\text{ran}(A) \subset D(A)$ and that $\sigma(A) \neq \mathbb{C}$. Prove that $A \in B(H)$.

Exercise 10.3.23. Let $A : L^2(\mathbb{R}) \to L^2(\mathbb{R})$ be an operator defined by

$$(Af)(x) = e^{-x^2} f(x-1)$$

where $f \in L^2(\mathbb{R})$.

(1) Show that A is bounded.
(2) Find A^* (the adjoint of A).
(3) Let $n \in \mathbb{N}$. Prove that
$$\|A^n\| = e^{-\frac{(n-1)n(n+1)}{12}}.$$
(4) Check that A is one-to-one and that $\operatorname{ran} A$ is dense in $L^2(\mathbb{R})$.
(5) Show that $r(A) = 0$. What is then $\sigma(A)$?
(6) Let B be an (unbounded) operator with domain $D(B) = \operatorname{ran} A$, such that
$$BAf = f, \; f \in L^2(\mathbb{R}).$$
Show that B is closed and that $\sigma(B) = \varnothing$.

Exercise 10.3.24. Define an (unbounded) closed operator A on a domain $D(A) \subset L^2[0,1]$ by

$$Af(x) = f'(x),$$

where

$$D(A) = \{f \in L^2[0,1] : \; f' \in L^2[0,1], \; f(0) = 0\}$$

(we wrote $f(0) = 0$ while f is in $L^2[0,1]$, is it Okay here?).

Show that $\sigma(A) = \varnothing$.

Remark.

The derivative here must be understood in the distributional sense. Otherwise, since absolutely continuous functions are differentiable almost everywhere, we may just replace $D(A)$ by

$$\widetilde{D}(A) = \{f \in \text{ AC[0,1]}, \, f' \in L^2[0,1] \text{ and } f(0) = 0\}.$$

10.4. Tests

Test 10.4.1. Let A be the multiplication operator defined on $L^2(\mathbb{R})$ by

$$Af(x) = xf(x)$$

with domain $C_0^\infty(\mathbb{R}) \subset L^2(\mathbb{R})$. Is A closed?

Test 10.4.2. Let A be a closed but not necessarily bounded operator. Let $D(A)$ be its domain (not necessarily a Hilbert space). Show that $\ker A$ is closed.

Test 10.4.3. Let A and B be two non necessarily bounded operators with domains $D(A)$ and $D(B)$ respectively. Assume that

$$\|Ax\| = \|Bx\|, \ \forall x \in D(A) = D(B)...(*).$$

Show that

$$A \text{ is closed} \iff B \text{ is closed.}$$

Test 10.4.4. Let $B \in B(H)$ and A be a non necessarily bounded operator such that $A + B$ is closed. Does it follow that A is closed?

Test 10.4.5. Let $B \in B(H)$ be a bounded normal operator and let A be closed on $D(A) \subset H$. Show, using two methods, that if BB^* is invertible, then BA is closed.

10.5. More Exercises

Exercise 10.5.1. Let A be a linear operator with a dense domain $D(A) \subset H$. Show that

$$\forall x \in D(A) : < Ax, x >= 0 \Longrightarrow Ax = 0, \ \forall x \in D(A).$$

Exercise 10.5.2. Let A, B be two linear operators on a Hilbert space H. Assume also that $B \in B(H)$. Assume further that A is closed on its domain $D(A)$ and that $A \subset B$. Show that

$$A = B \iff \overline{D(A)} = H.$$

Infer that if C is (boundedly) invertible, then

$$AC \subset B \Longrightarrow A = BC^{-1}$$

whenever $\overline{D(A)} = H$.

Exercise 10.5.3. Let A be defined by $Af(x) = -f'(x)$ on

$$D(A) = \{f \in L^2(0,1) : \ f' \in L^2(0,1)\}.$$

Show that A is closed.

Exercise 10.5.4. ([**163**]) On ℓ^2, define a *linear* operator A by

$$Ax = A(x_n) = (x_2, 0, 2x_4, 0, \cdots, \underbrace{nx_{2n}}_{2n-1}, \underbrace{0}_{2n}, \cdots)$$

on its domain

$$D(A) = \{x = (x_n) \in \ell^2 : \ (nx_{2n}) \in \ell^2\}.$$

(1) Check that $D(A)$ is dense in ℓ^2.
(2) Show that A is unbounded and closed.
(3) Compute A^2. What do you observe?

Exercise 10.5.5. Let A, B be two unbounded operators such that $AB = BA$. Assume that A and B are invertible (with A^{-1} and B^{-1} both in $B(H)$). Show that

$$A^{-1}B^{-1} = B^{-1}A^{-1}.$$

Remark. The reader should bear in mind that the previous result is not necessarily true if the inverses are not defined everywhere (even if stronger conditions are imposed on the operators A and B themselves). There is a famous counterexample by Schmüdgen. Details may be found in e.g. [**195**].

Exercise 10.5.6. Let A be an unbounded operator with a domain $D(A) \subset H$, where H is a Hilbert space. If $\rho(A) \neq \varnothing$, then show that

$$\overline{D(A)} = H \Longrightarrow \overline{D(A^2)} = H.$$

Exercise 10.5.7. Let A denote the operator $\frac{\partial}{\partial t} + \frac{\partial^3}{\partial x^3}$, defined on the domain

$$D(A) = \{u \in L^2(\mathbb{R}^2) : Lu, \text{ as a distribution, is an } L^2(\mathbb{R}^2)\text{-function}\}.$$

Then A is called the **Airy operator**.

(1) Show that A is not bounded.
(2) Show that A is closed in $D(A)$.
(3) It is shown in [**37**] that: *For all $a > 0$, there exists a $b > 0$ such that*

$$\|u\|_{L^8(\mathbb{R}^2)} \leq a\|Au\|_{L^2(\mathbb{R}^2)} + b\|u\|_{L^2(\mathbb{R}^2)}$$

for all $u \in D(A)$. Show that if $8 < p \leq \infty$, then there do not exist positive constants a and b such that

$$\|u\|_{L^p(\mathbb{R}^2)} \leq a\|Au\|_{L^2(\mathbb{R}^2)} + b\|u\|_{L^2(\mathbb{R}^2)}$$

for all $u \in D(A)$ as follows:

(a) Let $\delta > 0$. Consider

$$u_\delta(x,t) = \mathcal{F}^{-1}(g_\delta(\eta,\xi)) \text{ where } g_\delta(\eta,\xi) = \varphi(\delta\eta)V(\eta^3 + \xi)$$

where $\mathcal{F}^{-1}$ is the inverse L^2-Fourier transform, φ is a smooth function with compact support whereas V is a nonnegative smooth function of one variable with the following compact support $\{y : |y| \leq \frac{1}{2}\}$. Show, using the Plancherel Theorem, that $u_\delta \in D(A)$.

(b) Show that

$$\|u_\delta\|_2 = \delta^{-\frac{1}{2}}\|\varphi\|_{L^2(\mathbb{R})}\|V\|_{L^2(\mathbb{R})}.$$

(c) Compute $\|u_\delta\|_p$.

(d) Using Fatou's lemma, show that as $\delta \to 0$ we have

$$\frac{\|u_\delta\|_p}{\|u_\delta\|_2} \longrightarrow +\infty \text{ for } p > 8.$$

(e) Conclude.

Exercise 10.5.8. ([**162**]) Let A be a closed linear operator with domain $D(A)$, such that $\mathrm{ran}(A) \subset D(A)$ and

$$\forall x \in D(A): \ \|Ax\|^2 \leq \|x\|\|A^2x\|$$

(such an operator is called **paranormal** regardless of whether A is bounded or not). Prove that $A \in B(H)$.

Exercise 10.5.9. ([**151**]) Let A and B be two densely defined operators with respective domains $D(A)$ and $D(B)$. Assume that $D(A) \subset D(BA)$. Assume further that A is closed and that for some $a < 1$ and $b > 0$

$$\exists r > 0: \ \|(rB - I)A\varphi\| \leq a\|A\varphi\| + b\|\varphi\|, \ \forall \varphi \in D(A).$$

Show that BA is closed.

Hint: You may use Theorem 10.1.22.

CHAPTER 11

Functional Calculi

11.1. Basics

There are quite different statements and proofs of the spectral theorem in the literature and each form of it may be (and it is) called a **Spectral Theorem**. I think most mathematicians would agree that the spectral theorem is the most important result in Operator Theory. It was shown by Stone and von Neumann (independently) between 1929 and 1932.

Its proofs are well documented. We refer interested readers to [**31**], [**48**], [**52**], [**64**], [**74**], [**92**], [**96**], [**178**], [**190**] and [**195**]. See also [**23**] and [**212**].

Finally, we assume basic definitions and properties on how to integrate operator-valued (and vector-valued) functions. See e.g. [**114**] or [**190**].

11.1.1. Functional Calculus for Self-adjoint Operators. We have already defined polynomials of bounded operators. This construction did not require any particular condition on either the operator A or the polynomial p.

We also gave sense to e^A (even in a Banach space). Generalizing this, taking $f(x) = \sum_{n=0}^{\infty} \alpha_n x^n$ with radius of convergence R, we may then show that $\sum_{n=0}^{\infty} \alpha_n A^n$ converges absolutely whenever $\|A\| < R$ and thus it converges in $B(H)$. Therefore, we may set $f(A) = \sum_{n=0}^{\infty} \alpha_n A^n$ (cf. Theorem 11.1.50). This is part of the holomorphic functional calculus which is only referred to briefly in the subsection "A Little Digression".

We are primarily interested in defining $f(A)$ for a self-adjoint (or normal) $A \in B(H)$ where f is some bounded complex-valued function (continuous or Borel).

Based on Proposition 7.1.57, Theorem 1.1.32 and Theorem 2.1.104 we have the following important result about $f(A)$ (see Exercise 11.3.1 for a proof):

THEOREM 11.1.1. ***(Spectral Theorem for Self-adjoint Operators: Continuous Functional Calculus Form)*** *Let $A \in B(H)$ be* *a self-adjoint operator. Then there is a unique map* $\varphi : C(\sigma(A)) \to B(H)$*, defined by:* $\varphi(f) = f(A)$*, such that*

(1) $(f+g)(A) = f(A) + g(A)$, $(\alpha f)(A) = \alpha f(A)$ *for any complex* α, $f(A)g(A) = fg(A)$ *(*fg *being the pointwise product!),* $[f(A)]^* = \overline{f}(A)$, $1(A) = I$ *and* $id(A) = A$ *(where* $g \in C(\sigma(A))$ *and* $id(z) = z$*).*
(2) $\sigma(f(A)) = f(\sigma(A))$ *(**Spectral Mapping Theorem**).*
(3) $\|f(A)\|_{B(H)} = \|f\|_\infty$.

Corollary 11.1.2. *(see Test 11.4.1) Let* $A \in B(H)$ *be self-adjoint and let* $f : \sigma(A) \to \mathbb{C}$ *be continuous. Then*

(1) $f(A)$ *is self-adjoint iff* f *is real-valued.*
(2) $f(A) \geq 0$ *iff* $f \geq 0$.

Examples 11.1.3. *(cf. Exercise 4.3.33) Let* $A \in B(H)$ *be self-adjoint. Then:*

(1) e^{iA} *is unitary (see Exercise 11.3.15).*
(2) e^A *is self-adjoint as* e^x *is real-valued. In general,* α^A *is self-adjoint if* $\alpha > 0$*. We may write symbolically,*
$$\alpha^A = e^{(\log \alpha)A}.$$
(3) *In fact,* e^A *is positive for* e^x *is positive for each* $x \in \mathbb{R}$.
(4) *In fact,* e^A *is invertible. Indeed, the range of* $f(x) = e^x$ *is contained in* $\mathbb{R}_+^*$*. Hence* $0 \notin f(\sigma(A))$ *and so the spectral mapping theorem gives* $0 \notin \sigma(f(A)) = \sigma(e^A)$.

Example 11.1.4. *Let* $A \in B(H)$ *be positive with* $\sigma(A) \subset (0, \infty)$*. We can then define the function* log *on* $\sigma(A)$*. Therefore, it makes also sense to define* $\log(A)$*. We call it the **logarithm** of the operator* A*. It is also clear why* $\log(A)$ *is self-adjoint.*

Example 11.1.5. *Let* $A \in B(H)$ *be self-adjoint and such that* $\sigma(A) \subset [1, \infty)$*. Then (why?)*
$$\log(A) \geq 0.$$

The last property of Theorem 11.1.1 is of a great importance. Let's restate as a separate result.

Theorem 11.1.6. *Let* $A \in B(H)$ *be self-adjoint. If* f *is a continuous function defined on* $\sigma(A)$*, then there exists a sequence of polynomials* (p_n) *(defined on* $\sigma(A)$ *too) such that* $\|p_n - f\|_\infty \to 0$*. Therefore,*
$$\|p_n - f\|_\infty = \|p_n(A) - f(A)\|_{B(H)} \longrightarrow 0.$$

Corollary 11.1.7. *(cf. Exercise 11.3.10) Let $A, B \in B(H)$ where A is self-adjoint. If $AB = BA$, then $f(A)B = Bf(A)$ for any continuous function f on $\sigma(A)$.*

As a consequence of the previous corollary, we have:

Corollary 11.1.8. *(see Exercise 11.3.10) Let $A, B, C \in B(H)$ be such that A and C are self-adjoint. Assume that f is a continuous function on both $\sigma(A)$ and $\sigma(C)$. Then*

$$AB = BC \Longrightarrow f(A)B = Bf(C).$$

The previous corollary too has a direct and interesting consequence:

Corollary 11.1.9. *Let $A, B \in B(H)$ be self-adjoint and also similar, i.e.*

$$T^{-1}AT = B$$

for some invertible $T \in B(H)$. If f is continuous on both $\sigma(A)$ and $\sigma(B)$, then

$$T^{-1}f(A)T = f(B).$$

We now turn to more examples (mainly of type A^α).

Example 11.1.10. *Let $A \in B(H)$ be positive. Then $\sigma(A) \subset \mathbb{R}^+$. Let $f(x) = \sqrt{x}$ (defined for $x \geq 0$). Since f is continuous on $\sigma(A)$,*

$$[f(A)]^2 = A,$$

i.e. we have defined the square root of A. For another proof of uniqueness of the square root, see Exercise 11.3.3.

Example 11.1.11. *Let $A \in B(H)$ be positive. If $\alpha, \beta > 0$, then*

$$A^\alpha A^\beta = A^{\alpha+\beta}.$$

If $A, B \in B(H)$ are commuting, then we know that $(AB)^m = A^m B^m$ for all $m \in \mathbb{N}$. We also know from Corollary 5.1.36 that $(AB)^{\frac{1}{n}} = A^{\frac{1}{n}} B^{\frac{1}{n}}$ for any $n \in \mathbb{N}$ (if further $A, B \geq 0$). Hence $(AB)^{\frac{m}{n}} = A^{\frac{m}{n}} B^{\frac{m}{n}}$, i.e. $(AB)^q = A^q B^q$ for all $q \in \mathbb{Q}_+^*$ (we may therefore extend the result to real positive powers, how?).

Alternatively, as a direct consequence of the Functional Calculus, we have:

Example 11.1.12. *Let $A, B \in B(H)$ be positive and commuting. If $\alpha > 0$, then*

$$A^\alpha B^\alpha = (AB)^\alpha.$$

THEOREM 11.1.13. *(see Exercise 11.3.6, cf. Exercise 11.3.18) Let $A \in B(H)$ be positive. Then for any $\alpha > 0$, we have*

$$\|A^\alpha\| = \|A\|^\alpha.$$

REMARK. The previous result is no longer valid for $\alpha < 0$ even when A is invertible. Just for $\alpha = -1$, the result is untrue as already seen in the "True or False" Section of Chapter 2. Remember that back then we considered as a counterexample the *positive invertible* A defined by

$$Af(x) = (x+1)f(x)$$

on $L^2[0,1]$.

PROPOSITION 11.1.14. *(see Exercise 11.3.7, and Exercise 11.5.7 for a more general result) Let $A, B \in B(H)$ be positive and commuting. Then*

$$0 \leq A \leq B \Longrightarrow A^\alpha \leq B^\alpha$$

for any $\alpha \in (0, \infty)$.

Without commutativity, we "only" have:

THEOREM 11.1.15. ***(Löwner-Heinz Inequality)****Let $A, B \in B(H)$. If $0 \leq A \leq B$, then $A^\alpha \leq B^\alpha$ for any $\alpha \in [0,1]$.*

REMARK. For a fairly simple proof of the Löwner-Heinz Inequality, see [**167**]. A deep treatment of these concepts may be found in e.g. in [**27**], [**81**] and [**218**].

REMARK. There is a related definition to the previous theorem: *Let $A, B \in B(H)$ be self-adjoint. We say that a real-valued continuous function f (defined on some real interval containing $\sigma(A)$ and $\sigma(B)$) is* ***operator monotone*** *if*

$$A \leq B \Longrightarrow f(A) \leq f(B).$$

EXAMPLE 11.1.16. *By Löwner-Heinz Inequality the function given by $f(x) = x^\alpha$ is operator monotone when $0 \leq \alpha \leq 1$.*

Another famous inequality is the so-called **Furuta Inequality**:

THEOREM 11.1.17. *([**80**]) Let $A, B \in B(H)$ be such that $A \geq B \geq 0$. Then for each $r \geq 0$,*

$$(B^r A^p B^r)^{\frac{1}{q}} \geq B^{(p+2r)/q}$$

and

$$A^{(p+2r)/q} \geq (A^r B^p A^r)^{\frac{1}{q}}$$

hold for each p and q such that $p \geq 0$, $q \geq 1$ and $(1+2r)q \geq p + 2r$.

COROLLARY 11.1.18. *Let $A, B \in B(H)$ be such that $A \geq B \geq 0$. Then*

$$(B^r A^p B^r)^{\frac{1}{p}} \geq B^{(p+2r)/p}$$

and

$$A^{(p+2r)/p} \geq (A^r B^p A^r)^{\frac{1}{p}}$$

for each $r \geq 0$ and $p \geq 1$.

Although, $A \geq B \geq 0$ need not imply $A^2 \geq B^2$, as a particular case of the previous corollary we have:

COROLLARY 11.1.19. *Let $A, B \in B(H)$ be such that $A \geq B \geq 0$. Then*

$$(BA^2B)^{\frac{1}{2}} \geq B^2 \text{ and } A^2 \geq (AB^2A)^{\frac{1}{2}}.$$

We go back to more properties of the Functional Calculus. The next property concerns the operation of composition (for a proof, see Exercise 11.3.2):

PROPOSITION 11.1.20. *Let $A \in B(H)$ be self-adjoint. If f is a continuous real-valued function defined on $\sigma(A)$ and if g is a continuous function on $f(\sigma(A))$, then*

$$(g \circ f)(A) = g(f(A)).$$

EXAMPLE 11.1.21. *Let $A \in B(H)$ be positive. Then*

$$(A^\alpha)^\beta = A^{\alpha\beta},\ \forall \alpha, \beta > 0.$$

EXAMPLE 11.1.22. *Let $A \in B(H)$ be self-adjoint. Then*

$$\log(e^A) = A.$$

In particular,

$$\log I = \log(e^0) = 0.$$

If we further assume that A is positive with $0 \notin \sigma(A)$ (i.e. if A is invertible), then

$$e^{\log A} = A.$$

We may recover some known properties of "log". For instance, we have:

PROPOSITION 11.1.23. *(see Exercise 11.3.16) Let $A, B \in B(H)$ be both positive and such that $0 \notin \sigma(A) \cap \sigma(B)$. Then*

$$AB = BA \Longrightarrow \log(AB) = \log A + \log B.$$

COROLLARY 11.1.24. *(see also Exercise 11.3.16) Let $A \in B(H)$ be positive and invertible. Then*

$$\log(A^{-1}) = -\log A.$$

As a consequence of the Löwner-Heinz Inequality, we have:

PROPOSITION 11.1.25. *Let $A, B \in B(H)$ be positive and such that $A \geq B$. If B is assumed invertible, then (A is invertible):*

$$\log A \geq \log B.$$

REMARK. What about the converse? (Cf. Exercise 11.5.6).

Can we go beyond continuous functions in the definition of $f(A)$? The answer is yes! We can go up to Borel bounded functions on $\mathbb{R}$ (denoted by $\mathfrak{B}(\mathbb{R})$) and by keeping A self-adjoint.

This is a consequence of the Continuous Functional Calculus as well as the Riesz-Markov Theorem (see [**178**]). In particular, the key idea is to start by writing (for f continuous on $\sigma(A)$)

$$< f(A)x, x >= \int_{\sigma(A)} f(\lambda) d\mu_{x,x}$$

for some unique measure $\mu_{x,x}$ on the compact $\sigma(A)$ (this is possible as $< f(A)x, x >$ is a *positive* linear functional on $C(\sigma(A))$, that is, $< f(A)x, x > \geq 0$ whenever $f \geq 0$). In the end, we obtain (see e.g. [**178**] for more details):

THEOREM 11.1.26. ***(Spectral Theorem for Self-adjoint Operators: Bounded Borel Functional Calculus Form)*** *Let $A \in B(H)$ be a self-adjoint operator. Then there is a unique map $\varphi : \mathfrak{B}(\mathbb{R}) \to B(H)$, defined by: $\varphi(f) = f(A)$, such that*

(1) *$(f+g)(A) = f(A) + g(A)$, $(\alpha f)(A) = \alpha f(A)$ for any complex α, $f(A)g(A) = fg(A)$, $[f(A)]^* = \overline{f}(A)$, $1(A) = I$ and $id(A) = A$ (where $g \in \mathfrak{B}(\mathbb{R})$ and $id(z) = z$).*
(2) $\|f(A)\|_{B(H)} \leq \|f\|_\infty$.
(3) *Let (f_n) be a sequence of Borel and bounded functions on $\sigma(A)$ with $\|f_n\|_\infty$ is bounded. If $f_n(x) \to f(x)$ for all x, then*

$$\|f_n(A)x - f(A)x\| \to 0$$

for all x.
(4) *If $f \geq 0$, then $f(A) \geq 0$.*

We also have:

PROPOSITION 11.1.27. *If A is self-adjoint and f is Borel on $\sigma(A)$, then*

(1) *If λ is an eigenvalue for A, then $f(\lambda)$ is an eigenvalue for $f(A)$.*
(2) *If $B \in B(H)$ is such that $AB = BA$, then $f(A)B = Bf(A)$.*

Remark. Observe that we loose the Spectral Mapping Theorem by using bounded Borel functions.

Example 11.1.28. *Let $A \in B(H)$ be positive and invertible. We know that*

$$\lim_{\alpha \to 0} \frac{x^\alpha - 1}{\alpha} = \log x$$

(when $x > 0$). Therefore, by the previous theorem, we have

$$\lim_{\alpha \to 0} \frac{A^\alpha - 1}{\alpha} = \log A$$

(strongly).

As alluded to before the previous theorem, a self-adjoint operator gave rise to some measure. If we have some particular measure, can we obtain a bounded (self-adjoint) operator?

We introduce the concept of a spectral measure:

Definition 11.1.29. *Let X be a non empty set and let Σ be a sigma algebra of subsets of X. Let H be a Hilbert space. A* ***spectral measure*** *is a function $E : \Sigma \to B(H)$ such that:*

(1) *$E(\varnothing) = 0$ and $E(X) = I$ (I is the identity operator).*
(2) *For every $\Delta \in \Sigma$, $E(\Delta)$ is an orthogonal projection.*
(3) *For any $\Delta_1, \Delta_2 \in \Sigma$: $E(\Delta_1 \cap \Delta_2) = E(\Delta_1)E(\Delta_2)$.*
(4) *If $(\Delta_n)_{n \in \mathbb{N}} \in \Sigma$ are pairwise disjoint, then*

$$E\left(\bigcup_{n=1}^{\infty} \Delta_n\right) = \sum_{n=1}^{\infty} E(\Delta_n).$$

Remarks.

(1) By (3), we have
$$E(\Delta_1)E(\Delta_2) = E(\Delta_2)E(\Delta_1)$$
for any $\Delta_1, \Delta_2 \in \Sigma$. In particular, if $\Delta_1 \cap \Delta_2 = \varnothing$, then
$$E(\Delta_1)E(\Delta_2) = E(\Delta_2)E(\Delta_1) = 0,$$
i.e. $E(\Delta_1)$ and $E(\Delta_2)$ have orthogonal ranges.
(2) The limit in Property (4) is w.r.t. the Strong Operator Topology.
(3) If $(\Delta_n)_{n \in \mathbb{N}}$ constitutes a partition of X, then (strongly)
$$\sum_{n=1}^{N} E(\Delta_n) \longrightarrow E(X) = I$$
as $N \to \infty$.

Next, we give a link between a spectral measure and an ordinary measure.

Proposition 11.1.30. *(see Exercise 11.3.19) Let X be a non empty set and let Σ be a sigma algebra of subsets of X. Let H be a Hilbert space. If $x, y \in H$, then*

$$\mu_{x,y}(\Delta) = < E(\Delta)x, y >$$

is a countably additive measure on Σ.

Example 11.1.31. *(see Exercise 11.3.20) Let X be a compact set and let Σ be the Borel subsets of X. Set $H = L^2(\mu)$ where μ is a measure on Σ. If $\Delta \in \Sigma$, let*

$$E(\Delta)f(x) = \mathbb{1}_\Delta(x)f(x)$$

for any $f \in H$. Then E is a spectral measure.

Even if knowledge on integration of vector-valued functions is assumed, we would like to give a word or two on how to work with this type of integrals. If f is a simple function of the form

$$f = \sum_{i=1}^{n} \alpha_i \mathbb{1}_{\Delta_i}$$

where (Δ_i) are pairwise disjoint elements of some σ-algebra, then we define the **spectral integral** of (the simple) f by

$$\int f dE = \sum_{i=1}^{n} \alpha_i E(\Delta_i).$$

We can then show that

$$\left\| \int f dE \right\|_{B(H)} \leq \|f\|_\infty.$$

Since each Borel function f is the uniform limit of a sequence of simple functions (f_n), we can show that $\int f_n dE$ converges in norm to a limit in $B(H)$ which we denote by $\int f dE$. The latter then also verifies statements on $f(A)$ listed in Theorem 11.1.26 (see e.g. [**195**] for more details on these topics). In particular, if f is real-valued, then $\int f dE$ is self-adjoint (otherwise, $\int f dE$ is normal).

Summarizing everything:

Theorem 11.1.32. ***(Spectral Theorem for Self-adjoint Operators: Integral Form)*** *Let $A \in B(H)$ be self-adjoint. Then there* *exists a unique spectral measure E on the Borel subsets of $\sigma(A)$ such that:*

$$A = \int_{\sigma(A)} \lambda dE,$$

also written as (for all $x, y \in H$)

$$< Ax, y >= \int_{\sigma(A)} \lambda d\mu_{x,y},$$

where $\mu_{x,y}$ was introduced in Proposition 11.1.30. Also,

$$AE(\Delta) = E(\Delta)A$$

holds for any Borel set Δ and so $\operatorname{ran}(E(\Delta))$ reduces A. In fact, if $B \in B(H)$, then

$$BA = AB \Longleftrightarrow BE(\Delta) = E(\Delta)B, \text{ for all } \Delta.$$

Finally, if f is bounded and Borel on $\sigma(A)$, then

$$f(A) = \int_{\sigma(A)} f(\lambda) dE.$$

Remark. We may therefore say that there is a bijection between self-adjoint operators and the spectral measures.

Example 11.1.33. *Let $A \in B(H)$ be such that $0 \leq A \leq I$. If $\alpha \in [0, 1]$, then $A^\alpha \geq A$.*

Example 11.1.34. *Let $\varphi : \mathbb{R} \to \mathbb{R}$ is essentially bounded and consider the multiplication operator M_φ of Exercise 10.3.9. Then the spectral measure of M_φ is given by*

$$E(\Delta)f = M_{\mathbb{1}_{\varphi^{-1}(\Delta)}} f = \mathbb{1}_{\varphi^{-1}(\Delta)} f$$

for every Borel subset Δ of $\sigma(M_\varphi)$ (and $f \in L^2(\mathbb{R})$).

Remark. If $A, B \in B(H)$ are both self-adjoint with E and F as the corresponding spectral measures, then

$$AB = BA \Longleftrightarrow E(\Delta)F(\Delta') = F(\Delta')E(\Delta)$$

for all Borel sets Δ and Δ'.

We already know from Exercise 7.3.3 that a positive operator has a positive spectrum. The converse also holds and the proof is so simple using the Spectral Theorem.

Corollary 11.1.35. *(see Exercise 11.3.25) Let $A \in B(H)$ be self-adjoint such that $\sigma(A) \subset \mathbb{R}^+$. Then $A \geq 0$.*

Remark. The previous result will be generalized in Proposition 11.1.39 to normal operators.

There is yet another form of the Spectral Theorem, namely:

Theorem 11.1.36. *(**Spectral Theorem for Self-adjoint Operators: Multiplication Operator Form**, see [**178**]) Let $A \in B(H)$ be self-adjoint where H is separable. Then there exists a measure space (X, Σ, μ) (σ-finite), and a real-valued function φ in $L^\infty(X, \Sigma, \mu)$ such that A is unitarily equivalent to M_φ (the multiplication operator by φ) on $L^2(X, \Sigma, \mu)$.*

11.1.2. Functional Calculus for Normal Operators. Most of the results of the previous subsection remain valid with "A normal" instead of "A self-adjoint". We will be citing these results again to make the reader at ease (as there are courses which go up to the Spectral Theorem for normal operators and others which do not go that far). As above, the spectral measure gives rise to a normal operator $\int f dE$ in $B(H)$ (where f is defined on $\mathbb{C}$) which verifies the statements of Theorem 11.1.26.

We have:

Theorem 11.1.37. *(**Spectral Theorem for Normal Operators: Integral Form**, [**48**]) Let $A \in B(H)$ be normal. Then there exists a unique spectral measure E on the Borel subsets of $\sigma(A)$ such that:*

$$A = \int_{\sigma(A)} \lambda dE,$$

also written as (for all $x, y \in H$)

$$< Ax, y >= \int_{\sigma(A)} \lambda d\mu_{x,y},$$

where $\mu_{x,y}$ was introduced in Proposition 11.1.30.

If Y is a (non empty) relatively open subset of $\sigma(A)$, then $E(Y) \neq 0$. Also,

$$AE(\Delta) = E(\Delta)A$$

holds for any Borel set Δ and hence $\operatorname{ran}(E(\Delta))$ reduces A. More generally, if $B \in B(H)$, then

$$BA = AB \text{ and } BA^* = A^*B \Longleftrightarrow BE(\Delta) = E(\Delta)B, \text{ for all } \Delta.$$

Finally, if f is bounded and Borel on $\sigma(A)$, then

$$f(A) = \int_{\sigma(A)} f(\lambda) dE.$$

Remark. By the Fuglede Theorem, $BA = AB$ iff $BA^* = A^*B$ whenever $A \in B(H)$ is normal. So, we may then just drop the assumption $BA^* = A^*B$ in Theorem 11.1.37.

Remark. If $A, B \in B(H)$ are both normal with E and F as the corresponding spectral measures, then

$$AB = BA \Longleftrightarrow E(\Delta)F(\Delta') = F(\Delta')E(\Delta)$$

for all Borel sets Δ and Δ'.

Corollary 11.1.38. *Let $A \in B(H)$ be normal with E as its spectral measure. If $\sigma(A) = \{\lambda_1, \lambda_2, \cdots, \lambda_n\}$, then*

$$A = \sum_{k=1}^{n} \lambda_k E(\{\lambda_k\}) \text{ and } A^* = \sum_{k=1}^{n} \overline{\lambda_k} E(\{\lambda_k\}).$$

In particular (cf. Exercise 8.3.3):

$$\sigma(A) = \{\lambda\} \Longrightarrow A = \lambda I.$$

Remark. What is the corresponding formula for A in the case $\sigma(A) = \{\lambda_1, \lambda_2, \cdots, \lambda_n, \cdots\}$?

A non-trivial and interesting consequence of the Spectral Theorem is the following:

Proposition 11.1.39. *(see Exercise 11.3.28) Let $A \in B(H)$ be normal. Then*

(1) *A is self-adjoint iff $\sigma(A)$ is real.*
(2) *A is unitary iff $\sigma(A) \subset \{\lambda \in \mathbb{C} : |\lambda| = 1\}$.*
(3) *If $\alpha \in \mathbb{R}^+$, then $A \geq \alpha I$ iff $\sigma(A) \subset [\alpha, \infty)$. In particular, A is positive iff $\sigma(A)$ is positive.*
(4) *A is an orthogonal projection iff $\sigma(A) \subset \{0, 1\}$.*

Another great application concerns the Invariant Subspace Problem (this has been promised before):

Theorem 11.1.40. *(see Exercise 11.3.22) Any normal operator has nontrivial reducing subspaces. In fact, any non scalar normal operator has a nontrivial subspace which reduces every operator which commutes with it.*

Thanks to the Spectral Theorem, we may list the important properties of functions of normal A.

Theorem 11.1.41. *(**Spectral Theorem for Normal Operators: Functional Calculus Form**) Let $A \in B(H)$ be normal and let f and g be two bounded Borel functions on $\sigma(A)$. Then*

(1) $(f+g)(A) = f(A)+g(A)$ *and* $(\alpha f)(A) = \alpha f(A)$ *for any complex* α.
(2) $f(A)g(A) = fg(A)$, $[f(A)]^* = \overline{f}(A)$, $1(A) = I$ *and* $id(A) = A$.
(3) $\|f(A)\|_{B(H)} \leq \|f\|_\infty$, *and* $\|f(A)\|_{B(H)} = \|f\|_\infty$ *if* f *is continuous on* $\sigma(A)$.

Corollary 11.1.42. *Let* $A \in B(H)$ *be normal. If* (f_n) *is a sequence of bounded Borel functions on* $\sigma(A)$ *s.t.* $\lim_{n\to\infty} \|f_n - f\|_\infty = 0$ *for some bounded Borel function* f *on* $\sigma(A)$*, then*

$$\lim_{n\to\infty} \|f_n(A) - f(A)\|_{B(H)} = 0.$$

Theorem 11.1.43. *(**Spectral Mapping Theorem**, see e.g.* **[48]***) Let* $A \in B(H)$ *be normal and let* f *be a continuous function from* $\sigma(A)$ *into* $\mathbb{C}$*. Then*

$$\sigma(f(A)) = f(\sigma(A)).$$

Proposition 11.1.44. *Let* $A \in B(H)$ *be normal with a spectral measure* E*. Then for every Borel and bounded function* f *on* $\sigma(A)$*, we have*

$$\|f(A)x\|^2 = \int_{\sigma(A)} |f(\lambda)|^2 d\mu_{x,x},\ (x \in H),$$

where $\mu_{x,x}$ *was introduced in Proposition 11.1.30.*

Corollary 11.1.45. *Let* $A \in B(H)$ *be normal. If* (f_n) *is a sequence of bounded Borel functions on* $\sigma(A)$ *(with* $(\|f_n\|_\infty)_n$ *being bounded) such that* $\lim_{n\to\infty} f_n(t) = f(t)$ $(t \in \sigma(A))$ *for some bounded Borel function* f *on* $\sigma(A)$*, then*

$$\lim_{n\to\infty} \|f_n(A)x - f(A)x\|_H = 0$$

for each x.

The following result is a nice application of the Spectral Theorem.

Proposition 11.1.46. *(***[138]***, cf. Exercise 11.3.27) Let* $U \in B(H)$ *be unitary. Then there is a uniquely determined injective and positive operator* $A \in B(H)$ *with* $\sigma(A) \subset [0, 2\pi]$ *and such that*

$$U = e^{iA}.$$

Remark. Another important application of the Functional Calculus is the polar decomposition of normal operators. This was already given before. A proof is given in Exercise 11.3.26.

The next result generalizes Proposition 11.1.20.

PROPOSITION 11.1.47. *Let $A \in B(H)$ be normal. Let $f : \sigma(A) \to \mathbb{C}$ be continuous and let $g : f(\sigma(A)) \to \mathbb{C}$ be Borel and bounded. Then*

$$(g \circ f)(A) = g(f(A)).$$

If $A \in B(H)$ is normal, then one may ask what the spectral measure of the normal $f(A)$ is (if f is continuous, say)?

PROPOSITION 11.1.48. *(see Test 11.4.6) Assume that $A \in B(H)$ is normal with a spectral measure E. Let f be continuous on $\sigma(A)$. Then the spectral measure F of the normal $f(A)$ is given by*

$$F(\Delta) = E \circ f^{-1}(\Delta)$$

for each Borel subset Δ of $\sigma(f(A))$.

Finally, we present an analogue of Theorem 11.1.36 in the "normal case":

THEOREM 11.1.49. ***(Spectral Theorem for Normal Operators: Multiplication Operator Form, see* [48]*)** Let $A \in B(H)$ be normal. Then there exists a measure space (X, Σ, μ) and a complex-valued function φ in $L^\infty(X, \Sigma, \mu)$ such that A is unitarily equivalent to M_φ (the multiplication operator by φ) on $L^2(X, \Sigma, \mu)$. If H is separable, then (X, Σ, μ) may be taken to be σ-finite.*

11.1.3. A Little Digression. We would like to add a word on the **Riesz (-Dunford) Functional Calculus** without going into much detail though.

We do define $f(A)$ for an element A of a Banach algebra but f is taken to be an analytic (holomorphic) function in a neighborhood of $\sigma(A)$. More precisely, if $f : \Omega \to \mathbb{C}$ is analytic and $\sigma(A) \subset \Omega$ (Ω being an open subset of $\mathbb{C}$), then inspired by Cauchy's formula we define $f(A)$ as

$$f(A) = \frac{1}{2\pi i} \int_\Gamma f(\lambda)(\lambda I - A)^{-1} d\lambda$$

where Γ is any contour that surrounds $\sigma(A)$ in the sense given in Section 10.23 of **[190]**. Then it may be shown that $f(A)$ is well-defined (see **[48]** or **[63]** for more details). Besides, we have

THEOREM 11.1.50. *(see* **[48]***) Let $\mathcal{A}$ be a Banach algebra with identity and let $A \in \mathcal{A}$. Then*

1. *The map $f \mapsto f(A)$ from the space of analytic functions in a neighborhood of $\sigma(A)$ into $\mathcal{A}$ verifies:*
   - (a) *$(f + g)(A) = f(A) + g(A)$ and $(\alpha f)(A) = \alpha f(A)$ for any complex α.*
   - (b) *$f(A)g(A) = fg(A)$, $1(A) = I$ and $id(A) = A$.*

(2) *Let $\sigma(A) \subset \Omega$ and let f be analytic on Ω. If (f_n) is a sequence of analytic functions on Ω, and if $f_n(z) \to f(z)$ uniformly on compact subsets of Ω, then*

$$\lim_{n\to\infty} \|f_n(A) - f(A)\|_{\mathcal{A}} = 0.$$

(3) *If $f(z) = \sum_{n=0}^{\infty} \alpha_n z^n$ has radius of convergence $R > r(A)$ (the spectral radius of A), then f is analytic in a neighborhood of $\sigma(A)$ and*

$$f(A) = \sum_{n=0}^{\infty} \alpha_n A^n.$$

Other major results still hold and we state them separately.

Theorem 11.1.51. ***(Spectral Mapping Theorem)*** *Let A be an element of a Banach algebra and let f be an analytic function in a neighborhood of $\sigma(A)$. Then*

$$\sigma(f(A)) = f(\sigma(A)).$$

Theorem 11.1.52. *Let A, B be two elements of a Banach algebra and let f be an analytic function in a neighborhood of $\sigma(A)$. Then*

$$AB = BA \Longrightarrow f(A)B = Bf(A).$$

The following consequences are fundamental to prove some of our results on exponentials of normal operators.

Lemma 11.1.53. *Let A and B be two commuting normal operators, on a Hilbert space, having spectra contained in simply connected regions not containing 0 (cf.* **[47]***). Then*

$$A^i B^i = B^i A^i$$

where $i = \sqrt{-1}$ is the usual complex number.

Lemma 11.1.54. *Let A be a self-adjoint operator such that $\sigma(A) \subset (0, \pi)$ (or e.g. $(-\frac{\pi}{2}, \frac{\pi}{2})$). Then*

$$(e^{iA})^i = e^{-A}.$$

11.1.4. The Fuglede-Putnam Theorem Revisited. We have already introduced the Fuglede Theorem without proof. We exclusively devote this subsection to it. As quoted above, the Spectral Theorem is the most important result in Operator Theory. We can definitely say that, at least as far as normal operators are concerned, the Fuglede-Putnam Theorem is the second most important result!

THEOREM 11.1.55. *(**Fuglede-Putnam**, see Exercise 11.3.30) Let $A, B, T \in B(H)$ and assume that A and B are normal. Then*

$$TA = BT \Longleftrightarrow TA^* = B^*T.$$

Hence $TE(\Delta) = F(\Delta)T$ for any Borel set Δ (where E and F are the spectral measures associated with A and B respectively).

REMARK. The problem leading to this theorem was first raised by von Neumann in [**161**] who had already established the result in a finite dimensional setting (this proof is given in Exercise 11.3.29). Fuglede was the first one to answer this problem affirmatively in [**78**], where he proved the previous result in the case $A = B$. Then in 1951, Putnam [**168**] proved the result as it stands.

There are different proofs of the Fuglede-Putnam Theorem (e.g. the one in [**91**]). The elegant proof prescribed in Exercise 11.3.30 is due to Rosenblum (see [**184**]). Then came Berberian [**20**], who noted that the Fuglede's version is simply equivalent to the Putnam's version (see Exercise 11.3.31).

Another proof which is, unfortunately, not well-spread is due to Rehder in [**180**].

Notice also that the Fuglede-Putnam Theorem holds for non necessarily bounded operators (and keeping $T \in B(H)$). For new versions of the Fuglede-Putnam Theorem involving only unbounded operators, see [**142**], [**150**] and [**164**].

Among other generalizations, we note a generalization to the so-called "spectral operators" by Dunford [**62**] (and another proof of the latter by Radjavi-Rosenthal [**171**]):

THEOREM 11.1.56. *If $A \in B(H)$ is such that $A = \int \lambda dE$ (where E is a spectral measure) and it commutes with $B \in B(H)$, then*

$$E(\Delta)B = BE(\Delta)$$

for any Borel set Δ.

There is a particular terminology to the transformation which occurred in the Fuglede-Putnam Theorem.

DEFINITION 11.1.57. *We say that $T \in B(H)$ **intertwines** two operators $A, B \in B(H)$ if $TA = BT$.*

REMARK. With this definition, we may restate the Fuglede-Putnam theorem as follows: *If an operator intertwines two normal operators, then it intertwines their adjoints.*

Among the interesting consequences of the Fuglede-Putnam Theorem, we have (a proof is given in Exercise 11.3.36):

COROLLARY 11.1.58. *([**168**]) Any two similar normal operators are unitarily equivalent.*

There is a big number of research papers dealing with the Fuglede-Putnam Theorem and its generalizations. One of the interesting results is:

THEOREM 11.1.59. *Let $A_1, A_2, B_1, B_2 \in B(H)$ be normal operators such that $A_1A_2 = A_2A_1$ and $B_1B_2 = B_2B_1$. If $T \in B(H)$, then*

$$A_1TB_1 = A_2TB_2 \Longleftrightarrow A_1^*TB_1^* = A_2^*TB_2^*.$$

REMARK. The previous theorem can be used for example to prove the existence of a solution to the operator equation $AXB - XD = E$ under some extra conditions. See [**135**].

We finish with an application of the Fuglede-Putnam theorem. Kaplansky established the following remarkable result (cf. Exercise 12.3.9) on the normality of products. A proof is prescribed in Exercise 11.3.39.

THEOREM 11.1.60. *([**113**]) Let A and B be two bounded operators on a Hilbert space such that AB and A are normal. Then B commutes with AA^* iff BA is normal.*

11.1.5. The Beck-Putnam Theorem. A related result to the Fuglede-Putnam Theorem is the following result (due to Beck and Putnam):

THEOREM 11.1.61. *([**17**]) Let A be a bounded operator on a Hilbert space. Let N be a normal operator such that $AN = N^*A$. If, whenever z is not real, either z or its conjugate $\overline{z}$ does not lie in the spectrum of N, then $AN = NA$.*

Rehder [**180**] gave both a generalization and a different proof of the previous theorem. Rehder's version reads:

THEOREM 11.1.62. *Let $A, N, M \in B(H)$ such that N and M are normal such that $AN = MA$. Assume further that*

(1) *$\sigma(M^*) \subset \sigma(N)$, and*
(2) *whenever z is not real, either z or its conjugate $\overline{z}$ does not lie in the spectrum of N.*

Then $AN = M^*A$, and equivalently $AE(\Delta) = F(\Delta)A$ for any Borel set Δ (where E and F are the spectral measures associated with N and M^* respectively).

The previous theorem has been strengthened in [**146**]. Its proof, which is given in Exercise 11.3.41, is Fuglede-free and is a lot simpler than Rehder's:

THEOREM 11.1.63. *Let $A, M, N \in B(H)$ where N and M are also assumed to be normal and such that $AN = MA$. Assume further that, whenever z is not real, either z or its conjugate $\overline{z}$ does not lie in $\sigma(M^*) \cup \sigma(N)$. Then $AN = M^*A$.*

11.2. True or False: Questions

QUESTIONS. Comment on the following questions/statements and indicate those which are false and those which are true when this applies. Justify your answers.

(1) Let $A \in B(H)$ be self-adjoint and let f be a continuous function on $\sigma(A)$. If $x \in H$, then
$$Ax = \lambda x \Longrightarrow f(A)x = f(\lambda)x.$$
(2) Let $A \in B(H)$. If $f(A)$ is well-defined (and is bounded), then
$$f(\sigma_p(A)) = \sigma_p(f(A)).$$
(3) Let $A \in B(H)$ be self-adjoint. Then for all $\lambda \notin \sigma(A)$, we have
$$\|(A-\lambda)^{-1}\| = \frac{1}{d(\lambda, \sigma(A))}.$$
(4) A monotonic function on $\mathbb{R}^+$ is operator monotone.
(5) Let E be a spectral measure. Then E is "increasing", in the following sense
$$\Delta \subset \Delta' \Longrightarrow E(\Delta) \leq E(\Delta').$$
(6) Let X be a non empty set and let Σ be the sigma algebra of subsets of X. Let H be a Hilbert space. Then (for all $x \in H$)
$$\mu_{x,x}(\Delta) =< E(\Delta)x, x >$$
is a positive measure on Σ.
(7) Let $A \in B(H)$ be self-adjoint (or normal) and let E be the corresponding spectral measure. Then
$$E(\sigma(A)) = I.$$
(8) Let $A \in B(H)$ be positive and invertible. Then
$$\log A \geq 0.$$
(9) Let $A \in B(H)$ be self-adjoint. If we use the integral form of the Spectral Theorem to compute e^A, then we will obtain the exponential e^A as defined by a series in Example 2.1.69.

(10) Let $A \in B(H)$ be self-adjoint. Then A has a unique self-adjoint cube root.
(11) If $A \in B(H)$ is normal and f is Borel, bounded and real-valued on $\sigma(A)$. Then $f(A)$ is self-adjoint.
(12) If $A \in B(H)$ is normal and f is Borel and bounded on $\sigma(A)$. Then $f(A)$ is always normal.
(13) Let $A \in B(H)$ be normal. Then A has a unique square root.

11.3. Exercises with Solutions

Exercise 11.3.1. Let $A \in B(H)$ be self-adjoint. Show that there is a unique map $\varphi : C(\sigma(A)) \to B(H)$, defined by: $\varphi(f) = f(A)$, which verifies

(1) $(f+g)(A) = f(A)+g(A)$, $(\alpha f)(A) = \alpha f(A)$ for any complex α, $f(A)g(A) = fg(A)$, $[f(A)]^* = \overline{f}(A)$, $1(A) = I$ and $id(A) = A$ (where $g \in C(\sigma(A))$).
(2) $\|f(A)\|_{B(H)} = \|f\|_\infty$.
(3) $\sigma(f(A)) = f(\sigma(A))$.

Exercise 11.3.2. Let $A \in B(H)$ be self-adjoint. Prove that if f is a continuous real-valued function defined on $\sigma(A)$ and if g is a continuous function defined on $f(\sigma(A))$, then

$$(g \circ f)(A) = g(f(A))...(1).$$

Exercise 11.3.3. ([**112**]) Let $A, B \in B(H)$ be normal and let f be a homeomorphism on $\sigma(A)$ and $\sigma(B)$.

(1) Show that if $f(A) = f(B)$, then $A = B$.
(2) Deduce that if A and B are positive and such that $A^n = B^n$ ($n \in \mathbb{N}$), then $A = B$ (in particular, a positive operator has a unique square root).

Exercise 11.3.4. Let

$$A = \begin{pmatrix} 1 & 2 \\ 2 & 1 \end{pmatrix}.$$

(1) What is $\sigma(e^A)$?
(2) What is $\|e^A\|$? Infer that (for this particular A) we have (cf. Exercise 11.3.5)

$$\|e^A\| = e^{\|A\|}.$$

Exercise 11.3.5. Let $A \in B(H)$ be positive and let $B \in B(H)$ be self-adjoint.

(1) Show that, in general,

$$\|e^B\| \neq e^{\|B\|}.$$

(2) Show that

$$\|e^A\| = e^{\|A\|}.$$

Remark. See Exercise 11.3.18 for a generalization.

(3) Deduce that

$$\|e^{A^n}\| = e^{\|A\|^n} \text{ and } \|e^{B^{2n}}\| = e^{\|B\|^{2n}}$$

for all $n \in \mathbb{N}$.

(4) **Application:** Let V be the Volterra Operator on $L^2[0,1]$. Show that

$$\|e^{\operatorname{Re} V}\| = \sqrt{e}.$$

Exercise 11.3.6. Let $A \in B(H)$ be positive. Show that for any $\alpha \geq 0$, we have

$$\|A^\alpha\| = \|A\|^\alpha.$$

Remark. See Exercise 11.3.18 for a generalization.

Exercise 11.3.7. Let $A, B \in B(H)$ be positive and commuting. Show that

$$0 \leq A \leq B \Longrightarrow A^\alpha \leq B^\alpha$$

for any $\alpha \in (0, \infty)$.

Exercise 11.3.8. (cf. Exercise 11.5.14) Let $A \in B(H)$ be self-adjoint. Let $\lambda \in \sigma(A)$ be an isolated point in $\sigma(A)$. Show that $\lambda \in \sigma_p(A)$.

Remark. The previous result equally holds for normal operators (with the same idea of proof). In fact, it holds for "hyponormal" operators as well. See Page 510 of [**125**].

Exercise 11.3.9. Let $A, B \in B(H)$ be commuting and self-adjoint. Prove that

$$\|e^{iA} - e^{iB}\| \leq \|A - B\|.$$

Exercise 11.3.10. Let $A, B, C \in B(H)$ be such that A and C are self-adjoint. Assume that f is a continuous function on both $\sigma(A)$ and $\sigma(C)$. Show that

$$AB = BC \Longrightarrow f(A)B = Bf(C).$$

Exercise 11.3.11. ([**142**], cf. [**110**]) Let H be a Hilbert space. Let A be a closed operator with domain $D(A) \subset H$ and let $B \in B(H)$ be self-adjoint. Assume further that $BA \subset AB$.

(1) Show that
$$f(B)A \subset Af(B)$$
for any continuous function $f : \sigma(B) \to \mathbb{C}$. **Hint**: You may use Exercise 10.3.4.

(2) Infer that
$$\sqrt{B}A \subset A\sqrt{B}$$
if $B \geq 0$.

Exercise 11.3.12. Let $A \in B(H)$.

(1) If $A^3 \geq 0$, then does it follow that $A \geq 0$?
(2) Show that if A is self-adjoint, then $A^3 \geq 0$ does imply that $A \geq 0$.
(3) Show that if $A, B \in B(H)$ are self-adjoint such that
$$ABA = A^2 \text{ and } BAB = B^2,$$
then $A \geq 0$ and $B \geq 0$.

REMARK. Question (3) of the previous exercise is important for the solvability of the equations $FF^* = A$ and $F^*F = B$, where F is an idempotent operator. See **[209]**.

Exercise 11.3.13. Let $A, B \in B(H)$ be self-adjoint.

(1) Prove that
$$AB = BA \Longleftrightarrow e^A e^B = e^{A+B}.$$
(2) Prove that
$$AB = BA \Longleftrightarrow e^A e^B = e^B e^A.$$
(3) Show that
$$e^A = e^B \Longrightarrow AB = BA.$$
(4) (**[38]**) Generalize the result of Question (2) to the case of three operators as follows: Assume further that $C \in B(H)$ is self-adjoint. If
$$\begin{cases} \cosh A e^B = e^B \cosh A, \\ \sinh A e^C = e^B \sinh A, \\ e^C \cosh A = \cosh A e^C, \\ e^C \sinh A = \sinh A e^B, \end{cases}$$
then infer that
$$AC = BA.$$

Exercise 11.3.14. (see also Exercise 11.3.17) Let $A \in B(H)$ be self-adjoint such that $e^A = I$, where I designates the identity operator on H. Show that $A = 0$.

Exercise 11.3.15. Let $A, B \in B(H)$.

(1) Show that if A and B are both self-adjoint, then
$$e^A = e^B \Longleftrightarrow A = B.$$
(2) Show that if A is self-adjoint, then e^{iA} is unitary.
(3) If e^{iA} is unitary, then does it follow that A is self-adjoint?
(4) Show that if A is normal and e^{iA} is unitary then A is self-adjoint.
(5) Show that if A is normal, then
$$e^A \text{ is unitary} \Longleftrightarrow A = -A^*.$$
(6) If A is self-adjoint and $e^{iA} = e^B$, then show that $B^* = -B$ whenever B is normal.

Exercise 11.3.16. Let $A, B \in B(H)$ be both positive and such that $0 \notin \sigma(A) \cap \sigma(B)$. Show that
$$AB = BA \Longrightarrow \log(AB) = \log A + \log B.$$
Infer that
$$\log(A^{-1}) = -\log A.$$

Exercise 11.3.17. Let $A \in B(H)$ be self-adjoint and such that $A \geq I$. Let $B \in B(H)$ be self-adjoint.

(1) Show that
$$\log \|A\| = \|\log A\|.$$

REMARK. See Exercise 11.3.18 for a generalization.

(2) Deduce that
$$\|A\| = 1 \Longrightarrow A = I.$$
(3) Infer also that
$$e^B = I \Longrightarrow B = 0.$$

Exercise 11.3.18. Let $A \in B(H)$ be positive and consider an increasing homeomorphism $f : \sigma(A) \subset \mathbb{R}^+ \to \mathbb{R}^+$. Show that
$$f(\|A\|) = \|f(A)\|.$$

Exercise 11.3.19. Let X be a non empty set and let Σ be the sigma algebra of subsets of X. Let H be a Hilbert space. Let $x, y \in H$. Show that

$$\mu_{x,y}(\Delta) =< E(\Delta)x, y >$$

defines a countably additive measure on Σ.

Exercise 11.3.20. Let X be a compact set and let Σ be the sigma algebra of Borel subsets of X. Set $H = L^2(\mu)$ where μ is a measure on Σ. If $\Delta \in \Sigma$, let

$$E(\Delta)f(x) = \mathbb{1}_\Delta(x)f(x)$$

for any $f \in H$. Show that E is a spectral measure.

Exercise 11.3.21. Let $A \in B(H)$ be normal with a corresponding spectral measure E. Check that

$$AE(\Delta) = E(\Delta)A$$

(for any Borel set Δ) and that $\operatorname{ran}(E(\Delta))$ reduces A.

Exercise 11.3.22. ([**124**]) Show that any normal operator has nontrivial reducing subspaces. In fact, any non scalar normal operator has a nontrivial subspace which reduces every operator which commutes with it.

Exercise 11.3.23. ([**124**]) Let $A \in B(H)$ be satisfying

$$[A^*, A]A = (A^*A - AA^*)A = 0$$

(such an operator is named **quasinormal**). Show that A has a nontrivial invariant subspace.

Exercise 11.3.24. Let A be a normal operator with E as a corresponding spectral measure. Using Theorem 11.1.56, show that

$$AB = BA \Longleftrightarrow A^*B = BA^*.$$

Exercise 11.3.25. Let $A \in B(H)$ be self-adjoint such that $\sigma(A) \subset \mathbb{R}^+$. Show that $A \geq 0$.

Exercise 11.3.26. Let $A \in B(H)$ be normal. Show that $A = UP$, where P is positive and U is unitary. Moreover, all of A, P and U are pairwise commuting. What happens if A self-adjoint?

Exercise 11.3.27. Let $A \in B(H)$ be invertible.

(1) Show that there are self-adjoint operators B and C such that

$$A = e^{iC}e^B.$$

(2) Assume here that A is also normal. Show that $A = e^N$ for some normal $N \in B(H)$.

Exercise 11.3.28. Let $A \in B(H)$ be normal. Show that:

(1) A is self-adjoint iff $\sigma(A)$ is real.
(2) A is unitary iff $\sigma(A) \subset \{\lambda \in \mathbb{C} : |\lambda| = 1\}$.
(3) If $\alpha \in \mathbb{R}^+$, then $A \geq \alpha I$ iff $\sigma(A) \subset [\alpha, \infty)$. In particular, A is positive iff $\sigma(A)$ is positive.
(4) A is an orthogonal projection iff $\sigma(A) \subset \{0, 1\}$.

Exercise 11.3.29. Let A and B be two square matrices of order n.

(1) Show that

$$\|AB - BA\|_F^2 - \|AB^* - B^*A\|_F^2 = -\text{tr}[(A^*A - AA^*)(B^*B - BB^*)].$$

where $\|\cdot\|_F$ is the Frobenius Norm (introduced in Exercise 3.3.5).

(2) Infer that if either A or B is normal, then

$$AB = BA \Longleftrightarrow AB^* = B^*A.$$

Exercise 11.3.30. Let T, A, B be all in $B(H)$ such that both A and B are normal. Assume that $TA = BT$.

(1) Show that $Tp(A) = p(B)T$ where $p(z)$ is a complex polynomial in z and of degree n.
(2) Deduce that

$$Te^{i\overline{z}A} = e^{i\overline{z}B}T, \ \forall z \in \mathbb{C}.$$

(3) Set $f(z) = e^{-izB^*}Te^{izA^*}$. Show that f is bounded. What can you deduce?
(4) Infer that $TA^* = B^*T$.
(5) Show that if only T is normal (and $A, B \in B(H)$), then

$$TA = BT \not\Longleftrightarrow T^*A = BT^*.$$

Exercise 11.3.31. Via a matrix of operators trick, show that the Fuglede Theorem is actually equivalent to the Fuglede-Putnam Theorem.

Exercise 11.3.32.

(1) Let M and N be two normal operators, and let A and B be self-adjoint such that $AN = MB$. Does it follow that $AN^* = M^*B$?
(2) Let M, N, A and B be all *unitary* operators such that A and B are also self-adjoint such that $AN = MB$. Does it follow that $AN^* = M^*B$?
(3) Let M be an isometry and N be a co-isometry. Let A and B be such that $AN = MB$. Show that $BN^* = M^*A$.

(4) Give an example showing that the hypotheses M being an isometry and N being a co-isometry cannot be dropped.

Exercise 11.3.33. Let $A, B \in B(H)$ be self-adjoint such that AB is normal. Prove that

$$|AB| = |A||B|.$$

Show by examples that it may happen that $|AB| \neq |A||B|$ if (A and B are self-adjoint and AB is not normal) or if (A and B are not self-adjoint and AB is normal).

Exercise 11.3.34. (cf. Exercise 11.3.35) Show that if $A, B \in B(H)$ are self-adjoint, one of them is positive, such that AB is normal, then AB is self-adjoint.

Exercise 11.3.35. (cf. Exercise 13.2.11) Let $A, B \in B(H)$ be self-adjoint where one of them is positive. Using the Fuglede-Putnam Theorem show that AB is self-adjoint whenever AB is normal.

Exercise 11.3.36. Show that two similar normal operators are automatically unitarily equivalent.

Exercise 11.3.37. Let H be a complex Hilbert space. Let $\mathcal{N} \subset B(H)$ be constituted only of normal operators. Show that

$$\mathcal{N} \text{ is a vector space } \iff AB = BA,\ \forall A, B \in \mathcal{N}.$$

Exercise 11.3.38. ([**117**]) Let $B, N \in B(H)$ be commuting such that B is normal and $N^2 = 0$. Set $A = B + N$.

(1) Prove that if A is invertible and A^2 is normal, then A is normal.
(2) Show that the hypothesis A invertible cannot just be dispensed with.

Exercise 11.3.39. ([**113**]) Let A and B be two bounded operators on a Hilbert space such that AB and A are normal. Then B commutes with AA^* iff BA is normal.

Exercise 11.3.40 (Commutativity Up To A Factor). In all of this exercise: $A, B \in B(H)$, where H is a $\mathbb{C}$-Hilbert space, such that $AB = \lambda BA \neq 0$, where $\lambda \in \mathbb{C}$.

(1) Show that if A and B are both normal, then AB is normal for any λ.
(2) Deduce that $\lambda = 1$ if A and B are self-adjoint and one of them is positive.
(3) Deduce also that if both A and B are self-adjoint, then $\lambda \in \{-1, 1\}$
(4) Finally, show that if both A and B are normal, then $|\lambda| = 1$.

Remark. Commutation relations between self-adjoint operators in a complex Hilbert space are important in the interpretation of quantum mechanical observables. They also play an important role when analyzing their spectra. For more details see **[35]** (and the references therein), **[51]** and **[169]**. The problem of the previous exercise is of this sort.

Exercise 11.3.41. Prove Theorem 11.1.63.

11.4. Tests

Test 11.4.1. Let $A \in B(H)$ be self-adjoint and let $f : \sigma(A) \to \mathbb{C}$ be continuous. Show that

(1) $f(A)$ is self-adjoint iff f is real-valued.
(2) $f(A) \geq 0$ iff $f \geq 0$.

Test 11.4.2. (cf. Exercise 10.5.5) Let A, B be two unbounded operators such that $AB = BA$. Assume that B is invertible (with $B^{-1} \in B(H)$) and that $B^{-1} \geq 0$. Does it follow that $B^{-\frac{1}{2}}A \subset AB^{-\frac{1}{2}}$?

Test 11.4.3 (**[180]**). Let $T = A + iB$ be the Cartesian form of T (recall that A and B are self-adjoint). Show that if AB is normal and either A or B is positive, then T is normal.

Test 11.4.4. Assume that A and B are two self-adjoint operators such that $B \geq 0$.

(1) Show that if AB is an isometry, then it is self-adjoint.
(2) What can you deduce from the previous result?

Test 11.4.5. Show that the assumption of self-adjointness on A and B in Question (3) of Exercise 11.3.13 is important.

Test 11.4.6. Prove Proposition 11.1.48.

Test 11.4.7. Is the norm of any bounded self-adjoint operator always equal to the norm of some multiplication operator?

Test 11.4.8. (cf. **[29]**) Let $A, B \in B(H)$ be such that $A + B$ is invertible. If $p, q \in (0, \infty)$, then show that $|A|^p + |B|^q$ is invertible.

Test 11.4.9. Show that the group G of invertible operators on $B(H)$ is connected.

Test 11.4.10. Say why an isometry has a nontrivial invariant subspace.

Test 11.4.11. Let $A \in B(H)$ be normal and invertible. Using the Spectral Theorem, show that

$$|A^{-1}| = |A|^{-1}.$$

11.5. More Exercises

Exercise 11.5.1. Let $A \in B(H)$ be positive and invertible and let $\alpha > 0$. Does it follow that A^α is invertible? If yes, then does it follow that

$$\log(A^\alpha) = \alpha \log A?$$

Exercise 11.5.2. Let $A \in B(H)$ be such that $\|A - I\| < 1$. Show that the series

$$\sum_{n=1}^{\infty} \frac{(-1)^{n-1}}{n}(A - I)^n$$

converges in $B(H)$. As in the real case, we shall denote this series by $\log A$ (also called **logarithm** of an operator). If A is positive and $\|A - I\| < 1$, then does this definition of $\log(A)$ coincide with the one obtained by the functional calculus?

Exercise 11.5.3. Let $A \in B(H)$ be normal. Using the functional calculus, can we define $\log(A)$? If this is not always possible, when can we do that?

Exercise 11.5.4. Let $A \in B(H)$ be normal and such that $A^5 = A^4$. Does it follow that A is an orthogonal projection?

Exercise 11.5.5. ([**193**]) Let $A, B \in B(H)$ be such that $r(B) < 2\pi$ and A is positive and invertible. Assume that $e^B = A$. Show that $\sigma(B) \subset \mathbb{R}$.

REMARK. In fact, calling on a result by Hille [**103**], we may even show that B is self-adjoint.

Exercise 11.5.6. Let $A, B \in B(H)$ be such that $A \leq B$. Does it follow that $e^A \leq e^B$? **Hint:** According to Exercise 11.5.7, you need to avoid commuting A and B in case you think the answer is no.

Exercise 11.5.7. ([**48**]) Let $A, B \in B(H)$ be such that $AB = BA$ and $A \leq B$. Show that $f(A) \leq f(B)$ for each increasing and continuous function f on $\mathbb{R}$.

Exercise 11.5.8. ([**98**]) Let $A, B \in B(H)$ be such that $B \geq 0$ and $\|A\| \leq 1$. Assume that f be an operator monotone function defined on $\mathbb{R}^+$. Show that

$$A^* f(B) A \leq f(A^* B A) \text{ (Hansen Inequality)}.$$

REMARK. See [**99**] for the reverse inequality under different assumptions.

Exercise 11.5.9. (cf. [**84**]) Let $A \in B(H)$ be positive. Find the solution of the differential equation

$$\frac{d^2y}{dt^2} + Ay = 0$$

with boundary conditions $y(0) = y_0$ and $y'(0) = x_0$.

Exercise 11.5.10. ([**208**], cf. Exercise 5.3.5) Let $A, B \in B(H)$ be positive. Prove that $AB = BA$ if and only if $AB^n + B^nA \geq 0$ for all $n \in \mathbb{N}$.

Exercise 11.5.11 ([**145**]).

(1) Assume that N is unitary and that A and B are two bounded operators. Check that

$$NA = BN \Longrightarrow NA^* = B^*N.$$

(2) Let A and B be two bounded operators. Suppose N is a contraction such that

$$(1-N^*N)^{\frac{1}{2}}A = B(1-NN^*)^{\frac{1}{2}} = (1-N^*N)^{\frac{1}{2}}A^* = B^*(1-NN^*)^{\frac{1}{2}} = 0.$$

Prove that

$$NA = BN \Longrightarrow NA^* = B^*N.$$

Hint: Use the Julia Operator.

Exercise 11.5.12. Let H be an *infinite* dimensional space. Let $A \in B(H)$ be normal and *having only real eigenvalues.* Does it follow that A is self-adjoint?

Exercise 11.5.13. ([**190**]) Let $A \in B(H)$ be normal and compact.

(1) Show that A has an eigenvalue λ such that $|\lambda| = \|A\|$.
(2) Show that $f(A)$ is compact if f is continuous on $\sigma(A)$ and $f(0) = 0$.
(3) Also, show that $f(A)$ is not compact if f is continuous on $\sigma(A)$ and $f(0) \neq 0$, and $\dim H = \infty$.

Exercise 11.5.14. Let $A \in B(H)$ be normal and let E be its associated spectral measure. If $\lambda \in \sigma(A)$, then show that

(1) $\lambda \in \sigma_p(A)$ iff $E(\{\lambda\}) \neq 0$.
(2) If $\lambda \in \sigma_p(A)$, then $E(\{\lambda\})$ is the orthogonal projection of H onto the eigenspace $\ker(A - \lambda I)$.
(3) If λ is isolated, then $\lambda \in \sigma_p(A)$.

Exercise 11.5.15. Let A be a normal operator with a purely imaginary spectrum. Does it follow that A is skew-symmetric?

Exercise 11.5.16. ([**190**], **An Ergodic Theorem**) Let $U \in B(H)$ be unitary and let $x \in H$. Define a sequence (A_n) in $B(H)$ by

$$A_n x = \frac{1}{n}(x + Ux + \cdots + U^{n-1}x).$$

Show that (A_n) converges strongly. **Hint:** Use the Spectral Theorem for normal operators.

Exercise 11.5.17. ([**79**]) Let $A \in B(H)$. Show that

$$| < Ax, y > |^2 \leq < |A|^{2\alpha}x, x > < |A^*|^{2(1-\alpha)}y, y >$$

for all $x, y \in H$ and for any $\alpha \in [0, 1]$.

REMARK. This is also a **Generalized Cauchy-Schwarz Inequality**.

Exercise 11.5.18. Let $A, B \in B(H)$, where H is a $\mathbb{C}$-Hilbert space, such that $AB = \lambda BA \neq 0$, where $\lambda \in \mathbb{C}$.

(1) Show that if A or B is self-adjoint, then $\lambda \in \mathbb{R}$.
(2) Prove that if A or B is self-adjoint and the other is normal, then $\lambda \in \{-1, 1\}$.

Exercise 11.5.19. (cf. [**28**]) Let $A, B, X, Y \in B(H)$. Assume that the spectra of A and B are contained in the open right half plane and the open left half plane, respectively. Show that the solution of $AX - XB = Y$ may be expressed as

$$X = \int_0^\infty e^{-tA} Y e^{tB} dt.$$

CHAPTER 12

Hyponormal Operators

12.1. Basics

DEFINITION 12.1.1. *Let $A \in B(H)$. We say that the operator A is **hyponormal** if*

$$\forall x \in H: \ \|A^*x\| \leq \|Ax\|.$$

EXAMPLES 12.1.2.

(1) *Obviously, every normal operator is hyponormal.*
(2) *If A is hyponormal, then αA is hyponormal for all $\alpha \in \mathbb{C}$.*
(3) *The shift operator is hyponormal (cf. Exercise 5.3.2). The shift is the prominent example of a hyponormal operator which is not normal.*
(4) *More generally, every isometry is hyponormal (see Exercise 12.3.3).*
(5) *The Volterra Operator is not hyponormal (see Test 12.4.3).*
(6) *If A is hyponormal and B is an isometry, then BAB^* is hyponormal.*
(7) *If A is an invertible hyponormal operator, then A^{-1} is hyponormal (see Exercise 12.3.2 or Exercise 12.3.17 or Test 12.4.1).*

THEOREM 12.1.3. *(see Exercise 12.3.1) Let $A \in B(H)$. Then*

$$A \text{ is hyponormal} \iff AA^* \leq A^*A.$$

REMARK. Putnam in [**169**] used another (more general) terminology, namely an operator such that $AA^* \leq A^*A$ or $AA^* \geq A^*A$ is called **semi-normal**.

A practical characterization of hyponormal operators is the following (a proof may be found in Exercise 12.3.2):

PROPOSITION 12.1.4. *Let $A \in B(H)$. Then A is hyponormal iff there exists a contraction $K \in B(H)$ such that $A^* = KA$.*

REMARK. In [**61**], an interesting result (outlined in Exercise 5.5.6) includes the previous proposition as a particular case.

The Reid Inequality has yet another improvement (a proof can be found in Exercise 12.3.14):

THEOREM 12.1.5. *(cf. Exercise 12.5.8) Let $A, K \in B(H)$ be such that A is positive and $(AK)^*$ is hyponormal. Then*

$$|< AKx, x >| \leq ||K|| < Ax, x >$$

for all $x \in H$.

The proof of the previous theorem relies on the following:

PROPOSITION 12.1.6. *(**Kittaneh**, [**118**]) Let $A \in B(H)$ be hyponormal. Then*

$$|< Ax, x >| \leq < |A|x, x > .$$

Hyponormal operators have been well developed in the sixties and seventies of last century. Now, we gather what we see as the important facts.

First, we saw in Theorem 8.1.23 that the closure of the numerical range of a normal operator equals the convex hull of its spectrum. This holds for hyponormal operator too (first conjectured in [**22**]).

THEOREM 12.1.7. *([**192**] or [**202**]) Let $A \in B(H)$ be hyponormal. Then*

$$\overline{W(A)} = \operatorname{conv}(\sigma(A)).$$

We saw that a hyponormal operator need not be normal. Thus, the following questions are natural: When is a hyponormal operator normal? When is a hyponormal operator self-adjoint? When is a hyponormal operator unitary?

It turns out that we can get back many of similar results on normal operators by using the weaker assumption of hyponormality. This is certainly a plus to the theory of hyponormal operators.

THEOREM 12.1.8. *([**6**]) Every compact hyponormal operator is normal.*

THEOREM 12.1.9. *Let $A \in B(H)$ be hyponormal and such that $\sigma(A)$ is a smooth arc. Then A is normal.*

COROLLARY 12.1.10. *Let $A \in B(H)$ be hyponormal such that $\sigma(A) \subset \mathbb{R}$. Then A is self-adjoint.*

COROLLARY 12.1.11. *Let $A \in B(H)$ be hyponormal and such that $\sigma(A)$ lies on the unit circle. Then A is unitary.*

REMARK. The previous three results have all appeared in [**202**].

Now, we have a result linking the spectrum and the approximate point spectrum of the adjoint of a hyponormal operator. A proof is given in Exercise 12.3.18.

PROPOSITION 12.1.12. *([**22**]) Let $A \in B(H)$ be hyponormal. Then*

$$\sigma(A^*) = \sigma_a(A^*).$$

Another natural question is the following: Are hyponormal operators normaloid?

PROPOSITION 12.1.13. *(see Exercise 12.3.6) Hyponormal operators are normaloid.*

We finish with a generalization of the Fuglede-Putnam Theorem to hyponormal operators (cf. [**26**]):

THEOREM 12.1.14. *([**140**]) Let $A, B, T \in B(H)$ such that A^* and B are hyponormal. If $TA = BT$, then $TA^* = B^*T$.*

REMARK. The previous theorem need not hold if A and B are both hyponormal, even if they are equal. See Exercise 12.3.21.

12.2. True or False: Questions

QUESTIONS. Comment on the following questions/statements and indicate those which are false and those which are true when this applies. Justify your answers.

(1) Some of the definitions as self-adjoint or normal operators have already been met in an Advanced Linear Algebra Course. But hyponormal operators have not been met before. Why is this?
(2) Let A be a hyponormal operator. Then so is its adjoint.
(3) Let A be a hyponormal operator. Then so is its square.
(4) Can we expect to have a spectral theorem for hyponormal operators.
(5) If $A \in B(H)$ is hyponormal, then $-A^*$ too is hyponormal.
(6) We have seen that a left (or right) invertible normal operator is invertible (Exercise 6.3.26). This result also holds for hyponormal operators.
(7) We have seen (Exercise 7.3.20) that if $A, B \in B(H)$ and A (say) is normal, then $\sigma(AB) = \sigma(BA)$. The previous equality also holds if we replace normal by hyponormal.
(8) Let $A \in B(H)$ be hyponormal. Then A has a square root.
(9) Let $A, B \in B(H)$ be hyponormal. Then A and B are commuting in norm, i.e.

$$\|AB\| = \|BA\|.$$

(10) Let $A \in B(H)$. Then

$$A \text{ is hyponormal } \iff |A^*| \leq |A|.$$

12.3. Exercises with Solutions

Exercise 12.3.1. Let $A \in B(H)$.

(1) Show that

$$A \text{ is hyponormal iff } AA^* \leq A^*A.$$

(2) Apart from the shift, give an example of a hyponormal operator which is not normal.
(3) Show that if $\dim H < \infty$, then every hyponormal operator is normal.
(4) If A is hyponormal, then does it follow that A^2 is hyponormal?

Exercise 12.3.2. Let $A \in B(H)$.

(1) Show that

$$A \text{ is hyponormal } \iff \exists K \in B(H) \text{ contraction } : A^* = KA.$$

(2) Infer a simple proof that if A is invertible and hyponormal, then its inverse A^{-1} too is invertible.

Exercise 12.3.3.

(1) Show that αA is hyponormal whenever A is hyponormal and whichever $\alpha \in \mathbb{C}$.
(2) Show that every isometry is hyponormal.
(3) Show that if $A \in B(H)$ is hyponormal and invertible such that both A and A^{-1} are contractions, then A is normal.

Exercise 12.3.4. Let $A, B \in B(H)$ be unitarily equivalent. Show that A is hyponormal iff B is hyponormal.

Exercise 12.3.5. Show that the set of hyponormal operators on a Hilbert space is closed with respect to the strong operator topology.

Exercise 12.3.6 ([**6**], cf. Exercises 8.3.2 & 8.3.3). Let A be a hyponormal operator on H.

(1) Prove that

$$\|A^n\| = \|A\|^n, \ \forall n \in \mathbb{N}.$$

(2) What can you deduce from the previous result?

Exercise 12.3.7. Let $A, B \in B(H)$ be two hyponormal operators such that $AB^* = B^*A$. Show that AB is hyponormal.

Exercise 12.3.8. Let $A, B \in B(H)$.

(1) Show that if A^* is hyponormal, then $\|AB\| \leq \|A^*B\|$.
(2) Show that if A is hyponormal, then $\|A^*B\| \leq \|AB\|$. In particular,
$$\|A^2\| = \|A\|^2.$$
(3) Show that if B is hyponormal, then $\|A^*B\| \leq \|BA\|$.
(4) Show that if B^* is hyponormal, then $\|A^*B^*\| \leq \|A^*B\|$.
(5) Deduce that if A and B^* are both hyponormal, then
$$\|BA\| \leq \|AB\|$$
(hence if $AB = 0$, then $BA = 0$).
(6) Infer that if A^* and B are both hyponormal, then
$$\|AB\| \leq \|BA\|.$$

REMARKS.
(a) We could obtain more inequalities using the same method and/or the standard equality $\|T\| = \|T^*\|$ where $T \in B(H)$.
(b) Observe that if A, A^*, B and B^* are all hyponormal, that is, if A and B are normal, we obtain
$$\|AB\| = \|BA\|$$
(which is the result of Exercise 4.3.13).

Exercise 12.3.9 (**[19]**). *The aim of this exercise is to investigate to what extent the Kaplansky Theorem may be generalized to hyponormal operators.* Let A and B be both in $B(H)$ where H is a $\mathbb{C}$-Hilbert space. Assume that A is normal.
(1) Suppose that AB is hyponormal. Show that
$$AA^*B = BAA^* \Longrightarrow BA \text{ is hyponormal.}$$
(2) Show, by an example, that the reverse implication does not hold in the previous result.
(3) Show that if AB is hyponormal and that BA is co-hyponormal, then
$$AA^*B = BA^*A \Longleftrightarrow BA \text{ and } AB \text{ are normal.}$$

Exercise 12.3.10. Let $A \in B(H)$ be hyponormal.
(1) Show that for all $x \in H$, we have
$$\|Ax\|^2 \leq \|A^2x\|\|x\|.$$
REMARK. This shows that every hyponormal operator is paranormal, cf. Exercise 10.5.8.
(2) Deduce that if A^2 is compact, then so is A.

Exercise 12.3.11. Let $A \in B(H)$ be hyponormal. Show that
$$|\text{Re}A| \leq |A| \text{ and } |\operatorname{Im} A| \leq |A|.$$

Exercise 12.3.12. (cf. Exercise 6.5.10) Let $A \in B(H)$.

(1) Show that A is hyponormal iff
$$|A|^2 \geq (\text{Re}A)^2 + (\text{Im}A)^2.$$
(2) Show that if a hyponormal operator A satisfies $|A| \leq |\text{Re}A|$, then A is self-adjoint.

Exercise 12.3.13. Let $A \in B(H)$ be hyponormal.

(1) Is $A + \lambda A^*$ hyponormal for any $\lambda \in \mathbb{C}$?
(2) Show that if $|\lambda| \leq 1$, then $A + \lambda A^*$ hyponormal.

REMARK. The second question of the previous exercise is even true for unbounded hyponormal operators. This appeared in [**111**].

Exercise 12.3.14. Show that the Reid Inequality *does hold* if $(AK)^*$ is hyponormal and $A \geq 0$.

Exercise 12.3.15. Let $A \in B(H)$ be hyponormal and let λ be a complex number.

(1) Show that hyponormality is preserved under translation, that is, $A - \lambda I$ is hyponormal.
(2) Show that if $\lambda \in \sigma_p(A)$, then $\overline{\lambda} \in \sigma_p(A^*)$.
(3) Show that if λ and μ are two distinct eigenvalues for A with corresponding eigenvectors x and y respectively, then we must have $< x, y >= 0$.
(4) What is $\sigma_r(A^*)$?

Exercise 12.3.16. Let $A \in B(H)$ be hyponormal. Show that A is invertible iff A is right invertible.

Exercise 12.3.17. ([**202**]) Let $A \in B(H)$ be hyponormal and let $\lambda \in \rho(A)$.

(1) Prove that $(A - \lambda I)^{-1}$ is hyponormal.
(2) Let $x \in H$ be such that $\|x\| = 1$. Show that
$$\|(A - \lambda I)^{-1}x\| \leq \frac{1}{d(\lambda, \sigma(A))}$$
where $d(\lambda, \sigma(A)) = \inf\{|\lambda - \mu| : \mu \in \sigma(A)\}$.

Exercise 12.3.18. Let $A \in B(H)$ be hyponormal. Show that
$$\sigma(A^*) = \sigma_a(A^*).$$

Exercise 12.3.19. ([**75**]) Let $A \in B(H)$.

(1) Show that A^* is hyponormal iff $|A|^2 \leq (\operatorname{Re} A)^2 + (\operatorname{Im} A)^2$.
(2) Show that if there exists an $0 < \alpha < 1$ such that

$$|A|^2 \leq (\operatorname{Re} A)^2 + \alpha(\operatorname{Im} A)^2,$$

then A is self-adjoint.
(3) Infer that

$$A \text{ is self-adjoint } \Longleftrightarrow |A|^2 \leq (\operatorname{Re} A)^2.$$

Exercise 12.3.20. ([**116**]) Let $T = A+iB \in B(H)$ be hyponormal, where A and B are self-adjoint. Prove that if AB is hyponormal, then T is normal.

Exercise 12.3.21. Show that if $A, B, T \in B(H)$ are such that $TA = BT$ and A, B are *hyponormal*, then it *is not necessary* that $TA^* = B^*T$.

Exercise 12.3.22. Let $A \in B(H)$ be hyponormal. Consider

$$M = \{x \in H : \ \|Ax\| = \|A\|\|x\|\}.$$

(1) Show that M is a closed subspace of H.
(2) Show that M is an invariant subspace for A (notice that M may be trivial).
(3) Assume that A is normal. Show that M is reducing for A.

12.4. Tests

Test 12.4.1. We have already seen two different proofs of the inverse of an invertible hyponormal operator being hyponormal. Give a third proof!

Test 12.4.2. ([**155**]) Let $A, B \in B(H)$ be such that $AB = BA$. Show that if A is normal and B is hyponormal, then the following triangle inequality holds:

$$|A+B| \leq |A| + |B|.$$

Test 12.4.3. Is the Volterra Operator hyponormal?

Test 12.4.4. Show that if A and B are two hyponormal such that A commutes with B^*, then $A+B$ is hyponormal.

Remark. The interested reader is referred to [**1**] where a counterexample is provided showing that the sum of two commuting hyponormal operators need not be hyponormal.

Test 12.4.5. Let $A, B \in B(H)$ be self-adjoint, one of them is positive, such that AB is hyponormal. Does it follow that AB is self-adjoint?

12.5. More Exercises

Exercise 12.5.1.

(1) Show that every compact and hyponormal operator is automatically normal.
(2) Deduce that the Volterra operator cannot be hyponormal (cf. Test 12.4.3).

Exercise 12.5.2. ([**44**], you may use Exercise 12.3.8) Let $A, B \in B(H)$, where H is a $\mathbb{C}$-Hilbert space, such that $AB = \lambda BA \neq 0$, $\lambda \in \mathbb{C}$.

(1) Show that if A^* and B are both hyponormal, then $|\lambda| \leq 1$.
(2) Show that if A and B^* are both hyponormal, then $|\lambda| \geq 1$.

Exercise 12.5.3. (cf. Exercise 11.3.35, [**166**]) Let A and B be two hyponormal operators. Assume that A commutes with $|B|$ and that B commutes with $|A^*|$.

(1) Show that both AB and BA are hyponormal.
(2) Give an example which shows that the hypothesis "B commutes with $|A^*|$" cannot just be dispensed with.
(3) Deduce that if A and B are normal, then each of A and B commutes with the absolute value of the other *if and only if both* AB and BA are normal. **Hint:** You may need to use on Exercise 11.3.39.

Exercise 12.5.4. ([**155**]) Let $A, B \in B(H)$ be such that $AB = BA$. Assume that A is normal and B is hyponormal. Show that the following inequalities hold:

$$||A| - |B|| \leq |A \pm B|$$

and

$$\||A| - |B|\| \leq \|A \pm B\|.$$

Exercise 12.5.5. Let $n \in \mathbb{N}$. If $A \in B(H)$ is hyponormal, then does it follow that

$$|A^n| = |A|^n?$$

Exercise 12.5.6. Let $A \in B(H)$ be hyponormal such that A^n is normal ($n \in \mathbb{N}$). Show that A is normal.

Exercise 12.5.7. ([**30**])

(1) Show that an idempotent operator which is also hyponormal is an orthogonal projection.
(2) Deduce that tripotent hyponormal operators are self-adjoint.

Exercise 12.5.8. Does the Reid Inequality hold in case AK is hyponormal instead of normal and by keeping A positive?

CHAPTER 13

Similarities of Operators

13.1. Important Theorems

In this chapter, we give some theorems on similarity of operators.

13.1.1. The Berberian Theorem. First, we have a definition.

DEFINITION 13.1.1. *A unitary operator $U \in B(H)$ is said to be* ***cramped*** *if its spectrum is completely contained on some open semicircle (of the unit circle), that is*

$$\sigma(U) \subseteq \{e^{it} : \alpha < t < \alpha + \pi\}.$$

THEOREM 13.1.2 ([**24**]). *Let A be a bounded operator for which $0 \notin \overline{W(A)}$. Then A is invertible and the unitary operator $A(A^*A)^{-\frac{1}{2}}$ is cramped.*

An operator (even normal) which is unitarily equivalent to its adjoint need not be self-adjoint (counterexamples can be found e.g. in the setting of 2×2 matrices). Berberian, however, established the following remarkable result on (cramped) unitary equivalence and self-adjointness:

THEOREM 13.1.3. *(**Berberian**, [**21**]) Let U be a cramped operator and let A be a bounded operator such that $UAU^* = A^*$. Then A is self-adjoint.*

REMARK. In Exercise 13.2.14, we give a different proof from Berberian's. In fact, this proof was initially used to prove that this result holds even in case of an unbounded and closed A (see [**54**]).

13.1.2. The Williams Theorem. Here is another result on similarity (it has also been generalized to unbounded operators in [**54**]).

THEOREM 13.1.4. *(**Williams**, [**213**]) Let $A, B \in B(H)$ such that $A^{-1}BA = B^*$, where $0 \notin \overline{W(A)}$. Assume further that B is hyponormal. Then B is self-adjoint.*

After [**213**] quite a few papers on similarity had followed. Here is one of these results.

Theorem 13.1.5. *([**200**]) If $A \in B(H)$ is invertible and such that $A^* = U^*A^{-1}U$, where U is a cramped unitary operator, then A is unitary.*

13.1.3. The Embry Theorem. We finish this chapter by an important result due to Embry in [**68**]. It is powerful enough to be compared with the Fuglede-Putnam Theorem in some cases. See [**149**].

Theorem 13.1.6. *(**Embry Theorem**) Let $A, B \in B(H)$ be two commuting normal operators. Let $T \in B(H)$ be such that $0 \notin W(T)$. If $TA = BT$, then $A = B$.*

Remark. The proof of the previous result is fairly easy. It is mainly based on the Spectral Theorem as well as Putnam's version of the Fuglede-Putnam Theorem. Can you provide an algebraic proof of it? I.e. a proof which is not based on the Spectral Theorem?

Remark. The Embry, Berberian and Williams Theorems all generalize a result by Beck-Putnam in [**17**].

Corollary 13.1.7. *Let $A, B \in B(H)$ be two commuting normal operators. Let $T > 0$ be in $B(H)$ such that $TA = BT$. Then $A = B$.*

There are many interesting consequences of the Embry Theorem:

Corollary 13.1.8. *Let $A \in B(H)$ be such that $0 \notin W(A)$. If AA^* and A^*A commute, then A is normal.*

We already met an example of a non-normal A such that A^2 is normal (where?). However, we have:

Corollary 13.1.9. *Let $A \in B(H)$ be such that $0 \notin W(A)$. If A^2 is normal, then A too is normal.*

Corollary 13.1.10. *Let $A \in B(H)$ be such that $0 \notin W(A)$. Assume that $B \in B(H)$. If $AB = -BA$, then $B = 0$, whenever A or B is normal.*

Remark. Embry [**68**] gave similar results using the asymmetric condition $\sigma(A) \cap \sigma(-A) = \varnothing$ instead of $0 \notin W(A)$.

13.2. Exercises with Solutions

Exercise 13.2.1. ([**213**]) Let $A, B \in B(H)$ be such that $A^{-1}BA = B^*$, where $0 \notin \overline{W(A)}$. Assume further that B is hyponormal.

(1) Show that $\sigma(B) \subset \mathbb{R}$.
(2) Infer that B is self-adjoint.

Exercise 13.2.2. ([**148**]) Let N be a normal operator with Cartesian Decomposition $A+iB$ and such that $\sigma(B) \subset (0,\pi)$. Let S be a self-adjoint operator. Assume further that $e^Se^N = e^Ne^S$.

(1) Show that $e^Se^Ae^{iB} = e^{iB}(e^Se^A)^*$.
(2) Deduce that $AS = SA$.
(3) Infer that $SN = NS$.

Exercise 13.2.3. Let N and M be two normal operators with Cartesian Decompositions $A+iB$ and $C+iD$ respectively. Assume that $e^Me^N = e^Ne^M$ and that $\sigma(B), \sigma(D) \subset (0,\pi)$.

(1) Using Exercise 13.2.2 and the Fuglede Theorem, show that $CN = NC$.
(2) Show that
$$AC = CA,\ BC = CB,\ AM = MA \text{ and } AD = DA.$$
(3) Prove that $BM = MB$.
(4) Deduce that $NM = MN$.

Exercise 13.2.4. Let
$$A = \begin{pmatrix} 0 & \pi \\ -\pi & 0 \end{pmatrix} \text{ and } B = \begin{pmatrix} \pi & -2\pi \\ \pi & -\pi \end{pmatrix}.$$

(1) Check that A is normal and that $AB \neq BA$.
(2) Find $\operatorname{Im} A$ and $\sigma(\operatorname{Im} A)$.
(3) Find e^A and e^B.

Exercise 13.2.5. Let A be in $B(H)$. Let $N \in B(H)$ be normal such that $\sigma(\operatorname{Im} N) \subset (0,\pi)$ and $Ae^N = e^NA$.

(1) Show that $(\operatorname{Re}A)N = N(\operatorname{Re}A)$.
(2) Show that $(\operatorname{Im}A)N = N(\operatorname{Im}A)$ (so that $AN = NA$).
(3) Give an example showing that the assumption $\sigma(\operatorname{Im} N) \subset (0,\pi)$ may not just be dropped.

Exercise 13.2.6. Let A and B be both in $B(H)$. Assume that $A+B$ is normal such that $\sigma(\operatorname{Im}(A+B)) \subset (0,\pi)$. Show that if
$$e^Ae^B = e^Be^A = e^{A+B},$$
then $AB = BA$.

Exercise 13.2.7. Let $A, B \in B(H)$.

(1) Show that
$$e^Ae^{A^*} = e^{A^*}e^A = e^{A+A^*} \Longleftrightarrow A \text{ is normal}.$$

(2) Assume that A is normal such that $\sigma(\text{Im}A) \subset (0,\pi)$. Show also that
$$e^A = e^B \Longrightarrow AB = BA.$$
(3) Prescribe an example which shows that the hypothesis $\sigma(\text{Im}A) \subset (0,\pi)$ may not be omitted.

Exercise 13.2.8. Let A, B be two bounded operators satisfying: $A^{-1}B^*A = B$, $A^*AB = BA^*A$ and $0 \notin \overline{W(A)}$. Show that B is self-adjoint.

Exercise 13.2.9. Let A and B be two bounded operators such that $BA^*A = A^*AB$ and $A^{-1}B^*A = B^{-1}$, where $0 \notin \overline{W(A)}$ and where also B is invertible. Show that B is unitary.

Exercise 13.2.10. (mostly from [**68**]) Let $A, T \in B(H)$ be such that $0 \notin W(T)$.

(1) Show that if $TA = -AT$ and either A or T is normal, then $A = 0$.
(2) Show that if $TA = A^*T$ and $TA^* = AT$, then A is self-adjoint.
(3) Show that if $TA = A^*T$ and either A is normal or T is unitary, then A is self-adjoint.

Exercise 13.2.11.

(1) Let $T \in B(H)$ be self-adjoint such that $0 \notin W(T)$. Show that if $A, B \in B(H)$ are normal and verifying $TA = BT$, then $A = B$.
(2) Deduce that if $T \in B(H)$ is self-adjoint such that $0 \notin W(T)$, and if $A \in B(H)$ is normal and such that $TA = A^*T$, then A is self-adjoint.
(3) ([**147**]) Let $A, B \in B(H)$ be self-adjoint such that $0 \notin W(B)$. Show that if AB is normal, then it is self-adjoint (cf. Exercise 11.3.35).

Exercise 13.2.12. (cf. Theorem 13.1.6) Let $A, B, T \in B(H)$. Provide examples of normal A and normal B such that $TA = BT$, and

(1) A and B commute, but $A \neq B$ (this shows the importance of the assumption $0 \notin W(T)$);
(2) $0 \notin W(T)$ but $A \neq B$ (this shows the importance of the commutativity of A and B).

Exercise 13.2.13. Prove Corollary 13.1.9.

Exercise 13.2.14. Let $U \in B(H)$ be a cramped unitary operator. Let $A \in B(H)$ be such that $UA = A^*U$. We would like to show that A is self-adjoint:

(1) Show that $U^2A = AU^2$.
(2) Deduce that $AU = UA^*$.
(3) Show that A is normal (hint: set $S = \frac{1}{2}(U + U^*)$ and you may assume that $S > 0$ as done in [**213**]).
(4) Infer that A is self-adjoint.

13.3. Tests

Test 13.3.1. Let $A, B \in B(H)$ be self-adjoint ahere AB is normal and $A > 0$ (or $B > 0$). Using Embry Theorem, show that AB is self-adjoint.

Test 13.3.2. Prove Corollary 13.1.8.

13.4. More Exercises

Exercise 13.4.1. Prove Corollary 13.1.10. **Hint:** For one case use Theorem 11.1.55, and for the other use Theorem 13.1.6.

Exercise 13.4.2. Let $A \in B(H)$ be normal such that $0 \notin W(A)$. Then A and A^2 commute with exactly the same operators.

Part 2

SOLUTIONS

CHAPTER 1

Normed Vector Spaces. Banach Spaces

1.2. True or False: Answers

Answers.

(1) The converse is not always true. We give a counterexample. Take any non-empty set X which is not a vector space. Then we can define a metric on it (the discrete metric is the preferred counterexample in this case) but we cannot define a norm on it. Even if X is given the structure of a vector space (with $X \neq \{0\}$), we still cannot always define a norm on it (A vector space endowed with the discrete metric works again as a counterexample). For details see Exercises 1.3.3, 1.3.4 & 1.3.5.

If, however, $X \neq \{0\}$, then obviously $x \mapsto \|x\| = d(x, 0)$ defines a norm on X as it satisfies the conditions of Exercise 1.3.3.

(2) The answer is yes. To prove the equality, let $y \in x + rB(0_X, 1)$. Then, for some $t \in B(0_X, 1)$, $y = x + rt$. As a deduction,

$$\|y - x\| = \|rt\| = r\|t\| < r,$$

i.e. $y \in B(x, r)$.

Conversely, let $y \in B(x, r)$, i.e. $\|y - x\| < r$. Set $y - x = rt$. Then we have

$$r\|t\| = \|rt\| = \|y - x\| < r \Rightarrow \|t\| < 1,$$

i.e. $y \in x + rB(0_X, 1)$.

(3) In general, both equations do not hold. For the first one, take $A = \{1, 2\}$. Then $-A = \{-1, -2\}$. Accordingly,

$$A + \{-A\} = \{0, -1, 1\} \neq \{0\}.$$

As for the second equality, and for an arbitrary A, only $2A \subset A + A$ holds. Indeed, if $x \in 2A$, then for some $a \in A$, $x = 2a = a + a$ and hence $x \in A + A$.

If A is convex, then $A + A \subset 2A$ (i.e. equality holds). Let $x \in A + A$. Then for some a in A and some b in A, $x = a + b$.

But A is convex and hence for all $t \in [0,1]$, $(1-t)a + tb \in A$. Taking $t = \frac{1}{2}$, yields

$$\frac{1}{2}a + \frac{1}{2}b \in A \text{ or } x = 2\left(\frac{1}{2}a + \frac{1}{2}b\right) \in 2A.$$

It is worth recalling that if A is a *vector space*, then

$$A + \{-A\} = A + A = 2A = A.$$

(4) False! The first two axioms of a norm remain valid in this case but the Minkowski Inequality (the triangle inequality!) fails to hold. In fact, using the concavity of the function $x \mapsto x^p$ on $[0,\infty)$ and for $0 < p < 1$, we can show that

$$\forall f, g \in X : \ \|f+g\|_p > \|f\|_p + \|g\|_p.$$

Let us give a counterexample anyway. Consider

$$f(x) = \mathbb{1}_{(0,\frac{1}{2})}(x) \text{ and } g(x) = \mathbb{1}_{(\frac{1}{2},1)}(x).$$

Then

$$\|f+g\|_p = \sqrt[p]{\int_0^1 1^p dx} = 1$$

and

$$\|f\|_p + \|g\|_p = \frac{1}{2^{\frac{1}{p}}} + \frac{1}{2^{\frac{1}{p}}} = \frac{1}{2^{\frac{1}{p}-1}} < 1 = \|f+g\|_p$$

since $0 < p < 1$.

In fine, it is worth noting that for $0 < p < 1$, we can endow L^p with a distance, namely

$$d(f,g) = \|f-g\|_p^p$$

where $\|\cdot\|_p$ is the quantity defined in the question.

1.3. Solutions to Exercises

Solution 1.3.1. The answer is no. One reason is

$$\|(-2,3)\| = 0 \text{ while } (-2,3) \neq (0,0).$$

Solution 1.3.2. Let $x, y \in X$. We have

$$\|x\| = \|x-y+y\| \Longrightarrow \|x\| \le \|x-y\| + \|y\| \Longrightarrow \|x\| - \|y\| \le \|x-y\|.$$

Inverting the roles of x and y gives us

$$\|y\| - \|x\| \le \|y-x\| = \|x-y\|.$$

Thus

$$|\|x\| - \|y\|| \le \|x-y\|.$$

Replacing y by $-y$ yields

$$|\|x\| - \|-y\|| \leq \|x-(-y)\| \iff |\|x\| - \|y\|| \leq \|x+y\|.$$

SOLUTION 1.3.3. First, since d is a distance, we immediately deduce that

$$\forall x \in X : \ \|x\| = d(x,0) \geq 0.$$

Now we check the other axioms of a norm (the two hypotheses and the fact that d is a metric will be used!).

(1) Let $x \in X$. Then

$$\|x\| = 0 \iff d(x,0) = 0 \iff x = 0.$$

(2) Let $x \in X$ and $\alpha \in \mathbb{F}$. We have

$$\|\alpha x\| = d(\alpha x, 0) = |\alpha| d(x,0) = |\alpha|\|x\|.$$

(3) Let $x, y, z \in X$. We have

$$\|x+y\| = d(x+y,0) \leq d(x+y,y) + d(y,0).$$

But

$$d(x+y,y) = d(x+y-y, y-y) = d(x,0).$$

Hence

$$\|x+y\| \leq d(x,0) + d(y,0) = \|x\| + \|y\|.$$

SOLUTION 1.3.4. If X is a normed vector space with norm $\|\cdot\|$, then the associated metric is given by $d(x,y) = \|x-y\|$ for all $x, y \in X$. Let us check that this metric does satisfy the listed properties of the preceding exercise. Let $x, y, a \in X$ and let $\alpha \in \mathbb{K}$. We have

$$d(x+a, y+a) = \|x+a-(y+a)\| = \|x-y\| = d(x,y).$$

Also,

$$d(\alpha x, \alpha y) = \|\alpha x - \alpha y\| = \|\alpha(x-y)\| = |\alpha|\|x-y\| = |\alpha| d(x,y)$$

(since $\|\cdot\|$ is a norm).

SOLUTION 1.3.5. It will be normed if its metric satisfies the properties of Exercise 1.3.3. But for $x \neq 0$ (hence $3x \neq 0$), we have

$$d(3x, 0) = 1 \neq 3d(x,0) = 3.$$

Thus a *non-trivial* vector space endowed with a discrete metric cannot be normed.

REMARK. What we did in the previous exercise was enough for proving that X cannot be normed. It is, however, worth noting that the other property (in Exercise 1.3.3) is satisfied since in a vector space X we have for all $x, y, a \in X$

$$x + a = y + a \Longleftrightarrow x = y$$

and hence

$$d(x + a, y + a) = d(x, y).$$

SOLUTION 1.3.6. Let $x, y \in X$ be satisfying $\|x + y\| = \|x\| + \|y\|$ and let $\alpha, \beta \geq 0$. WLOG, we may assume that $\alpha \geq \beta \geq 0$.

First, we obviously have

$$\|\alpha x + \beta y\| \leq \|\alpha x\| + \|\beta y\| = |\alpha|\|x\| + |\beta|\|y\| = \alpha\|x\| + \beta\|y\|.$$

Now we prove the reverse inequality. We have

$$\begin{aligned}\|\alpha x + \beta y\| =& \|\alpha(x + y) - (\alpha - \beta)y\| \\ \geq& \|\alpha(x + y)\| - \|(\alpha - \beta)y\| \text{ (by Exercise 1.3.2)} \\ \geq& |\alpha|\|x + y\| - |\alpha - \beta|\|y\| \\ \geq& \alpha\|x + y\| - (\alpha - \beta)\|y\| \\ =& \alpha(\|x\| + \|y\|) - (\alpha - \beta)\|y\| \text{ (by the assumption of the exercise)} \\ =& \alpha\|x\| + \beta\|y\|,\end{aligned}$$

as required.

SOLUTION 1.3.7.

(1) (a) Since f is continuous, so is $|f|$ and from here $\|f\|_1$ is a well-defined positive quantity.
Now, we show that $\|\cdot\|_1$ is a norm on X.
(i) Let $f, g \in X$. We have

$$f = 0 \Longrightarrow \|f\|_1 = 0 \text{ (this is obvious!)}.$$

Now since $|f|$ is a *continuous* and *positive* function, we have

$$\int_0^1 |f(x)|dx = 0 \Longrightarrow |f(x)| = 0,\ \forall x \in [0, 1] \Longleftrightarrow f = 0.$$

(ii) It is plain that for all f in X and all $\lambda \in \mathbb{R}$, one has

$$\|\lambda f\|_1 = \int_0^1 |\lambda f(x)|dx = \int_0^1 |\lambda||f(x)|dx = |\lambda| \int_0^1 |f(x)|dx = |\lambda|\|f\|_1.$$

(iii) Let us show now the triangle inequality. Let $f, g \in X$. We have for all $x \in [0,1]$

$$|f(x) + g(x)| \leq |f(x)| + |g(x)|.$$

Hence

$$\int_0^1 |f(x) + g(x)|dx \leq \int_0^1 |f(x)|dx + \int_0^1 |g(x)|dx,$$

i.e.

$$\|f + g\|_1 \leq \|f\|_1 + \|g\|_1.$$

(b) To prove $\|\cdot\|_2$ is a norm on X, we first observe that for all $f \in X$, $\|f\|_2 \geq 0$. Now we check the other axioms of a norm.

(i) Let $f \in X$. If $f = 0$, then $\|f\|_2 = \sqrt{\int_0^1 0^2 dx} = 0$. Conversely, if $\|f\|_2 = 0$, then by the continuity and the positivity of $|f|^2$ we get that $|f|^2 = 0$ or $f = 0$.

(ii) Let $(f, \lambda) \in X \times \mathbb{R}$. It is plain that

$$\|\lambda f\|_2 = |\lambda| \|f\|_2.$$

(iii) To prove the triangle inequality, we need the Cauchy-Schwarz Inequality for integrals. Let $f, g \in X$. We have

$$\begin{aligned}
\|f+g\|_2^2 &= \int_0^1 |f(x) + g(x)|^2 \\
&= \int_0^1 |f(x)|^2 dx + 2\int_0^1 |f(x)g(x)|dx + \int_0^1 |g(x)|^2 dx \\
&\leq \int_0^1 |f(x)|^2 dx + 2 \underbrace{\sqrt{\int_0^1 |f(x)|^2 dx}}_{\|f\|_2} \underbrace{\sqrt{\int_0^1 |g(x)|^2 dx}}_{\|g\|_2} + \int_0^1 |g(x)|^2 dx \\
&= (\|f\|_2 + \|g\|_2)^2.
\end{aligned}$$

Therefore,

$$\|f + g\|_2 \leq \|f\|_2 + \|g\|_2.$$

(c) (i) If $f = 0$, then obviously $\|f\|_\infty = 0$. Conversely, if $\|f\|_\infty = 0$, then for all $x \in [0,1]$ we have

$$0 \leq |f(x)| \leq \sup_{x \in [0,1]} |f(x)| = 0.$$

Thus $f = 0$.

(ii) For all $f \in X$ and all $\lambda \in \mathbb{R}$, we have

$$\|\lambda f\|_\infty = \sup_{x\in[0,1]} |\lambda f(x)| = |\lambda|\|f\|_\infty.$$

(iii) Let $f, g \in X$. Then we have for all $x \in [0,1]$

$$|f(x)+g(x)| \le |f(x)| + |g(x)| \le \|f\|_\infty + \|g\|_\infty.$$

Thus

$$\|f+g\|_\infty \le \|f\|_\infty + \|g\|_\infty.$$

(2) Let $f \in X$. Then

$$\begin{aligned}
\|f\|_1 &= \int_0^1 |f(x)|dx \\
&= \int_0^1 1\cdot|f(x)|dx \\
&\le \left(\int_0^1 1^2dx\right)^{\frac{1}{2}}\left(\int_0^1 |f(x)|^2dx\right)^{\frac{1}{2}} \text{ (by "Cauchy-Schwarz")} \\
&\le \left(\int_0^1 |f(x)|^2dx\right)^{\frac{1}{2}} \\
&= \|f\|_2
\end{aligned}$$

and we are half-way through. To prove the other inequality, we have

$$\forall x \in [0,1]: \ |f(x)|^2 \le \|f\|_\infty^2.$$

Hence

$$\|f\|_2 \le \|f\|_\infty.$$

Therefore, for all $f \in X$,

$$\|f\|_1 \le \|f\|_2 \le \|f\|_\infty.$$

(3) First, we show that $\|\cdot\|_\infty$ is not equivalent to $\|\cdot\|_1$. Take $f(x) = x^n$, defined on $[0,1]$, where $n \in \mathbb{N}$. Then

$$\|f\|_1 = \frac{1}{n+1} \text{ and } \|f\|_\infty = 1.$$

If $\|\cdot\|_1$ were equivalent to $\|\cdot\|_\infty$, then we would have

$$\exists a > 0, \forall n \in \mathbb{N}: \ a \le \frac{1}{n+1}$$

which would then imply that $\mathbb{N}$ is bounded! Thus, we have reached a contradiction. Hence, $\|\cdot\|_1$ and $\|\cdot\|_\infty$ are not equivalent.

Now, we show that $\|\cdot\|_1$ is not equivalent to $\|\cdot\|_2$. Let

$$f_n(x) = \begin{cases} 1 - nx, & 0 \le x < \frac{1}{n}, \\ 0 & \frac{1}{n} \le x \le 1. \end{cases}$$

Then

$$\|f_n\|_1 = \int_0^{\frac{1}{n}} (1 - nx)dx = \frac{1}{2n}$$

and

$$\|f_n\|_2 = \sqrt{\int_0^{\frac{1}{n}} (1 - nx)^2 dx} = \frac{1}{\sqrt{3n}}.$$

Since

$$\frac{\|f_n\|_2}{\|f_n\|_1} = \frac{2}{\sqrt{3}}\sqrt{n} \longrightarrow \infty,$$

as $n \to \infty$, we conclude that the two norms cannot be equivalent.

(4) Since the norms $\|\cdot\|_1$ and $\|\cdot\|_2$, *defined on the same vector space*, are not equivalent, we immediately deduce that $\dim X = \infty$.

Solution 1.3.8. First, we show that $\|f\| = |f(0)| + \|f'\|_\infty$ defines a norm on X.

(1) Let $f \in X$. We have

$$\begin{aligned} \|f\| = 0 &\iff \begin{cases} f'(x) = 0, \ \forall x \in [0,1] \\ f(0) = 0 \end{cases} \\ &\iff \begin{cases} f(x) = C \text{ (a constant)}, \ \forall x \in [0,1] \\ f(0) = 0 \end{cases} \\ &\iff f(x) = 0, \ \forall x \in [0,1]. \\ &\iff f = 0. \end{aligned}$$

(2) Let $f \in X$ and let $\lambda \in \mathbb{R}$. We have

$$\|\lambda f\| = |\lambda f(0)| + \|(\lambda f)'\|_\infty = |\lambda|(|f(0)| + \|f'\|_\infty) = |\lambda|\|f\|.$$

(3) Let $f, g \in X$. We have

$$\begin{aligned} \|f + g\| &= |f(0) + g(0)| + \|f' + g'\|_\infty \\ &\le |f(0)| + |g(0)| + \|f'\|_\infty + \|g'\|_\infty \\ &= |f(0)| + \|f'\|_\infty + |g(0)| + \|g'\|_\infty \\ &= \|f\| + \|g\|. \end{aligned}$$

Therefore, $f \mapsto \|f\|$ is in effect a norm on X. Let us show its equivalence to $\|f\|_\infty + \|f'\|_\infty$. Let $f \in X$. We have

$$\|f\| = |f(0)| + \|f'\|_\infty \leq \|f\|_\infty + \|f'\|_\infty.$$

On the other hand and by means of the mean value theorem, for some t in $(0, x)$ $(\subset (0,1))$, we have

$$f(x) = f(0) + xf'(t).$$

Hence

$$|f(x)| \leq |f(0)| + |xf'(t)| \leq |f(0)| + |xf'(t)| \leq |f(0)| + \|f'\|_\infty, \ \forall x \in [0,1]$$

and so

$$\|f\|_\infty \leq |f(0)| + \|f'\|_\infty = \|f\|.$$

This yields

$$\begin{aligned} \|f\|_\infty + \|f'\|_\infty &\leq |f(0)| + 2\|f'\|_\infty \\ &\leq 2(|f(0)| + \|f'\|_\infty) \\ &= 2\|f\|. \end{aligned}$$

Thus

$$\forall f \in X: \ \|f\| \leq \|f\|_\infty + \|f'\|_\infty \leq 2\|f\|,$$

quod erat demonstrandum.

Solution 1.3.9. We already know from a Topology Course (see e.g. [**153**]) that $B_c(a,r) \supset \overline{B(a,r)}$. So, let $x \in B_c(a,r)$ and consider $x_n = \frac{1}{n}a + \left(1 - \frac{1}{n}\right)x$. We have

$$\begin{aligned} \|a - x_n\| &= \left\| a - \frac{1}{n}a - \left(1 - \frac{1}{n}\right)x \right\| \\ &= \left(1 - \frac{1}{n}\right)\|a - x\| \\ &\leq \left(1 - \frac{1}{n}\right) r < r. \end{aligned}$$

So $x_n \in B(a,r)$. We need only check that $x_n \to x$ in X. We have

$$\|x_n - x\| = \frac{1}{n}\|a - x\| \leq \frac{r}{n}, \ \forall n \geq 1.$$

Passing to the limit as n tends to infinity yields $x_n \to x$ in X. Thus $x \in \overline{B(a,r)}$.

SOLUTION 1.3.10. The given vector space c_{00} is not closed. Consider the sequence (x^k) defined by

$$x^k = \left(1, \frac{1}{2^2}, \frac{1}{3^2}, \cdots, \frac{1}{k^2}, 0, 0, \cdots\right).$$

Then (x^k) obviously does belong to c_{00}. It has the following limit with respect to $\|\cdot\|_1$

$$x = \left(1, \frac{1}{2^2}, \frac{1}{3^2}, \cdots, \frac{1}{k^2}, \frac{1}{(k+1)^2}, \frac{1}{(k+2)^2}, \cdots\right)$$

for

$$\|x^k - x\|_1 = \sum_{n=k+1}^{\infty} \frac{1}{n^2} \longrightarrow 0 \text{ as } k \longrightarrow \infty,$$

because this is a remainder of a convergent series. Since, clearly $x \notin c_{00}$, c_{00} is not closed.

SOLUTION 1.3.11. Let $x, y \in \overline{Y}$ and let $\alpha, \beta \in \mathbb{C}$. Since $x, y \in \overline{Y}$, $\exists x_n, y_n \in Y$ such that $x_n \to x$ and $y_n \to y$.

As Y is a vector space, then $\alpha x_n + \beta y_n \in Y$ for all $n \in \mathbb{N}$. We also know that the multiplication by a scalar and the function sum are both continuous and so

$$\alpha x + \beta y = \lim_{n\to\infty} (\alpha x_n + \beta y_n) \in \overline{Y}.$$

Therefore, $\overline{Y}$ is indeed a vector space.

SOLUTION 1.3.12.

(1) We prove that if Y is open, then so will be $Y + Z$. One can write the following

$$Y + Z = \bigcup_{z\in Z} (Y + \{z\}).$$

Since the arbitrary union of open sets is open, we only have to show that $Y + \{z\}$ are open for each $z \in Z$. Let $z \in Z$. Let $x \in Y + \{z\}$. Then there exists $y \in Y$ such that $x = y + z$. Since Y is open, we know that for some $r > 0$, $B(y, r) \subset Y$. Whence

$$B(x, r) = B(y + z, r) \subset Y + \{z\}.$$

Thus $Y + \{z\}$ is open.

(2) Let $f : X \times X \to X$ defined by $f(x, y) = x + y$. Then f is continuous. Since Y and Z are compact, so is $Y \times Z$ and hence so is $f(Y \times Z) = Y + Z$ too.

(3) The answer is exactly the same as the foregoing one with the word "connected" instead of "compact".

(4) A counterexample is the following: In $\mathbb{R}^2$ endowed with the Euclidian norm, let

$$Y = \left\{ \left(x, \frac{1}{x} \right) : \ x > 0 \right\} \text{ and } Z = \{(y, 0) : \ y \in \mathbb{R}\}.$$

Then Y and Z are both closed while their sum $Y + Z$ is not closed (do the details or see [**153**]).

(5) Let $x \in \overline{Y + Z}$. Then there exist y_n in Y and z_n in Z such that $y_n + z_n \to x$. But, Y is compact and so there exists a subsequence (y_{nk}) of (y_n) such that $y_{nk} \to y \in Y$.

Hence, the subsequence $(y_{nk} + z_{nk})$ of $(y_n + z_n)$ converges to x. Thus $(z_{nk}) = (y_{nk} + z_{nk} - y_{nk})$ converges too. Since Z is closed, (z_{nk}) has a limit z belonging to Z. Therefore,

$$x = y + z \in Y + Z,$$

establishing the closedness of $Y + Z$.

Solution 1.3.13.

(1) We can show the "Cauchyness" of (f_n) either by doing some arithmetic or geometrically by observing that $\|f_n - f_{n+p}\|_1$ is actually the area of the triangle of apexes $\left(\frac{1}{2} - \frac{1}{n}, 0\right)$, $\left(\frac{1}{2} - \frac{1}{n+p}\right)$ and $\left(\frac{1}{2}, 0\right)$ where $p, n \in \mathbb{N}$. Ergo

$$\begin{aligned}\|f_n - f_{n+p}\|_1 =& \frac{1}{2} \times 1 \times \left[\frac{1}{2} - \frac{1}{n+p} - \left(\frac{1}{2} - \frac{1}{n} \right) \right] \\ =& \frac{1}{2} \left(\frac{1}{n} - \frac{1}{n+p} \right) \\ \leq& \frac{1}{2n}.\end{aligned}$$

Hence

$$\forall \varepsilon > 0, \exists N = \left[\frac{\varepsilon}{2} \right] + 1 \in \mathbb{N}, \forall n, p \in \mathbb{N} \ (n \geq N \Rightarrow \|f_n - f_{n+p}\|_1 < \varepsilon),$$

proving the "Cauchyness" of (f_n).

(2) The pointwise limit of (f_n) is

$$f(x) = \begin{cases} 0, & 0 \leq x < \frac{1}{2}, \\ 1, & \frac{1}{2} \leq x \leq 1. \end{cases}$$

Hence

$$\begin{aligned}\|f_n - f\|_1 &= \int_0^1 |f_n(x) - f(x)|dx \\ &= \int_0^{\frac{1}{2}-\frac{1}{n}} 0dx + \int_{\frac{1}{2}-\frac{1}{n}}^{\frac{1}{2}} \left[nx + \left(1 - \frac{1}{2}n\right)\right] dx + \int_{\frac{1}{2}}^1 0dx \\ &= \frac{1}{2n} \longrightarrow 0\end{aligned}$$

as $n \to \infty$.

Now, to show $(X, \|\cdot\|_1)$ is not complete, assume that there is a continuous function g on $[0,1]$ such that $\|f_n - g\|_1 \to 0$ and we will reach a contradiction. Since

$$|f(x) - g(x)| \le |f(x) - f_n(x)| + |f_n(x) - g(x)|, \ \forall x \in [0,1],$$

we get

$$0 \le \int_0^1 |f(x) - g(x)|dx \le \int_0^1 |f(x) - f_n(x)|dx + \int_0^1 |f_n(x) - g(x)|dx.$$

Passing to the limit and using our hypotheses yield

$$\int_0^1 |f(x) - g(x)|dx = 0$$

and hence $f(x) = g(x)$ at each x for which $f - g$ is continuous, i.e.

$$f(x) = g(x), \ \forall x \in [0,1] \setminus \left\{\frac{1}{2}\right\}.$$

But, g is assumed to be continuous on $[0,1]$ and so it must be continuous at $\frac{1}{2}$ and we would have

$$\lim_{x \to \frac{1}{2}^-} g(x) = \lim_{x \to \frac{1}{2}^+} g(x) = g\left(\frac{1}{2}\right).$$

However,

$$\lim_{x \to \frac{1}{2}^-} g(x) = \lim_{x \to \frac{1}{2}^-} f(x) = 0 \neq \lim_{x \to \frac{1}{2}^+} g(x) = \lim_{x \to \frac{1}{2}^+} f(x) = 1.$$

Therefore, such a function g cannot exist. Thus $(X, \|\cdot\|_1)$ is not a Banach space.

Solution 1.3.14. Let (f_n) be a Cauchy sequence in $(X, \|\cdot\|_\infty)$. Then

$$\forall \varepsilon > 0, \exists N \in \mathbb{N}, \forall n, m \in \mathbb{N} \ (n, m \ge N \Longrightarrow \|f_n - f_m\|_\infty < \varepsilon)$$

or

$$\forall \varepsilon > 0, \exists N \in \mathbb{N}, \forall n, m \in \mathbb{N}\ (n, m \geq N \Longrightarrow \sup_{x \in [0,1]} |f_n(x) - f_m(x)| < \varepsilon).$$

This implies that for each particular x in $[0, 1]$ one has

$$|f_n(x) - f_m(x)| < \varepsilon \text{ for all } n, m \geq N.$$

Thus for each $x \in [0, 1]$, $(f_n(x))_n$ is a Cauchy sequence in usual $\mathbb{R}$. Since the latter is a Banach space, we deduce that $(f_n(x))_n$ must converge to some $f(x) \in \mathbb{R}$ (for each $x \in [0, 1]$). Thus we have a function f defined for all $x \in [0, 1]$. Fixing n in the last displayed equation, and letting m tend to infinity yield

$$|f_n(x) - f(x)| < \varepsilon \text{ whenever } n \geq N$$

and for each x where N is independent of x. Thus

$$\|f_n - f\|_\infty = \sup_{x \in [0,1]} |f_n(x) - f(x)| < \varepsilon \text{ for all } n \geq N.$$

Therefore, we have shown that (f_n) converges in the supremum norm to f, i.e. that (f_n) converges uniformly to f.

To summarize, the (arbitrary!) Cauchy sequence (f_n) is convergent in $(X, \|\cdot\|_\infty)$ to a *continuous* function f. In other language, $(X, \|\cdot\|_\infty)$ is complete.

Solution 1.3.15. The answer is no. Take any continuous function f on $[0, 1]$ which *does not* belong to X (e.g. $x \mapsto |x|$). The Weierstrass Theorem (Theorem 1.1.30) then implies that there exists a polynomial sequence (P_n) such that

$$\|P_n - f\|_\infty \to 0 \text{ as } n \to \infty.$$

Since each P_n is a polynomial, $(P_n) \subset X$. Since (P_n) converges to f in $(X, \|\cdot\|_\infty)$, (P_n) is Cauchy in $(X, \|\cdot\|_\infty)$. But $f \notin X$ and thus $(X, \|\cdot\|_\infty)$ is not complete.

Solution 1.3.16.

(1) To show that $x \mapsto e^x$ is not a polynomial we use a contradiction argument. Assume that $x \mapsto e^x$ is a polynomial of degree m, say, i.e. for some real $a_0, a_1, \cdots, a_m$ we have

$$e^x = a_0 + a_1 x + \cdots + a_m x^m.$$

Then by differentiating $(m+1)$-times both sides of the previous displayed equation we get

$$(e^x)^{(m+1)} = e^x = (a_0 + a_1 x + \cdots + a_m x^m)^{(m+1)} = 0$$

which is of course absurd. Thus the exponential function is not a polynomial.

(2) Obviously $P_n(x) = \left(1+\frac{x}{n}\right)^n$ (for $n \geq 1$) is a polynomial sequence. So, it only remains to show that $\|P_n - e^x\|_\infty \to 0$ as $n \to \infty$, and by uniqueness of the limit e^x, is the *only* limit of (P_n).

Let us show now that $P_n \to e^x$ in $(X, \|\cdot\|_\infty)$. To this end, we compute $\sup_{x\in[0,1]} |P_n(x) - e^x|$. We can write

$$\sup_{x\in[0,1]} |P_n(x) - e^x| \leq e^1 \sup_{x\in[0,1]} |P_n(x)e^{-x} - 1|.$$

Hence

$$0 \leq |P_n(x)e^{-x} - 1| \leq 1 - P_n(1)e^{-1} = 1 - \left(1+\frac{1}{n}\right)^n e^{-1}.$$

Taking the supremum and passing to the limit as $n \to \infty$ yield

$$0 \leq \|P_n - e^x\|_\infty \leq e\left(1 - \left(1+\frac{1}{n}\right)^n e^{-1}\right) \to 0.$$

(3) Take any continuous function f which is not a polynomial (e.g. e^x, $\sin x$, etc...) and hence $f \notin X$. By Theorem 1.1.30, we know that there is a polynomial sequence $(P_n)_n$ which converges uniformly to f, i.e. $P_n \to f$ with respect to $\|\cdot\|_\infty$. Hence $(P_n)_n$ is Cauchy but converging to f which is outside X. Thus $(X, \|\cdot\|_\infty)$ is not complete.

SOLUTION 1.3.17.

(1) Let $1 \leq p < \infty$. First we treat *real* ℓ^p spaces. We must show that there exists a countable set X such that $\overline{X} = \ell^p$. Let

$$X_n = \{(q_1, q_2, \cdots, q_n, 0, 0, \cdots) : q_1, \cdots, q_n \in \mathbb{Q}\}.$$

Then each X_n is countable and so is their union, which we denote by X. We are done as soon as we show that X is dense in ℓ^p, i.e. $\ell^p \subset \overline{X}$.

Let $x = (x_n) = (x_1, x_2, \cdots, x_n, x_{n+1}, \cdots) \in \ell^p$. We show that $x \in \overline{X}$. Let $q = (q_1, q_2, \cdots, q_N, 0, 0, \dots) \in X_N$, $N \in \mathbb{N}$. We ought to show that $\|x - q\|_p$ is arbitrarily small. Since $\mathbb{Q}$ is dense in usual $\mathbb{R}$, we know that for all $x_1, \cdots, x_N$ and all $\varepsilon > 0$ there are $q_1, \cdots, q_N \in \mathbb{Q}$ such that

$$\sum_{i=1}^N |x_i - q_i|^p = |x_1 - q_1| + \cdots + |x_N - q_N| < \varepsilon$$

making $\sum_{i=1}^N |x_i - q_i|^p$ as small as we wish.

Now, for $i \geq N+1$, we have

$$\|x-q\|_p^p = \sum_{i=N+1}^{\infty} |x_i|^p \longrightarrow 0 \text{ as } N \longrightarrow \infty$$

for this is the tail of a convergent series! Thus (real!) ℓ^p is separable.

To see that complex ℓ^p is separable, just use Gaussian rational numbers $p+iq$, where $p,q \in \mathbb{Q}$, in the definition of X_n and the latter remains countable. All the rest stays unchanged.

(2) A very similar method may be applied. Everything as before (mutatis mutandis), we will have in the end that

$$\|x-q\|_\infty = \sup_{i\geq N+1} |x_i| \longrightarrow 0 \text{ as } N \longrightarrow \infty$$

(otherwise we would get a contradiction with (x_n) tending to zero). Thus $(c_0, \|\cdot\|_\infty)$ is separable.

Solution 1.3.18.

(1) Since $(\ell^\infty, \|\cdot\|_\infty)$ is a Banach space, to show that $(c_0, \|\cdot\|_\infty)$ is complete, it suffices to show that it is closed, i.e. $\overline{c_0} \subset c_0$. So, let $x \in \overline{c_0}$ (remember that $x = (x_n) \in \ell^\infty$ for the moment). Let $\varepsilon > 0$. Hence there is a sequence y in c_0 such that

$$\|x-y\|_\infty < \varepsilon.$$

Since $y \in c_0$, $y = (y_n)$ with $y_n \to 0$. Hence for some $N \in \mathbb{N}$ we have $|y_n| < \varepsilon$ when $N \geq n$. Accordingly, for $n \geq N$ we have

$$|x_n| = |x_n - y_n + y_n| \leq |x_n - y_n| + |y_n| \leq \|x-y\|_\infty + |y_n| < 2\varepsilon.$$

Thus, $x_n \to 0$, i.e. $x \in c_0$, i.e. $\overline{c_0} \subset c_0$, as needed.

(2) Let

$$x = \left(\frac{1}{\ln 2}, \frac{1}{\ln 3}, \cdots, \frac{1}{\ln n}, \cdots\right).$$

Then $x \in c_0$ because $\frac{1}{\ln n} \to 0$. Assume now for the sake of contradiction that x is in some ℓ^p, where $1 \leq p < \infty$. Then this would mean that $\sum_{n\geq 2} \frac{1}{(\ln n)^p}$ converges. But, this is a particular case of a Bertrand's series, i.e. $\sum_{n\geq 2} \frac{1}{n^q (\ln n)^p}$, which diverges if $q < 1$ regardless the value of p (which is our case!). Thus, x is not in any ℓ^p, where $p \geq 1$.

(3) Let $x \in \ell^p$. Then $\sum_{n=1}^{\infty} |x_n|^p$ converges. This implies that $|x_n|^p \to 0$ when $n \to \infty$ or simply $|x_n| \to 0$ or $x_n \to 0$, that is, $x \in c_0$.

(4) We already know that $(c_0, \|\cdot\|_\infty)$ is closed. By the foregoing question, we may then write

$$\ell^p \subset c_0 \Longrightarrow \overline{\ell^p} \subset \overline{c_0} = c_0$$

where the closures are with respect to the topology of $\|\cdot\|_\infty$.

So, it only remains to show that $c_0 \subset \overline{\ell^p}$. To achieve this, let $x = (x_n) \in c_0$, i.e. x_n tends to zero. Take now $x^k = (x_1, x_2, \cdots, x_k, 0, 0, \cdots)$ (the superscript is a label and not a power!) which is clearly in ℓ^p. Hence

$$\|x - x^k\|_\infty = \sup_{n \geq k+1} |x_n| \longrightarrow 0,$$

proving that $x \in \overline{\ell^p}$, i.e. $c_0 \subset \overline{\ell^p}$ so that $c_0 = \overline{\ell^p}$ (w.r.t. the topology of $\|\cdot\|_\infty$).

(5) It is clear that $c_{00} \subset \ell^p$ so that $\overline{c_{00}} \subset \overline{\ell^p} = \ell^p$ (w.r.t. the topology of ℓ^p). So we need only show that $\ell^p \subset \overline{c_{00}}$. Let $x \in \ell^p$. Consider (x^k) where $x^k = (x_1, x_2, \cdots, x_k, 0, 0, \cdots)$, i.e. (x^k) is in c_{00}. Now, we have

$$\|x - x^k\|_p^p = \sum_{n=k+1}^{\infty} |x_n|^p \longrightarrow 0$$

and this completes the proof.

(6) No, c_{00} is not dense in ℓ^∞, that is, $\ell^\infty \not\subset \overline{c_{00}}$. To see this, let $x = (1, 1, 1, \cdots) \in \ell^\infty$. Then this x cannot be in $\overline{c_{00}}$ as for any (x^k) in c_{00}, then $\|x - x^k\|_\infty$ cannot be arbitrarily small as it will at least be bigger than one (why?).

We claim that $\overline{c_{00}} = c_0$ (w.r.t. the topology of ℓ^∞). We already know that c_0 is closed so that

$$c_{00} \subset c_0 \Longrightarrow \overline{c_{00}} \subset c_0.$$

So let $x = (x_n)$ be in c_0. Consider now (x^k) where $x^k = (x_1, x_2, \cdots, x_k, 0, 0, \cdots)$. Remembering that $x_n \to 0$, we get

$$\|x - x^k\|_\infty = \sup_{n \geq k+1} |x_n| \longrightarrow 0,$$

i.e. $x \in \overline{c_{00}}$, and the proof is over.

SOLUTION 1.3.19. Let $X \subset \mathbb{N}$. Define a sequence (x_n) as follows

$$x_n = \mathbb{1}_X(n) = \begin{cases} 1, & n \in X, \\ 0, & n \notin X. \end{cases}$$

Then the set all of all sequences of this type, denoted by $\{0,1\}^{\mathbb{N}}$, is uncountable (see e.g. **[153]**).

It is clear that every x_n is in ℓ^∞. Now, let $X, Y \subset \mathbb{N}$ be such that $X \neq Y$. Then if $m \in \mathbb{N}$, then either $m \in X \setminus Y$ or $m \in Y \setminus X$. For example, assume that $m \in X \setminus Y$ (the other case can be dealt with similarly). Hence

$$\|\mathbb{1}_X - \mathbb{1}_Y\|_\infty = \sup_{n\in\mathbb{N}} |\mathbb{1}_X(n) - \mathbb{1}_Y(n)| \geq |\mathbb{1}_X(m) - \mathbb{1}_Y(m)| = |1-0| = 1.$$

Therefore, by Proposition 1.1.29, we infer that ℓ^∞ is not separable.

Solution 1.3.20. We use again Proposition 1.1.29. Let

$$f_a(x) = \mathbb{1}_{(0,a)}(x) \text{ and } f_b(x) = \mathbb{1}_{(0,b)}(x),$$

where $a, b \in (0,1)$. Then for $a \neq b$,

$$\|f_a - f_b\|_\infty = 1.$$

Since $\{f_a : a \in (0,1)\}$ is clearly uncountable, Proposition 1.1.29 does the remaining job, i.e. it tells us that $L^\infty(0,1)$ is not separable.

Solution 1.3.21. We treat two separate cases.

(1) The case $p = \infty$: First, recall that if $f \in L^\infty(\mathbb{R}^n)$, then

$$|f(x)| \leq \|f\|_\infty, \text{ for almost every } x \in \mathbb{R}^n.$$

Now, let (f_n) be a Cauchy sequence in $L^\infty(\mathbb{R}^n)$. Hence we have a set $X_{m,n} \subset \mathbb{R}$ of strictly positive measure such that for all $x \in X_{m,n}$

$$|f_n(x) - f_m(x)| \leq \|f_n - f_m\|_\infty$$

for all $n, m \in \mathbb{N}$.

Set $X = \cup_{n,m} X^c_{n,m}$, where $X^c_{m,n}$ is the complement of $X_{m,n}$ and X has measure zero. This implies that $(f_n(x))$ is a Cauchy sequence outside X (with respect to the supremum norm!). Define

$$f(x) = \begin{cases} 0, & \text{if } x \in X, \\ \lim_{n\to\infty} f_n(x), & \text{otherwise.} \end{cases}$$

Then f is measurable and (f_n) converges uniformly to f outside X. Hence $\|f_n - f\|_\infty \to 0$ too. To finish the proof, we need to show that $f \in L^\infty(\mathbb{R}^n)$. This easily comes from

$$|f(x)| \leq |f(x) - f_n(x)| + |f_n(x)| \leq \|f - f_n\|_\infty + \|f_n\|_\infty$$

for almost every x, i.e. $f \in L^\infty(\mathbb{R}^n)$.

(2) The case $1 \leq p < \infty$: Let (f_n) be a Cauchy sequence in $L^p(\mathbb{R}^n)$, that is, for all $\varepsilon > 0$, there is $N \in \mathbb{N}$ such that for all $n, m \in \mathbb{N}$

$$\|f_n - f_m\|_p < \varepsilon.$$

when $n, m \in \geq N$. In particular, for $\varepsilon = \frac{1}{2}$, we have an N_1; and for $\varepsilon = \frac{1}{2^2}$, we have an N_2, etc... Hence we have a subsequence $(f_{n_k})_k$ with the following property:

$$\|f_{n_{k+1}} - f_{n_k}\|_p < \frac{1}{2^k}, \ \forall k \in \mathbb{N}.$$

Now consider the following two series (whose convergence is yet to be justified)

$$f(x) = f_{n_1}(x) + \sum_{k=1}^{\infty}(f_{n_{k+1}}(x) - f_{n_k}(x)),$$

$$g(x) = |f_{n_1}(x)| + \sum_{k=1}^{\infty} |f_{n_{k+1}}(x) - f_{n_k}(x)|,$$

and their corresponding partial sums

$$S_K[f(x)] = f_{n_1}(x) + \sum_{k=1}^{K}(f_{n_{k+1}}(x) - f_{n_k}(x)),$$

$$S_K[g(x)] = |f_{n_1}(x)| + \sum_{k=1}^{K} |f_{n_{k+1}}(x) - f_{n_k}(x)|.$$

Then $S_K g$ is measurable and increasing. By the Minkwoski Inequality we have

$$\begin{aligned}
\|S_K(g)\|_p &\leq \|f_{n_1}\|_p + \sum_{k=1}^{K} \|f_{n_{k+1}} - f_{n_k}\|_p \\
&\leq \|f_{n_1}\|_p + \sum_{k=1}^{K} \frac{1}{2^k} \\
&\leq \|f_{n_1}\|_p + 1 \\
&:= \alpha.
\end{aligned}$$

Whence $([S_K g]^p)$ is also increasing (in K) and that

$$\int_{\mathbb{R}^n} [S_K g(x)] dx \leq \alpha^p.$$

By the Monotone Convergence Theorem applied to $S_K(g)$ we know that

$$\int_{\mathbb{R}^n} [g(x)]^p dx = \lim_{K\to\infty} \int_{\mathbb{R}^n} [S_K g(x)]^p dx \leq \alpha^p < \infty.$$

Hence by a simple result of Measure Theory, g^p must be finite almost everywhere and therefore g is finite almost everywhere. Thus the series intervening in the definition of g is absolutely convergent almost everywhere, and hence so is the (telescoping) series appearing in the definition of $f(x)$ and hence by a simple computation

$$\lim_{k\to\infty} f_{n_k}(x) = f(x)$$

(holding almost everywhere). Since $|f| \leq g$, we also have that $f \in L^p(\mathbb{R}^n)$. We will be done as soon as we show that (f_{n_k}) converges to f w.r.t. the topology of $L^p(\mathbb{R}^n)$.

We clearly have for all k (as also $|f_{n_k}| \leq g$ for all k) and almost everywhere

$$|f_{n_k}(x) - f(x)|^p \leq (|f_{n_k}(x)| + |f(x)|)^p \leq 2^p |g(x)|^p.$$

Since $\lim_{k\to\infty} |f_{n_k}(x) - f(x)|^p = 0$ a.e., and $|g|^p \in L^1(\mathbb{R}^n)$, the Dominated Convergence Theorem yields

$$\lim_{k\to\infty} \int_{\mathbb{R}^n} |f_{n_k}(x) - f(x)|^p dx = 0, \text{ that is, } \lim_{k\to\infty} \|f_{n_k} - f\|_p = 0$$

In the end, since (f_n) is a Cauchy sequence which has a convergent subsequence, namely (f_{n_k}), Proposition 1.1.24 tells us that in fact

$$\lim_{n\to\infty} \|f_n - f\|_p = 0,$$

completing the proof that $L^p(\mathbb{R}^n)$ is a Banach space.

Solution 1.3.22.

(1) None of the inclusions $L^1(\mathbb{R}) \subset L^2(\mathbb{R})$ and $L^2(\mathbb{R}) \subset L^2(\mathbb{R})$ holds. Take

$$f(x) = \frac{1}{\sqrt{x}} \mathbb{1}_{(0,1)}(x) \text{ and } g(x) = \frac{1}{x} \mathbb{1}_{[1,\infty)}(x).$$

Then

$$\int_{\mathbb{R}} |f(x)| dx = \int_{\mathbb{R}} \frac{1}{\sqrt{x}} \mathbb{1}_{(0,1)}(x) dx = 2, \int_{\mathbb{R}} |f(x)|^2 dx = \int_{\mathbb{R}} \frac{1}{x} \mathbb{1}_{(0,1)}(x) dx = \infty$$

and hence $f \in L^1(\mathbb{R})$ but $f \notin L^2(\mathbb{R})$, i.e.

$$L^1(\mathbb{R}) \not\subset L^2(\mathbb{R}).$$

Similarly, we may show that

$$\int_{\mathbb{R}} |g(x)|dx = \infty, \int_{\mathbb{R}} |g(x)|^2 dx = 1$$

and hence $g \in L^2(\mathbb{R})$ but $g \notin L^1(\mathbb{R})$, i.e.

$$L^2(\mathbb{R}) \not\subset L^2(\mathbb{R}).$$

(2) Let $f \in L^2(\Omega)$. We have

$$\|f\|_{L^1(\Omega)} = \int_\Omega |f(x)|dx = \int_\Omega 1 \times |f(x)|dx \leq |\Omega|^{\frac{1}{2}} \|f\|_{L^2(\Omega)}$$

by the Cauchy-Schwarz Inequality.

(3) From the assumption, $|f(x)| \leq K$ a.e. x we may write

$$|f(x)| \leq K \Longrightarrow |f(x)|^2 \leq K|f(x)| \Longrightarrow \int_{\mathbb{R}} |f(x)|^2 dx \leq K \int_{\mathbb{R}} |f(x)|dx$$

Therefore,

$$\|f\|_{L^2(\mathbb{R})} \leq K^{\frac{1}{2}} \|f\|_{L^1(\mathbb{R})}^{\frac{1}{2}}.$$

Solution 1.3.23. Let

$$f(x) = \frac{1}{x^{\frac{1}{4}}}, \ x > 0.$$

Clearly $f \notin L^\infty(0,\infty)$. It may also be shown that $f \notin L^2(0,\infty)$. However, we can write (can't we?)

$$f(x) = \underbrace{\mathbb{1}_{(0,1)}(x)\frac{1}{x^{\frac{1}{4}}}}_{f_1(x)} + \underbrace{\mathbb{1}_{[1,\infty)}(x)\frac{1}{x^{\frac{1}{4}}}}_{f_2(x)}.$$

Then clearly $f_1 \in L^2(0,\infty)$ and $f_2 \in L^\infty(0,\infty)$ and this means that $f \in L^2(0,\infty) + L^\infty(0,\infty)$.

Solution 1.3.24. Fix $-\frac{1}{2} < \alpha < 0$ and define φ in each interval $k < t \leq k+1$, $k \in \mathbb{Z}$, by

$$\varphi(t) = (t-k)^\alpha.$$

Then φ is not essentially bounded on $\mathbb{R}$. Moreover one has

$$\int_k^{k+1} |\varphi(t)|^2 dt = \int_k^{k+1} (t-k)^{2\alpha} dt = \frac{1}{2\alpha+1}.$$

Hence

$$\sup_{k\in\mathbb{Z}} \int_k^{k+1} |\varphi(t)|^2 dt = \frac{1}{2\alpha+1} < \infty.$$

But

$$\sum_{k=-\infty}^{\infty}\frac{1}{2\alpha+1}=\infty.$$

SOLUTION 1.3.25. Let $1 \leq p, q \leq \infty$ and let $\frac{1}{p}+\frac{1}{q}=\frac{1}{r}$ and $\frac{1}{p}+\frac{1}{q}\leq 1$. Let $f \in L^p(\mathbb{R}^n)$ and $g \in L^q(\mathbb{R}^n)$.

Observe that

$$|f(x)g(x)|^r = |f(x)|^r|g(x)|^r.$$

Also, observe that

$$\frac{1}{p}+\frac{1}{q}=\frac{1}{r}\Longrightarrow \frac{1}{\frac{p}{r}}+\frac{1}{\frac{q}{r}}=1.$$

Now apply the Hölder Inequality (Proposition 1.1.63) to $|f|^r$ and $|g|^r$ with respect to the exponents $\frac{p}{r}$ and $\frac{q}{r}$. We then obtain

$$\int_{\mathbb{R}^n}|f(x)g(x)|^r dx \leq \left(\int_{\mathbb{R}^n}|f(x)|^{\frac{rp}{r}}dx\right)^{\frac{r}{p}}\left(\int_{\mathbb{R}^n}|g(x)|^{\frac{rq}{r}}dx\right)^{\frac{r}{q}}$$

or merely

$$\left(\int_{\mathbb{R}^n}|f(x)g(x)|^r dx\right)^{\frac{1}{r}} \leq \left(\int_{\mathbb{R}^n}|f(x)|^p dx\right)^{\frac{1}{p}}\left(\int_{\mathbb{R}^n}|g(x)|^q dx\right)^{\frac{1}{q}},$$

that is,

$$\|fg\|_r \leq \|f\|_p\|g\|_q,$$

as required.

SOLUTION 1.3.26. Let $f \in L^p(\mathbb{R}^n)$. Also, let

$$E_\lambda = \{x \in \mathbb{R}^n : |f(x)| \geq \lambda\}.$$

Using

$$|E_\lambda| = \int\limits_{E_\lambda} dx,$$

where $|\cdot|$ denotes Lebesgue's measure, we then obtain

$$p\int_0^\infty \lambda^{p-1}|E_\lambda|d\lambda = p\int_0^\infty \lambda^{p-1}\int\limits_{E_\lambda} dxd\lambda.$$

Since everything is positive, we obtain (using the Fubini Theorem),

$$p\int_0^\infty \lambda^{p-1}|E_\lambda|d\lambda = p\int\limits_{\mathbb{R}^n}\left(\int_0^{|f(x)|}\lambda^{p-1}d\lambda\right)dx = \int\limits_{\mathbb{R}^n}|f(x)|^p dx = \|f\|_p^p,$$

as suggested.

Solution 1.3.27. ([**141**]) Let

$$E_\lambda = \{x \in \mathbb{R}^n : |f(x)| \geq \lambda\}.$$

Let $f \in L^r_w(\mathbb{R}^n)$. By definition

$$|E_\lambda| = |\{x \in \mathbb{R}^n : |f(x)| \geq \lambda\}| \leq c_r \lambda^{-r},$$

Also when $f \in L^s_w(\mathbb{R}^n)$, then

$$|E_\lambda| = |\{x \in \mathbb{R}^n : |f(x)| \geq \lambda\}| \leq c_s \lambda^{-s}$$

(here c_s and c_r denote $\|\cdot\|_{s,w}$ and $\|\cdot\|_{r,w}$ respectively). So

$$\begin{aligned}
\|f\|^p_{L^p(\mathbb{R}^n)} &= p\int_0^\infty \lambda^{p-1}|E_\lambda|d\lambda \\
&= p\int_0^1 \lambda^{p-1}|E_\lambda|d\lambda + p\int_1^\infty \lambda^{p-1}|E_\lambda|d\lambda \\
&\leq pc_r\int_0^1 \lambda^{p-r-1}d\lambda + pc_s\int_1^\infty \lambda^{p-s-1}d\lambda.
\end{aligned}$$

Hence

$$\|f\|^p_{L^p(\mathbb{R}^n)} \leq pc_r\left[\frac{\lambda^{p-r}}{p-r}\right]_0^1 + pc_s\left[\frac{\lambda^{p-s}}{p-s}\right]_1^\infty$$

which is finite if $r < p < s$. Therefore

$$\|f\|^p_{L^p(\mathbb{R}^n)} \leq \frac{p}{p-r}\|f\|^r_{r,w} + \frac{p}{s-p}\|f\|^s_{s,w}.$$

Thus

$$\|f\|_{L^p(\mathbb{R}^n)} \leq \tilde{c}\|f\|^{\frac{r}{p}}_{r,w} + \tilde{c}\|f\|^{\frac{s}{p}}_{s,w} ...(1)$$

for some constant $\tilde{c}$ depending on p, r and s.

Now we proceed to make all the powers in (1) equal to one. We replace f by cf where c is a constant to be determined. We then have

$$\|f\|_{L^p(\mathbb{R}^n)} \leq \tilde{c}c^{\frac{r-p}{p}}\|f\|^{\frac{r}{p}}_{r,w} + \tilde{c}c^{\frac{s-p}{p}}\|f\|^{\frac{s}{p}}_{s,w}.$$

Minimizing the quantity on the right hand side with respect to c shows that

$$\|f\|_{L^p(\mathbb{R}^n)} \leq \tilde{c}\left(\|f\|^{\frac{s(r-p)}{p(r-s)}}_{s,w}\|f\|^{\frac{r(r-p)}{(s-r)p}+\frac{r}{p}}_{r,w}\right) + \tilde{c}\left(\|f\|^{\frac{s(s-p)}{(r-s)p}+\frac{s}{p}}_{s,w}\|f\|^{\frac{r(s-p)}{p(s-r)}}_{r,w}\right).$$

Now since the sum of the powers in each part of the right hand side is one, Young's inequality (see e.g. [**153**]) shows that

$$\|f\|_p \leq a\|f\|_{r,w} + b\|f\|_{s,w},$$

quod erat demonstrandum.

Solution 1.3.28.

(1) The sequence (e_n) belongs to the closed ball in ℓ^1 as

$$\|e_n - 0\| = \|e_n\| = 1.$$

The sequence (e_n) cannot be Cauchy as for $n > m$ say, one has

$$\|e_n - e_m\| = 2.$$

It also follows that (e_n) cannot have any Cauchy subsequence and hence none of its subsequences converges. Thus $B_c(0,1)$ is not compact.

(2) As $B_c(0,1)$ is not compact, then ℓ^1 is infinite-dimensional.

(3) We can use the method of the previous question to prove ℓ^2 and ℓ^∞ are also infinite-dimensional...

SOLUTION 1.3.29.

(1) Denote the closed unit ball in X by $B_c(0,1)$.

(a) Assume that $B_c(0,1)$ is compact and let $x \in X$. Define a function $f : B_c(0,1) \to B_c(x,1)$, for all $y \in B_c(0,1)$, by: $f(y) = x + y$. Then f is continuous and onto and hence

$$f(B_c(0,1)) = B_c(x,1)$$

is compact. It certainly contains a neighborhood of x (for example, $B(x,1)$, the open ball!). This proves that X is locally compact.

(b) Now, suppose that X is locally compact. Then there exists a compact set K which contains a neighborhood of 0, i.e.

$$\exists r > 0,\ B(0,r) \subset K.$$

Whence

$$\overline{B(0,r)} = B_c(0,r) \subset K$$

for K is closed. This also tells us that $B_c(0,r)$ is compact. To finish the proof, define a function

$$f : B_c(0,r) \longrightarrow B_c(0,1),$$

for all $x \in B_c(0,r)$, by: $f(x) = \frac{1}{r}x$. Since f is continuous and onto, we immediately deduce that

$$f(B_c(0,r)) = B_c(0,1)$$

is compact.

(2) Let

$$B(0,1) = \{x \in X : \|x\| \leq 1\}.$$

Let $x_1 \in B(0,1)$. Since $\dim X = \infty$, surely $\operatorname{span}\{x_1\} \neq X$. Also $\operatorname{span}\{x_1\}$ is closed for it is a finite dimensional vector space. By Theorem 1.1.40, and choosing $\alpha = \frac{1}{2}$, we know that

$$\exists x_2 \in B(0,1) : \|x_2 - y\| \geq \frac{1}{2}$$

for all $y \in \operatorname{span}\{x_1\}$, that is for all $\alpha \in \mathbb{F}$, with $y = \alpha x_1$. Hence (setting $\alpha = 1$)

$$\|x_2 - x_1\| \geq \frac{1}{2}.$$

Similarly, $\operatorname{span}\{x_1, x_2\} \neq X$, and it is also closed for it is a finite dimensional vector space. By Theorem 1.1.40 again,

$$\exists x_3 \in B(0,1) : \|x_3 - y\| \geq \frac{1}{2}$$

for all $y \in \operatorname{span}\{x_1, x_2\}$, that is for all $\alpha, \beta \in \mathbb{F}$, with $y = \alpha x_1 + \beta y_2$. Hence setting respectively $\alpha = 1, \beta = 0$ and $\alpha = 0, \beta = 1$, we obtain

$$\|x_3 - x_1\| \geq \frac{1}{2} \text{ and } \|x_3 - x_2\| \geq \frac{1}{2}.$$

Proceeding inductively, we obtain

$$\exists x_{n+1} \in B(0,1) : \|x_{n+1} - y\| \geq \frac{1}{2}$$

for all $y \in \operatorname{span}\{x_1, x_2, \ldots, x_n\}$. Thus, we obtain a sequence (x_n) such that

$$\|x_n - x_m\| \geq \frac{1}{2}, \ \forall n \neq m,$$

and such a sequence cannot have any convergent subsequence and hence $B(0,1)$ is not sequentially compact, that is, it is not compact.

Remark. The same proof may be used to show that $S(0,1)$ is not compact.

1.4. Hints/Answers to Tests

Solution 1.4.1. No! It is certainly complete but it is not a vector space...

Solution 1.4.2. No! We can show that its complement, i.e. $\mathcal{P}^c$ is not closed in $(C[-1,1], \|\cdot\|_\infty)$. It suffices to find a non polynomial sequence which converges to a polynomial. Let

$$f(x) = x \text{ and } f_n(x) = \frac{1}{n}e^x + x.$$

Then f is a polynomial, i.e. $f \notin \mathcal{P}^c$ and for all n, $f_n \in \mathcal{P}^c$. Finally,

$$\|f_n - f\|_\infty = \sup_{x\in[-1,1]} \left|\frac{1}{n}e^x\right| \longrightarrow 0$$

as n approaches ∞.

Solution 1.4.3. A is not closed...

CHAPTER 2

Bounded Linear Operators on Banach Spaces

2.2. True or False: Answers

ANSWERS.

(1) True! We know from Theorem 2.1.78 that a Banach space with a Schauder basis is automatically separable. Since ℓ^∞ is a Banach space without being separable (by Exercise 1.3.19), it does not have a Schauder basis.

(2) In this particular example, it depends on what we mean by boundedness. Obviously, A is linear and besides for all $x \in \mathbb{R}$, we have

$$\|Ax\| = |Ax| = 2|x| = 2\|x\|$$

so that A is bounded in the sense of Definition 2.1.2.

If, however, we talk about boundedness in the sense of functions from $\mathbb{R}$ into $\mathbb{R}$, then obviously A is not bounded. In fact, we do know that just in $\mathbb{R}$ (let alone other cases) $A : \mathbb{R} \to \mathbb{R}$ is linear iff $Ax = \alpha x$, where $\alpha \in \mathbb{R}$. With this observation, all linear "functions" are never bounded in the classical sense (unless $\alpha = 0$)! So readers should forget about the boundedness in the classical sense in this book and should remember what we mean by a bounded operator (as in Definition 2.1.2!).

(3) No! Take

$$Af(x) = (x+1)f(x)$$

defined on $L^2[0,1]$. Then A is invertible and we have

$$A^{-1}f(x) = \frac{1}{x+1}f(x).$$

Hence (see Exercise 2.3.3)

$$\|A\| = 2 \text{ and } \|A^{-1}\| = 1.$$

Therefore,

$$\|A^{-1}\| \neq \frac{1}{\|A\|}.$$

What we always have is

$$\|A\| \geq \frac{1}{\|A^{-1}\|} \text{ or } \|A^{-1}\| \geq \frac{1}{\|A\|},$$

both coming from

$$1 = \|I\| = \|AA^{-1}\| \leq \|A\|\|A^{-1}\|.$$

Remark. Even in a finite dimensional setting, this is not always true. The reader may look for a counterexample.

(4) True! Let $A : X \to X$ be linear. We know by Theorem 2.1.4 that A is continuous everywhere iff A is continuous at 0. Thanks to the linearity of A, we may easily see that A is continuous everywhere iff A is continuous at x_0, where $x_0 \in X$. Therefore, a linear operator is either continuous everywhere or discontinuous everywhere.

(5) False in general! If $\dim X < \infty$, then the result holds simply because $S(0,1)$ (or $B(0,1)$), the unit sphere, is compact (Corollary 1.1.42), and since also $f : X \to \mathbb{R}^+$ defined by $x \mapsto f(x) = \|Ax\|$ is continuous as whenever $x_n \to x$, we have by Theorem 1.1.5

$$0 \leq |f(x_n) - f(x)| = |\|Ax_n\| - \|Ax\|| \leq \|Ax_n - Ax\| \leq \|A\|\|x_n - x\| \longrightarrow 0$$

as $n \to \infty$. Thus the supremum is attained, that is, it is a maximum.

If $\dim X = \infty$, then we know that neither $B(0,1)$ nor $S(0,1)$ is compact. So we can no more argue as before. A counterexample may be found in Exercise 2.3.18.

(6) The answer is no in general. For instance, let I be the identity mapping from X into Y where $X = C[0,1]$ is equipped with $\|f\|_1 = \int_0^1 |f(x)|dx$ and $Y = C[0,1]$ is equipped with $\|f\|_\infty = \sup_{x\in[0,1]} |f(x)|$. Then I is not continuous (see Exercise 2.3.48). Nevertheless,

$$\ker I = \{f \in X : \; If(x) = f(x) = 0, \; \forall x \in [0,1]\} = \{0\},$$

being a singleton in a separated space, is closed in X.

Of course, if A is continuous, then $\ker A$ is closed. It is also worth noting that if A is a linear *functional*, then $\ker A$ is closed iff A is continuous (see Exercise 2.3.20).

(7) The answer is yes except for the trivial case when $X = \{0\}$. Indeed, in that event, $B(X,Y) = B(\{0\},Y)$ reduces to the vector space constituted of the zero operator. It is finite and

so it is a Banach space. However, Y can be any non necessarily complete normed vector space.

If $X \neq \{0\}$, then the result is true, i.e. $B(X,Y)$ being Banach does imply that Y is a Banach space. A proof (borrowed from [**50**]) may be found in Exercise 2.3.43.

(8) True. Since $A^2 = I$, $AA = I$, and hence A is invertible with inverse $A^{-1} = A$.

(9) The right-to-left implication does not hold in general, i.e. $I - A$ may well be invertible even when $\|A\| \geq 1$. For a counterexample, let

$$Af(x) = (e^x + 1)f(x) \text{ be defined for all } f \in C([0,1],\mathbb{R})$$

with respect to the supremum norm. Then it is easy to see that $I - A$ is invertible with inverse given by

$$(I - A)^{-1}f(x) = -e^{-x}f(x)$$

whereas

$$\|A\| = \max_{0\leq x\leq 1} |e^x + 1| = e + 1 \geq 1.$$

Obviously, the left-to-right implication is always true (the Neumann Series!).

(10) True. If $\|Ax\| \leq \|x\|$ for all $x \in X$, then clearly $\|A\| \leq 1$. Conversely, if $\|A\| \leq 1$, then for all $x \in X$ we have

$$\|Ax\| \leq \|A\|\|x\| \leq \|x\|.$$

2.3. Solutions to Exercises

Solution 2.3.1.

(1) We need to show that A is a continuous linear functional on X. The linearity of A is left to readers. Let us show the continuity of A. Let $f \in X$. We have

$$\begin{aligned}
|Af| &\leq |f(-2)| + |f(-1)| + 3|f(0)| + 2|f(1)| + 3|f(2)| + 2|f(3)| \\
&\leq \sup_{x\in[-2,3]} |f(x)| + \sup_{x\in[-2,3]} |f(x)| + 3\sup_{x\in[-2,3]} |f(x)| \\
&+ 2\sup_{x\in[-2,3]} |f(x)| + 3\sup_{x\in[-2,3]} |f(x)| + 2\sup_{x\in[-2,3]} |f(x)| \\
&= 12\sup_{x\in[-2,3]} |f(x)| \\
&= 12\|f\|_\infty.
\end{aligned}$$

Hence A is continuous and besides

$$\|A\| = \sup_{\|f\|_\infty=1} \frac{|Af|}{\|f\|_\infty} \leq 12.$$

(2) We claim that $\|A\| = 12$. We already know from the previous question that $\|A\| \leq 12$. So, we need only show that $\|A\| \geq 12$. We need to find a *continuous* function g such that $\|g\|_\infty = 1$ and

$$\|A\| = \sup_{\|f\|_\infty=1} |Af| \geq |Ag| = 12.$$

For the last purpose, it will be sufficient that g satisfies

$$g(-2) = g(0) = g(1) = g(2) = 1 \text{ and } g(-1) = g(3) = -1.$$

The following function

$$x \mapsto g(x) = \begin{cases} -2x-3, & -2 \leq x < -1, \\ 2x+1, & -1 \leq x < 0, \\ 1, & 0 \leq x < 2, \\ -2x+5, & 2 \leq x < 3 \end{cases}$$

will do, i.e. g is continuous on $[-2,3]$, $\|g\|_\infty = 1$ and $Ag = 12$. Thus $\|A\| = 12$.

SOLUTION 2.3.2.

(1) The linearity of A is left to the reader. We need to show that $Ax \in \ell^p$, whenever $x \in \ell^p$. Let $1 \leq p < \infty$. We have for all n

$$|\alpha_n x_n|^p \leq \|\alpha\|_\infty^p |x_n|^p.$$

Then the series $\sum_n |\alpha_n x_n|^p$ converges by the comparison test as so does the series $\|\alpha\|_\infty^p \sum_n |x_n|^p$. This proves that $Ax \in \ell^p$. By the previous inequality, we deduce that

$$\|A\| \leq \|\alpha\|_\infty.$$

Now, if $e_n = (0,0,\cdots,\underbrace{1}_{n},0,\cdots)$ (which is clearly in ℓ^p), then for all n

$$\|A\| = \sup_{x\neq 0} \frac{\|Ax\|}{\|x\|} \geq \frac{\|Ae_n\|_p}{\|e_n\|_p} = |\alpha_n|.$$

Thus

$$\|A\| \geq \|\alpha\|_\infty \text{ or } \|A\| = \|\alpha\|_\infty.$$

(2) By the previous question we do know that $(\alpha_n x_n)$ is in ℓ^1, showing that the series intervening in the definition of B is convergent, that is, it is an element of $\mathbb{C}$. Hence B is well-defined and, as above, linearity is left to the interested reader. Next, we have for all n

$$|\alpha_n x_n| \le \|\alpha\|_\infty |x_n|$$

and hence

$$|Bx| = \left|\sum_{n=1}^{\infty} \alpha_n x_n\right| \le \sum_{n=1}^{\infty} |\alpha_n x_n| \le \|\alpha\|_\infty \sum_{n=1}^{\infty} |x_n| = \|\alpha\|_\infty \|x\|_1.$$

Therefore

$$\|B\| \le \|\alpha\|_\infty.$$

Conversely, let $e_n = (0, 0, \cdots, \underbrace{1}_{n}, 0, \cdots)$. Then

$$\|B\| \ge \frac{|Be_n|}{\|e_n\|_1} = |\alpha_n|$$

which holds for all n.

Thus

$$\|B\| \ge \|\alpha\|_\infty \text{ or } \|B\| = \|\alpha\|_\infty.$$

Solution 2.3.3.

(1) Let $f, g \in L^2([0,1])$ and let $\alpha, \beta \in \mathbb{C}$. We have

$$\begin{aligned} A(\alpha f + \beta g)(x) &= \varphi(x)(\alpha f(x) + \beta g(x)) \\ &= \alpha\varphi(x)f(x) + \beta\varphi(x)g(x) \\ &= \alpha(Af)(x) + \beta(Ag)(x). \end{aligned}$$

Accordingly, A is linear.

Since φ is continuous on a compact set, it is bounded and so

$$\|\varphi\|_\infty = \sup_{x\in[0,1]} |\varphi(x)| < \infty.$$

As $\forall x \in [0,1] : |\varphi(x)|^2 \le \|\varphi\|_\infty^2$, then

$$\int_0^1 |\varphi(x)f(x)|^2 dx \le \|\varphi\|_\infty^2 \int_0^1 |f(x)|^2 dx.$$

Hence

$$\|Af\|_2 \le \|\varphi\|_\infty \|f\|_2$$

for all $f \in L^2([0,1])$. This shows that A is bounded and that

$$\|A\| \le \|\varphi\|_\infty.$$

(2) We claim that $\|A\| = \|\varphi\|_\infty$, and so we only show

$$\|A\| \geq \|\varphi\|_\infty.$$

Since φ is continuous on the compact $[0,1]$, we know that

$$\|\varphi\|_\infty = |\varphi(a)|$$

for some $a \in [0,1]$. Consider the sequence of functions (f_n) defined by

$$f_n(x) = \begin{cases} \sqrt{\frac{n}{2}}, & a - \frac{1}{n} \leq x \leq a + \frac{1}{n} \\ 0, & \text{otherwise.} \end{cases}$$

As one can check, $\|f_n\|_2 = 1$ for all n. Finally, invoking the Mean Value Theorem for integrals and the continuity of φ, we obtain

$$\|A\|^2 \geq \|Af_n\|^2 = \frac{n}{2}\int_{a-\frac{1}{n}}^{a+\frac{1}{n}} |\varphi(x)|^2 dx \longrightarrow |\varphi(a)|^2$$

as $n \to \infty$. Consequently,

$$\|A\| \geq |\varphi(a)| = \|\varphi\|_\infty,$$

as needed.

SOLUTION 2.3.4. Let (A_n) and (B_n) be two convergent sequences to A and B respectively with respect to the operator norm, that is,

$$\lim_{n\to\infty} \|A_n - A\| = 0 \text{ and } \lim_{n\to\infty} \|B_n - B\| = 0.$$

Hence $(\|B_n\|)$ is bounded by some $M \geq 0$, say, and also

$$\begin{aligned} 0 \leq \|A_nB_n - AB\| &= \|A_nB_n - AB_n + AB_n - AB\| \\ &\leq \|A_nB_n - AB_n\| + \|AB_n - AB\| \\ &= \|(A_n - A)B_n\| + \|A(B_n - B)\| \\ &\leq \|A_n - A\|\|B_n\| + \|A\|\|B_n - B\| \\ &\leq \|A_n - A\|M + \|A\|\|B_n - B\| \longrightarrow 0 \end{aligned}$$

as n tends to ∞. So much for the case of convergence in norm.

Now, assume that (A_n) and (B_n) converge strongly to A and B, i.e.

$$\lim_{n\to\infty} \|A_nx - Ax\| = 0 \text{ and } \lim_{n\to\infty} \|B_nx - Bx\| = 0$$

for all $x \in X$.

So, let $x \in X$. Then as before, we obtain

$$0 \leq \|A_nB_nx - ABx\| \leq \|(A_n - A)Bx\| + \|A_n\|\|B_nx - Bx\|.$$

But since $(\|A_n x\|)$ is bounded (why?), Theorem 2.1.107 tells us that also $(\|A_n\|)$ is bounded. Hence

$$\lim_{n\to\infty} \|A_n B_n x - ABx\| = 0,$$

as required.

Solution 2.3.5. Let (A_n) be a Cauchy sequence in $B(X,Y)$. Hence (A_n) is bounded, i.e.

$$\exists M > 0,\ \forall n \in \mathbb{N}:\ \|A_n\| \leq M.$$

The idea now is to place some Cauchy sequence in Y and then exploit its completeness. Let $x \in X$. We have (for all $n, m \in \mathbb{N}$)

$$\|A_n x - A_m x\| = \|(A_n - A_m)x\| \leq \|A_n - A_m\|\|x\|.$$

Since (A_n) is Cauchy, it is clear that so is $(A_n x)$ in Y. Since Y is a Banach space, $(A_n x)$ converges to something in Y. We may then define $A : X \to Y$ by

$$Ax = \lim_{n\to\infty} A_n x$$

(remember that this means $\|A_n x - Ax\| \to 0$). What remains to show is: A is in $B(X,Y)$ and that $\|A_n - A\| \to 0$ as n tends to infinity.

(1) $A \in B(X,Y)$:

(a) $A \in L(X,Y)$: Let $x, y \in X$ and $\alpha, \beta \in \mathbb{K}$. Then by the linearity of each A_n we may write

$$\begin{aligned} A(\alpha x + \beta x) &= \lim_{n\to\infty} A_n(\alpha x + \beta y) \\ &= \lim_{n\to\infty} [A_n(\alpha x) + A_n(\beta y)] \\ &= \lim_{n\to\infty} (\alpha A_n x + \beta A_n y) \\ &= \lim_{n\to\infty} \alpha A_n x + \lim_{n\to\infty} \beta A_n y \\ &= \alpha \lim_{n\to\infty} A_n x + \beta \lim_{n\to\infty} A_n y \\ &= \alpha Ax + \beta Ay. \end{aligned}$$

Thus A is linear.

(b) A is bounded: Since $\|A_n x - Ax\| \to 0$, it follows by Theorem 1.1.5 that $\|A_n x\| \to \|Ax\|$. Since (A_n) is bounded (with respect to n) and each A_n is in $B(X,Y)$, we have

$$\|A_n x\| \leq \|A_n\|\|x\| \leq M\|x\|$$

so that by passing to the limit we get

$$\|Ax\| = \lim_{n\to\infty} \|A_n x\| \leq M\|x\|.$$

This settles the proof that $A \in B(X,Y)$.

(2) $\|A_n - A\| \to 0$: First, we use again the assumption that (A_n) is Cauchy, i.e. for any $\varepsilon > 0$, there is $N \in \mathbb{N}$ such that for $n, m \geq N$, we have

$$\|A_n - A_m\| < \varepsilon.$$

Before carrying on, we choose the norm $\|A\| = \sup_{\|x\|=1} \|Ax\|$. So if $x \in X$ is such that $\|x\| = 1$, then

$$\|A_n x - A_m x\| \leq \|A_n - A_m\| \|x\| < \varepsilon$$

for all $n, m \geq N$.

Now, keeping n fixed, and letting m goes to infinity, we obtain for all $x \in X$ such that $\|x\| = 1$

$$\|A_n x - Ax\| \leq \varepsilon$$

for all $n \geq N$.

Passing to the supremum over the unit sphere in X gives us in the end (for all $n \geq N$)

$$\|A_n - A\| \leq \varepsilon,$$

meaning that $\|A_n - A\| \to 0$.

Solution 2.3.6. First, BA is linear as it is the composition of linear mappings. Now, for any $x \in X$

$$\|BAx\| = \|B(Ax)\| \leq \|B\| \|Ax\| \leq \|B\| \|A\| \|x\|.$$

Hence BA is bounded, and passing to the supremum over $\|x\| = 1$ gives

$$\|BA\| \leq \|B\| \|A\|,$$

as required.

Solution 2.3.7. We use a proof by induction. The statement is obviously true for $n = 0$ as

$$\|A^0\| = \|I\| = 1 = \|A\|^0.$$

Now we assume that $\|A^n\| \leq \|A\|^n$ and we show that $\|A^{n+1}\| \leq \|A\|^{n+1}$. Clearly,

$$\|A^{n+1}\| = \|A^n A\| \text{ and hence } \|A^n A\| \leq \|A^n\| \|A\|$$

from the previous exercise. The induction hypothesis can be easily used now to establish

$$\|A^{n+1}\| \leq \|A\|^{n+1}.$$

SOLUTION 2.3.8. *Recall the following **Hölder Inequality**:*

$$\sum_{k=1}^{n} a_k b_k \leq \left(\sum_{k=1}^{n}(a_k)^p\right)^{\frac{1}{p}} \left(\sum_{k=1}^{n}(b_k)^q\right)^{\frac{1}{q}}$$

holding for positive numbers $a_1, \cdots, a_n$ *and* $b_1, \cdots, b_n$ *with* $\frac{1}{p} + \frac{1}{q} = 1$.

Clearly,

$$\left\|\sum_{k=1}^{n} A_k B_k\right\| \leq \sum_{k=1}^{n} \|A_k B_k\| \leq \sum_{k=1}^{n} \|A_k\| \|B_k\|.$$

The proof of the desired inequality then follows by invoking the above Hölder Inequality and by observing that $\|A_k\|$ and $\|B_k\|$ are positive numbers...

SOLUTION 2.3.9. Set $\alpha = \inf_{n\in\mathbb{N}} \|A^n\|^{\frac{1}{n}}$. Then, clearly

$$\alpha \leq \liminf_{n\to\infty} \|A^n\|^{\frac{1}{n}}.$$

By definition of "inf", we know that for any $\varepsilon > 0$, there is some $N \in \mathbb{N}$ for which

$$\|A^N\|^{\frac{1}{N}} \leq \alpha + \varepsilon.$$

On the other hand, we can express each n as $n = Np + q$, where p and q both depend on n and $0 \leq q < N$. Therefore,

$$\|A^n\| = \|A^{Np}A^q\| \leq \|A^{Np}\|\|A^q\| = \|(A^N)^p\|\|A^q\| \leq \|A^N\|^p \|A\|^q$$

and so

$$\|A^n\|^{\frac{1}{n}} \leq \|A^N\|^{\frac{p}{n}} \|A\|^{\frac{q}{n}}.$$

Passing to "lim sup" (reason: We don't know yet that $\|A^n\|^{\frac{1}{n}}$ converges!), and observing that $\frac{q}{n} \to 0$ and $\frac{p}{n} \to \frac{1}{N}$, we get

$$\limsup_{n\to\infty} \|A^n\|^{\frac{1}{n}} \leq \|A^N\|^{\frac{1}{N}} \leq \alpha + \varepsilon$$

which is valid for all $\varepsilon > 0$. Accordingly,

$$\limsup_{n\to\infty} \|A^n\|^{\frac{1}{n}} \leq \alpha \leq \liminf_{n\to\infty} \|A^n\|^{\frac{1}{n}}.$$

Since the reverse inequality always holds, we finally infer that

$$\lim_{n\to\infty} \|A^n\|^{\frac{1}{n}} = \alpha = \inf_{n\in\mathbb{N}} \|A^n\|^{\frac{1}{n}},$$

as needed.

SOLUTION 2.3.10.

(1) We use a proof by induction. The statement clearly holds for $n = 0$. Assume that $(A - B)^{2n+1} = A - B$. Then we have
$$(A - B)^{2n+3} = (A - B)^{2n+1}(A - B)^2 = (A - B)(A - B)^2.$$
But as $AB = BA$, $A^2 = A$ and $B^2 = B$, then
$$(A - B)(A - B)^2 = A^3 - 3A^2B + 3AB^2 - B^3 = A - B,$$
completing the proof.

(2) Since $(A - B)^{2n+1} = A - B$ for all n, we may write for all n
$$\|A - B\| = \|(A - B)^{2n+1}\| \leq \|A - B\|^{2n+1}.$$
It thus becomes clear that if $\|A - B\| < 1$, then passing to the limit as $n \to \infty$, we obtain $A = B$.

SOLUTION 2.3.11. The answer is no. Let us show by induction that for any $n \in \mathbb{N}$:
$$A^nB - BA^n = nA^{n-1}.$$
The statement is true for $n = 1$ since
$$A^1B - BA^1 = AB - BA = I = 1A^{1-1}.$$
Assume now that $A^nB - BA^n = nA^{n-1}$ and we will show that
$$A^{n+1}B - BA^{n+1} = (n + 1)A^n.$$
We have
$$\begin{aligned} A^{n+1}B - BA^{n+1} &= A^n(AB - BA) + (A^nB - BA^n)A \\ &= A^nI + nA^{n-1}A \\ &= A^n + nA^n \\ &= (n + 1)A^n. \end{aligned}$$
This finishes the proof by induction.

Now we have
$$\|nA^{n-1}\| = \|A^nB - BA^n\| \leq \|A^nB\| + \|BA^n\| \leq 2\|B\|\|A\|\|A^{n-1}\|.$$
Two cases are to be examined.

(1) If $\|A^{n-1}\| \neq 0$, then one gets
$$2\|B\|\|A\| \geq n, \ \forall n \in \mathbb{N}.$$
This contradicts the hypotheses A and B being both bounded.

(2) If $\|A^{n-1}\| = 0$, then the relation $\|nA^{n-1}\| \leq 2\|B\|\|A^n\|$ (by changing n by $n - 1$) would imply that $\|A^{n-2}\| = 0$. Carrying on this process we arrive at $A = 0$, which is clearly a contradiction. Hence this case cannot occur either.

Therefore,

$$\nexists A, B \in B(X) : AB - BA = I.$$

SOLUTION 2.3.12. By hypothesis,

$$\exists \alpha > 0, \ \forall x \in X : \ \|Ax\| \geq \alpha \|x\| ...(1).$$

To show that $\operatorname{ran}A$ is closed, we need only prove that $\overline{\operatorname{ran}A} \subset \operatorname{ran}A$. Let $y \in \overline{\operatorname{ran}A}$. Then there exists a sequence (y_n) in $\operatorname{ran}A$ such that $\|y_n - y\|_Y \to 0$ as n goes to ∞. Since y_n is in $\operatorname{ran}A$, for some $x_n \in X$, $y_n = Ax_n$. Hence (Ax_n) converges in Y and so it is Cauchy still in Y. This implies that (x_n) too is Cauchy (in X!) since by Inequality (1) we have for all $n, m \in \mathbb{N}$

$$\|x_n - x_m\| \leq \frac{1}{\alpha} \|Ax_n - Ax_m\|.$$

Since X is complete, (x_n) converges to some $x \in X$. Since A is continuous, Ax_n converges to Ax. By the uniqueness of the limit, $Ax = y$, so that $y \in \operatorname{ran}(A)$, as needed.

SOLUTION 2.3.13.

(1) Let

$$\|x\|_{G(A)} = \sqrt{\|x\|_X^2 + \|Ax\|_Y^2}.$$

It defines a norm on X since:

(a) For all $x \in X$, $\|x\|_{G(A)} \geq 0$.

(b) If $\|x\|_{G(A)} = 0$, then $\|x\|_X = \|Ax\|_Y = 0$ so that $x = 0$, and if $x = 0$, then clearly $\|x\|_X = \|Ax\|_Y = 0$, that is, $\|x\|_{G(A)} = 0$.

(c) For all $\lambda \in \mathbb{F}$ and all $x \in X$:

$$\|\lambda x\|_{G(A)} = (\|\lambda x\|_X^2 + \|A(\lambda x)\|_Y^2)^{\frac{1}{2}} = (|\lambda|^2 \|x\|_X^2 + |\lambda|^2 \|A(x)\|_Y^2)^{\frac{1}{2}} = |\lambda| \|x\|_{G(A)}$$

for A is linear and $\|\cdot\|_X$ and $\|\cdot\|_Y$ are norms.

(d) For all $x, y \in X$,

$$\begin{aligned} \|x+y\|_{G(A)}^2 &= \|x+y\|_X^2 + \|A(x+y)\|_Y^2 \\ &= \|x+y\|_X^2 + \|Ax+Ay\|_Y^2 \text{ (as } A \text{ is linear)} \\ &\leq (\|x\|_X + \|y\|_X)^2 + (\|Ax\|_Y + \|Ay\|_Y)^2 \end{aligned}$$

because $\|\cdot\|_X$ and $\|\cdot\|_Y$ are norms. Now, taking square roots in the previous inequality leads to

$$\|x+y\|_{G(A)} \leq \|x\|_X + \|y\|_X + \|Ax\|_Y + \|Ay\|_Y$$

signifying that

$$\|x+y\|_{G(A)} \leq \|x\|_{G(A)} + \|y\|_{G(A)}.$$

(2) The proof that $\|\cdot\|_1$ is a norm on X is very similar to the one for $\|\cdot\|_{G(A)}$ and so we leave it to readers.
(3) The norms $\|\cdot\|_{G(A)}$ and $\|\cdot\|_1$ are equivalent. To see this, it suffices to establish the following double inequality:
$$\exists \alpha, \beta > 0, \ \forall a, b \geq 0: \ \alpha(a+b) \leq \sqrt{a^2+b^2} \leq \beta(a+b).$$
These inequalities are easily deduced by taking square roots once we check that
$$(a+b)^2 \geq a^2+b^2$$
and
$$2a^2+2b^2 \geq (a+b)^2$$
(so that $\alpha = 2^{-\frac{1}{2}}$ and $\beta = 1$).

Solution 2.3.14.

(1) Let $A \in L(X,Y)$. We already saw in Exercise 2.3.13 that
$$\|x\|_1 = \|x\|_X + \|Ax\|_Y \ (\geq \|Ax\|_Y)$$
defines a norm on X. Since X is finite dimensional, Proposition 1.1.34 tells us that $\|\cdot\|_1$ is equivalent to $\|\cdot\|_X$. Hence
$$\exists \alpha > 0, \ \forall x \in X: \ \|x\|_1 \leq \alpha \|x\|_X$$
and so
$$\exists \alpha > 0, \ \forall x \in X: \ \|Ax\|_Y \leq \alpha \|x\|_X,$$
i.e. $A \in B(X,Y)$.
(2) It suffices to exhibit a sequence (p_n) in X (of norm one!) such that $\|Ap_n\|$ goes to infinity. Consider $p_n(x) = x^n$, where $x \in [0,1]$. Then
$$\|p_n\|_\infty = \sup\{x^n : \ x \in [0,1]\} = 1,$$
whereas
$$\|Ap_n\| = |Ap_n| = |p_n'(1)| = |n| = n \longrightarrow \infty$$
as $n \to \infty$. Therefore, A is unbounded.

Solution 2.3.15.

(1) It is clear, using basic properties of integrals, that both A and B linear. The boundedness of A follows from
$$\forall f \in X, \ |Af| = \left|\int_0^1 f(x)dx\right| \leq \int_0^1 |f(x)|dx \leq \|f\|_\infty \int_0^1 dx = \|f\|_\infty.$$
Hence
$$\|A\| = \sup_{\|f\|_\infty = 1} |Af| \leq 1.$$

To prove that B is bounded, a similar method may be used with a few more steps though. Let $f \in X$ and $x \in [0,1]$. Then

$$\begin{aligned}|[Bf](x)| &= \left|\int_0^x f(y)dy\right| \\ &\leq \int_0^x |f(y)|dy \\ &\leq \max_{0\leq y\leq x} |f(y)| \int_0^x dy \\ &= \max_{0\leq y\leq x} |f(y)|(x-0) \\ &\leq \|f\|_\infty\end{aligned}$$

since $x \leq 1$ and $\max_{0\leq y\leq x}|f(y)| \leq \max_{0\leq y\leq 1}|f(y)|$. Therefore,

$$\|B\| = \sup\{\|Bf\|_\infty : \|f\|_\infty = 1\} = \sup\{\max_{0\leq x\leq 1} |[Bf](x)| : \|f\|_\infty = 1\} \leq 1.$$

(2) To show that $\|A\| = \|B\| = 1$, based on the previous question, it then suffices to prove that $\|A\| \geq 1$ and $\|B\| \geq 1$.

Let's start with A. Let $f_0(x) = 1$ for all $x \in [0,1]$. Then $\|f_0\|_\infty = 1$ and

$$\|A\| = \sup_{\|f\|_\infty = 1} |Af| \geq |Af_0| = \int_0^1 1dx = 1,$$

proving that $\|A\| = 1$.

As for B, consider again $f_0(x) = 1$ for all $x \in [0,1]$. Then

$$\|B\| = \sup_{\|f\|_\infty = 1} \|Bf\|_\infty \geq \|Bf_0\|_\infty.$$

But

$$Bf_0(x) = \int_0^x dy = x \text{ and so } \|Bf_0\|_\infty = \sup_{0\leq x\leq 1} |Bf_0(x)| = 1.$$

Finally, $\|B\| \geq 1$ and thus $\|B\| = 1$.

Solution 2.3.16. First, note that both A and B are obviously linear.

(1) The functional A is continuous since for any $f \in X$, one has

$$|A(f)| = |f(0)| \leq \sup_{0\leq x\leq 1} |f(x)| = \|f\|_\infty.$$

(2) The functional B cannot be continuous. For if we consider the following sequence of *continuous* functions defined by

$$f_n(x) = \begin{cases} n(1-nx), & 0 \le x \le \frac{1}{n}, \\ 0, & \frac{1}{n} \le x \le 1, \end{cases}$$

then we easily see that $\|f_n\|_1 = \frac{1}{2}$ and hence

$$\frac{|Bf_n|}{\|f_n\|_1} = \frac{|f_n(0)|}{\|f_n\|_1} = 2n,$$

which cannot be bounded!

SOLUTION 2.3.17.

(1) The linearity of φ can easily be deduced from that of the integral. Let us show now the boundedness of φ. Let $f \in X$. Since $|f(x)| \le \|f\|_\infty$, $\forall x \in [0,1]$, one has

$$|\varphi(f)| = \left|\int_0^1 f(x)w(x)dx\right| \le \int_0^1 |f(x)w(x)|dx \le \|f\|_\infty \int_0^1 |w(x)|dx.$$

(2) From the previous question we see that

$$\|\varphi\| = \sup_{\|f\|_\infty = 1} |\varphi(f)| \le \int_0^1 |w(x)|dx$$

and we have not needed so far the fact that w has a constant sign. Now, a known result from Calculus says that if an integrable function w has a constant sign over I, then $\int_I |w(x)|dx = |\int_I w(x)dx|$. Thus, for $f = 1$ we have

$$\|\varphi\| \ge |\varphi(g)| = \left|\int_0^1 w(x)dx\right| = \int_0^1 |w(x)|dx,$$

leading to $\|\varphi\| = \int_0^1 |w(x)|dx$.

(3) The answer is yes. The best proof among many ones existing in the literature is due to Maddox (see [**134**]). It goes as follows:

Let $n \geq 2$. Then

$$\begin{aligned}
\int_0^1 |w(x)|dx &= \int_0^1 \frac{(1+n|w(x)|)|w(x)|}{1+n|w(x)|}dx \\
&= \int_0^1 \frac{|w(x)|}{1+n|w(x)|}dx + \int_0^1 \frac{n|w(x)|^2}{1+n|w(x)|}dx \\
&= \int_0^1 \frac{|w(x)|}{1+n|w(x)|}dx + \varphi\left(\frac{nw}{1+n|w|}\right) \\
&\leq \int_0^1 \frac{1}{n}dx + \|\varphi\| \left\|\frac{nw}{1+n|w|}\right\|_\infty \\
&\leq \frac{1}{n} + \|\varphi\|
\end{aligned}$$

since $1+n|w(x)| \geq |w(x)|$ for all x and since $\left\|\frac{nw}{1+n|w|}\right\|_\infty \leq 1$. Therefore, for all $n \geq 2$

$$\|\varphi\| \geq \int_0^1 |w(x)|dx - \frac{1}{n}.$$

Sending n to infinity yields $\|\varphi\| \geq \int_0^1 |w(x)|dx$. Thus

$$\|\varphi\| = \int_0^1 |w(x)|dx.$$

SOLUTION 2.3.18.

(1) The linearity of the integral implies that of φ. Let us show that φ is continuous. Let $f \in X$. We have

$$|\varphi(f)| \leq \int_{-1}^0 |f(x)|dx + \int_0^1 |f(x)|dx = \int_{-1}^1 |f(x)|dx \leq \sup_{-1\leq x\leq 1} |f(x)| \int_{-1}^1 dx.$$

Hence

$$|\varphi(f)| \leq 2\|f\|_\infty \Longrightarrow \|\varphi\| \leq 2.$$

(2) The given sequence (g_n) is continuous and evidently $\|g_n\|_\infty = 1$ for all n. We also have

$$\begin{aligned}
\varphi(g_n) &= \int_{-1}^{\frac{-1}{2n}} dx + \int_{\frac{-1}{2n}}^0 (-2nx)dx + \int_0^{\frac{1}{2n}} 2nxdx - \int_{\frac{1}{2n}}^1 (-1)dx \\
&= -\frac{1}{2n} + 1 + 1 - \frac{1}{2n} \\
&= 2 - \frac{1}{2n}
\end{aligned}$$

and hence

$$\|\varphi\| = \sup_{\|f\|_\infty=1} |\varphi(f)| \geq \varphi(g_n) = 2 - \frac{1}{2n}, \ \forall n \in \mathbb{N}$$

which yields $\|\varphi\| \geq 2$. Thus $\|\varphi\| = 2$.

(3) If there were an element $g \in X$ with $\|g\|_\infty = 1$ and $2 = \|\varphi\| = \varphi(g)$, i.e.

$$\int_{-1}^{0} g(x)dx - \int_{0}^{1} g(x)dx = 2,$$

then we would have

$$2 = \left|\int_{-1}^{0} g(x)dx - \int_{0}^{1} g(x)dx\right| \leq \int_{-1}^{0} |g(x)|dx + \int_{0}^{1} |g(x)|dx \leq 2$$

since

$$\int_{-1}^{0} |g(x)|dx \leq \|g\|_\infty \int_{-1}^{0} dx = 1 \text{ and } \int_{0}^{1} |g(x)|dx \leq \|g\|_\infty \int_{0}^{1} dx = 1.$$

But, this would imply that

$$\int_{-1}^{0} g(x)dx = 1 \text{ and } \int_{0}^{1} g(x)dx = -1$$

or

$$\int_{-1}^{0} (1 - g(x))dx = 0 \text{ and } \int_{0}^{1} (g(x) + 1)dx = 0.$$

However, for all $x \in [0, 1] : |g(x)| \leq \|g\|_\infty = 1$. Hence $1 - g$ and $1 + g$ are positive functions. They are also continuous and hence

$$g(x) = 1 \text{ on } [-1, 0] \text{ and } g(x) = -1 \text{ on } [0, 1]$$

and this would mean that g has two different values at zero!! Thus, we arrived at a contradiction and hence the norm is not attained.

Solution 2.3.19.

(1) We have already done that in Exercise 2.3.17. We then found

$$\|\varphi\| = \int_{0}^{1} |x|dx = \frac{1}{2}.$$

(2) Yes, the norm is attained at $f(x) = 1$. This also follows from Exercise 2.3.17.

(3) (a) Since $Y \subset X$, we deduce that $\|\psi\| \leq \|\varphi\|$. Let us show now that $\|\psi\| \geq \|\varphi\| = \frac{1}{2}$. Let

$$f_n(x) = \begin{cases} 1, & 0 \leq x \leq 1 - \frac{1}{n}, \\ n - nx, & 1 - \frac{1}{n} \leq x \leq 1. \end{cases}$$

Then for all n, $f_n \in Y$, $\|f_n\|_\infty = 1$ and

$$\|\psi\| = \|\psi\| \|f_n\|_\infty \geq |\psi(f_n)|.$$

Thus

$$\|\psi\| \geq \lim_{n \to \infty} |\psi(f_n)| = \frac{1}{2}$$

which follows from an easy calculation.

(b) We need to show that there is no f in Y for which $|\psi(f)| = \|\psi\|$ $(= \frac{1}{2})$. Assume the contrary, that is, there is such an f. Since $f(1) = 0$ and f is continuous, there is $\varepsilon > 0$ (and smaller than one) such that $f(x) = 0$ for all x in $(1-\varepsilon, 1] = I_\varepsilon$. Then we would have

$$\begin{aligned} \frac{1}{2} =& |\psi(f)| \\ =& \left| \int_{I_\varepsilon} x f(x) dx + \int_{I_\varepsilon^c} x f(x) dx \right| \\ \leq& \int_{I_\varepsilon} |x f(x)| dx + \int_{I_\varepsilon^c} |x f(x)| dx \\ \leq& 0 + \underbrace{\|f\|_\infty}_{1} \int_0^{1-\varepsilon} x dx \\ =& \frac{(1-\varepsilon)^2}{2} \end{aligned}$$

which does not hold for any $0 < \varepsilon \leq 1$, i.e. we reached a contradiction.

SOLUTION 2.3.20.

(1) If f is continuous, then it is plain that $\ker f = f^{-1}(\{0\})$ is closed.

Let's now show the other direction. Suppose that $\ker f$ is closed. Since $f \neq 0$, there is $a \in X$ such that $f(a) = 1$ (if not take $\frac{a}{f(a)}$ instead). Assume that f is not continuous, i.e. it has an infinite norm. Let $x_n \in X$ be such that $\|x_n\| = 1$ and $|f(x_n)| \to \infty$. Let $y_n = a - \frac{x_n}{f(x_n)}$. Then

$$f(y_n) = f\left(a - \frac{x_n}{f(x_n)}\right) = f(a) - \frac{f(x_n)}{f(x_n)} = 0, \text{ i.e. } y_n \in \ker f.$$

Hence

$$0 \leq \|y_n - a\| = \left\| \frac{x_n}{f(x_n)} \right\| = \frac{\|x_n\|}{|f(x_n)|} = \frac{1}{|f(x_n)|} \to 0,$$

giving $y_n \to a$. Hence $a \in \overline{\ker f} = \ker f$ (since $\ker f$ is closed) and so we obtain $f(a) = 0$ contradicting $f(a) = 1$! Thus, f must be continuous.

Remark. This result also holds for $f = 0$, in which case $\ker f = X$, i.e. it stays closed.

(2) One implication is deduced from the previous answer, namely if $\ker f$ is dense, then f is not continuous (indeed, if $\ker f$ were dense, then it would not be closed). Assume now that f is not continuous and hence it has an infinite norm, i.e. there is some $x_n \in X$ such that $\|x_n\| = 1$ and $|f(x_n)| \to \infty$. Let $y_n = x - \frac{f(x)}{f(x_n)} x_n$ where $x \in X$. Then $y_n \in \ker f$ and besides

$$0 \leq \|y_n - x\| = \left\| \frac{f(x)}{f(x_n)} x_n \right\| = \frac{|f(x)|}{|f(x_n)|} \longrightarrow 0.$$

Thus $y_n \to x$ yielding the density of $\ker f$.

(3) A is closed with respect to the supremum norm and it is not with respect to the integral norm (it is dense in this case). The answer is based on Exercise 2.3.16 and the first two questions of this exercise.

Solution 2.3.21. The answer is yes. Taking into account the results of the preceding exercise and in order to establish the closedness of Y, it suffices to show that $f \mapsto \int_{-1}^{1} f(x)dx$ (call it φ in the sequel) is a continuous linear functional on X.

It is clear that φ is into $\mathbb{R}$. The linearity of φ is an immediate consequence of that of the integral. Let us check its continuity. Let $f \in X$. We have

$$|\varphi(f)| \leq \int_{-1}^{1} |f(x)|dx = \|f\|_1.$$

Hence φ is continuous. Thus,

$$Y = \left\{ f \in X : \ \varphi(f) = \int_{-1}^{1} f(x)dx = 0 \right\}$$

is closed.

Solution 2.3.22.

(1) Left to interested readers...

(2) Proving $\|\cdot\|$ is a norm is a routine by now and should be easily established by the reader.
(3) Let $f : \ell^1 \to X$ be defined by

$$f(x) = \left(\frac{x_n}{n^2}\right)_{n\geq 1}.$$

The mapping f is onto since for any $y = (y_n) \in X$, there exists $x = (n^2 y_n) \in \ell^1$ such that $y = f(x)$.
    Let $x \in \ell^1$. One has

$$\|f(x)\|_X = \left\|\left(\frac{x_n}{n^2}\right)_{n\geq 1}\right\| = \sum_{n=1}^{\infty} n^2 \left|\frac{x_n}{n^2}\right| = \sum_{n=1}^{\infty} |x_n| = \|x\|_{\ell^1}.$$

Thus f is a surjective isometry on ℓ^1. Therefore, since ℓ^1 is complete, so is X.

Solution 2.3.23. Let $(X, \|\cdot\|)$ be a normed vector space and let

$$B(0,1) = \{x \in X : \ \|x\| < 1\}.$$

Let us show that these two sets are homeomorphic, that is, we need to find a homeomorphism $f : (X, \|\cdot\|) \to (B(0,1), \|\cdot\|)$. Let

$$f(x) = \frac{x}{1 + \|x\|}$$

and let $g : (B(0,1), \|\cdot\|) \to (X, \|\cdot\|)$ be defined by

$$g(x) = \frac{x}{1 - \|x\|}.$$

To see that f is continuous, we use the definition of a continuous (not necessarily linear) function between normed vector spaces. So we should utilize either the "$\varepsilon - \delta$" definition or sequential continuity. We choose the latter. Let (x_n) be a sequence in X which converges to $x \in X$ (remember that we must show that $f(x_n) \to f(x)$). Hence $\|x_n\| \to \|x\|$, and (x_n) is bounded by a certain M, say. Now, we have

$$\begin{aligned}
\|f(x_n)-f(x)\| &= \left\| \frac{x_n}{1+\|x_n\|} - \frac{x}{1+\|x\|} \right\| \\
&= \left\| \frac{x_n+\|x\|x_n-x-\|x_n\|x}{(1+\|x_n\|)(1+\|x\|)} \right\| \\
&\leq \|x_n+\|x\|x_nx-\|x_n\|x\| \text{ (as } \frac{1}{(1+\|x_n\|)(1+\|x\|)} \leq 1) \\
&= \|x_n+\|x\|x_n-x-\|x_n\|x+\|x\|x-\|x\|x\| \\
&\leq \|x_n-x\|+\|x\|\|x_n-x\|+|\|x\|-\|x_n\||\ \|x\| \\
&\longrightarrow 0 \text{ (as } n \text{ tends to } \infty).
\end{aligned}$$

Thus f is continuous.

In a very similar way, we may establish the continuity of g. Finally f is bijective, and to see this, we will show that

$$f \circ g = \mathrm{id}_{B(0,1)} \text{ and } g \circ f = \mathrm{id}_X.$$

Let $x \in B(0,1)$ (i.e. $\|x\| < 1$), then

$$(f \circ g)(x) = f[g(x)] = \frac{g(x)}{1+\|g(x)\|} = \frac{\frac{x}{1-\|x\|}}{1+\frac{\|x\|}{1-\|x\|}} = x$$

and if $x \in X$, then

$$(g \circ f)(x) = g[f(x)] = \frac{f(x)}{1-\|f(x)\|} = \frac{\frac{x}{1+\|x\|}}{1-\frac{\|x\|}{1+\|x\|}} = x.$$

Accordingly, f is continuous, bijective, and its inverse g is also continuous, that is, f is a homeomorphism.

Solution 2.3.24.

(1) We have to show that there is an isometric isomorphism between ℓ^∞ and $(\ell^1)'$. Let $A : \ell^\infty \to (\ell^1)'$ be defined by

$$A\alpha = f_\alpha,$$

where $f_\alpha : \ell^1 \to \mathbb{C}$ is defined by

$$f_\alpha(x) = f_\alpha(x_n) = \sum_{n=1}^{\infty} \alpha_n x_n$$

with $\alpha = (\alpha_n) \in \ell^\infty$.

First, we must show that $f_\alpha \in (\ell^1)'$. It is very easy to see that f_α is linear and that f_α is well-defined and takes its values in $\mathbb{C}$. By Exercise 2.3.2, f_α is bounded and we have

$$\|f_\alpha\| \leq \|\alpha\|_\infty.$$

Now, we go back to A. We need to show that A is onto, so let $g \in (\ell^1)'$. We must find an $\alpha \in \ell^\infty$ such that $A\alpha = g$. Let $e_n = (0, 0, \cdots, 0, 1, 0, \cdots)$ and set $\alpha_n = g(e_n)$. Let $\alpha = (\alpha_n)$. Before showing that $A\alpha = g$, we have to check that α is indeed in ℓ^∞. For all n we have (as g is bounded)

$$|\alpha_n| = |g(e_n)| \leq \|g\|$$

so that

$$\|\alpha\|_\infty \leq \|g\|.$$

We are done with the surjectivity of A once we show that $A\alpha = f_\alpha = g$. Let $x \in \ell^1$. As c_{00} is dense in ℓ^1, then there is an (x^k) in c_{00} such that $\|x^k - x\|_1 \to 0$, where as usual $x^k = (x_1, x_2, \cdots, x_k, 0, \cdots) = \sum_{n=1}^k e_n x_n$. Then

$$\begin{aligned} f_\alpha(x^k) - g(x^k) &= \sum_{n=1}^k \alpha_n x_n - g\left(\sum_{n=1}^k e_n x_n\right) \\ &= \sum_{n=1}^k \alpha_n x_n - \sum_{n=1}^k x_n g(e_n) \\ &= \sum_{n=1}^k \alpha_n x_n - \sum_{n=1}^k \alpha_n x_n \\ &= 0. \end{aligned}$$

Hence f_α coincides with g on a dense set, namely c_{00}. By an elementary result, we know that f_α then coincides with g on $\overline{c_{00}} = \ell^1$, showing that f_α is onto.

Finally, we show that A is an isometry, i.e. $\|A\alpha\| = \|f_\alpha\| = \|\alpha\|_\infty$ for each $\alpha \in \ell^\infty$. In fact, there is nothing to prove as we have already shown above that

$$\|f_\alpha\| \leq \|\alpha\|_\infty \text{ and } \|\alpha\|_\infty \leq \|g\| = \|f_\alpha\|,$$

leading to

$$\forall \alpha \in \ell^\infty : \ \|A\alpha\| = \|\alpha\|_\infty,$$

finishing the proof.

(2) We omit details. Consider $A : \ell^1 \to (c_0)'$ defined by

$$A\alpha = f_\alpha,$$

where $f_\alpha : c_0 \to \mathbb{C}$ is given by

$$f_\alpha(x_n) = \sum_{n=1}^\infty \alpha_n x_n.$$

Then prove that f_α is well-defined, linear and bounded and that A is an isometric isomorphism.

SOLUTION 2.3.25. Let $A, B \in B(L^2[0,1])$ be defined by

$$Af(x) = (1+x)f(x) \text{ and } Bf(x) = xf(x).$$

(1) A is invertible: Let $Sf(x) = \frac{1}{1+x}f(x)$. Then $S \in B(L^2[0,1])$ and for all $f \in B(L^2[0,1])$

$$ASf = SAf = f,$$

proving that A is invertible.

(2) B is not invertible (certainly not because $Tf(x) = \frac{1}{x}f(x)$ is not in $B(L^2[0,1])$ and $TB = BT = I$). One way of seeing this is to use Corollary 2.1.65. Let $n \in \mathbb{N}$ and set $f_n(x) = \sqrt{n}\mathbb{1}_{[0,\frac{1}{n}]}(x)$. Then for all n, $f_n \in L^2[0,1]$ and $\|f_n\| = 1$ (do it!). In the end, we have

$$\|Bf_n\|^2 = \int_0^1 x^2 n\mathbb{1}_{[0,\frac{1}{n}]}(x)dx = \frac{1}{3n^2}$$

so that $Bf_n \to 0$ as $n \to \infty$. Thus B is not invertible.

SOLUTION 2.3.26. We claim that A is invertible iff $\varphi(x) \neq 0$ for all $x \in [a,b]$.

(1) If $\varphi(x) \neq 0$ for all $x \in [a,b]$, then clearly $\frac{1}{\varphi(x)}$ is continuous and bounded and so the *bounded* operator

$$Bf(x) = \frac{1}{\varphi(x)}f(x)$$

works perfectly as the inverse of A.

(2) Assume that A is invertible. We know that

$$\|Af\| \geq \alpha\|f\|$$

for some $\alpha > 0$ and all f. Suppose that $\varphi(y) = 0$ for some $y \in [a,b]$. Then by the continuity of φ, we know that for any $\varepsilon > 0$, there is a $\beta > 0$ such that $|\varphi(x)| < \varepsilon$ whenever $|x-y| < \beta$. Take $f(x) = \mathbb{1}_{[y-\beta,y+\beta]}(x)$, i.e. the indicator function of the interval $[y-\beta, y+\beta]$. Then

$$\alpha^2\|f\|^2 \leq \|Af\|^2 = \int_{y-\beta}^{y+\beta} |\varphi(x)|^2|f(x)|^2 \leq \varepsilon^2\|f\|^2.$$

If (the arbitrary!) ε is such that $\varepsilon < \alpha$, then the previous inequality is not consistent with the inequality $\|Af\| \geq \alpha\|f\|$! Therefore, we must have $\varphi(x) \neq 0$ for all $x \in [a,b]$.

Remark. With this result, the non-invertibility of the operator B of Exercise 2.3.25 is obvious now. Indeed, B is a multiplication by the continuous function $\varphi(x) = x$ on $[0,1]$. Since $\varphi(y) = 0$ for $y = 0 \in [0,1]$, B is not invertible.

Solution 2.3.27. The operator A is not invertible. To see this, we use again Corollary 2.1.65. Let $e_n = (0, \cdots, 0, 1, 0, \cdots)$. Then $e_n \in \ell^2$, $\|e_n\| = 1$ and

$$Ae_n = \left(0, \cdots, 0, \frac{1}{n}, 0, \cdots\right)$$

so that

$$\lim_{n\to\infty} \|Ae_n\| = \lim_{n\to\infty} \frac{1}{n} = 0.$$

Solution 2.3.28. First, suppose that X is a Banach space, and let $\sum_n x_n$ be an absolutely convergent series in X meaning that the series $\sum_n \|x_n\|$ is convergent, signifying that its sequence of partial sums (denoted by (S_n)) converges (hence it is Cauchy!).

To establish the convergence of $\sum_n x_n$, we must establish the convergence of the sequence of partial sums, which we denote by (s_n). We first show that it is Cauchy. Let $\varepsilon > 0$. Then for $n, m \in \mathbb{N}$ such that $n \geq m$

$$\|s_n - s_m\| = \left\|\sum_{k=m}^{n} x_k\right\| \leq \sum_{k=m}^{n} \|x_k\| < \varepsilon$$

and hence (s_n) is Cauchy (in X) as so is (S_n). Since X is a Banach space, it follows that (s_n) converges or that $\sum_n x_n$ converges, as required.

Now, assume that every absolutely convergent series is convergent. To show that X is a Banach space, let (x_n) be a Cauchy sequence in X. Then

$$\forall \varepsilon > 0,\ \exists N \in \mathbb{N},\ \forall m, n \in \mathbb{N}\ (m, n \geq N \Longrightarrow \|x_n - x_m\| < \varepsilon).$$

Hence for $\varepsilon = \frac{1}{2}$, we have an N_1; for $\varepsilon = \frac{1}{2^2}$, we have an N_2;..., and so forth. WLOG, one may assume that $N_1 < N_2 < \cdots$. Now, for all k, we have that $n_{k+1} > N_{k+1} > N_k$ so that

$$\|x_{n_{k+1}} - x_{n_k}\| < \frac{1}{2^k}.$$

Hence

$$\sum_{k=1}^{\infty} \|x_{n_{k+1}} - x_{n_k}\| \leq \sum_{k=1}^{\infty} \frac{1}{2^k} = 1.$$

This tells us that $\sum_{k=1}^{\infty} \|x_{n_{k+1}} - x_{n_k}\|$ is convergent, which, by assumption, implies that $\sum_{k=1}^{\infty}(x_{n_{k+1}} - x_{n_k})$ too is convergent. This means that its sequence of partial sums (denoted by (s_m)) converges to a limit x, say. Writing this explicitly

$$x \longleftarrow s_m = \sum_{k=1}^{m}(x_{n_{k+1}} - x_{n_k}) = x_{n_{m+1}} - x_{n_1}$$

(a telescoping series!). Therefore, we see that the subsequence $(x_{n_{m+1}})_m$ converges (to $x + x_{n_1} \in X$!). By Proposition 1.1.24, (x_n) converges in X. Consequently, X is a Banach space.

SOLUTION 2.3.29.

(1) Let us show that c_{00} is not a Banach space (we already treated a similar case before). Consider $S_n = \left(1, \frac{1}{2^2}, \frac{1}{3^2}, \cdots, \frac{1}{n^2}, 0, 0, \cdots\right)$. Then (S_n) is in c_{00} and besides it is a Cauchy sequence. For if $m \geq n$, then

$$\begin{aligned}\|S_n - S_m\|_\infty &= \left\|\left(0,0,\cdots,0,\frac{1}{(n+1)^2},\cdots,\frac{1}{m^2},0,0,\cdots\right)\right\|_\infty\\ &= \max_{1\leq k\leq m-n}\frac{1}{(n+k)^2}\\ &= \frac{1}{(n+1)^2} \longrightarrow 0\end{aligned}$$

as $n, m \to \infty$. Hence (S_n) is Cauchy. Its limit is

$$S = \left(1, \frac{1}{2^2}, \frac{1}{3^2}, \cdots, \frac{1}{n^2}, \frac{1}{(n+1)^2}, \frac{1}{(n+2)^2}, \cdots\right)$$

because

$$\begin{aligned}\|S_n - S\|_\infty &= \left\|\left(0,0,\cdots,0,\frac{1}{n^2},\frac{1}{(n+1)^2},\frac{1}{(n+2)^2},\cdots\right)\right\|_\infty\\ &= \max_{k\in\mathbb{N}}\frac{1}{(n+k)^2} = \frac{1}{(n+1)^2} \to 0 \text{ as } n \to \infty.\end{aligned}$$

Therefore, c_{00} is not complete.

(2) Let

$$X_n = \left(0, 0, \cdots, 0, \frac{1}{n^2}, 0, \cdots\right) \in c_{00}.$$

Then $\|X_n\|_\infty = \frac{1}{n^2}$. Hence $\sum_{n\geq 1}\|X_n\|_\infty$ is a convergent series, i.e. $\sum_{n\geq 1} X_n$ is absolutely convergent. However, it is not convergent as its sequence of partial sums (S_n) does not converge in $(c_{00}, \|\cdot\|_\infty)$ since if it were, we would have the convergence of

$$S_n = \sum_{k=1}^{n} X_k = X_1 + X_2 + \cdots + X_n$$
$$= \left(1, \frac{1}{2^2}, \frac{1}{3^2}, \cdots, \frac{1}{n^2}, 0, 0, \cdots\right).$$

But, we saw in the first answer that this sequence does not converge in $(c_{00}, \|\cdot\|_\infty)$.

(3) There is nothing wrong with the previous question. The series $\sum_{n\geq 1} X_n$ is absolutely convergent but it is not convergent. The reason is that $(c_{00}, \|\cdot\|_\infty)$ is not a Banach space.

Solution 2.3.30. Since X is a Banach space, so is $B(X)$. As $\|A\| < 1$, then the (real) series $\sum_{n=0}^{\infty} \|A\|^n$ converges.

Now, by Proposition 2.1.27, we know that $\|A^n\| \leq \|A\|^n$ for any n. Hence the series $\sum_{n=0}^{\infty} \|A^n\|$ converges. That is, the series $\sum_{n=0}^{\infty} A^n$, which is in $B(X)$, converges absolutely and so it converges by calling on Theorem 2.1.68. So much for the proof of the convergence of the series.

To show that this sum is actually the inverse of $I - A$, set

$$B = \sum_{n=0}^{\infty} A^n \text{ and } B_N = \sum_{n=0}^{N} A^n$$

(we have just shown that $B_N \to B$ in $B(X)$). We clearly have

$$0 \leq \|(I - A)B_N - I\| = \|I - A^{N+1} - I\| = \|A^{N+1}\| \leq \|A\|^{N+1} \longrightarrow 0$$

as $N \to \infty$ because $\|A\| < 1$. Hence $(I - A)B_N \to I$. Consequently,

$$(I - A)B = \lim_{N\to\infty} (I - A)B_N = I.$$

Similarly, we can show that $B(I - A) = I$. Thus, $I - A$ is invertible with the inverse given by

$$(I - A)^{-1} = B = \sum_{n=0}^{\infty} A^n.$$

For the last estimate on the norm of $(I - A)^{-1}$, it follows from Theorem 2.1.68 as we all know that

$$\sum_{n=0}^{\infty} \|A\|^n = \frac{1}{1 - \|A\|}.$$

SOLUTION 2.3.31. Since A is nilpotent, there exists $N \in \mathbb{N}$ such that $A^N = 0$. Using the idea of the Neumann Series, we claim that $I - A$ is invertible with inverse given by $\sum_{n=0}^{N} A^n$. Indeed, we have

$$\begin{aligned}(I-A)\sum_{n=0}^{N} A^n =& I + A + A^2 + \cdots + A^N - A - A^2 - \cdots - A^N - A^{N+1} \\ =& I\end{aligned}$$

since $A^N = 0$ yields $A^{N+1} = 0$.

Similarly, we can verify that

$$\left(\sum_{n=0}^{N} A^n\right)(I-A) = I.$$

Therefore $I - A$ is invertible and

$$(I-A)^{-1} = I + A + A^2 + \cdots + A^{N-1}.$$

Similarly, we may show that $I + A$ is invertible (find its inverse!).

SOLUTION 2.3.32.

(1) The given series converges absolutely as for all n one has

$$\left\|\frac{A^n}{n!}\right\| = \frac{\|A^n\|}{n!} \leq \frac{\|A\|^n}{n!}$$

and the series $\sum_{n\geq 0} \frac{\|A\|^n}{n!}$ converges and its sum is $e^{\|A\|}$. Hence $\sum_{n\geq 0} x_n$ converges absolutely.

(2) Since X is a Banach, so is $B(X) = B(X,X)$. Thus $\sum_{n\geq 0} x_n$ also converges by Theorem 2.1.68. From the previous question we observe that

$$\|e^A\| \leq e^{\|A\|}.$$

(3) Since A and B commute, we can easily apply the Binomial Theorem to compute $(A+B)^n$. We then obtain:

$$\frac{1}{n!}(A+B)^n = \sum_{p+q=n} \frac{n!}{p!q!} A^p B^q.$$

Then we have

$$e^{A+B} = \sum_{n=0}^{\infty} \frac{1}{n!}(A+B)^n = \sum_{n=0}^{\infty} \frac{1}{n!} \sum_{p+q=n} \frac{n!}{p!q!} A^p B^q.$$

It is clear that we will be summing over all possible values of the integers p and q. This leads to

$$e^{A+B} = \sum_{p=0}^{\infty}\sum_{q=0}^{\infty} \frac{1}{p!q!} A^p B^q.$$

Separating the p-terms from the q-terms allows us to obtain

$$e^{A+B} = \left(\sum_{p=0}^{\infty} \frac{1}{p!} A^p\right)\left(\sum_{q=0}^{\infty} \frac{1}{q!} B^q\right) = e^A e^B.$$

Similarly,

$$e^{A+B} = e^{B+A} = e^B e^A,$$

that is,

$$e^{A+B} = e^A e^B = e^B e^A.$$

(4) If A and B do not commute, then the previous relation does not necessarily hold as shown by the following example. Consider the following matrices

$$A = \begin{pmatrix} 0 & 1 \\ 0 & 0 \end{pmatrix} \text{ and } B = \begin{pmatrix} 0 & 0 \\ 1 & 0 \end{pmatrix}.$$

The reader can easily check that A and B do not commute. Since for all $n \geq 2$, $A^n = 0$ and $B^n = 0$ (the null matrices!), one has

$$e^A = I + A = \begin{pmatrix} 1 & 1 \\ 0 & 1 \end{pmatrix} \text{ and } e^B = I + B = \begin{pmatrix} 1 & 0 \\ 1 & 1 \end{pmatrix}.$$

Then

$$e^A e^B = \begin{pmatrix} 2 & * \\ * & * \end{pmatrix} \neq e^B e^A = \begin{pmatrix} 1 & * \\ * & * \end{pmatrix}.$$

Now, set

$$A + B := C = \begin{pmatrix} 0 & 1 \\ 1 & 0 \end{pmatrix}.$$

Then

$$C^2 = I,\ C^3 = C,\ C^4 = I, \dots \text{ etc.}$$

Thus

$$\begin{aligned} e^{A+B} =& I + C + \frac{I}{2!} + \frac{C}{3!} + \frac{I}{4!} + \cdots \\ =& \left(1 + \frac{1}{2!} + \frac{1}{4!} + \cdots\right) I + \left(1 + \frac{1}{3!} + \frac{1}{5!} + \cdots\right) C \\ =& \begin{pmatrix} \cosh 1 & 0 \\ 0 & \cosh 1 \end{pmatrix} + \begin{pmatrix} 0 & \sinh 1 \\ \sinh 1 & 0 \end{pmatrix} \\ =& \begin{pmatrix} \cosh 1 & \sinh 1 \\ \sinh 1 & \cosh 1 \end{pmatrix}. \end{aligned}$$

Therefore,

$$e^{A+B} \neq e^A e^B \neq e^B e^A.$$

(5) Since $A(-A) = (-A)A$, we have

$$e^A e^{-A} = e^{-A} e^A = e^{A-A} = e^{(0x)} = I + 0 + 0 + \cdots = I.$$

This means that e^A invertible and that

$$(e^A)^{-1} = e^{-A}.$$

(6) Let

$$A = \begin{pmatrix} 1 & -1 & -1 \\ 2 & 5 & 2 \\ -2 & -3 & 0 \end{pmatrix}$$

The given system can be written "matricially" as

$$X'(t) = \frac{dX}{dt} = \begin{pmatrix} x'(t) \\ y'(t) \\ z'(t) \end{pmatrix} = A \begin{pmatrix} x(t) \\ y(t) \\ z(t) \end{pmatrix}.$$

Then, we easily see that A is diagonalizable with 1, 2 and 3 as its eigenvalues, and

$$A = PDP^{-1}$$

where

$$D = \begin{pmatrix} 1 & 0 & 0 \\ 0 & 2 & 0 \\ 0 & 0 & 3 \end{pmatrix} \text{ and } P = \begin{pmatrix} -1 & -1 & 0 \\ 1 & 0 & 1 \\ -1 & 1 & -1 \end{pmatrix}.$$

Hence, we rewrite the system as

$$Y'(t) = DY(t)$$

with $Y(t) = P^{-1}X(t)$. Therefore

$$Y(t) = e^{tD}\begin{pmatrix} a \\ b \\ c \end{pmatrix}$$

or

$$Y(t) = \begin{pmatrix} ae^t \\ be^{2t} \\ ce^{3t} \end{pmatrix}$$

where $a, b, c \in \mathbb{R}$. Thus

$$X(t) = \begin{pmatrix} x(t) \\ y(t) \\ z(t) \end{pmatrix} = PY(t) = \begin{pmatrix} -ae^t - be^{2t} \\ ae^t + ce^{3t} \\ -ae^t + be^{2t} - ce^{3t} \end{pmatrix}.$$

SOLUTION 2.3.33. The answer is no in general. Take

$$A = 2\pi i I \neq 4\pi i I = B$$

where i is the complex number and I is the identity operator on X. Then

$$\begin{aligned} e^A = e^{2\pi i I} &= I + (2\pi i)I + \frac{(2\pi i)^2 I^2}{2!} + \cdots \\ &= \left(1 + (2\pi i) + \frac{(2\pi i)^2}{2!} + \cdots\right) I \\ &= e^{2\pi i} I = I \end{aligned}$$

and similarly we obtain

$$e^B = e^{4\pi i I} = I.$$

Thus $e^A = e^B$ whereas $A \neq B$.

REMARK. This result is true if both of the operators A and B are *self-adjoint* on a *Hilbert* space. For a proof and more related results, see Exercise 11.3.13. See also Exercise 11.3.15.

SOLUTION 2.3.34. Let

$$S_n = \sum_{k=0}^{n} \frac{A^n}{n!} \text{ and } T_n = \left(I + \frac{1}{n}A\right)^n$$

and

$$s_n = \sum_{k=0}^{n} \frac{\|A\|^n}{n!} \text{ and } t_n = \left(I + \frac{1}{n}\|A\|\right)^n$$

where (S_n) and (T_n) are sequences in $B(X)$; and (s_n) and (t_n) are sequences in $\mathbb{R}^+$ (also remember that $\lim_{n\to\infty} s_n = \lim_{n\to\infty} t_n = e^{\|A\|}$). We already know that (in $B(X)$)

$$e^A = \lim_{n\to\infty} S_n = \lim_{n\to\infty} \sum_{k=0}^{n} \frac{A^n}{n!}.$$

Let's show that (S_n) and (T_n) have the same limit in $B(X)$. Observe first that (since I commutes with $\frac{1}{n}A$)

$$\begin{aligned}\left(I + \frac{1}{n}A\right)^n &= \sum_{k=0}^{n} C_n^k \left(\frac{1}{n^k}\right) A^k \\ &= I + A + \frac{n-1}{n}\frac{1}{2!}A^2 + \cdots + \frac{1}{n^n}A^n\end{aligned}$$

where

$$C_n^k = \frac{n!}{(n-k)!k!}.$$

But

$$\frac{1}{n^n} = \frac{1}{n^{n-1}}\frac{(n-1)!}{n!} = \frac{1}{n!}\left(1 - \frac{1}{n}\right)\left(1 - \frac{2}{n}\right)\cdots\left(1 - \frac{n-1}{n}\right)$$

and so

$$T_n = I + \frac{1}{1!}A + \left(1 - \frac{1}{n}\right)\frac{1}{2!}A^2 + \cdots + \left(1 - \frac{1}{n}\right)\left(1 - \frac{2}{n}\right)\cdots\left(1 - \frac{n-1}{n}\right)\frac{1}{n!}A^n.$$

Hence

$$S_n - T_n = \sum_{k=2}^{n} c_k A^k,$$

where $c_k \geq 0$ (for $k \in \{2, 3, \cdots, n\}$). It turns out that the same c_k appears in $s_n - t_n$. More precisely, we have

$$0 \leq \|S_n - T_n\| \leq \sum_{k=2}^{n} c_k \|A^k\| \leq \sum_{k=2}^{n} c_k \|A\|^k = s_n - t_n.$$

Since the previous holds for all n, passing to the limit as n approaches infinity, finally leads (in $B(X)$!) to:

$$\lim_{n\to\infty} \left(I + \frac{1}{n}A\right)^n = \lim_{n\to\infty} T_n = \lim_{n\to\infty} S_n = \lim_{n\to\infty} \sum_{k=0}^{n} \frac{A^n}{n!} = e^A.$$

SOLUTION 2.3.35. Yes, we have! We have (don't we?)

$$A^{-1}e^BA = A^{-1}\left(I + B + \frac{B^2}{2!} + \cdots + \frac{B^n}{n!} + \cdots\right)A$$

so that

$$A^{-1}e^BA = I + A^{-1}BA + \frac{A^{-1}B^2A}{2!} + \cdots + \frac{A^{-1}B^nA}{n!} + \cdots.$$

But e.g.

$$A^{-1}B^2A = A^{-1}BAA^{-1}BA = (A^{-1}BA)^2.$$

Therefore,

$$A^{-1}e^BA = I + A^{-1}BA + \frac{(A^{-1}BA)^2}{2!} + \cdots + \frac{(A^{-1}BA)^n}{n!} + \cdots$$

so that finally

$$A^{-1}e^BA = e^{A^{-1}BA},$$

as required.

SOLUTION 2.3.36. We claim that the solution is $y(t) = e^{tA}x$. To see this, we must show that $y'(t) = Ay(t)$ in $\|\cdot\|_X$. We have

$$\begin{aligned}
\left\|\frac{y(t+h)-y(t)}{h} - Ay(t)\right\| &= \left\|\frac{e^{(t+h)A}-e^{tA}}{h}x - Ae^{tA}x\right\| \\
&= \left\|e^{tA}\left(\frac{e^{hA}-I}{h} - A\right)x\right\| \\
&= \left\|e^{tA}A\sum_{n=1}^{\infty}\frac{h^nA^n}{(n+1)!}x\right\| \\
&\leq \|e^{tA}\|\|A\|\|x\|\sum_{n=1}^{\infty}\frac{|h|^n\|A\|^n}{(n+1)!} \\
&\leq e^{|t|\|A\|}\|\|A\|\|x\|\sum_{n=1}^{\infty}\frac{|h|^n\|A\|^n}{(n+1)!} \\
&\leq e^{|t|\|A\|}\|A\|\|x\|(e^{|h|\|A\|}-1) \\
&\longrightarrow 0
\end{aligned}$$

as $h \to 0$. In the end, clearly $y(0) = e^0x = x$. So much for the existence of a solution. To prove uniqueness, assume that there were another solution $z(t)$ say. Then $z'(t) = Az(t)$ and $z(0) = x$. Setting $u(t) = e^{-tA}z(t)$ yields

$$u'(t) = -e^{-tA}Az(t) + e^{-tA}z'(t) = 0$$

and so $u(t)$ is constant (as $\mathbb{R}$ is connected! See e.g. [**42**]). Hence

$$u(t) = u(0) = e^0 z(0) = x \text{ and so } z(t) = e^{tA}x$$

exactly like $y(t)$! Therefore, the solution is unique.

SOLUTION 2.3.37. It is no obvious to find the solution. It is not so hard either! A not bad idea is to look at the analogous equation in $\mathbb{R}$, namely

$$t - atb = c,$$

where $t, a, b, c \in \mathbb{R}$. Then clearly $t = (1 - ab)^{-1}c$ as long as $1 \neq ab$. So if $|ab| < 1$, then this solution may be expressed as

$$t = \sum_{n=0}^{\infty} (ab)^n c \ldots (1).$$

It is true that we did use commutativity which is not available all the time in $B(X)$. One may therefore conjecture that the solution in $B(X)$ is an adaptation of one of the variants of (1) in $B(X)$. Let us check that the sought solution is given by:

$$T = \sum_{n=0}^{\infty} A^n C B^n.$$

The first thing to do is to verify the convergence of the series. Since $B(X)$ is Banach space (as X is), it suffices to show that this series converges absolutely. For all n, we have

$$\|A^n C B^n\| \leq \|A^n\| \|C\| \|B^n\| \leq \|C\| \|A\|^n \|B\|^n.$$

Since $\|A\| < 1$ and $\|B\| < 1$, the real (positive) series $\|C\| \sum_{n \geq 0} (\|A\| \|B\|)^n$ clearly converges and hence so does $\sum_{n \geq 0} A^n C B^n$.

In the end, let us check that this sum does satisfy the equation: $T - ATB = C$. We have

$$\sum_{n=0}^{\infty} A^n C B^n - A \left(\sum_{n=0}^{\infty} A^n C B^n \right) B = \sum_{n=0}^{\infty} A^n C B^n - \sum_{n=0}^{\infty} A^{n+1} C B^{n+1}.$$

But

$$\sum_{n=0}^{\infty} A^n C B^n - \sum_{n=0}^{\infty} A^{n+1} C B^{n+1} = C + \sum_{n=1}^{\infty} A^n C B^n - \sum_{n=1}^{\infty} A^n C B^n = C,$$

i.e. $T = \sum_{n=0}^{\infty} A^n C B^n$ is in effect a solution of $T - ATB = C$.

Solution 2.3.38. As in the solution of Exercise 2.3.37, the reader may check that

$$T = \sum_{n=0}^{\infty} A^{-n-1} C B^n$$

is a convergent series which satisfies the equation $AT - BT = C$...

Solution 2.3.39.

(1) Yes, $B_i(X, Y)$ may well be empty (of course it is still open!). For instance, if $X = \mathbb{R}^p$ and $Y = \mathbb{R}^q$ (hence they are Banach spaces) and $p \neq q$.

(2) To show that $B_i(X, Y)$ is open in $B(X, Y)$, we need to find for any A in $B_i(X, Y)$, a strictly positive r such that $B(A, r) \subset B_i(X, Y)$ where $B(A, r)$ denotes the open ball in $B(X, Y)$.

Let A be in $B_i(X, Y)$, i.e. A is invertible. Take $r = \|A^{-1}\|^{-1}$. Let $B \in B(A, r)$, i.e. $\|B - A\| < r$. We have

$$\|(A - B)A^{-1}\| \leq \|A - B\|\|A^{-1}\| < \|A^{-1}\|^{-1}\|A^{-1}\| = 1.$$

Hence $I - (A - B)A^{-1}$ is invertible (why?). But

$$I - (A - B)A^{-1} = I - I + BA^{-1} = BA^{-1}.$$

Therefore, BA^{-1} is invertible and since A is invertible, so is $BA^{-1}A = B$, i.e. $B \in B_i(X, Y)$. Thus $B_i(X, Y)$ is open in either case.

(3) The answer is no (even if $X = Y$!). Consider the sequence (A_n) defined by

$$A_n = \frac{1}{n} I, \ n \in \mathbb{N},$$

where I is the identity operator. Then each A_n is clearly invertible, that is, $A_n \in B_i(X)$ for all $n \in \mathbb{N}$ and yet

$$\lim_{n \to \infty} \|A_n - 0\| = \lim_{n \to \infty} \frac{1}{n}\|I\| = \lim_{n \to \infty} \frac{1}{n} = 0,$$

that is, (A_n) converges w.r.t. the operator norm to the *zero* operator which is *not* invertible. This proves that $B_i(X)$ is not closed in $B(X)$.

Solution 2.3.40.

(1) Let (A_n) be a sequence of $B_i(X)$ which converges to $A \in B_i(X)$, i.e.

$$\lim_{n \to \infty} \|A_n - A\| = 0.$$

Hence $\lim_{n \to \infty} \|A^{-1}A_n - I\| = 0$ for

$$0 \leq \|A^{-1}A_n - I\| = \|A^{-1}(A_n - A)\| \leq \|A^{-1}\|\|A_n - A\| \to 0.$$

Hence for all $\varepsilon > 0$, and in particular for $\varepsilon = 1$, there exists an N_1 such that for all $n \geq N_1$,

$$\|A^{-1}A_n - I\| < 1.$$

By Theorem 2.1.88, $A^{-1}A_n$ is invertible. By Corollary 2.1.89, we know that the inverse of $A^{-1}A_n$ is given by

$$(A^{-1}A_n)^{-1} = A_n^{-1}A = I + \sum_{k=1}^{\infty}(I - A^{-1}A_n)^k$$

or simply

$$A_n^{-1}A - I = \sum_{k=1}^{\infty}(I - A^{-1}A_n)^k.$$

Hence

$$\begin{aligned}
0 \leq \|I - A_n^{-1}A\| &= \|\sum_{k=1}^{\infty}(I - A^{-1}A_n)^k\| \\
&\leq \sum_{k=1}^{\infty}\|(I - A^{-1}A_n)^k\| \\
&= \sum_{k=1}^{\infty}\|A^{-k}(A - A_n)^k\| \\
&\leq \sum_{k=1}^{\infty}\|A^{-1}\|^k\|A - A_n\|^k.
\end{aligned}$$

But $\|A_n - A\| \to 0$, hence we can take $\|A - A_n\|^k$ as small as we wish to make $\sum_{k=1}^{\infty}\|A^{-1}\|^k\|A - A_n\|^k$ goes to zero as n approaches infinity. Thus

$$\|I - A_n^{-1}A\| \longrightarrow 0 \text{ or } \|(A^{-1} - A_n^{-1})A\| \longrightarrow 0 \text{ or } \|A^{-1} - A_n^{-1}\| \longrightarrow 0,$$

that is, $A_n^{-1} \to A^{-1}$, proving the continuity of the map $A \mapsto A^{-1}$.

(2) Yes for the simple reason that the inverse of the map $A \mapsto A^{-1}$ is itself, hence continuous too!

Solution 2.3.41. Let

$$(Au)(x) = \int_0^1 e^{x-t}u(t)dt$$

be defined on $C([0,1], \mathbb{R})$ with the standard norm. The given equation may be re-written as

$$(I - \lambda A)u = f.$$

Now, to use the Neumann Series, we need to find powers of A. First, we have for all x

$$(A^2u)(x)=\int_0^1 e^{x-t}Au(t)dt=\int_0^1 e^{x-t}\int_0^1 e^{t-r}u(r)drdt=\int_0^1 e^{x-r}u(r)dr,$$

i.e. for all x

$$(A^2u)(x)=(Au)(x).$$

Hence for all $n\in\mathbb{N}$, $A^n=A$. Now, if we choose λ such that $\|\lambda A\|<1$, then the Neumann Series converges and gives

$$u=(I-\lambda A)^{-1}f=\sum_{n=0}^{\infty}(\lambda A)^n f=f+\lambda Af+\lambda^2A^2f+\lambda^3A^3f+\cdots.$$

Hence

$$u=f+\lambda Af+\lambda^2Af+\lambda^3Af+\cdots.$$

Thus by a simple results on real series, we finally obtain

$$u=f+\frac{\lambda}{1-\lambda}Af,$$

that is,

$$u(x)=f(x)+\frac{\lambda}{1-\lambda}\int_0^1 e^{x-t}f(t)dt.$$

Remark. The last expression of $u(x)$ remains valid as a solution of the given integral equation for any $\lambda\neq 1$ as one can check.

Solution 2.3.42.

(1) There are many examples. Here is a very simple one:

$$x=(x_n)=(1,-1,0,0,\cdots)$$

Then $x\in X_p$ for $x\in\ell^p$ and $\sum_{n=0}^{\infty}x_n=1-1+0+\cdots=0$.

(2) Left to the interested reader...

(3) To show that X_1 is closed in ℓ^1, one of the ways of doing it is via the function $f:\ell^1\to\mathbb{R}$ defined by

$$f(x)=\sum_{n=0}^{\infty}x_n.$$

Indeed, if we show that f is continuous, then X_1, being equal to $\ker f$ will then be closed. Let $x\in\ell^1$. Then

$$|f(x)|=\left|\sum_{n=0}^{\infty}x_n\right|\leq\sum_{n=0}^{\infty}|x_n|=\|x\|_1.$$

Then f is bounded (and linear). Therefore, X_1 is closed in ℓ^1.

Let $1 < p < \infty$. To show that X_p is not closed in ℓ^p, we exhibit a sequence (x_n) in X_p whose limit w.r.t. ℓ^p leaves X_p. Let

$$\left(-1, \frac{1}{n}, \cdots, \frac{1}{n}, 0, 0, \cdots\right)$$

where $n \geq 1$. It is clearly in X_p and it converges to the x given by $(-1, 0, \cdots, 0, \cdots)$ because

$$\|x_n - x\|_p = \left[0 + \frac{1}{n^p} + \cdots + \frac{1}{n^p} + \cdots\right]^{\frac{1}{p}} = \left(\frac{n}{n^p}\right)^{\frac{1}{p}} = n^{\frac{1}{p}-1} \longrightarrow 0$$

exactly when $p > 1$. Finally, as $x \notin X_p$, then X_p is not closed in ℓ^p, when $1 < p < \infty$.

(4) Let $f : \ell^p \to \mathbb{R}$ be continuous and such that $f = 0$ on X_p. If we come to show that $f = 0$ on all of ℓ^p, then invoking Proposition 2.1.102 (and the remark below it!) we obtain that $\overline{X_p} = \ell^p$.

Since $f \in (\ell^p)'$, and the latter is isometrically isomorphic to ℓ^q, where $\frac{1}{p} + \frac{1}{q} = 1$, we know that there exists a $y = (y_n) \in \ell^q$ such that for all $x = (x_n) \in \ell^p$

$$f(x) = \sum_{n=0}^{\infty} x_n y_n.$$

As f vanishes on X_p, and $\alpha = (\underbrace{-1}_{0}, 0, \cdots, 0, \underbrace{1}_{n}, 0 \cdots,) \in X_p$, we then have that $f(\alpha) = 0$, from which we easily get

$$y_n = y_0, \quad \text{for } n = 1, 2, 3, \cdots.$$

Since $\alpha \in X_p$, we know that such a (y_n) does exist and it belongs to ℓ^q. Hence

$$(y_0, y_1, \cdots, y_n, \cdots) = (y_0, y_0, \cdots, y_0, \cdots) \in \ell^q$$

only if $y_0 = 0$, forcing the vector y to vanish. Thus for all $x \in \ell^p$

$$f(x) = \sum_{n=0}^{\infty} x_n y_n = \sum_{n=0}^{\infty} x_n \times 0 = 0,$$

i.e. $f = 0$ on ℓ^p, establishing the result.

Solution 2.3.43. Let $x_0 \in X$. WLOG we may consider $\|x_0\| = 1$ (why?). By Corollary 2.1.101, we know that there is $f_0 \in X'$ such that

$$\|f_0\| = \|x_0\| = 1 \text{ and } f_0(x_0) = \|x_0\|^2 = 1.$$

To show that Y is a Banach space, let (y_n) be a Cauchy sequence in Y and define a sequence of operators (A_n) by

$$A_n : X \to Y \text{ with } A_n(x) = f_0(x)y_n.$$

It is clear that each A_n is linear and bounded, i.e. $A_n \in B(X,Y)$. Now, we have for all $n, m \in \mathbb{N}$,

$$\begin{aligned}
\|A_n - A_m\| &= \sup_{\|x\|\leq 1} \|A_n(x) - A_m(x)\| \\
&= \sup_{\|x\|\leq 1} \|f_0(x)(y_n - y_m)\| \\
&= \sup_{\|x\|\leq 1} |f_0(x)|\|y_n - y_m\| \\
&= \|y_n - y_m\| \sup_{\|x\|\leq 1} |f_0(x)| \\
&= \|y_n - y_m\| \underbrace{\|f_0\|}_{=1} \\
&= \|y_n - y_m\|,
\end{aligned}$$

so that (A_n) is Cauchy in $B(X,Y)$ because (y_n) is so (in Y). Since by hypothesis $B(X,Y)$ is a Banach space, we know that (A_n) converges to some $A \in B(X,Y)$.

To finish the proof, we need to show that (y_n) has a limit in Y. So, let $Ax_0 = y_0$ $(\in Y)$. Then we have

$$\begin{aligned}
\|y_n - y_0\| &= \|A_n(x_0) - A(x_0)\| \\
&= \|(A_n - A)x_0\| \\
&\leq \|(A_n - A)\|\|x_0\| \\
&= \|A_n - A\|,
\end{aligned}$$

and this shows that (y_n) converges to y_0.

Solution 2.3.44.

(1) Let $Y \subsetneq X$. Assume for the sake of contradiction that Y has a nonempty interior, that is, $\overset{\circ}{Y} \neq \varnothing$, i.e. $\overset{\circ}{Y}$ contains at least one element a. Hence

$$\exists r > 0, \ B(a,r) \subset Y.$$

Now, for any $x \in X$ such that $x \neq 0$, we have

$$y = a + \frac{r}{2}\frac{x}{\|x\|} \in B(a,r)$$

because

$$\|y-a\| = \left\|\frac{r}{2}\frac{x}{\|x\|}\right\| = \frac{r}{2}\left\|\frac{x}{\|x\|}\right\| = \frac{r}{2} < r.$$

Consequently, $y \in Y$. Since Y is a linear subspace (and $a, y \in Y$), we have

$$x = \frac{2}{r}(y-a)\|x\| \in Y,$$

which means that $X \subset Y$ or $X = Y$, contradicting the hypothesis Y being a *proper* subspace of X.

Finally, if Y is closed and a proper linear subspace of X, then $\overset{\circ}{\overline{Y}} = \overset{\circ}{Y} = \varnothing$.

(2) Let $Y \subset X$ be a linear subspace. Assume that Y is not dense in X, that is, $\overline{Y} \neq X$, and let us show that $\overset{\circ}{\overline{Y}} = \varnothing$. We argue by contradiction and so assume that $\overset{\circ}{\overline{Y}} \neq \varnothing$, that is, there is at least a $y \in \overset{\circ}{\overline{Y}}$, i.e.

$$\exists r > 0,\ B(a,r) \subset \overline{Y}.$$

Proceeding as in the previous question, we may obtain that $X \subset \overline{Y}$ and hence $\overline{Y} = X$! But this contradicts $\overline{Y} \neq X$. The proof is therefore complete.

(3) Assume that X (where $\dim X = \infty$) is generated by the (countable family!) of vectors $x_1, x_2, \cdots$. For each $n \in \mathbb{N}$, set

$$X_n = \mathrm{sp}\{x_1, x_2, \cdots, x_n\}.$$

It is then clear that each X_n is closed and that

$$\bigcup_{n=1}^{\infty} X_n = X.$$

Since X is a complete metric space, by the Baire Category Theorem, it follows that

$$\exists n_0 \in \mathbb{N} :\ \overset{\circ}{X}_{n_0} \neq \varnothing$$

or as each X_n is closed,

$$\exists n_0 \in \mathbb{N} :\ \overset{\circ}{\overline{X_{n_0}}} \neq \varnothing,$$

and this contradicts the result of the first question, completing the proof.

SOLUTION 2.3.45. The hypothesis

$$\exists k > 0,\ \|x\|_1 \leq k\|x\|_2,\ \forall x \in X$$

means that the identity operator $I : (X, \|\cdot\|_2) \to (X, \|\cdot\|_1)$ (which is of course linear and bijective) is continuous. The Banach Isomorphism Theorem implies that $I^{-1} = I : (X, \|\cdot\|_1) \to (X, \|\cdot\|_2)$ is also continuous and hence

$$\exists m > 0,\ \|x\|_2 \leq m\|x\|_1,\ \forall x \in X.$$

Thus $\|\cdot\|_1$ and $\|\cdot\|_2$ are equivalent norms.

SOLUTION 2.3.46.

(1) To show that $G(A)$ is closed in $X \times Y$, let $(x_n, y_n) \in G(A)$ be a converging sequence to (x, y) in $X \times Y$ (we must show that $(x, y) \in G(A)$). Then (x_n) converges to x in X and (y_n) converges to y in Y. As $(x_n, y_n) \in G(A)$, then we have $y_n = Ax_n$ for every $n \in \mathbb{N}$. Since A is continuous, we obviously have

$$y = \lim_{n\to\infty} y_n = \lim_{n\to\infty} Ax_n = Ax.$$

Thus $(x, y) = (x, Ax) \in G(A)$, establishing the closedness of $G(A)$.

(2) Consider two norms on X:

$$\|x\|_1 = \|x\|_X + \|Ax\|_Y \text{ and } \|x\|_2 = \|x\|_X.$$

Since $G(A)$ is closed, X is a Banach space with respect to $\|\cdot\|_1$ (show it or cf. Exercise 10.3.6).

On the other hand, it is clear that for all $x \in X$,

$$\|x\|_2 = \|x\|_X \leq \|x\|_X + \|Ax\|_Y = \|x\|_1.$$

Since X is a Banach space with respect to $\|\cdot\|_2$ too, Exercise 2.3.45 applies and gives the equivalence of the norms $\|\cdot\|_1$ and $\|\cdot\|_2$, i.e.

$$\exists \alpha > 0,\ \forall x \in X :\ \|x\|_1 \leq \alpha\|x\|_2.$$

Therefore,

$$\exists \beta > 0,\ \forall x \in X :\ \|Ax\|_Y \leq \beta\|x\|_X,$$

that is, A is bounded.

SOLUTION 2.3.47. Let $A \in B(X, Y)$ (X and Y are two Banach spaces) be surjective. By assumption, there is an $\alpha > 0$ such that

$$B'(0, \alpha) \subset A[B(0, 1)],$$

where $B'(0, \alpha)$ is the open ball of center 0 and radius α in Y, and $B(0, 1)$ is the unit ball in X.

Let us then show that A is an open map. To this end, let U be an open set in X (remember that we need to show that $A(U)$ is open in Y). Let $y \in A(U)$ so that $y = Ax$ for some $x \in U$. Since U is open, we know that for some $r > 0$ such that $B(x,r) \subset U$.

Hence (by Proposition 1.1.10)

$$x + B(0,r) \subset U$$

so that

$$y + A[B(0,r)] = A[x + B(0,r)] \subset A(U).$$

Since by assumption $B'(0,\alpha) \subset A(B(0,1))$, we get that $B'(0,r\alpha) \subset A(B(0,r))$ (why?) and hence

$$B'(y,r\alpha) = y + B'(0,r\alpha) \subset y + A[B(0,r)] \subset A(U).$$

Therefore, $A(U)$ is open.

Solution 2.3.48.

(1) To show that $I : (X, \|\cdot\|_\infty) \to (X, \|\cdot\|_1)$ is bounded, we need to find a positive M such that for all f in X:

$$\|f\|_1 = \|If\|_1 \le M\|f\|_\infty.$$

Let $f \in X$. We have

$$|f(x)| \le \sup_{x\in[0,1]} |f(x)| = \|f\|_\infty, \ \forall x \in [0,1]$$

and hence

$$\|If\|_1 = \|f\|_1 = \int_0^1 |f(x)|dx \le \|f\|_\infty \int_0^1 dx = \|f\|_\infty.$$

Thus $I : (X, \|\cdot\|_\infty) \to (X, \|\cdot\|_1)$ is bounded. The identity operator is obviously bijective and hence onto. The mapping $I : (X, \|\cdot\|_\infty) \to (X, \|\cdot\|_1)$ cannot be open. Since if it were, we would have (taking into account the fact that $I^{-1} = I$) that $I : (X, \|\cdot\|_1) \to (X, \|\cdot\|_\infty)$ is bounded. Hence $\|\cdot\|_\infty$ and $\|\cdot\|_1$ would be equivalent norms which is not true (see Exercise 1.3.7).

(2) The conclusion of the previous question does not contradict the Open Mapping Theorem since $(X, \|\cdot\|_1)$ is not a Banach space (see Exercise 1.3.13).

Solution 2.3.49.

(1) That A is linear can easily be established by the reader. Let us show now that A is unbounded. If A were bounded, then we would have

$$\exists M \ge 0, \ \forall f \in X : \ \|f'\|_\infty \le M\|f\|_\infty.$$

But if $f(x) = x^n$ (which is in X), then

$$\|f'\|_\infty = \sup_{x\in[0,1]} |nx^{n-1}| = n \text{ and } \|f\|_\infty = \sup_{x\in[0,1]} |x^n| = 1.$$

Thus there cannot exist an $M \geq 0$ such that $\|f'\|_\infty \leq M\|f\|_\infty$ for all $f \in X$ (otherwise $\mathbb{N}$ would be bounded!).

(2) Denote the graph of A by $G(A)$. Let us show that $G(A)$ is closed. Take any sequence (x_n, x_n') in $G(A)$ such that $(x_n, x_n') \to (x, y)$. Then $x_n \to x$ in X and $x_n' \to y$ in Y, i.e. $x_n \to x$ uniformly and $x_n' \to y$ uniformly. By a standard result from an Advanced Calculus Course, we obtain $y = x'$ and $x \mapsto x'$ is continuous. Accordingly, $(x, x') \in G(A)$, i.e. $G(A)$ is closed.

(3) No, the conclusion of the previous question does not violate the Closed Graph Theorem for the simple reason is that X is not a Banach space (see Exercise 1.3.15).

Solution 2.3.50.

(1) Proving the linearity of A is straightforward. Let us check its boundedness. Let $x = (x_n)_n \in c_{00}$. We clearly have

$$\|Ax\| \leq \|x\|,$$

showing the boundedness of A.

(2) To prove that A is bijective, it suffices to exhibit its "formal" inverse without caring about the continuity at this stage. Observe that the mapping defined by

$$B(x_n)_n = ((n+1)x_n)_n$$

is the "formal" inverse of A since for all $x \in c_{00}$ one has

$$BAx = ABx = x.$$

(3) Although we have found the "formal" inverse of A previously, it cannot be considered as the actual inverse of A in terms of operators because B is not bounded. For if it were, then one would have

$$\exists M \geq 0,\ \forall x \in c_{00} : \ \|Bx\| \leq M\|x\|.$$

But, if $x = e_n = (0, 0, \cdots, 0, 1, 0 \cdots) \in c_{00}$, then $\|e_n\| = 1$ and

$$\|Be_n\| = n + 1 \longrightarrow +\infty.$$

Hence, B is not bounded. But, *we are not done yet*! Nothing makes us sure that there does not exist another bounded operator C, say, such that $CA = AC = I$ (the inverse is unique *when* it exists!).

Going back to our example, let $x_n = (0, 0, \cdots, 1, 0, \cdots)$. Then $\|x_n\| = 1$ for all $n \in \mathbb{N}$. Besides

$$A(x_n) = \left(0, 0, \cdots, \frac{1}{n+1}, 0 \cdots\right)$$

leading to

$$\lim_{n\to\infty} \|A(x_n)\| = \lim_{n\to\infty} \frac{1}{n+1} = 0.$$

Therefore, A cannot be bounded below and thus it is not invertible by Proposition 2.1.62.

(4) The result in the previous question does not contradict Banach Isomorphism Theorem since c_{00} is not complete (cf. Exercise 1.3.18).

Solution 2.3.51. This is a consequence of the Hahn-Banach Theorem. Let $Y = \text{span}\{a\}$. Let $f : Y \to \mathbb{C}$ be a function defined by $f(x) = f(\lambda a) = \lambda\|a\|$ for all $x \in Y$ where $\lambda \in \mathbb{C}$. Then f is linear since for any $\alpha, \beta \in \mathbb{C}$ and any $x, y \in Y$ (i.e. $x = \lambda a$ and $y = \mu a$ for some λ and some μ), one has

$$\begin{aligned} f(\alpha x + \beta y) &= f(\alpha\lambda a + \beta\mu a) \\ &= f((\alpha\lambda + \beta\mu)a) \\ &= (\alpha\lambda + \beta\mu)\|a\| \\ &= \alpha\lambda\|a\| + \beta\mu\|a\| \\ &= \alpha f(x) + \beta f(y). \end{aligned}$$

It is also continuous since Y is finite dimensional. Besides, one has $\|f(a)\| = \|a\|$ and hence $\|f\| = 1$. According to the Hahn-Banach Theorem (more precisely Corollary 2.1.100), f admits a continuous extension denoted, by F, and which verifies $F(a) = \|F\|\|a\|$. This completes the proof.

Solution 2.3.52. As usual, no proof is given for the linearity of A. Now, since $(A_n(x))_n$ converges, it is bounded. The Uniform Boundedness Principle then implies that the sequence $(\|A_n\|)_n$ is bounded, by some $M > 0$, say. Hence

$$\|A_n x\| \leq \|A_n\|\|x\| \leq M\|x\|, \ \forall (x, n) \in X \times \mathbb{N}.$$

Taking the limit as n tends to infinity (and observing that $A_n x \to Ax$ implies $\|A_n x\| \to \|Ax\|$) yields

$$\|Ax\| \leq M\|x\|, \ \forall x \in X,$$

i.e. A is bounded.

2.4. Hints/Answers to Tests

SOLUTION 2.4.1. The proof is as in Exercise 2.3.15. You should find using Cauchy-Schwarz Inequality that $\|A\| = 1$.

SOLUTION 2.4.2. Both Y and Z are closed...

SOLUTION 2.4.3. If A has a left inverse B and a right inverse C, then clearly

$$B = B(AC) = (BA)C = C...$$

SOLUTION 2.4.4. As BA is nilpotent, there is some $n \in \mathbb{N}$ such that $(BA)^n = 0$. Now compute $(AB)^{n+1}$ using the associativity of the product of operators...

SOLUTION 2.4.5. No! Use properties of the trace...

SOLUTION 2.4.6. Yes! This follows from $A^2 = I$...

SOLUTION 2.4.7. Yes, V is an isometry. It is linear... Now, set $V(x) = f$ and let $t \geq 0$. Then write (since each $|f_n(t)|^p = |f(t)|^p \mathbb{1}_{[n-1,n)}(t)$ is measurable and positive, and the sets $\{[n-1,n)\}_{n\geq 1}$ are pairwise disjoint)

$$\begin{aligned}
\|Vx\|_{L^p}^p =& \int_0^\infty |f(t)|^p dt \\
=& \int_0^\infty \sum_{n=1}^\infty |f(t)|^p \mathbb{1}_{[n-1,n)}(t) dt \\
=& \sum_{n=1}^\infty \int_0^\infty |f(t)|^p \mathbb{1}_{[n-1,n)}(t) dt \\
=& \int_0^1 |f(t)|^p dt + \int_1^2 |f(t)|^p dt + \cdots \\
=& \int_0^1 |x_1|^p dt + \int_1^2 |x_2|^p dt + \cdots \\
=& |x_1|^p \int_0^1 dt + |x_2|^p \int_1^2 dt + \cdots \\
=& \sum_{n=1}^\infty |x_n|^p \\
=& \|x\|_{\ell^p}^p,
\end{aligned}$$

for all $x \in \ell^p$...

Solution 2.4.8. Write the given system as

$$X'(t) = AX(t),$$

where

$$A = \begin{pmatrix} 0 & 1 \\ 1 & 0 \end{pmatrix}.$$

Thus it will have a solution given by

$$X(t) = e^{tA}K \text{ where } K = {}^t(k_1, k_2)$$

is a constant in $\mathbb{R}^2$. We can find that for any $t \in \mathbb{R}$,

$$e^{tA} = \begin{pmatrix} \cosh t & \sinh t \\ \sinh t & \cosh t \end{pmatrix}$$

Hence

$$X(t) = \begin{pmatrix} x(t) \\ y(t) \end{pmatrix} = \begin{pmatrix} k_1 \cosh t + k_2 \sinh t \\ k_1 \sinh t + k_2 \cosh t \end{pmatrix} = \begin{pmatrix} \alpha e^t + \beta e^{-t} \\ \alpha e^t - \beta e^{-t} \end{pmatrix}$$

where α and β are two real constants.

Solution 2.4.9. Just apply Theorem 2.1.99 to $p(x) = \|g\|_{Y'}\|x\|$...

CHAPTER 3

Hilbert Spaces

3.2. True or False: Answers

ANSWERS.

(1) The answer depends on the book you are using. Most texts use the linearity of the inner product with respect to the first coordinate and the conjugate-linearity with respect to the second. ***We have also adopted this convention in the present book.*** However, the reader should be aware that some books utilize the other definition, i.e. the linearity with respect to the *second* vector and the conjugate-linearity with respect to the *first* (see e.g. [**178**]). This difference does not change any major result in the theory of Hilbert spaces.

(2) The answer is yes and this is a simple application of the Cauchy-Schwarz Inequality. By assumption $fg \geq 1$. Hence $\sqrt{fg} \geq 1$ and so

$$1 = \int_0^1 1dx \leq \int_0^1 \sqrt{f(x)}\sqrt{g(x)}dx \leq \left(\int_0^1 f(x)dx\right)^{\frac{1}{2}} \left(\int_0^1 g(x)dx\right)^{\frac{1}{2}}.$$

Thus

$$\int_0^1 f(x)dx \int_0^1 g(x)dx \geq 1,$$

as required.

(3) The answer is yes and we actually have an equivalence, i.e.

$$\|x+y\|^2 = \|x\|^2 + \|y\|^2 \Longleftrightarrow x \perp y.$$

The proof easily follows from

$$\begin{aligned} \|x+y\|^2 &= < x+y, x+y > \\ &= \|x\|^2 + < x,y > + < y,x > + \|y\|^2 \\ &= \|x\|^2 + 2 < x,y > + \|y\|^2 \end{aligned}$$

(as $< x, y >$ is real).

(4) In a complex inner product space, the answer is not true anymore. Consider $x = (1,0)$ and $y = (i,0)$ (where $i = \sqrt{-1}$).

Then

$$\|x+y\|^2 = 2 = 1+1 = \|x\|^2 + \|y\|^2$$

whereas $x \not\perp y$ as $< x, y >= i \neq 0$.

(5) For any $Z \subset X$, it is known that $Z^\perp$ is a ***closed vector space***. Hence if Y is not closed and/or it is not a vector space, then $Y = Y^{\perp\perp}$ need not hold.

(6) By the Bessel Inequality, the series $\sum_{n\geq 1} |< x, e_n >|^2$ converges (for each $x \in X$) and so its general term necessarily tends to zero.

(7) False! See Exercise 3.3.21.

(8) False! On ℓ^2, let (e_n) be the standard orthonormal basis. If $x = (1, \frac{1}{2}, \cdots, \frac{1}{n}, \cdots)$, then clearly

$$x = \sum_{n=1}^{\infty} \frac{1}{n} e_n.$$

By Corollary 3.1.39, the series $\sum_{n=1}^{\infty} \frac{1}{n} e_n$ converges if $(\frac{1}{n}) = x$ is in ℓ^2, and this clearly holds. However, $\sum_{n=1}^{\infty} \|\frac{1}{n} e_n\|$ does not converge for

$$\sum_{n=1}^{\infty} \left\| \frac{1}{n} e_n \right\| = \sum_{n=1}^{\infty} \frac{1}{n} \|e_n\| = \sum_{n=1}^{\infty} \frac{1}{n} = \infty.$$

(9) True if the linear operator in question is bounded! Nonetheless, if an operator is not bounded, then its adjoint does not have to be unique (for a counterexample, see e.g. [**210**]). If, however, its domain is dense, then the adjoint is also unique in this case. The reader need not bother about this in this book as we only deal with adjoints of bounded operators.

(10) True. In fact $A = 0$ iff $A^* = 0$ and this comes from the known property

$$\|A\| = \|A^*\|.$$

(11) False! We give a counterexample. Let

$$A = \begin{pmatrix} 0 & 1 \\ 0 & 0 \end{pmatrix} \neq \begin{pmatrix} 0 & 0 \\ 0 & 0 \end{pmatrix}.$$

Then

$$A^2 = \begin{pmatrix} 0 & 0 \\ 0 & 0 \end{pmatrix}.$$

Hence,

$$\|A^2\| = 0 \neq \|A\|^2$$

as if it were true, then this would lead to $A = 0$!

(12) The reason is simple: $(\ker A)^\perp$ is always closed while $\operatorname{ran}(A^*)$ is not always closed. Let us give a counterexample anyway. Consider

$$A(x_1, x_2, x_3, \cdots) = \left(x_1, \frac{1}{2}x_2, \frac{1}{3}x_3, \cdots\right)$$

as a bounded operator from ℓ^2 into ℓ^2. Then A is injective so that

$$\ker(A) = \{0\} \text{ or } [\ker(A)]^\perp = \{0\}^\perp = \ell^2.$$

It is also easy to see that $A = A^*$. Now, if $y = (1, \frac{1}{2}, \frac{1}{3}, \cdots)$ (which is in ℓ^2), then obviously there cannot exist any $x \in \ell^2$ such that $A^*x = y$. This means that A^* is not surjective, i.e. $\operatorname{ran}A^* \neq \ell^2$. Therefore,

$$[\ker(A)]^\perp \neq \operatorname{ran}(A^*).$$

(13) True. This comes from the following observation

$$\|A_n - A\| = \|(A_n - A)^*\| = \|A_n^* - A^*\|.$$

(14) False! By Exercise 3.3.41, the sequence (A_n) strongly converges to 0 but (A_n^*) does not converge strongly to 0 (where $A_n^* = B_n$!).

(15) True. Let $x, y \in H$. Then the following tells us why this is true:

$$< Ax, y > \longleftarrow < A_nx, y > = < x, A_n^*y >$$

and hence

$$< x, A_n^*y > \longrightarrow < x, A^*y > ...$$

3.3. Solutions to Exercises

SOLUTION 3.3.1. The answer is no since $(1,1) \neq (0,0)$ but

$$< (1,1), (1,1) > = 1 - 1 = 0.$$

SOLUTION 3.3.2. No! Let $f(x) = g(x) = 2$. Then $f, g \in X$,

$$< f, f > = \int_0^1 0dx = 0$$

whereas $f \neq 0$.

SOLUTION 3.3.3.

(1) Let $x \in \mathbb{R}^n$. Then $< x, x > = \sum_{k=1}^{n} x_i^2$ is obviously real and positive.

(2) Let $x \in \mathbb{R}^n$. We have

$$\begin{aligned} < x, x >= 0 &\Longleftrightarrow \sum_{k=1}^{n} x_i^2 = 0 \\ &\Longleftrightarrow x_i^2 = 0, \ \forall i = 1, 2, \cdots, n \\ &\Longleftrightarrow x_i = 0, \ \forall i = 1, 2, \cdots, n \\ &\Longleftrightarrow x = 0_{\mathbb{R}^n}. \end{aligned}$$

(3) Let $x, y, z \in \mathbb{R}^n$ and let $\alpha, \beta \in \mathbb{R}$. We have

$$< \alpha x + \beta y, z >= \sum_{k=1}^{n} (\alpha x_i + \beta y_i) z_i = \alpha \sum_{k=1}^{n} x_i z_i + \beta \sum_{k=1}^{n} y_i z_i$$

and hence

$$< \alpha x + \beta y, z >= \alpha < x, z > + \beta < y, z > .$$

(4) We have for all $x, y \in \mathbb{R}^n$

$$< x, y >= \sum_{k=1}^{n} x_i y_i = \sum_{k=1}^{n} y_i x_i =< y, x > .$$

Thus $< \cdot, \cdot >$ is a real inner product on $\mathbb{R}^n$.

Solution 3.3.4.

(1) First, we need to justify the existence of the integral. Since f and g are both continuous on $[0, 1]$, so is $f\overline{g}$. Thus the integral $\int_0^1 f(x)\overline{g(x)}dx$ is finite. Let us now turn to the verification of the axioms of an inner product.
   (a) Let $f \in X$. Then

$$< f, f >= \int_0^1 f(x)\overline{f(x)}dx = \int_0^1 |f(x)|^2 dx$$

   is a positive real number.
   (b) If $f = 0$, then $< 0, 0 >= \int_0^1 0\overline{0}dx = 0$. Conversely, let $f \in X$ be such that $< f, f >= 0$. Whence $\int_0^1 |f(x)|^2 dx = 0$. But, $|f|^2$ is a positive and continuous function whose integral vanishes. Hence $|f|^2$ must also vanish, leading to $f = 0$.

(c) Let $f, g, h \in X$ and $\alpha, \beta \in \mathbb{C}$. We have

$$\begin{aligned}
<\alpha f + \beta g, h> &= \int_0^1 (\alpha f(x) + \beta g(x))\overline{h(x)}dx \\
&= \int_0^1 \alpha f(x)\overline{h(x)}dx + \int_0^1 \beta g(x)\overline{h(x)}dx \\
&= \alpha < f, h > + \beta < g, h > .
\end{aligned}$$

(d) Let $f, g \in X$. We have

$$\begin{aligned}
\overline{< g, f >} &= \overline{\int_0^1 g(x)\overline{f(x)}dx} \\
&= \int_0^1 \overline{\overline{f(x)}}\,\overline{g(x)}dx \\
&= \int_0^1 f(x)\overline{g(x)}dx \\
&= < f, g > .
\end{aligned}$$

(2) Basically, the proof is very similar to that of the previous question (and should be verified by the reader). There are some different points which we emphasize.
- The integral intervening in the definition of $< f, g >$ converges. This is a simple application of the Cauchy-Schwarz inequality. Alternatively, this follows from

$$2|f\overline{g}| \leq |f|^2 + |g|^2.$$

- Now we show that $< f, f >= 0 \Rightarrow f = 0$. We have

$$< f, f >= \int_{\mathbb{R}} |f(x)|^2 dx = 0 \Longrightarrow f = 0, \text{ a.e.}$$

because $|f|^2$ is positive and measurable. Thus $f = 0$ on $L^2(\mathbb{R})$.

(3) First, we note that $< f, f >$ is a real and nonnegative number. Second, if $< f, f >= 0$ and since $|f'|$ is positive and continuous, one has $f'(x) = 0$, $\forall x \in [0, 1]$ and $f(0) = 0$. Hence $f = 0$. The remaining properties may be easily checked by the reader.

(4) First both integrals are finite since $f\overline{g}$ and $f'\overline{g'}$ are continuous on $[0, 1]$.

(a) Let $f \in X$. Then

$$< f, f >= \int_0^1 |f(x)|^2 dx + \int_0^1 |f'(x)|^2 dx \geq 0.$$

(b) It is plain that $f = 0$ implies that $< f, f >= 0$. Conversely, if $< f, f >= 0$, then

$$\int_0^1 |f(x)|^2 dx = 0 \text{ and } \int_0^1 |f'(x)|^2 dx = 0.$$

Since $|f|^2$ and $|f'|^2$ are continuous and positive, we obtain $f = 0$ and $f' = 0$. Hence $f = 0$.

(c) Let $f, g, h \in X$ and let $\alpha, \beta \in \mathbb{C}$. We have

$$\begin{aligned} < \alpha f+\beta g, h >= & \int_0^1 (\alpha f(x)+\beta g(x))\overline{h(x)}dx+\int_0^1 (\alpha f'(x)+\beta g'(x))\overline{h'(x)}dx \\ =& \alpha\left(\int_0^1 f(x)\overline{h(x)}dx+\int_0^1 f'(x)\overline{h'(x)}dx\right) \\ &+\beta\left(\int_0^1 g(x)\overline{h(x)}dx+\int_0^1 g'(x)\overline{h'(x)}dx\right) \\ =& \alpha < f, h > +\beta < g, h > . \end{aligned}$$

(d) Finally, one easily verifies that for any $f, g \in X$:

$$< f, g >= \overline{< g, f >}.$$

(5) Observe first that the integral in question is finite since the function $x \mapsto x^2 f(x)g(x)$ is continuous on $[-1, 1]$. Everything should be readily verified except perhaps for the first property. Let $f \in X$ be such that $< f, f >= 0$. Then

$$\int_{-1}^1 x^2 f^2(x) dx = 0.$$

But $x \mapsto x^2 f^2(x)$ is continuous and positive. Hence

$$x^2 f^2(x) = 0 \text{ or } f(x) = 0 \text{ for all } x \in [-1, 1] \setminus \{0\}.$$

However, f is continuous on $[-1, 1]$ and so

$$f(0) = \lim_{x \to 0} f(x) = 0$$

yielding $f = 0$.

(6) The only two non obvious things which need to be taken care of are:

(a) the convergence of the integral in question.

(b) the property $< P, P >= 0 \Rightarrow P = 0$.

To show the convergence of the integral, we note that since $P(x)Q(x)$ is a polynomial, it suffices to prove that the integrals $\int_{-\infty}^{\infty} x^n e^{-x^2} dx$ converge for all n. We instead show that those integrals converge *absolutely*. A change of variable shows it is enough to establish the convergence of $\int_0^\infty x^n e^{-x^2} dx$. Now, observe that for x large enough (say for $x > A > 1$), $x^n e^{-x^2} \leq \frac{1}{x^2}$. Thus and since $x \mapsto x^n e^{-x^2}$ is continuous on $[0, A]$, one has

$$\begin{aligned}\int_0^\infty x^n e^{-x^2} dx &= \int_0^A x^n e^{-x^2} dx + \int_A^\infty x^n e^{-x^2} dx \\ &\leq \int_0^A x^n e^{-x^2} dx + \int_A^\infty \frac{dx}{x^2} < \infty.\end{aligned}$$

Now, let P be such that $< P, P >= 0$. Then $\int_{-\infty}^\infty e^{-x^2} P^2(x) dx = 0$. Since $x \mapsto e^{-x^2} P^2(x)$ is continuous and nonnegative, one concludes that $e^{-x^2} P^2(x) = \frac{P^2(x)}{e^{x^2}} = 0$, $\forall x \in \mathbb{R}$ which implies that $P = 0$.

SOLUTION 3.3.5. Recall that the product of A^t with A is always a square matrix whose diagonal is constituted of squares (of the moduli over $\mathbb{C}$) of entries of A.

First, observe that $< A, B >$ is well-defined as $A^t B$ is always a square matrix (of order m) and hence we can compute its trace.

(1) (a) Let $A \in X$. Then

$$< A, A >= \mathrm{tr}(A^t A) \geq 0$$

as it is the sum of *squares* of all entries of A.

(b) If $A = 0$, then $< A, A >= \mathrm{tr}(0_X^t 0_X) = 0$. Conversely, assume that $< A, A >= 0$. Then the sum of the squares of all entries is zero. Hence, the entries must all vanish and thus $A = 0_X$.

(c) Let $A, B, C \in X$ and let $\alpha, \beta \in \mathbb{R}$. We have by the linearity of the trace:

$$< \alpha A + \beta B, C >= \mathrm{tr}((\alpha A + \beta B)^t C) = \alpha \mathrm{tr}(A^t C) + \beta \mathrm{tr}(B^t C).$$

Therefore, for all $A, B, C \in X$ and all $\alpha, \beta \in \mathbb{R}$

$$< \alpha A + \beta B, C >= \alpha < A, C > + \beta < B, C > .$$

(d) Lastly, let $A, B \in X$. Since for any square matrix C: $\mathrm{tr}\, C^t = \mathrm{tr} C$, we can write

$$< A, B >= \mathrm{tr}(A^t B) = \mathrm{tr}((A^t B)^t) = \mathrm{tr}(B^t A) =< B, A > .$$

(2) Since $(X, < \cdot, \cdot >)$ is an inner product space, Corollary 3.1.7 says that

$$A \mapsto \|A\|_F = \sqrt{\mathrm{tr}(A^t A)}$$

is a norm on $X = \mathcal{M}_{n,m}(\mathbb{R})$.

SOLUTION 3.3.6. The right-to-left implication is obvious.

Conversely, since $<x,y>=0$ holds for all $y \in X$, it certainly holds for $y=x$. Then $\|x\|^2=0$ yielding $x=0$, proving the other implication.

SOLUTION 3.3.7.

(1) Observe first that if $<x,y>=0$, then the inequality is satisfied. So, assume that $<x,y>\neq 0$ and let $x,y\in X$ and $\alpha\in\mathbb{C}$. Then

$$<x-\alpha y, x-\alpha y>=\|x-\alpha y\|^2 \geq 0.$$

But

$$\begin{aligned}<x-\alpha y, x-\alpha y> &= \|x\|^2-\overline{\alpha}<x,y> -\alpha<y,x> +\alpha\overline{\alpha}\|y\|^2\\ &= \|x\|^2-2\text{Re}\alpha<y,x> +|\alpha|^2\|y\|^2.\end{aligned}$$

Substituting $\alpha=\frac{\|x\|^2}{<y,x>}$ in the previous equality yields

$$0\leq \|x\|^2-2\text{Re}\alpha<y,x> +|\alpha|^2\|y\|^2=\|x\|^2-2\|x\|^2+\left|\frac{\|x\|^2}{<y,x>}\right|^2\|y\|^2$$

or

$$-\|x\|^2+\frac{\|x\|^4\|y\|^2}{|<y,x>|^2}\geq 0,$$

i.e.

$$|<y,x>|=|\overline{<y,x>}|=|<x,y>|\leq \|x\|\|y\|,$$

as desired.

(2) Assume that $|<x,y>|=\|x\|\|y\|$ and x and y both different from zero (otherwise, x and y are clearly linearly dependent). Then $<x,y>\neq 0$ and $<y,y>\neq 0$. Taking as above $\alpha=\frac{\|x\|^2}{<y,x>}$ $(\neq 0)$, we obtain

$$\|x-\alpha y\|^2=<x-\alpha y, x-\alpha y>=0,$$

i.e. $x=\alpha y$, that is, x and y are linearly dependent.

Conversely, assume that x and y are linearly dependent, i.e. $x=\alpha y$ for some $\alpha\neq 0$. Hence

$$|<x,y>|=|<\alpha y,y>|=|\alpha|<y,y>=|\alpha|\|y\|^2=|\alpha|\|y\|\|y\|=\|\alpha y\|\|y\|,$$

i.e.

$$|<x,y>|=\|x\|\|y\|.$$

(3) Let $x,y\in X$ and $\alpha\in\mathbb{F}$. Then as a consequence of the properties of an inner product space, we have:

(a) $\|x\|\geq 0$ (clear!).

(b) $\|x\|=0 \Leftrightarrow <x,x>=0 \Leftrightarrow x=0$.

(c)

$$\|\alpha x\| = \sqrt{< \alpha x, \alpha x >} = \sqrt{|\alpha|^2 < x, x >} = |\alpha|\sqrt{< x, x >} = |\alpha|\|x\|.$$

(d) Finally,

$$\begin{aligned}\|x+y\|^2 &=< x+y, x+y > \\ &=< x,x > + < x,y > + < y,x > + < y,y > \\ &=< x,x > + < x,y > + \overline{< x,y >} + < y,y > \\ &= \|x\|^2 + 2\mathrm{Re} < x,y > + \|y\|^2 \\ &\leq \|x\|^2 + 2\|x\|\|y\| + \|y\|^2 \text{ (by the first question)} \\ &= (\|x\| + \|y\|)^2.\end{aligned}$$

Therefore,

$$\|x+y\| \leq \|x\| + \|y\|,$$

as needed.

SOLUTION 3.3.8. For all $x, y \in X$ (in particular for those s.t. $\|x\| = 1$), we have by the Cauchy-Schwarz Inequality

$$| < x,y > | \leq \|x\|\|y\| = \|y\|.$$

Thereupon,

$$\sup_{\|x\|=1} | < x,y > | \leq \|y\|.$$

Assume now that $y \neq 0$. Since $\|\frac{y}{\|y\|}\| = 1$, we get

$$\sup_{\|x\|=1} | < x,y > | \geq \left| < \frac{y}{\|y\|}, y > \right| = \frac{| < y,y > |}{\|y\|} = \frac{\|y\|^2}{\|y\|} = \|y\|,$$

and since the required equality does hold for $y = 0$, we infer that (for all y)

$$\sup_{\|x\|=1} | < x,y > | = \|y\|.$$

SOLUTION 3.3.9. Let $x, y \in X$.

(1) We have

$$\|x+y\|^2 =< x,x > + < x,y > + < y,x > + < y,y >$$

and

$$\|x-y\|^2 =< x,x > - < x,y > - < y,x > + < y,y > .$$

Hence

$$\|x+y\|^2 = \|x\|^2 + 2\mathrm{Re} < x,y > + \|y\|^2 ...(1)$$

and

$$\|x-y\|^2 = \|x\|^2 - 2\text{Re} < x,y > +\|y\|^2 ...(2)$$

so that

$$\|x+y\|^2 + \|x+y\|^2 = 2\|x\|^2 + 2\|y\|^2.$$

(2) Since X real, Re $< x,y >=< x,y >$. Making (1)–(2) gives

$$< x,y >= \frac{1}{4}(\|x+y\|^2 - \|x-y\|^2).$$

(3) To establish the desired equality, we need to find $\|x+iy\|^2$ and $\|x-iy\|^2$. We have

$$\begin{aligned}\|x+iy\|^2 &= < x+iy, x+iy > \\ &= < x,x > +i < y,x > -i < x,y > -i^2 < y,y >,\end{aligned}$$

that is,

$$\|x+iy\|^2 = \|x\|^2 + i < y,x > -i < x,y > +\|y\|^2.$$

Similarly, we find that

$$\|x-iy\|^2 = \|x\|^2 - i < y,x > +i < x,y > +\|y\|^2.$$

Consequently,

$$\|x+y\|^2 - \|x-y\|^2 + i\|x+iy\|^2 - i\|x-iy\|^2 = 4 < x,y >,$$

as required.

Solution 3.3.10. We will be showing that none of these norms satisfies the parallelogram law.

(1) Let $x = (1,0,0,\cdots,0)$ and $y = (0,1,0,\cdots,0)$. Then

$$\|x-y\|_1^2 + \|x+y\|_1^2 = 4+4 \neq 2\|x\|_1^2 + 2\|y\|_1^2 = 2+2.$$

Now, let $x = (1,0,0,\cdots,0)$ and $y = (-1,1,0,\cdots,0)$. Then

$$\|x-y\|_\infty^2 + \|x+y\|_\infty^2 = 4+1 \neq 2\|x\|_\infty^2 + 2\|y\|_\infty^2 = 2+2 = 4.$$

(2) Let $f(x) = x$ and $g(x) = 1-x$ (both defined on $[0,1]$). Then

$$\|f-g\|_1^2 + \|f+g\|_1^2 = \frac{1}{4} + 1 \neq 2\|f\|_1^2 + 2\|g\|_1^2 = \frac{1}{2} + \frac{1}{2} = \frac{1}{4}.$$

Let $f(x) = 1$ and $g(x) = x$ both defined on $[0,1]$. They are obviously continuous. Moreover,

$$\|f-g\|_\infty^2 + \|f+g\|_\infty^2 = 1+4 \neq 2\|f\|_\infty^2 + 2\|g\|_\infty^2 = 2+2 = 4.$$

(3) Let

$$A = \begin{pmatrix} 1 & 0 \\ 0 & -1 \end{pmatrix} \text{ and } B = \begin{pmatrix} -1 & 0 \\ 0 & 0 \end{pmatrix}.$$

Then

$$\|A\| = 1,\ \|B\| = 1,\ \|A+B\| = 1 \text{ and } \|A-B\| = 2$$

and hence

$$\|A-B\|^2 + \|A+B\|^2 = 5 \neq 2\|A\|^2 + 2\|B\|^2 = 4.$$

(4) We cannot take the functions of Answer 2 as $1, 1-x \notin L^1(\mathbb{R})$. We can, however, take

$$f(x) = x\mathbb{1}_{(0,1)}(x) \text{ and } g(x) = (1-x)\mathbb{1}_{(0,1)}(x).$$

Then $f, g \in L^1(\mathbb{R})$ and we leave details to be checked by readers...

SOLUTION 3.3.11.

(1) Let (x_n) be a sequence in X converging strongly to a point x, i.e. $\|x_n - x\| \to 0$ as $n \to \infty$. Let $y \in X$. The Cauchy-Schwarz Inequality then implies that

$$0 \leq |< x_n - x, y >| \leq \underbrace{\|x_n - x\|}_{\searrow 0} \cdot \|y\|$$

and hence $< x_n - x, y > \to 0$, i.e. for all y we have

$$< x_n, y > \to < x, y >,$$

establishing the weak convergence of (x_n).

(2) The converse is false. For a counterexample, take any orthonormal sequence (x_n) in X. This implies that $\|x_n\| = 1$, for all $n \in \mathbb{N}$. Since $\|x_n - 0\| = 1 \not\to 0$, x_n does not converge strongly to 0. In fact, such a sequence does not converge to any point in X. For

$$\|x_n - x_m\|^2 = 2$$

which implies that (x_n) cannot be Cauchy.

Nevertheless, (x_n) converges weakly to 0 since by the Bessel Inequality we have

$$\sum_{n=1}^{\infty} |< x_n, y >|^2 \leq \|y\|^2$$

and we must necessarily have

$$\lim_{n\to\infty} < x_n, y >= 0 =< 0, y >,\ \forall y \in X.$$

(3) Since (x_n) converges weakly to x, $< x_n, y > \to < x, y >$ for all $y \in X$. Hence $< x_n, x > \to < x, x >$. Then

$$\begin{aligned} \|x_n - x\|^2 &= < x_n - x, x_n - x > \\ &= \|x_n\|^2 - < x_n, x > - < x, x_n > - \|x\|^2 \\ &\longrightarrow \|x\|^2 - \|x\|^2 - \|x\|^2 + \|x\|^2 \\ &= 0. \end{aligned}$$

Thus (x_n) converges strongly to x.

Solution 3.3.12. Assume that X has dimension m with basis $\mathcal{B} = \{e_1, e_2, \cdots, e_m\}$. By the Gram-Schmidt Algorithm, we can construct from $\mathcal{B}$ an orthonormal sequence which we keep on noting by $\mathcal{B} = \{e_1, e_2, \cdots, e_m\}$. Then one can write

$$x = < x, e_1 > e_1 + \cdots + < x, e_m > e_m.$$

Since (x_n) converges weakly to x, $< x_n, y > \to < x, y >$ for all $y \in X$ and so $< x_n, e_k > \to < x, e_k >$, for all $k = 1, 2, \cdots, m$ (as n tends to infinity). Now we have

$$\begin{aligned} \|x - x_n\|^2 &= < x - x_n, x - x_n > \\ &= \sum_{k=1}^{m} | < x, e_k > - < x_n, e_k > |^2 \\ &\longrightarrow 0, \end{aligned}$$

i.e. (x_n) converges strongly to x.

The conclusion is: In a finite-dimensional inner product space a sequence converges weakly iff it converges strongly (to the same limit).

Solution 3.3.13.

(1) X is an inner product space which is not complete (Exercise 1.3.13). Thus X is not a Hilbert space.
(2) Since $\mathbb{R}^n$ is finite dimensional, it is complete. Hence $\mathbb{R}^n$ is a Hilbert space.
(3) By Theorem 1.1.73, $L^{\pi}(\mathbb{R})$ is a Banach space. By Exercise 3.5.4, it cannot be a Hilbert space.
(4) ℓ^2 is a Hilbert space since ℓ^2 is an inner product space, which is complete (Exercise 1.5.6).

Solution 3.3.14. We already know that ℓ^2 is a Hilbert space. Let $1 \le p < \infty$ and assume that ℓ^p is a Hilbert space (and we will force p to equal 2). Then it must satisfy the parallelogram law

$$\|x + y\|_p^2 + \|x - y\|_p^2 = 2\|x\|_p^2 + 2\|y\|_p^2 \cdots (1)$$

for all $x, y \in \ell^p$. In particular for (e_n) and (e_m) $(n \neq m)$ the usual bases. So we must have

$$\|e_n + e_m\|_p^2 + \|e_n - e_m\|_p^2 = 2\|e_n\|_p^2 + 2\|e_m\|_p^2$$

WLOG, take $m > n$. Then $\|e_n\|_p^2 = \|e_m\|_p^2 = 1$ and

$$e_n + e_m = (0, \cdots, 0, \underbrace{1}_{n}, 0 \cdots, \underbrace{1}_{m}, 0 \cdots)$$

and

$$e_n - e_m = (0, \cdots, 0, \underbrace{1}_{n}, 0 \cdots, \underbrace{-1}_{m}, 0 \cdots).$$

Hence

$$\|e_n + e_m\|_p^2 = \|e_n - e_m\|_p^2 = 2^{\frac{2}{p}}$$

so that Equation (1) becomes

$$2^{\frac{2}{p}} + 2^{\frac{2}{p}} = 2 + 2 \text{ or } 2^{\frac{2}{p}+1} = 2^2, \text{ i.e. } p = 2.$$

Solution 3.3.15.

(1) That Y is a vector space can be easily established by the reader. Now, we need to check that Y is a closed set in X. Set $\varphi : C[0,1] \to C[0,\frac{1}{2}]$ defined for all $f \in X$ by $\varphi(f) = f \mid [0, \frac{1}{2}]$. Then φ is linear (clear) and continuous since for all $f \in X$ we have

$$\int_0^{\frac{1}{2}} |\varphi(f)(x)|^2 dx = \int_0^{\frac{1}{2}} |f(x)|^2 dx \leq \int_0^1 |f(x)|^2 dx.$$

Thus $\ker \varphi$, being equal to Y, must then be closed.

(2) Let us find $Y^\perp$. Remember that

$$Y^\perp = \{f \in X :< f, g >= 0, \ \forall g \in Y\}.$$

Whence $\int_{\frac{1}{2}}^1 f(x)g(x)dx = 0$ for all continuous functions g. Setting $f = g$, gives $\int_{\frac{1}{2}}^1 f^2(x)dx = 0$ which, since f^2 is continuous and positive, gives $f(x) = 0$ for $x \in \left[\frac{1}{2}, 1\right]$. Therefore,

$$Y^\perp = \left\{f \in X : \ f(x) = 0 \text{ for all } \frac{1}{2} \leq x \leq 1\right\}.$$

(3) First, $Y \cap Y^\perp = \{0\}$ since Y is a vector space.

Now we proceed to show that $Y + Y^\perp = Z$. Let $f \in Z$. Define

$$g(x) = \begin{cases} 0, & 0 \leq x \leq \frac{1}{2}, \\ f(x), & \frac{1}{2} \leq x \leq 1 \end{cases} \text{ and } h(x) = \begin{cases} f(x), & 0 \leq x \leq \frac{1}{2}, \\ 0, & \frac{1}{2} \leq x \leq 1. \end{cases}$$

The functions g and h are both well-defined and continuous thanks to the fact that $f \in Z$. It is also plain that

$$f(x) = g(x) + h(x),\ \forall x \in [0,1].$$

Lastly, $g \in Y$ and $h \in Y^{\perp}$ and hence $f = g + h \in Y + Y^{\perp}$.

Conversely, let $f \in Y + Y^{\perp}$. Then there exist g in Y and h in $Y^{\perp}$ such that $f(x) = g(x) + h(x)$. The functions g and h being continuous, we deduce that f is also continuous. Finally,

$$f\left(\frac{1}{2}\right) = g\left(\frac{1}{2}\right) + h\left(\frac{1}{2}\right) = 0,$$

i.e. $f \in Z$ and clearly $Z \neq X$ (for example $x \mapsto e^x \in X$ but it is not in Z).

(4) The reason is that X is not complete, i.e. it is not a Hilbert space despite the fact that Y is closed. In other words, the completion of the space cannot be simply dispensed with.

Solution 3.3.16.

(1) Remember that $\text{span}\{y\} = \{\alpha y\}_{\alpha \in \mathbb{C}}$. Now we have

$$(\text{span}\{y\})^{\perp} = \{x \in X :< x, \alpha y >= 0,\ \forall \alpha \in \mathbb{C}\}.$$

But

$$< x, y >= 0 \Longleftrightarrow \overline{\alpha} < x, y >= 0,\ \forall \alpha \in \mathbb{C} \Longleftrightarrow < x, \alpha y >= 0,\ \forall \alpha \in \mathbb{C}.$$

Thus $Y = (\text{span}\{y\})^{\perp}$.

(2) Yes, Y is closed as e.g. it is equal to the orthogonal of some set.

(3) Since $\text{span}\{y\}$ is a *finite-dimensional* vector space, it is closed. Therefore,

$$Y^{\perp} = (\text{span}\{y\})^{\perp\perp} = \overline{\text{span}\{y\}} = \text{span}\{y\}.$$

(4) We can alternatively write Y as

$$Y = \{f \in X : < f, \mathbf{1} >= 0\}$$

where $\mathbf{1}$ denotes the constant function taking the value 1 on $[0,1]$. The previous question yields $Y^{\perp} = \{\alpha\}$, i.e. the constant functions on $[0,1]$.

Solution 3.3.17.

(1) Yes, Y is certainly closed...

(2) First, recall that c_{00} is not complete. Using the previous exercise, we have $Y^\perp = \{\alpha x\}_{\alpha\in\mathbb{C}}$. We give an example of a vector in c_{00} which is not in $Y \oplus Y^\perp$. Let

$$y = (1, 1, 0, \cdots, 0, \cdots),\ a = (\frac{1}{2}, -1, 0 \cdots, 0, \cdots) \text{ and}$$

$$b = (\frac{1}{2}, 2, 0 \cdots, 0, \cdots).$$

Obviously $a + b = y \in c_{00}$. Also $a \in Y$ since

$$< (\frac{1}{2}, -1, 0 \cdots, 0, \cdots), (1, \frac{1}{2}, \cdots, \frac{1}{n}, \cdots) >= 0.$$

In the end, $b \notin Y^\perp$ as

$$\exists x = a \in Y : < a, b >= -\frac{7}{4} \neq 0.$$

(3) It does not contradict the Orthogonal Decomposition Theorem because c_{00} is not complete (even if Y is closed). In other words, the completeness of the vector space cannot merely be dropped.

Solution 3.3.18. Consider $H = \ell^2$ and $Y = c_{00}$. We know, from Exercise 1.3.18 that c_{00} is dense and non-closed in ℓ^2. Hence

$$(c_{00})^\perp = (\overline{c_{00}})^\perp = (\ell^2)^\perp = \{0\}.$$

Identifying $c_{00} \oplus \{0\}$ with c_{00}, it then follows that the trueness of $c_{00} \oplus (c_{00})^\perp = \ell^2$ would lead to $c_{00} = \ell^2$, which is impossible!

Solution 3.3.19. We need to show that X is complete. Since X is a normed vector space, it has a completion which we denote by $\tilde{X}$. So to show that X is a Hilbert space, it then suffices to show that $X = \tilde{X}$. Let $\tilde{x} \in \tilde{X}$ be such that $\tilde{x} \neq 0$. Define a linear functional $f : \tilde{X} \to \mathbb{C}$ such that

$$f(x) =< x, \tilde{x} > .$$

Then f is continuous and non-identically zero. It then follows that the restriction of f to X remains continuous. Since X is dense in $\tilde{X}$, we may easily see that $f : X \to \mathbb{C}$ is non-identically zero.

Now, let

$$Y = \ker f = \{x \in X : f(x) = 0\}.$$

Then Y is closed and $Y \subsetneq X$ and $Y^\perp \neq \{0\}$. Indeed if $Y^\perp = \{0\}$, then we would have

$$Y = Y^{\perp\perp} = \{0\}^\perp = X$$

which is a contradiction. Since $Y^\perp \neq \{0\}$, there is a $y \in Y^\perp$ ($y \neq 0$) such that $\|y\| = 1$ (otherwise normalize it!). Set $z = \overline{f(y)}y$ which is in X. Since $f(x)y - f(y)x \in Y$ and $y \in Y^\perp$, we have

$$< f(x)y - f(y)x, y >= 0 \text{ or } f(x) < y, y >= f(y) < x, y > .$$

Hence for all $x \in X$

$$< x, \tilde{x} >= f(x) = f(x) < y, y >= f(y) < x, y >=< x, \overline{f(y)}y >=< x, z >.$$

Since X is dense in $\tilde{X}$, we infer that for all $x \in \tilde{X}$ (why?)

$$< x, \tilde{x} >=< x, z > .$$

Thus $\tilde{x} = z \in X$, proving that $\tilde{X} \subset X$ so that finally $X = \tilde{X}$.

SOLUTION 3.3.20.

(1) It is very easy to see that $\{e_n\}$ is an orthonormal sequence. To see that it is an orthonormal basis, remember that it must satisfy either of the statements of Theorem 3.1.40. We use here the first statement. Let $x = (x_n) \in \ell^2$ be such that $< x, e_n >= 0$ for all $n \in \mathbb{N}$. Then clearly

$$x_n = 0, \ \forall n \in \mathbb{N}, \text{ that is, } x = 0.$$

(2) ([**210**]) We first check that the given sequence is orthonormal. It is clear that

$$< e_n, e_m >= \int_0^1 e^{2\pi i n x} e^{-2\pi i m x} dx = \begin{cases} 1, & n = m, \\ 0, & n \neq m, \end{cases}$$

proving that the given sequence is orthonormal. To show that this sequence is actually a Hilbert basis, we use the second property of Theorem 3.1.40. Set

$$X = \{f \in C[0,1] : \ f(0) = f(1)\}.$$

By Fejér's theorem, for each $f \in X$, there is a sequence of (f_n) of trigonometric polynomials (implying that $f_n \in \operatorname{span}\{e_n : n \in \mathbb{Z}\}$ such that

$$\|f_n - f\|_\infty \longrightarrow 0$$

as $n \to \infty$. Hence (why?)

$$\|f_n - f\|_{L^2[0,1]} \longrightarrow 0$$

as $n \to \infty$. The proof is over as soon as we show that X is dense in $C[0,1]$. To see this, for any $f \in C[0,1]$, define

$$f_n(x) = \begin{cases} f(x), & 0 \le x \le 1 - \frac{1}{n}, \\ f(0) + (1-x)n(f(1-\frac{1}{n}) - f(0)), & 1 - \frac{1}{n} < x \le 1. \end{cases}$$

Then clearly, $f_n \in X$ for all n and $\|f_n - f\|_\infty \to 0$. The result now follows from the fact that $C[0,1]$ is dense in $L^2[0,1]$.

Solution 3.3.21. It is clear that $\{e_{2n} : n \in \mathbb{N}\}$ remains an orthonormal sequence in ℓ^2 as by assumption $\{e_n : n \in \mathbb{N}\}$ is orthonormal. But, $\{e_{2n} : n \in \mathbb{N}\}$ is not a Hilbert basis as if it were then we would have for $x \in \ell^2$

$$< x, e_{2n} >= 0, \ \forall n \in \mathbb{N} \Longrightarrow x = 0.$$

But $e_1 = (1, 0, 0, \cdots, 0, \cdots) \in \ell^2$ and clearly

$$< e_1, e_{2n} >= 0, \ \forall n \in \mathbb{N}$$

and yet $e_1 \neq 0_{\ell^2}$!

Solution 3.3.22. ([**50**])

(1) Let $f : \mathbb{D} \to \mathbb{C}$ be such that $f(z) = \sum_{k=0}^{\infty} a_k z^k$. Let $0 \le r < 1$ and $\theta \in [0, 2\pi]$. Set

$$f_r(\theta) = f(re^{i\theta}).$$

Then

$$\begin{aligned} \left| f_r(\theta) - \sum_{k=0}^{n} a_k r^k e^{i\theta k} \right| &= \left| \sum_{k=n+1}^{\infty} a_k r^k e^{i\theta k} \right| \le \sum_{k=n+1}^{\infty} |a_k r^k e^{i\theta k}| \\ &= \sum_{k=n+1}^{\infty} |a_k| r^k \end{aligned}$$

and the latter is a tail of a convergent series hence it goes to zero (uniformly) in θ. Before carrying on the proof, we recall that

$$\int_0^{2\pi} e^{i(k-l)} d\theta = \begin{cases} 2\pi, & k = l, \\ 0, & k \neq l. \end{cases}$$

Hence, we have

$$\frac{1}{2\pi}\int_0^{2\pi}|f(re^{i\theta})|^2 d\theta = \lim_{n\to\infty}\frac{1}{2\pi}\int_0^{2\pi}\left|\sum_{k=0}^n a_k r^k e^{i\theta k}\right|^2 d\theta$$

$$= \lim_{n\to\infty}\frac{1}{2\pi}\int_0^{2\pi}\sum_{k=0}^n a_k r^k e^{i\theta k}\sum_{l=0}^n \overline{a_l} r^l e^{-i\theta l} d\theta$$

$$= \lim_{n\to\infty}\frac{1}{2\pi}\sum_{k,l=0}^n a_k\overline{a_l}r^{k+l}\int_0^{2\pi} e^{i(k-l)}d\theta$$

$$= \lim_{n\to\infty}\sum_{k=0}^n a_k\overline{a_k}r^{2k}$$

$$= \lim_{n\to\infty}\sum_{k=0}^n |a_k|^2 r^{2k}$$

$$= \sum_{k=0}^{\infty}|a_k|^2 r^{2k},$$

that is (for $r \in [0,1)$),

$$\|f_r\|_2 = \sqrt{\sum_{n=0}^{\infty}|a_n|^2 r^{2n}}$$

(where $\|\cdot\|_2$ is the usual L^2-norm). Now, we compute the norm of f w.r.t. $H^2(\mathbb{D})$. We have (as the sequence of partial sums appearing just above is increasing)

$$\|f\|_{H^2(\mathbb{D})} = \sup_{0\le r<1}\|f_r\|_2 = \sup_{0\le r<1}\sup_{n\ge 0}\sqrt{\sum_{k=0}^n |a_k|^2 r^{2k}} = \sup_{n\ge 0}\sup_{0\le r<1}\sqrt{\sum_{k=0}^n |a_k|^2 r^{2k}}.$$

But,

$$\sup_{0\le r<1}\sqrt{\sum_{k=0}^n |a_k|^2 r^{2k}} = \lim_{r\to 1}\sqrt{\sum_{k=0}^n |a_k|^2 r^{2k}} = \sqrt{\sum_{k=0}^n |a_k|^2}$$

so that

$$\|f\|_{H^2(\mathbb{D})} = \sup_{n\ge 0}\sqrt{\sum_{k=0}^n |a_k|^2} = \sqrt{\sum_{n=0}^{\infty}|a_n|^2} = \|(a_n)\|_{\ell^2}.$$

To prove the other question, observe first that φ is linear and by the previous calculations, we obtain that (for all $f \in$

$H^2(\mathbb{D})$)

$$\|\varphi(f)\|_{\ell^2} = \|f\|_{H^2(\mathbb{D})}$$

signifying that φ is an isometry. It only remains to show that φ is surjective. To this end, let $(a_n) \in \ell^2$. Then by considering $f : \mathbb{D} \to \mathbb{C}$ defined by $f(z) = \sum_{n=0}^{\infty} a_n z^n$, we may show that f is a well defined analytic function on $\mathbb{D}$ which belongs to $H^2(\mathbb{D})$ and $\varphi(f) = (a_n)$, showing that φ is surjective.

(2) By Exercise 1.5.7, $H^2(\mathbb{D})$ is a complete normed vector space. To show that $H^2(\mathbb{D})$ is a Hilbert space, it suffices to show that it does satisfy the parallelogram law. Let $f, g \in H^2(\mathbb{D})$, $f(z) = \sum_{n=0}^{\infty} a_n z^n$ and $g(z) = \sum_{n=0}^{\infty} b_n z^n$. We then have (by the foregoing question)

$$\begin{aligned}\|f+g\|^2_{H^2(\mathbb{D})} + \|f-g\|^2_{H^2(\mathbb{D})} &= \|(a_n+b_n)\|^2_{\ell^2} + \|(a_n-b_n)\|^2_{\ell^2}\\ &= 2\|(a_n)\|^2_{\ell^2} + 2\|(b_n)\|^2_{\ell^2} \text{ (as } \ell^2 \text{ is a Hilbert space)}\\ &= 2\|f\|^2_{H^2(\mathbb{D})} + 2\|g\|^2_{H^2(\mathbb{D})}\end{aligned}$$

Remark. The associated inner product space is given by

$$< f, g > = \lim_{r\to 1} \frac{1}{2\pi} \int_0^{2\pi} f(re^{i\theta})\overline{g(re^{i\theta})} d\theta.$$

(3) First, we show that $\{z^n : n \geq 0\}$ is an orthonormal sequence. We have

$$< z^n, z^n > = \lim_{r\to 1} \frac{1}{2\pi} \int_0^{2\pi} r^n e^{i\theta n} r^n e^{-i\theta n} d\theta = \lim_{r\to 1} r^{2n} \frac{1}{2\pi} \int_0^{2\pi} d\theta = 1$$

and for all $n \neq m$ we have

$$< z^n, z^m > = \lim_{r\to 1} \frac{1}{2\pi} \int_0^{2\pi} r^n e^{i\theta n} r^m e^{-i\theta m} d\theta = \frac{1}{2\pi} \int_0^{2\pi} e^{i\theta(n-m)} d\theta = 0$$

proving that $\{z^n : n \geq 0\}$ is an orthonormal sequence.

To prove that $\{z^n : n \geq 0\}$ is an orthonormal basis if e.g. it satisfies the Parseval Formula (Theorem 3.1.40), i.e. for all $f \in H^2(\mathbb{D})$

$$\|f\|^2_{H^2(\mathbb{D})} = \sum_{n=0}^{\infty} | < f, z^n > |^2.$$

So let $f \in H^2(\mathbb{D})$. By the first question

$$\|f\|^2_{H^2(\mathbb{D})} = \|\varphi(f)\|^2_{\ell^2} = \|(a_n)\|^2_{\ell^2} = \sum_{n=0}^{\infty} |a_n|^2.$$

On the other hand,

$$\begin{aligned}
< f, z^n > &= \lim_{r\to 1} \frac{1}{2\pi} \int_0^{2\pi} \sum_{m=0}^{\infty} a_m r^m e^{i\theta m} r^n e^{-i\theta n} d\theta \\
&= \lim_{r\to 1} \frac{1}{2\pi} \int_0^{2\pi} \sum_{m=0}^{\infty} a_m r^{m+n} e^{i\theta(m-n)} d\theta \\
&= \lim_{r\to 1} \sum_{m=0}^{\infty} \frac{1}{2\pi} \int_0^{2\pi} a_m r^{m+n} e^{i\theta(m-n)} d\theta \text{ (why?)} \\
&= \lim_{r\to 1} \sum_{m=0}^{\infty} a_m r^{m+n} \frac{1}{2\pi} \int_0^{2\pi} e^{i\theta(m-n)} d\theta \\
&= \lim_{r\to 1} a_n r^{n+n} \frac{1}{2\pi} \int_0^{2\pi} d\theta \\
&= a_n.
\end{aligned}$$

Consequently,

$$\sum_{n=0}^{\infty} |< f, z^n >|^2 = \sum_{n=0}^{\infty} |a_n|^2 = \|f\|^2_{H^2(\mathbb{D})},$$

as required.

Solution 3.3.23.

(1) Let $x \in X$. Since $\{a\}^\perp$ is a closed vector space, the Projection Theorem says that there exists a unique $b \in \{a\}^\perp$ such that

$$\|x - b\| = \inf_{t\in\{a\}^\perp} \|x - t\| = d(x, \{a\}^\perp).$$

One can also write $x = a + c$ (where $c \in \{a\}^\perp$) and hence

$$|< x, a >| = |< a + c, a >| =< a, a >= \|a\|^2$$

and so

$$\frac{|< x, a >|}{\|a\|} = \|a\|.$$

So we are, in fact, required to show that

$$\|x - b\| = \|a\|.$$

This first follows from:

$$\|x - b\| = \inf_{t\in\{a\}^\perp} \|x - t\| \leq \|x - c\| = \|a\|.$$

Then, if $t \in \{a\}^\perp$, we get

$$|< x - t, a >| = |< x, a >| = \|a\|^2,$$

and so by invoking the Cauchy-Schwarz Inequality, we get

$$\|x-t\| \geq \|a\|.$$

Hence

$$\|x-b\| = \inf_{t\in\{a\}^\perp} \|x-t\| \geq \|a\|,$$

as wanted.

(2) (a) The closedness of Y is a direct consequence of Exercise 3.3.16. Alternatively, we can use the following method. Let (f_n) be a sequence in Y converging in $L^2[0,1]$ to some f. We need to check that $f \in Y$. We first note that $f \in L^2[0,1]$ since $L^2[0,1]$ is complete. It remains to show that $\int_0^1 f(x)dx = 0$. Since $f_n \in Y$, we have

$$\begin{aligned} 0 \leq \left|\int_0^1 f(x)dx\right| &= \left|\int_0^1 f(x)dx - \int_0^1 f_n(x)dx\right| \\ &= \left|\int_0^1 (f(x)-f_n(x))dx\right| \\ &\leq \int_0^1 |f(x)-f_n(x)|dx \\ &\leq \sqrt{\int_0^1 |f(x)-f_n(x)|^2 dx} \longrightarrow 0 \end{aligned}$$

where we used the Cauchy-Schwarz Inequality in the last inequality. Hence $\int_0^1 f(x)dx = 0$, i.e. $f \in Y$. Finally, the reader can easily check that Y is a vector subspace of $L^2[0,1]$.

(b) We clearly see that $Y = [\{a\}_{a\in\mathbb{C}}]^\perp = [\text{span}\{1\}]^\perp$. Hence

$$Y^\perp = \{a\}^{\perp\perp} = \overline{\{a\}} = \{a\},$$

i.e. it is constituted of constant functions.

(c) Let $f(x) = x$. Since Y is closed, $Y = (Y^\perp)^\perp$ and using Question 1, we are led to

$$d(f,Y) = d(f,(Y^\perp)^\perp) = d(f,\{a\}^\perp) = \frac{|<f,a>|}{\|a\|} = \frac{|a|\int_0^1 xdx}{|a|}.$$

Thus

$$d(f,Y) = \frac{1}{2}.$$

SOLUTION 3.3.24.

(1) (a) Let $p \neq q$. We have

$$< e_p, e_q >= \frac{1}{2\pi} \int_0^{2\pi} e_p(x)\overline{e_q(x)} dx = \frac{1}{2\pi} \int_0^{2\pi} e^{ipx} e^{-iqx} dx$$
$$= \left[\frac{1}{2\pi(p-q)} e^{i(p-q)x} \right]_0^{2\pi} = 0.$$

(b) Let $p = q$. We have

$$\|e_p\|^2 =< e_p, e_p >= \frac{1}{2\pi} \int_0^{2\pi} e^{ipx} e^{-ipx} dx = \frac{1}{2\pi} \int_0^{2\pi} dx = 1.$$

(2) Since $(e_p = e^{ipx})_{p \in \mathbb{N}}$ is an orthonormal sequence, the Bessel Inequality then implies that

$$\sum_{p=1}^{\infty} | < f, e_p > |^2 = \sum_{p=1}^{\infty} \left| \frac{1}{2\pi} \int_0^{2\pi} f(x) e^{-ipx} dx \right|^2 \leq \frac{1}{2\pi} \int_0^{2\pi} |f(x)|^2 dx < \infty.$$

Thus the series on the left hand converges and hence its general term tends to zero, i.e.

$$\lim_{p \to \infty} \frac{1}{2\pi} \int_0^{2\pi} f(x) e^{-ipx} dx = 0.$$

SOLUTION 3.3.25.

(1) The linearity of f_x follows easily from the properties of an inner product. Now let $y \in X$. The Cauchy-Schwarz Inequality then yields

$$|f_x(y)| = | < x, y > | \leq \|x\| \|y\|.$$

Thus f_x is continuous.

(2) We have

$$Y^\perp = \{y \in X :< y, x >= 0, \ \forall x \in Y\} = \bigcap_{x \in Y} \ker f_x.$$

Hence $Y^\perp$ is a vector space since all $\ker f_x$ are so. Also, as f_x is continuous, $\ker f_x$ is closed. Hence $Y^\perp$, being an arbitrary intersection of closed sets, is itself closed.

SOLUTION 3.3.26. Denote the norm on $L^2[-1,1]$ by $\|\cdot\|_2$. Recall that for $f \in L^2[-1,1]$

$$\|f\|_2 = \sqrt{\int_{-1}^1 |f(x)|^2 dx}.$$

We want to construct from $1, x, x^2$ an orthonormal set which we denote by $\{e_1, e_2, e_3\}$. The Gram-Schmidt Algorithm gives

$$e_1 = \frac{1}{\|1\|_2} = \frac{1}{\sqrt{2}}.$$

Now

$$y_2 := x - < x, e_1 > e_1 = x - \frac{1}{2}\int_{-1}^{1} x dx = x.$$

Then

$$e_2 = \frac{y_2}{\|y_2\|_2} = \sqrt{\frac{3}{2}} x.$$

Setting

$$y_3 = x^2 - < x^2, e_1 > e_1 - < x^2, e_2 > e_2$$

yields

$$y_3 = x^2 - \frac{1}{3} \text{ and } \|y_3\|_2 = \sqrt{\frac{8}{45}} = \frac{\sqrt{8}}{3\sqrt{5}}.$$

Therefore,

$$e_3 = \frac{y_3}{\|y_3\|_2} = \sqrt{\frac{5}{8}}(3x^2 - 1).$$

Solution 3.3.27. The question amounts to finding the best approximation of x^3 by a polynomial of degree 2 in $L^2[-1, 1]$. We will use the Projection Theorem. Let $f(x) = x^3$ be defined on $[-1, 1]$.

Let $Y = \mathbb{R}_2[X]$ (hence Y is a complete vector space). We need to find

$$\inf_{q \in Y} \|f - q\|_2^2 \text{ where } q(x) = ax^2 + bx + c.$$

Since Y is complete, there exists a **unique** $p \in Y$ such that

$$\|f - p\|_2^2 = \inf_{q \in Y} \|f - q\|_2^2.$$

Let $p(x) = \alpha x^2 + \beta x + \gamma$ where α, β and γ are to be found. We also know that p has to satisfy

$$< f - p, q >= 0, \ \forall q \in Y, \text{ i.e.}$$

$$\int_{-1}^{1} (x^3 - \alpha x^2 - \beta x - \gamma)(ax^2 + bx + c) dx, \ \forall a, b, c \in \mathbb{R}.$$

The monomials having odd powers are odd functions over the symmetric interval $[-1, 1]$. Hence we are left with

$$\int_{-1}^{1} [b(x^4 - \beta x^2) + a(-\alpha x^4 - \gamma x^2) + c(-\alpha x^2 - \gamma)] dx = 0, \ \forall a, b, c \in \mathbb{R}.$$

Hence

$$b\left[\frac{x^5}{5}-\frac{\beta x^3}{3}\right]_{-1}^{1}+a\left[-\frac{\alpha x^5}{5}-\frac{\gamma x^3}{3}\right]_{-1}^{1}+c\left[-\frac{\alpha x^3}{3}-\gamma x\right]_{-1}^{1}=0$$

for all $a, b, c \in \mathbb{R}$. Thus, we must have

$$\begin{cases} \frac{\alpha}{5}+\frac{\gamma}{3}=0, \\ \frac{1}{5}-\frac{\beta}{3}=0, \\ \frac{\alpha}{3}+\gamma=0, \end{cases} \quad \text{or} \quad \begin{cases} \alpha=\gamma=0, \\ \beta=\frac{3}{5}. \end{cases}$$

The required polynomial is $p(x)=\frac{3}{5}x$. The corresponding infimum is then

$$\|f-p\|_2^2=\int_{-1}^{1}\left|x^3-\frac{3}{5}x\right|^2 dx=\frac{8}{175}.$$

Remark. We can use a different method to find the inf in this exercise, namely the one used to find an extremum point for functions of several variables.

Solution 3.3.28. We can use the same method as in the foregoing exercise. We only give the important results. The polynomial q is defined for every $x \in \mathbb{R}$ by

$$q(x)=x-\frac{1}{6}.$$

Thus

$$\inf_{a,b\in\mathbb{R}}\int_0^1 (x^2-ax-b)^2 dx=\frac{1}{180}.$$

Solution 3.3.29.

(1) Since $\dim Y=1<\infty$, Y is complete.

(2) Since Y is a complete vector space, the Projection Theorem may be applied to write each $g \in X$ as

$$g=\alpha(1-x)+h,\ \alpha \in \mathbb{R} \text{ and } h \in Y^{\perp}.$$

The orthogonal projection of $f(x)=1$ is then $\alpha(1-x)$. It remains to find the right α. We have

$$< g, 1-x >=< \alpha(1-x)+h, 1-x >= \alpha\|1-x\|^2$$

yielding

$$\alpha=\frac{< g, 1-x >}{\|1-x\|^2}.$$

Now take $g=1$ and since

$$< 1, 1-x >=\int_0^1 (1-x)dx=1-\frac{1}{2}=\frac{1}{2}$$

and

$$\|1-x\|^2 = \int_0^1 (1-x)^2 dx = 1 - 1 + \frac{1}{3} = \frac{1}{3},$$

we deduce that the orthogonal projection of 1 onto Y is given by $\frac{3}{2}(1-x)$.

SOLUTION 3.3.30. Since H is separable, H possesses a countable dense subset which we denote by X (hence $X = \{x_0, x_1, \cdots\}$). Let $\mathcal{O}$ be an orthonormal set in H.

Since X is dense in H, we know that for each $e \in \mathcal{O}$, there is an $n \in \mathbb{N}$ such that $\|x_n - e\| < \frac{1}{3}$ (why?). We may then define a function $\varphi : \mathcal{O} \to \mathbb{N}$ by $\varphi(e) = n$. To prove that $\mathcal{O}$ is countable, we need only show that φ is injective. Let $e \neq e'$ where $e, e' \in \mathcal{O}$. Then

$$\|e - e'\|^2 =< e - e', e - e' >=< e, e > + < e', e' >= 2.$$

If $\varphi(e) = \varphi(e')$, we would have $\sqrt{2} < \frac{2}{3}$! Thus $\varphi(e) \neq \varphi(e')$ and so φ is injective.

SOLUTION 3.3.31.

(1) The properties of an inner product are easy to check, so we should only worry about the existence of the limit. It suffices to check that the limit exists for u_r and u_s. If $r = s$, then

$$\frac{1}{2n} \int_{-n}^{n} e^{irx} e^{-isx} dx = 1.$$

If $r \neq s$, then

$$\frac{1}{2n} \int_{-n}^{n} e^{irx} e^{-isx} dx = \frac{1}{2ni(r-s)} (e^{i(r-s)n} - e^{-i(r-s)n}) = \frac{\sin[(r-s)n]}{(r-s)n}$$

and this goes to zero as n goes to infinity. Thus the limit exists.

(2) The reader is asked in Exercise 3.5.3 to prove that X is not a Hilbert space. Let H be its completion then. We know from the first question that $\{u_r(x) : x \in \mathbb{R}\}$ is an uncountable orthonormal set in X. Thus, we have an uncountable orthonormal set in H, proving that H cannot be separable (by Exercise 3.3.30).

REMARK. We could have also proved the non-separability of H using Proposition 1.1.29.

SOLUTION 3.3.32.

(1) Left to the reader...

(2) First, by Cauchy's integral formula, we know that if f is an analytic function within and on a simple closed contour γ (in the positive sense), and if a is any complex interior to γ, then

$$f(a)=\frac{1}{2\pi i}\int_{\gamma}\frac{f(z)dz}{z-a}.$$

Applying this formula to $f\in\mathcal{A}^2$ (with the usual parametrization) yields (for any $s\in(0,r)$)

$$f(a)=\frac{1}{2\pi i}\int_{\gamma}\frac{f(z)dz}{z-a}=\frac{1}{2\pi i}\int_0^{2\pi}\frac{f(a+se^{i\theta})}{se^{i\theta}}sie^{i\theta}d\theta=\frac{1}{2\pi}\int_0^{2\pi}f(a+se^{i\theta})d\theta.$$

Hence

$$\iint\limits_{B_c(a,r)} f(x+iy)dxdy=\int_0^r\int_0^{2\pi}f(a+se^{i\theta})sdsd\theta=\int_0^r 2\pi f(a)sds=\pi r^2 f(a).$$

Thus

$$f(a)=\frac{1}{\pi r^2}\iint\limits_{B_c(a,r)} f(x+iy)dxdy.$$

(3) Since f is analytic, f^2 too is analytic so that by the previous question

$$f^2(a)=\frac{1}{\pi r^2}\iint\limits_{B_c(a,r)} f^2(x+iy)dxdy.$$

Hence

$$|f(a)|^2\le\frac{1}{\pi r^2}\iint\limits_{B_c(a,r)}|f(x+iy)|^2dxdy\le\frac{1}{\pi r^2}\|f\|^2$$

or

$$|f(a)|\le\frac{1}{\sqrt{\pi}r}\|f\|.$$

(4) Let $a\in K$. As $B_c(a,R)\subset\Omega$, then by the foregoing question

$$\pi R^2|f(a)|^2\le\iint\limits_{B_c(a,R)}|f(x+iy)|^2dxdy\le\iint\limits_{\Omega}|f(x+iy)|^2dxdy=\|f\|^2.$$

Thus

$$\sup_{z\in K}|f(z)|\le\frac{1}{\sqrt{\pi}R}\|f\|.$$

(5) We show that $\mathcal{A}^2(\Omega)$ is closed in the Hilbert space $L^2(\Omega)$ (with respect to the usual Lebesgue's measure restricted to Ω). Let (f_n) be a sequence in $\mathcal{A}^2(\Omega)$ which converges to f w.r.t. the norm of $L^2(\Omega)$. By the previous question, we have

$$\sup_{z\in K}|f_n(z)-f(z)| \leq \frac{1}{\sqrt{\pi}R}\|f_n-f\|$$

which tells us that (f_n) converges uniformly to f on every compact of Ω. By the hint, f is analytic (and obviously also $f\in L^2(\Omega)$), i.e. $f\in\mathcal{A}^2(\Omega)$.

Solution 3.3.33.

(1) Let (e_n) be an orthonormal sequence in H. Consider now all sequences $\alpha=(\alpha_n)$ with values in $\{0,1\}$. This is an uncountable set (see e.g. [**153**]). Define

$$A_\alpha x = \sum_{n=1}^{\infty} \alpha_n < x, e_n > e_n.$$

Then A_α is obviously linear. It is also bounded as for any $x\in H$, we have

$$\begin{aligned}
\|A_\alpha x\|^2 &=< A_\alpha x, A_\alpha x > \\
&=< \sum_{n=1}^{\infty}\alpha_n < x,e_n > e_n, \sum_{m=1}^{\infty}\alpha_m < x,e_m > e_m > \\
&= \sum_{n=1}^{\infty}\sum_{m=1}^{\infty}\alpha_n\overline{\alpha_m} < x,e_n > \overline{< x,e_m >} < e_n,e_m > \\
&= \sum_{n=1}^{\infty}\alpha_n\overline{\alpha_n} < x,e_n > \overline{< x,e_n >}\underbrace{< e_n,e_n >}_{=1} \text{ ((}e_n\text{) is orthonormal)} \\
&= \sum_{n=1}^{\infty}|\alpha_n|^2| < x,e_n > |^2 \\
&\leq \sum_{n=1}^{\infty}| < x,e_n > |^2 \text{ (for } |\alpha_n|^2\leq 1) \\
&\leq \|x\|^2 \text{ (by the Bessel Inequality).}
\end{aligned}$$

Hence $\|A_\alpha\|\leq 1$. This tells us that $\{A_\alpha\}_\alpha$ is an uncountable family in $B(H)$. To finish the proof, we calculate $\|A_\alpha - A_\beta\|$.

We have for any $x \in H$

$$A_\alpha x - A_\beta x = \sum_{n=1}^{\infty} (\alpha_n - \beta_n) < x, e_n > e_n.$$

Arguing as above, and taking into account that $|\alpha_n - \beta_n|$ cannot exceed one whatsoever n, we obtain

$$\|A_\alpha - A_\beta\| \leq 1.$$

On the other hand, since $\|e_m\| = 1$, we have

$$\|A_\alpha - A_\beta\| \geq \|(A_\alpha - A_\beta)e_m\| = \|\sum_{n=1}^{\infty}(\alpha_n - \beta_n) < e_m, e_n > e_n\| = |\alpha_m - \beta_m|$$

Whence (why?)

$$\|A_\alpha - A_\beta\| \geq 1 \text{ or } \|A_\alpha - A_\beta\| = 1$$

for all $\alpha \neq \beta$. By Proposition 1.1.29, $B(H)$ is not separable.

(2) Recall that a compact metric space is necessarily separable. Since $B(H)$ is non separable (and a normed vector space), it cannot be compact.

(3) Since H is a Hilbert space, it is a Banach space, and so $B(H)$ is a Banach space. To become "Hilbert", it must therefore satisfy the parallelogram law! This is not always the case. Indeed, consider $H = L^2[0,1]$ and consider

$$Af(x) = xf(x) \text{ and } Bf(x) = (1-x)f(x).$$

Then (cf. Exercise 2.5.1) $A, B \in B(L^2[0,1])$ and

$$\|A\| = \sup_{0\leq x\leq 1} |x| = 1, \ \|B\| = \sup_{0\leq x\leq 1} |1-x| = 1,$$

$$\|A + B\| = 1 \text{ and } \|A - B\| = 1,$$

so that

$$\|A+B\|^2 + \|A-B\|^2 = 1 + 1 = 2 \neq 2\|A\|^2 + 2\|B\|^2 = 2 + 2 = 4.$$

Remark. In **[50]**, it is shown that $B(H)$ is a Hilbert space iff $\dim H = 1$.

Solution 3.3.34. A first thought would be to do the following (using the Cauchy-Schwarz Inequality): For all $x \in H$,

$$\|Ax\| = \| < x, b > a + < x, a > b\| \leq \|a\|\|x\|\|b\| + \|b\|\|x\|\|a\|$$

Hence

$$\|Ax\| \leq 2\|a\|\|b\|\|x\|$$

so that

$$\|A\| \leq 2\|a\|\|b\|.$$

So according to the question, we have somehow lost some precious information doing it this way (besides, we have not used the fact that a and b are orthogonal!).

So here is a more careful way: Since a and b are orthogonal, we have

$$<< x,b > a, < x,a > b >=< x,b > \overline{< x,a >} < a,b >= 0$$

so that applying the Pythagorean Theorem yields

$$\|Ax\|^2 = \|a\|^2 | < x,b > |^2 + | < x,a > |^2 \|b\|^2$$

and hence

$$\|Ax\|^2 = \|a\|^2\|b\|^2 \left| < x, \frac{b}{\|b\|} > \right|^2 + \|a\|^2\|b\|^2 \left| < x, \frac{a}{\|a\|} > \right|^2.$$

Using the Bessel Inequality applied to the orthonormal set $\{\frac{a}{\|a\|}, \frac{b}{\|b\|}\}$, gives

$$\|Ax\|^2 = \|a\|^2\|b\|^2 \left(\left| < x, \frac{b}{\|b\|} > \right|^2 + \left| < x, \frac{a}{\|a\|} > \right|^2 \right) \le \|a\|^2\|b\|^2\|x\|^2$$

from which we deduce that

$$\|A\| \le \|a\|\|b\|.$$

This is on the one hand. On the other hand, we have

$$\|A\|\|b\| \ge \|Ab\| = \| < b,b > a\| = \|a\|\|b\|^2 \text{ or } \|A\| \ge \|a\|\|b\|.$$

Consequently,

$$\|A\| = \|a\|\|b\|.$$

SOLUTION 3.3.35.

(1) That f is linear can easily be established by the reader. Now let $x \in c_{00}$. By the Cauchy-Schwarz Inequality one has

$$\left| \sum_{n=1}^{\infty} \frac{x_n}{n} \right| \le \sqrt{\sum_{n=1}^{\infty} |x_n|^2} \sqrt{\sum_{n=1}^{\infty} \frac{1}{n^2}}$$

and so

$$\|f\| \le \sqrt{\sum_{n=1}^{\infty} \frac{1}{n^2}} = \frac{\pi}{\sqrt{6}}.$$

(2) Let $e_k = (0, 0, \cdots, 1, 0, \cdots)$. If there were a $y \in c_{00}$ such that for all $x \in c_{00}$: $f(x) =< x, y >$, then one would have

$$f(e_k) = \frac{1}{k} =< y, e_k >= y_k.$$

However, this y does not belong to c_{00} as it is of the form

$$y = \left(1, \frac{1}{2}, \cdots, \frac{1}{n}, \frac{1}{n+1}, \cdots\right).$$

(3) This does not contradict the Riesz-Fréchet Representation Theorem since c_{00} is not a Hilbert space.

Solution 3.3.36. The answer is no. To see that, let $e_n = (0, \cdots, 0, 1, 0, \cdots)$. Then

$$\|A\| \geq \|A(e_n)\|_{\ell^2} = n, \ \forall n \in \mathbb{N}.$$

Thus A cannot be bounded on ℓ^2, i.e. A is unbounded.

Solution 3.3.37. Since A is *linear*, by the Closed Graph Theorem, it suffices to prove that $G(A)$ (the graph of A) is closed. Let (x_n, Ax_n) be a sequence of $G(A)$ converging to (x, y). Since H is complete, $x \in H$. We need only show that $y = Ax$. Let $z \in H$. We have by our assumption and by the continuity of the inner product

$$< z, y > \longleftarrow < z, Ax_n >=< Az, x_n > \longrightarrow < Az, x >=< z, Ax >$$

and hence $Ax = y$, as required.

Solution 3.3.38. We have

$$A^* = \begin{pmatrix} -i & 2i \\ 1 & 3 \end{pmatrix} \text{ and } B^* = \begin{pmatrix} 1 & 0 \\ -1 & 2 \end{pmatrix}.$$

Solution 3.3.39. Let $f, g \in L^2([0,1])$. Then

$$< Af, g >= \int_0^1 \varphi(x) f(x) \overline{g(x)} dx = \int_0^1 f(x) \overline{\overline{\varphi(x)} g(x)} dx =< f, A^* g >$$

where $(A^* f)(x) = \overline{\varphi(x)} f(x)$.

Solution 3.3.40.

(1) Since K is continuous on a compact set, it is bounded and hence $\|K\|_\infty = \sup_{0\le x,y\le 1} |K(x,y)|$ is finite. Let $f \in H$. We have

$$\begin{aligned}
\|Af\|^2 &= \int_0^1 \left|\int_0^1 K(x,y)f(y)dy\right|^2 dx \\
&\le \int_0^1 \left(\int_0^1 |K(x,y)f(y)|dy\right)^2 dx \\
&\le \|K\|_\infty^2 \int_0^1 \left(\int_0^1 |f(y)|dy\right)^2 dx \\
&= \|K\|_\infty^2 \left(\int_0^1 |f(y)|dy\right)^2 \\
&\le \|K\|_\infty^2 \int_0^1 |f(y)|^2 dy \\
&= \|K\|_\infty^2 \|f\|^2
\end{aligned}$$

where we have used the Cauchy-Schwarz Inequality in the ultimate inequality.

(2) Let $f, g \in H$. By the Fubini Theorem

$$\begin{aligned}
< Af, g > &= \int_0^1 \left(\int_0^1 K(x,y)f(y)dy\right)\overline{g(x)}dx \\
&= \int_0^1 \int_0^1 K(x,y)f(y)\overline{g(x)}dxdy \\
&= \int_0^1 \int_0^1 f(y)\overline{\overline{K(x,y)}g(x)}dxdy \\
&= \int_0^1 f(y)\left(\int_0^1 \overline{\overline{K(x,y)}g(x)}dx\right)dy \\
&= \int_0^1 f(y)\overline{\left(\int_0^1 \overline{K(x,y)}g(x)dx\right)}dy \\
&= < f, A^*g >
\end{aligned}$$

where

$$A^*g(y) = \int_0^1 \overline{K(x,y)}g(x)dx.$$

Therefore, A^* is given, for all $f \in H$, by

$$A^*f(x) = \int_0^1 \overline{K(y,x)}f(y)dy.$$

Solution 3.3.41.

(1) It is very easy to check the linearity of S. To prove its boundedness, let $x \in \ell^2$. Then

$$\|S(x)\|_{\ell^2}^2 = 0^2 + |x_1|^2 + |x_2|^2 + \cdots + |x_n|^2 + \cdots = \|x\|_{\ell^2}^2.$$

Whence S is bounded and besides

$$\|S\| = \sup_{\|x\|=1} \|S(x)\|_{\ell^2} = \sup_{\|x\|=1} \|x\| = 1.$$

(2) Yes! ran$S$ is closed as $\|Sx\| = \|x\|$ for all x and by a glance at Proposition 2.1.63.

(3) Let $x, y \in \ell^2$. Then

$$\begin{aligned} < Sx, y > &= 0\overline{y_1} + x_1\overline{y_2} + x_2\overline{y_3} + \cdots + x_n\overline{y_n} + \cdots \\ &= x_1\overline{y_2} + x_2\overline{y_3} + \cdots + x_n\overline{y_n} + \cdots \\ &= < x, S^*y > \end{aligned}$$

where

$$S^*(y_1, y_2, \cdots, y_n, \cdots) = (y_2, \cdots, y_n, \cdots).$$

That is,

$$S^*(x) = (x_2, \cdots, x_n, \cdots).$$

(4) It is evident that $\ker S^* = \text{span}\{e_1\}$ (check it!).

(5) As known, the shift acts on the standard basis $\{e_n : n \in \mathbb{N}\}$ as

$$Se_n = e_{n+1},\ n \in \mathbb{N}.$$

One of the ways of answering this question is the following: Since (e_n) is a Hilbert basis, we know that

$$\ell^2 = \overline{\text{span}\{e_1, e_2, e_3, \cdots\}}.$$

Hence as span$\{e_1\}$ is compact

$$\overline{\text{span}\{e_1, e_2, e_3, \cdots\}} = \overline{\text{span}\{e_2, e_3, \cdots\}} \oplus \underbrace{\text{span}\{e_1\}}_{\ker S^*}.$$

But, as $\ker S^*$ is closed, we know that

$$\ell^2 = (\ker S^*)^\perp \oplus \ker S^*$$

and so

$$\text{ran}S = \overline{\text{ran}S} = (\text{ran}S)^{\perp\perp} = (\ker S^*)^\perp = \overline{\text{span}\{e_2, e_3, \cdots\}}$$

as suggested.

(6) Let us first find A_n. Let $x = (x_n)_n \in \ell^2$. We can easily establish, using a proof by induction, that

$$A_n(x_n)_n = (S^*)^n(x_n)_n = (x_{n+1}, x_{n+2}, \cdots)$$

and whence

$$\|A_n x\|_{\ell^2}^2 = \sum_{k=n+1}^{\infty} |x_k|^2,$$

which, being a remainder of a convergent series, necessarily goes to zero. Hence $A_n x \to 0$ for all $x \in \ell^2$, i.e. (A_n) converges strongly to 0. The previous displayed equation also implies that $\|A_n\| = 1$ (why?), for all $n \in \mathbb{N}$. Therefore, (A_n) does not converge to zero in norm.

(7) First we have

$$B_n(x_1, x_2, \cdots) = (\underbrace{0, 0, \cdots, 0}_{n \text{ terms}}, x_1, x_2, \cdots).$$

Hence

$$\forall x \in \ell^2 : \ \|B_n x\| = \|x\|$$

and this tells us that (B_n) does not converges to zero in the strong operator topology and hence it cannot converge to zero in norm either. Now, we show that (B_n) weakly converges to zero. Let $x, y \in \ell^2$. Then

$$< B_n x, y >= x_1\overline{y_{n+1}} + x_2\overline{y_{n+2}} + \cdots = \sum_{k=n+1}^{\infty} x_{k-n}\overline{y_k}$$

so that

$$| < B_n x, y > | = \left| \sum_{k=n+1}^{\infty} x_{k-n}\overline{y_k} \right| \leq \sum_{k=n+1}^{\infty} |x_{k-n}||y_k|.$$

By the Cauchy-Schwarz Inequality, we obtain

$$| < B_n x, y > | \leq \sqrt{\sum_{k=n+1}^{\infty} |x_{k-n}|^2} \sqrt{\sum_{k=n+1}^{\infty} |y_k|^2}.$$

But

$$\sum_{k=n+1}^{\infty} |x_{k-n}|^2 = \sum_{n=1}^{\infty} |x_n|^2 = \|x\|^2.$$

Thus

$$0 \leq |< B_n x, y >| \leq \|x\| \sqrt{\sum_{k=n+1}^{\infty} |y_k|^2} \longrightarrow 0$$

as n tends to ∞, proving that (B_n) weakly converges to zero.

Solution 3.3.42.

(1) Linearity is trivial. As for boundedness, for all $x \in \ell^2(\mathbb{Z})$

$$\|Rx\|_{\ell^2(\mathbb{Z})} = \|x\|_{\ell^2(\mathbb{Z})} = \sqrt{\sum_{n=-\infty}^{\infty} |x_n|^2}$$

so that $\|R\| = 1$.

(2) For all $x, y \in \ell^2(\mathbb{Z})$, we may write

$$\begin{aligned} < Rx, y > &= \sum_{n=-\infty}^{\infty} R(x_n)_n \overline{y_n} \\ &= \cdots + x_{-1}\overline{y_0} + x_0\overline{y_1} + x_1\overline{y_2} + \cdots \\ &=< (\cdots, x_{-1}, \boldsymbol{x_0}, x_1, \cdots), (\cdots, y_0, \boldsymbol{y_1}, y_2, \cdots) >, \end{aligned}$$

so that

$$R^*(\cdots, x_{-1}, \boldsymbol{x_0}, x_1, \cdots) = (\cdots, x_{-1}, x_0, \boldsymbol{x_1}, x_2, \cdots).$$

Solution 3.3.43.

(1) We already showed the right to left implication in Exercise 2.3.2 and that

$$\|A\| \leq \|\alpha\|_\infty.$$

Let us turn now to the other implication. So assume that A is bounded (i.e. $\|A\|$ is finite). Hence

$$\forall x \in \ell^2 : \ \|Ax\|_2 \leq \|A\| \|x\|_2.$$

If we choose $x = e_n = (0, 0, \cdots, \underbrace{1}_{nth}, 0, \cdots)$, then we obtain

$$|\alpha_n| = \|Ae_n\|_2 \leq \|A\| \|e_n\|_2 = \|A\|,$$

i.e. we have proved that (α_n) is bounded.

Finally, we show that $\|A\| \geq \|\alpha\|_\infty$. We easily see that for all $n \in \mathbb{N}$

$$\|A\| = \sup_{\|x\|=1} \|Ax\| \geq \|Ae_n\| = |\alpha_n|$$

where $e_n = (0, 0, \cdots, 0, 1, 0, \cdots)$. Hence

$$\|A\| \geq \sup_{n\in\mathbb{N}} |\alpha_n| = \|\alpha\|_\infty \text{ and thus } \|A\| = \|\alpha\|_\infty.$$

(2) Let $x, y \in \ell^2$. We have

$$< Ax, y >= \sum_{n=1}^{\infty} \alpha_n x_n \overline{y_n} = \sum_{n=1}^{\infty} x_n \overline{\overline{\alpha_n} y_n}.$$

Hence

$$A^*(x_n) = (\overline{\alpha_n} x_n) = (\overline{\alpha_1} x_1, \overline{\alpha_2} x_2, \cdots).$$

(3) The "formal" inverse of A, denoted by B, is defined as

$$B(x_1, x_2, \cdots, x_n, \cdots) = \left(\frac{x_1}{\alpha_1}, \frac{x_2}{\alpha_2}, \cdots, \frac{x_n}{\alpha_n}, \cdots\right)$$

since $ABx = BAx = x$, $\forall x \in \ell^2$. Nonetheless, it cannot be considered as the inverse (in terms of operators) of A unless B is a bounded operator which is realized if and only if $\sup_{n\in\mathbb{N}} \frac{1}{|\alpha_n|} < \infty$. This shows that if $\sup_{n\in\mathbb{N}} \frac{1}{|\alpha_n|} < \infty$, then A is invertible. Let us pass to the converse. Assume that A is invertible, then for all $x \in \ell^2$, we have (Proposition 2.1.62)

$$\|Ax\| \geq \|A^{-1}\|^{-1}\|x\|.$$

In particular, for $e_n = (0, \cdots, 0, 1, 0, \cdots)$,

$$\|Ae_n\| \geq \|A^{-1}\|^{-1}\|e_n\|,$$

that is, for all n

$$|\alpha_n| \geq \|A^{-1}\|^{-1} \text{ or } \frac{1}{|\alpha_n|} \leq \|A^{-1}\|,$$

i.e. $\sup_{n\in\mathbb{N}} \frac{1}{|\alpha_n|} < \infty$, and this finishes the proof.

(4) The answer is no. By Question 1 and an easy result from the course of calculus we have

$$\|A^{-1}\| = \sup_{n\in\mathbb{N}} \frac{1}{|\alpha_n|} = \frac{1}{\inf_{n\in\mathbb{N}} |\alpha_n|} \neq \frac{1}{\sup_{n\in\mathbb{N}} |\alpha_n|} = \frac{1}{\|A\|}$$

(for instance, consider a bounded real-valued sequence (α_n) such that $\inf_{n\in\mathbb{N}} \alpha_n = 1$ and $\sup_{n\in\mathbb{N}} \alpha_n = 2$, say).

Solution 3.3.44. The answer is no! Let ran$A$ denote the range of A. We may proceed directly by finding a sequence in ran$A$ whose ℓ^2-limit lies outside ran$A$.

Alternatively, we may do the following: It is clear that A is equal to its adjoint A^* (by the previous exercise!). This yields

$$\{0\} = \ker A^* = (\operatorname{ran} A)^{\perp} \Longrightarrow \overline{\operatorname{ran} A} = (\operatorname{ran} A)^{\perp\perp} = \{0\}^{\perp} = \ell^2.$$

Hence $\operatorname{ran} A$ is dense in ℓ^2. So if $\operatorname{ran} A$ were closed, A would be onto. But this is not the case as

$$\exists y = \left(\frac{1}{2}, \frac{1}{3}, \cdots, \frac{1}{n+1}, \cdots\right) \in \ell^2, \forall x = (x_1, x_2, \cdots, x_n, \cdots) \in \ell^2$$

we have $Ax \neq y$ since otherwise the series "$\sum_{n\geq 1} 1$" would be convergent! Therefore, $\operatorname{ran} A$ is not closed.

Solution 3.3.45.

(1) Let $x, y \in H$. Since H is a vector space, $x + y \in H$ and hence our assumption implies that

$$\begin{aligned} 0 &= < A(x+y), x+y > \\ &= \underbrace{< Ax, x >}_{0} + < Ax, y > + < Ay, x > + \underbrace{< Ay, y >}_{0} \\ &= < Ax, y > + < Ay, x > . \end{aligned}$$

Substituting y by iy combined with the linearity of A yield

$$-i < Ax, y > +i < Ay, x >= 0.$$

Multiplying the previous equation by i and adding it to the one before it give

$$< Ax, y >= 0, \ \forall x, y \in H.$$

Setting $y = Ax \in H$, we obtain $\|Ax\|^2 = 0$ for all $x \in H$, i.e. $Ax = 0$ for each $x \in H$, i.e. $A = 0$.

To get an answer to the remaining question, it suffices to apply what we have just proved to the operator $B - A$.

(2) No, the result is no longer true if the field of scalars is $\mathbb{R}$. For if one takes

$$A = \begin{pmatrix} 0 & 1 \\ -1 & 0 \end{pmatrix}$$

on $\mathbb{R}^2$ (considered as an $\mathbb{R}$-Hilbert space), then one sees immediately that for all $x = (x_1, x_2)$

$$< Ax, x >= x_2 x_1 - x_1 x_2 = 0$$

while $A \neq \begin{pmatrix} 0 & 0 \\ 0 & 0 \end{pmatrix}$.

SOLUTION 3.3.46. The implication "$\Leftarrow$" is trivial. To prove the other implication, assume that $\forall x \in H$: $\|Ax\| \le \|Bx\|$.

Now, define $Ky = Ax$ for each $y = Bx$. Then K unambiguously defines a linear operator on ran(B). Indeed, if $x' \in H$ and Bx' is such that $Bx = Bx'$, then clearly

$$\|A(x - x')\| \le \|B(x - x')\| = 0$$

and so $Ax = Ax'$, i.e. $KBx = KBx'$, i.e. K is well-defined on ran(B).

Let's turn to linearity. Let $y, z \in$ ran(B) and let $\alpha, \beta \in \mathbb{C}$. Then

$$\begin{aligned} K(\alpha y + \beta z) &= K(\alpha Br + \beta Bs),\ r, s \in H \\ &= KB(\alpha r + \beta s) \text{ (for } B \text{ is linear)} \\ &= A(\alpha r + \beta s) \\ &= \alpha Ar + \beta As \text{ (for } A \text{ is linear)} \\ &= \alpha KBr + \beta KBs \\ &= \alpha Ky + \beta Kz, \end{aligned}$$

as needed.

Moreover, for all $y \in$ ran(B) we have:

$$\|Ky\| = \|Ax\| \le \|Bx\| = \|y\|.$$

This tells us that K is uniformly continuous on ran(B) and so we can extend it to $\overline{\text{ran}(B)}$ (and K remains a contraction!). Finally, taking $K = 0$ on $[\text{ran}(B)]^\perp$, gives us the required contraction K for which $A = KB$.

SOLUTION 3.3.47. Let $A \in B(H)$ and let A^* be its adjoint. Hence for all $y \in H$, we have

$$\lim_{n\to\infty} < Ax_n, y >= \lim_{n\to\infty} < x_n, A^*y >=< x, A^*y >=< Ax, y >,$$

and this completes the proof.

SOLUTION 3.3.48. Assume that ran(A) is closed. Then we may write, as $\ker(A^*) = [\text{ran}(A)]^\perp$,

$$H = \text{ran}A \oplus \ker(A^*).$$

To show that ran(A^*) is closed, let $y \in \overline{\text{ran}(A^*)}$. Then there exists an $y_n \in \text{ran}(A^*)$ such that $y_n = A^*x_n \to y$ where $x_n \in H$ (the limit is a strong one).

WLOG, we may assume that $x_n \in$ ran(A). Indeed, as $H = \text{ran}A \oplus \ker(A^*)$, then $x_n \in H$ may be decomposed uniquely as $a_n + b_n$ where $a_n \in$ ran(A) and $b_n \in \ker(A^*)$. Then

$$A^*x_n = A^*a_n + A^*b_n = A^*a_n.$$

Now, since $y_n = A^*x_n \to y$ strongly, we deduce that $y_n = A^*x_n \to y$ weakly, i.e.

$$\forall z \in H : < A^*x_n, z > \longrightarrow < y, z >$$

or

$$\forall z \in H : < x_n, Az > \longrightarrow < y, z > .$$

This implies that the complex sequence $(< x_n, t >)$ converges for any $t \in H$. Indeed, if $t = a + b \in \operatorname{ran}(A) \oplus \ker(A^*)$ (remember that $a = Aw$ for some $w \in H$ and $A^*b = 0$), then

$$\begin{aligned}
< x_n, t > &= < x_n, a > + < x_n, b > \\
&= < x_n, Aw > + < x_n, b > \\
&= < x_n, Aw > \text{ (for } x_n \in \operatorname{ran}(A) \text{ and } b \in \ker(A^*) = [\operatorname{ran}(A)]^\perp) \\
&\longrightarrow < y, w >,
\end{aligned}$$

i.e. $(< x_n, t >)$ converges for all t to a complex number. We then deduce that (x_n) weakly converges to x so that equally (A^*x_n) converges weakly to A^*x by Exercise 3.3.47. Finally, as (A^*x_n) converges strongly to y, then it also converges to y weakly. But weak limits are unique, we deduce that $y = A^*x$ meaning that $y \in \operatorname{ran}(A^*)$ and establishing the closedness of $\operatorname{ran}(A^*)$.

The other inclusion is now an easy consequence of the first one. Indeed, if $\operatorname{ran}(A^*)$ is closed, by the previous proof, we know that $\operatorname{ran}(A^{**})$ is closed, i.e. $\operatorname{ran}(A)$ is closed.

Finally, if S is the shift operator, then $\operatorname{ran}S^*$ is closed as S^* is surjective. Therefore, $\operatorname{ran}S$ too is closed, as wished for.

Solution 3.3.49. Since A is surjective, $\operatorname{ran}(A) = H$. Hence

$$\ker A^* = [\operatorname{ran}(A)]^\perp = H^\perp = \{0\},$$

i.e. A^* is injective. Hence $A^* : H \to \operatorname{ran}(A^*)$ is bijective. Also H is a Banach space, and so is $\operatorname{ran}(A^*)$. Indeed, it is closed (by Proposition 3.1.68) in a Banach space. By the Banach Isomorphism Theorem, $(A^*)^{-1} : \operatorname{ran}(A^*) \to H$ is bounded, that is,

$$\exists \alpha > 0, \ \forall y \in \operatorname{ran}(A^*) : \ \|(A^*)^{-1}y\| \leq \alpha\|y\|.$$

Letting $y = A^*x$, we see that for all $x \in H$,

$$\|A^*x\| \geq \frac{1}{\alpha}\|x\|,$$

signifying that A^* is bounded below.

3.4. Hints/Answers to Tests

Solution 3.4.1. The answer is no!...

Solution 3.4.2. Apply the Bessel Inequality to some orthonormal set (more precisely, to an orthonormal singleton!)...

Solution 3.4.3. Take $f(z) = 1$ and $g(z) = z$ and investigate when the parallelogram law is satisfied...

Solution 3.4.4. Yes! $Y^\perp$ is a closed linear subspace of a Hilbert space...

Solution 3.4.5. The answer is

$$Y^\perp = \{(x_n) \in \ell^2 : \ x_{2n-1} = 0, \ \forall n \in \mathbb{N}\}...$$

Solution 3.4.6. No! $B(H)$ is a Banach space which is not separable...

Solution 3.4.7. The answer is no! One reason is if I is the identity matrix on H (with $\dim H = n \geq 2$), then we know that $\|I\|_{B(H)} = 1$ while

$$\|I\|_F = \sqrt{\operatorname{tr}(I^t I)} = \sqrt{\operatorname{tr} I} = \sqrt{n}...$$

CHAPTER 4

Classes of Linear Operators on Hilbert Spaces

4.2. True or False: Answers

Answers.

(1) False! In general, only the left-to-right implication is true. The converse may not hold if H is over $\mathbb{R}$. In this case $< Ax, x >$ is always real anyway, and for all x and regardless of what A can be! If, however, H is over $\mathbb{C}$, then the right-to-left implication becomes true. Details may be found in Exercise 4.3.5.

(2) True. Indeed, if $T \in B(H)$, then $\|T\| = \|T^*\|$. So, since A and B are self-adjoint, we clearly get

$$\|AB\| = \|(AB)^*\| = \|B^*A^*\| = \|BA\|.$$

This type of questions appears on several occasions in this book. Let's give it a name:

We say that $A, B \in B(H)$ are ***commuting in norm*** *if* $\|AB\| = \|BA\|$.

Remark. Any commuting operators are commuting in norm. The converse need not hold. Indeed, take any two non commuting self-adjoint operators and we have just seen that self-adjoint operators are commuting in norm.

Notice also that any operator commutes in norm with its adjoint.

(3) True again! This is a consequence of Exercise 4.3.13.

(4) True! We prove that for any self-adjoint operators A and B and for any *real* number α, $\alpha A + B$ remains self-adjoint. This becomes obvious once we observe that

$$(\alpha A + B)^* = \alpha A^* + B^* = \alpha A + B.$$

(5) False! The reason is simple, that is, if A is self-adjoint, then αA need not be self-adjoint anymore since α is complex (e.g. take $A = I$ and $\alpha = i$).

(6) If A is unitary, then it is clearly invertible and normal. But, a *normal and invertible operator need not be unitary.* Simply,

consider $A = 2I$. Then A is invertible and even self-adjoint. It is, however, not unitary as

$$AA^* = A^*A = 4I \neq I.$$

(7) Clearly, every isometry is a contraction but not vice versa. Indeed, let $A = \frac{1}{2}I$, where I is the usual identity on H. Then it is plain that A is a contraction which is not an isometry for

$$A^*A = \frac{1}{4}I \neq I.$$

(8) No in general, and yes if $P \neq 0$ (what we are always sure of is $\|P\| \leq 1$ for any projection P).

Obviously, if $P = 0$, then $\|P\| = 0$; and if $P = I$, then $\|P\| = 1$.

In case $P \neq 0$, the proof is simple. Let $x \in H$. Then

$$\|Px\|^2 =< Px, Px >=< P^2x, x >=< Px, x >\leq \|Px\|\|x\|$$

so that $\|Px\| \leq \|x\|$ (by the Cauchy-Schwarz Inequality). This leads to $\|P\| \leq 1$. Conversely,

$$\|Px\| = \|P^2x\| \leq \|P\|\|Px\|$$

and hence $\|P\| \geq 1$.

Another much easier way of seeing this is the following: Since $P = P^*$, we have

$$\|P\| = \|P^2\| = \|PP^*\| = \|P\|^2.$$

Hence $\|P\| = 1$.

Remark. See Test 8.4.3 for yet another proof.

(9) The answer is no in general! It is true if the basis is orthonormal and in both directions! Let us then give counterexamples. First, let

$$A(x, y) = (y, x)$$

be defined on $\mathbb{C}^2$, say. Then A is clearly self-adjoint as for all $(x, y) \in \mathbb{C}^2$

$$\begin{aligned} < A(x,y), (x,y) > &= < (y,x), (x,y) >= y\overline{x} + x\overline{y} \\ &= x\overline{y} + y\overline{x} =< (x,y), A(x,y) > . \end{aligned}$$

However, the associated matrix with respect to the *non-orthonormal* basis $\{(1, 0), (1, 1)\}$ is easily seen to be

$$\begin{pmatrix} -1 & 0 \\ 1 & 1 \end{pmatrix},$$

which is not symmetric!

Let now exhibit a non self-adjoint transformation whose associated matrix is symmetric. Define on $\mathbb{R}_2[X]$ an inner product

$$< p, q >= \int_0^1 p(x)q(x)dx.$$

Define a linear transformation A on $\mathbb{R}_2[X]$ by

$$A(p) := A(ax^2 + bx + c) = bx.$$

Then A is not self-adjoint since (for example)

$$< A1, x >=< 0, x >= 0 \neq < 1, Ax >=< 1, x >= \int_0^1 x dx = \frac{1}{2}.$$

But, the associated matrix with respect to the *non-orthonormal* standard basis $\{1, x, x^2\}$ is given by

$$\begin{pmatrix} 0 & 0 & 0 \\ 0 & 1 & 0 \\ 0 & 0 & 0 \end{pmatrix},$$

which is clearly symmetric.

4.3. Solutions to Exercises

SOLUTION 4.3.1.

(1) (a) Since

$$(A + A^*)^* = A^* + A^{**} = A^* + A,$$

$A + A^*$ is self-adjoint. Since

$$(iA)^* = -iA^*,$$

iA is not necessarily self-adjoint. As a counterexample, consider the (*self-adjoint*!) matrix:

$$A = \begin{pmatrix} 1 & 0 \\ 0 & 0 \end{pmatrix}.$$

Then

$$iA = \begin{pmatrix} i & 0 \\ 0 & 0 \end{pmatrix}$$

is not self-adjoint (*anymore*!).

As for $iA - iA^*$, it is self-adjoint for

$$(iA - iA^*)^* = -iA^* + iA^{**} = iA - iA^*.$$

Both A^*A and AA^* are self-adjoint. For instance, AA^* is self-adjoint because

$$(AA^*)^* = A^{**}A^* = AA^*.$$

As for A^2A^*, it is not self-adjoint. For example,

$$A = \begin{pmatrix} 0 & 1 \\ 2 & 0 \end{pmatrix}.$$

Then

$$A^2A^* = \begin{pmatrix} 0 & 4 \\ 2 & 0 \end{pmatrix}$$

is obviously not self-adjoint.

Since

$$(A - A^*)^* = A^* - A^{**} = A^* - A = -(A - A^*),$$

it is not guaranteed that $A - A^*$ is self-adjoint. To confirm that, consider

$$A = \begin{pmatrix} 1 & 2 \\ 0 & 0 \end{pmatrix}.$$

Then

$$A - A^* = \begin{pmatrix} 0 & 2 \\ -2 & 0 \end{pmatrix}$$

is not self-adjoint.

Finally, $I + A$ is not self-adjoint! As a counterexample, consider any non self-adjoint matrix A.

(b) The non self-adjoint matrices are: iA, A^2A^*, $A - A^*$ and $I + A$. Then iA and $I + A$ are not normal (it suffices to consider the same non normal matrix for both cases). As for A^2A^*, the same example as above works.

Finally, $A - A^*$ is normal as it is skew-symmetric.

(2) Let A be self-adjoint. Then iA is not always self-adjoint (see above). It is, however, normal because

$$iA(iA)^* = iA(-iA) = A^2 = (-iA)(iA) = (iA)^*iA$$

(or because it is skew-symmetric!).

Now, $A + A^2$ is self-adjoint (and hence normal!) for

$$(A + A^2)^* = A^* + (A^2)^* = A + (AA)^* = A + A^*A^* = A + A^2.$$

(3) All operators are normal in this case. Indeed, A^2 is normal because

$$A^2A^{2*} = A^2A^{*2} = AAA^*A^* = AA^*AA^* = AA^*A^*A = A^*AA^*A = A^*A^*AA$$

and hence

$$A^2A^{2*} = A^{2*}A^2$$

(where we have used on several occasions the normality of A).

The proof that A^2A^* is normal is left to readers. Finally, iA is normal for

$$(iA)^*iA = -iA^*iA = A^*A = AA^* = iA(-iA^*) = iA(iA)^*.$$

Solution 4.3.2. Let

$$A = \begin{pmatrix} a & b \\ c & d \end{pmatrix},$$

where $a, b, c, d \in \mathbb{R}$. Then

$$A^* = \begin{pmatrix} a & c \\ b & d \end{pmatrix}$$

so that

$$AA^* = \begin{pmatrix} a^2+b^2 & ac+bd \\ ac+bd & c^2+d^2 \end{pmatrix} \text{ and } A^*A = \begin{pmatrix} a^2+c^2 & ab+cd \\ ab+cd & b^2+d^2 \end{pmatrix}.$$

In order that A be normal, the entries must obey

$$b^2 - c^2 = 0 \text{ and } (b-c)(a-d) = 0.$$

So basically there are two cases to look at:

(1) If $b = c$, then A is clearly symmetric (hence normal) regardless of the other values.
(2) If $b = -c$, then $a = d$.

Accordingly, the 2×2 real normal matrices are **either** *symmetric* **or** of the form

$$A = \begin{pmatrix} a & b \\ -b & a \end{pmatrix}.$$

Remark. It is quite useful to keep this result in mind mainly as a source of counterexamples. Indeed, when searching for counterexamples in Operator Theory we usually start by investigating 2×2 matrices.

Solution 4.3.3.

(1) No! Let

$$A = \begin{pmatrix} 0 & 0 \\ 1 & 0 \end{pmatrix} \neq \begin{pmatrix} 0 & 0 \\ 0 & 0 \end{pmatrix}$$

and yet

$$A^2 = \begin{pmatrix} 0 & 0 \\ 1 & 0 \end{pmatrix}\begin{pmatrix} 0 & 0 \\ 1 & 0 \end{pmatrix} = \begin{pmatrix} 0 & 0 \\ 0 & 0 \end{pmatrix}.$$

(2) One of the ways of seeing this is to write (for any $x \in H$)

$$\|Ax\|^2 =< Ax, Ax >=< A^2x, x >= 0,$$

that is, $A = 0$.

Solution 4.3.4. Let $x, y \in H$. Then as in the first question of Exercise 3.3.45 we have

$$0 = < A(x+y), x+y > = < Ax, y > + < Ay, x > .$$

However, A is self-adjoint and hence $< Ay, x > = < y, Ax >$. Also, since we are dealing with a real inner product,

$$< y, Ax > = < Ax, y > .$$

We obtain,

$$< Ax, y > = 0, \ \forall x, y \in H,$$

and so $A = 0$ as we did in the solution of Exercise 3.3.45.

Solution 4.3.5. Assume $< Ax, x > \in \mathbb{R}$ for all $x \in H$. We have

$$< Ax, x > = < x, A^*x > = \overline{< A^*x, x >} = < A^*x, x >$$

for all $x \in H$. Exercise 3.3.45 gives us $A = A^*$, establishing the self-adjointness of A.

Conversely, if A is self-adjoint, then for all $x \in H$

$$< Ax, x > = < x, A^*x > = < x, Ax > = \overline{< Ax, x >},$$

i.e. $< Ax, x >$ is real-valued.

Solution 4.3.6.

(1) Since V is an isometry, we clearly have

$$\|VA\| = \sup_{\|x\|=1} \|VAx\| = \sup_{\|x\|=1} \|Ax\| = \|A\|.$$

Hence

$$\|BV^*\| = \|(BV^*)^*\| = \|VB^*\| = \|B^*\| = \|B\|.$$

Now, by the preceding two equalities, and since V is an isometry, we have

$$\|VAV^*\| = \|AV^*\| = \|A\|.$$

Finally, as U is unitary, then U^* is an isometry so that by the previous result we obtain

$$\|U^*BU\| = \|B\|.$$

(2) The equality $\|BV\| = \|B\|$ need not always hold. Indeed, let

$$V(x_1, x_2, \cdots) = (x_1, 0, x_2, 0, \cdots)$$

and

$$B(x_1, x_2, \cdots) = (0, x_2, 0, x_4, 0, \cdots),$$

both defined on ℓ^2. The adjoint of V is given by

$$V^*(x_1, x_2, \cdots) = (x_1, x_3, x_5, \cdots).$$

Hence it easily follows that

$$V^*V(x_1, x_2, \cdots) = (x_1, x_2, \cdots) = I(x_1, x_2, \cdots),$$

that is, V is an isometry.

Now,

$$BV(x_1, x_2, \cdots) = B(x_1, 0, x_2, 0, \cdots) = (0, 0, 0, \cdots).$$

Hence $\|BV\| = 0$. It may also easily be checked that $\|B\| = 1$ so that finally

$$\|BV\| = 0 \neq 1 = \|B\|.$$

Remark. If $\dim H < \infty$, then $\|BV\| = \|B\|$ does hold since in such case, V becomes a *unitary* operator.

(3) We have

$$(VW)^*VW = W^* \underbrace{V^*V}_{=I} W = W^*IW = W^*W = I,$$

i.e. VW is an isometry.

(4) We use a proof by induction. The statement is true (by assumption) for $n = 1$. Now, suppose that $\|V^n\| = 1$. We then have (for all x)

$$\|V^{n+1}x\| = \|V(V^nx)\| = \|V^nx\| = \|x\|,$$

quod erat demonstrandum.

Consequently, for all n

$$\|V^n\| = \sup_{\|x\|=1} \|V^nx\| = \sup_{\|x\|=1} \|x\| = 1.$$

Solution 4.3.7. Let $x, y \in H$. Then

$$\begin{aligned}
\left| < \sum_{k=1}^{n} A_k B_k x, y > \right| &= \left| \sum_{k=1}^{n} < A_k B_k x, y > \right| \\
&= \left| \sum_{k=1}^{n} < B_k x, A_k^* y > \right| \\
&\leq \sum_{k=1}^{n} | < B_k x, A_k^* y > | \\
&\leq \sum_{k=1}^{n} \|B_k x\| \|A_k^* y\| \\
&\leq \sqrt{\sum_{k=1}^{n} \|B_k x\|^2} \sqrt{\sum_{k=1}^{n} \|A_k^* y\|^2} \\
&= \sqrt{\sum_{k=1}^{n} < B_k^* B_k x, x >} \sqrt{\sum_{k=1}^{n} < A_k A_k^* y, y >} \\
&= \sqrt{< \sum_{k=1}^{n} B_k^* B_k x, x >} \sqrt{< \sum_{k=1}^{n} A_k A_k^* y, y >}.
\end{aligned}$$

Then pass to the supremum over $\|x\| = 1$ and $\|y\| = 1$, by invoking Propositions 3.1.51 & 4.1.3, to get the desired inequality

$$\left\| \sum_{k=1}^{n} A_k B_k \right\|^2 \leq \left\| \sum_{k=1}^{n} A_k A_k^* \right\| \left\| \sum_{k=1}^{n} B_k^* B_k \right\|.$$

Solution 4.3.8. This is a simple application of the Fuglede Theorem. We have

$$\begin{aligned}
B^*A^* = A^*B^* &\Longrightarrow BA = AB \text{ (by taking adjoints)} \\
&\Longrightarrow BA^* = A^*B \text{ (by the Fuglede Theorem as } A \text{ is normal)} \\
&\Longrightarrow AB^* = B^*A \text{ (by taking adjoints)} \\
&\Longrightarrow B^*A^* = A^*B^* \text{ (by the Fuglede Theorem as } A \text{ is normal)}.
\end{aligned}$$

Solution 4.3.9.

(1) The answer is no for both questions. For instance, let

$$A = \begin{pmatrix} -1 & 0 \\ 0 & 2 \end{pmatrix} \text{ and } B = \begin{pmatrix} 0 & 1 \\ 1 & 0 \end{pmatrix}.$$

Then A and B are both self-adjoint (hence normal). Moreover,

$$AB = \begin{pmatrix} 0 & -1 \\ 2 & 0 \end{pmatrix}$$

We clearly see that AB is not normal and hence it cannot be self-adjoint either.

(2) Let A and B be two self-adjoint operators. Since,

$$(AB)^* = B^*A^* = BA,$$

we clearly have

$$AB \text{ is self-adjoint} \iff AB = BA.$$

(3) Let A and B be two normal operators such that $AB = BA$. Hence $B^*A^* = (AB)^* = (BA)^* = A^*B^*$. By the Fuglede Theorem,

$$AB = BA \Longrightarrow AB^* = B^*A \text{ or } A^*B = BA^*.$$

With these equalities on our hands (remembering also that $BB^* = B^*B$ and $AA^* = A^*A$), we may write the following

$$\begin{aligned}(AB)^*AB &= B^*A^*AB \\ &= B^*AA^*B \\ &= AB^*A^*B \\ &= AB^*BA^* \\ &= ABB^*A^* \\ &= AB(AB)^*.\end{aligned}$$

So much for the proof of the normality of AB.

SOLUTION 4.3.10. Define T and S on $B(H \oplus H)$ by

$$T = \begin{pmatrix} B & 0 \\ 0 & C \end{pmatrix} \text{ and } S = \begin{pmatrix} 0 & A \\ A & 0 \end{pmatrix}.$$

It is easy to see that since A, B and C are normal, so are S and T. Moreover,

$$ST = \begin{pmatrix} 0 & AC \\ AB & 0 \end{pmatrix} \text{ and } TS = \begin{pmatrix} 0 & BA \\ CA & 0 \end{pmatrix}.$$

Since we have assumed that $AC = BA$ and $AB = CA$, we get $TS = ST$. By Exercise 4.3.9, ST is normal, that is, $(ST)^*ST = ST(ST)^*$. Writing down what this means yields

$$(AB)^*AB = AC(AC)^* \text{ and } (AC)^*AC = AB(AB)^*,$$

as suggested.

SOLUTION 4.3.11.

(1) Let A and B be two self-adjoint operators. It is then obvious that
$$(A+B)^* = A^* + B^* = A + B,$$
that is, $A+B$ is self-adjoint.

(2) Here is a counterexample: Consider the matrices A and B defined as
$$A = \begin{pmatrix} 1 & -1 \\ 1 & 1 \end{pmatrix},\ B = \begin{pmatrix} -1 & 2 \\ 2 & 1 \end{pmatrix}.$$
Obviously A and B are both normal (B is even self-adjoint). However,
$$A+B = \begin{pmatrix} 0 & 1 \\ 3 & 2 \end{pmatrix}$$
is not normal (cf. Exercise 4.3.2).

(3) Let A and B be two normal operators such that $AB = BA$. We must show that $A+B$ is normal. We have on the one hand,
$$(A+B)(A+B)^* = AA^* + AB^* + BA^* + BB^*,$$
and on the other hand,
$$(A+B)^*(A+B) = A^*A + A^*B + B^*A + B^*B.$$
But, A and B are normal, that is, $AA^* = A^*A$ and $BB^* = B^*B$. So, we need only compare $AB^* + BA^*$ and $A^*B + B^*A$. Since $AB = BA$, the Fuglede Theorem gives $AB^* = B^*A$. Taking adjoints yields $BA^* = A^*B$. This leads to
$$AB^* + BA^* = A^*B + B^*A.$$
Consequently,
$$(A+B)(A+B)^* = (A+B)^*(A+B).$$
So much for the proof of the normality of $A+B$.

(4) The answer is no! Let
$$A = \begin{pmatrix} 2 & 0 \\ 0 & -2 \end{pmatrix} \text{ and } B = \begin{pmatrix} 0 & 1 \\ -1 & 0 \end{pmatrix}.$$
Then A and B are both normal (in fact, A is self-adjoint and B is unitary!). They anti-commute because
$$AB = -BA = \begin{pmatrix} 0 & 2 \\ 2 & 0 \end{pmatrix},$$

but

$$A+B=\begin{pmatrix} 2 & 1 \\ -1 & -2 \end{pmatrix}$$

is not normal.

(5) As A^*B and AB^* are self-adjoint, then

$$A^*B=(A^*B)^*=B^*A \text{ and } AB^*=(AB^*)^*=BA^*.$$

But A, B and $A+B$ are normal. Hence

$$AB^*+BA^*=A^*B+B^*A.$$

Hence

$$2BA^*=2A^*B \text{ or merely } BA^*=A^*B.$$

An application of the Fuglede Theorem to the normal A yields $BA=AB$. The proof is therefore complete.

(6) Let S be the unilateral shift. Set $A=SS^*$ and $B=S+S^*$. Then A and B are obviously self-adjoint (hence normal!). Their sum is self-adjoint and so it is normal. Now

$$A^*B=AB^*=AB=SS^*(S+S^*)=SS^*S+S(S^*)^2=S+S(S^*)^2.$$

Hence it is plain that neither A^*B nor AB^* is self-adjoint which means that $BA\neq AB$, that is, A does not commute with B.

Solution 4.3.12.

(1) We know that the normality of A means that $AA^*=A^*A$ or

$$\|A^*x\|=\|Ax\|,\ \forall x\in H.$$

It is therefore obvious that

$$\ker A=\ker A^*.$$

We deduce that a normal operator A is injective iff its adjoint A^* is injective.

(2) The answer is no. We give an example of a *non-normal* A such that $\ker A=\ker A^*$. Let

$$A=\begin{pmatrix} 0 & 1 \\ 2 & 0 \end{pmatrix}.$$

Then it can easily be seen that A and A^* are invertible so that their associated linear mappings have trivial kernels. Hence $\ker A=\ker A^*$. Finally, it is plain that A is not normal as

$$AA^*=\begin{pmatrix} 1 & 0 \\ 0 & 4 \end{pmatrix}\neq A^*A=\begin{pmatrix} 4 & 0 \\ 0 & 1 \end{pmatrix}.$$

(3) We have

$$\begin{aligned} A \text{ is surjective } &\Longleftrightarrow \text{ran}A = H \\ &\Longrightarrow (\text{ran}A)^\perp = H^\perp = \{0\} \\ &\Longleftrightarrow \ker A^* = \{0\} \\ &\Longleftrightarrow \ker A = \{0\} \text{ (by the first question).} \end{aligned}$$

Hence A is injective so that it becomes invertible. Whence so is A^*. This entails that A^* is surjective.

Conversely, if A^* is surjective, then by the previous reasoning $A^{**} = A$ is surjective.

SOLUTION 4.3.13.

(1) Let $x \in H$. Then by the normality of A, we get

$$\|ABx\| = \|A^*Bx\|.$$

Hence

$$\|AB\| = \|A^*B\|.$$

(2) As before! Otherwise, we may do

$$\begin{aligned} \|BA\|^2 &= \|(BA)^*BA\| \\ &= \|A^*B^*BA\| \\ &= \|A^*BB^*A\| \text{ (as } B \text{ is normal)} \\ &= \|A^*B(A^*B)^*\| \\ &= \|A^*B\|^2. \end{aligned}$$

SOLUTION 4.3.14. Since N is normal and invertible, one has

$$NN^* = N^*N \text{ and hence } N^{-1}N^* = N^*N^{-1}.$$

Then we have

$$(N(N^*)^{-1})(N(N^*)^{-1})^* = N(N^*)^{-1}N^{-1}N^* = N(N^*)^{-1}N^*N^{-1} = NN^{-1} = I.$$

Similarly, we find that $(N(N^*)^{-1})^*(N(N^*)^{-1}) = I$. Thus $N(N^*)^{-1}$ is unitary.

SOLUTION 4.3.15.

(1) We clearly have

$$(UV)^*UV = V^*U^*UV = V^*V = I$$

and

$$UV(UV)^* = UVV^*U^* = UU^* = I = (UV)^*UV.$$

Thus UV is unitary.

(2) No! Let $U \in B(H)$ be unitary and set $V = U$ (so V too is unitary!). Then

$$U + V = U + U = 2U$$

is not unitary for

$$(2U)(2U)^* = 4UU^* = 4I \neq I!$$

Solution 4.3.16. Let $y \in U(X^\perp)$. Then for some $x \in X^\perp$: $y = Ux$. Let $t \in X$. We have

$$< y, Ut >=< Ux, Ut >=< x, U^*Ut >=< x, t >= 0.$$

Hence $y \in (U(X))^\perp$. Thus we have proved that $U(X^\perp) \subset (U(X))^\perp$.

Next, let $y \in (U(X))^\perp$. Using the previous inclusion yields

$$y \in (U(X))^\perp \Longrightarrow U^*y \in U^*\left((U(X))^\perp\right) \subset \left(\underbrace{U^*U}_{I}(X)\right)^\perp = X^\perp.$$

So

$$y = UU^*y \in U(X^\perp),$$

i.e. we have proved that $U(X^\perp) \supset (U(X))^\perp$. Therefore, the equality holds.

Solution 4.3.17. Since U is unitary, $U^{-1} = U^*$. Let us first show that B is normal whenever A is normal. We have

$$\begin{aligned} BB^* =& U^*AU(U^*AU)^* \\ =& U^*AUU^*A^*U^{**} \\ =& U^*AA^*U \\ =& U^*A^*AU \text{ (as } A \text{ is normal)} \\ =& U^*A^*UU^*AU \\ =& (U^*AU)^*U^*AU \\ =& B^*B, \end{aligned}$$

which shows that B is normal. To prove the converse implication, apply a very similar argument to $A = UBU^*$.

Solution 4.3.18.

(1) Setting $\sqrt{x} = y$ yields $dy = \frac{dx}{2\sqrt{x}}$ so that

$$I = 2\int_0^\infty e^{-y^2} dy = \sqrt{\pi}.$$

(2) Let $f \in L^2(\mathbb{R}_+^*)$. We may do the following

$$\begin{aligned}|Lf(s)|^2 &= \left(\int_0^\infty f(t)e^{-st}dt\right)^2\\ &= \left(\int_0^\infty f(t)t^{\frac{1}{4}}e^{-\frac{st}{2}}t^{-\frac{1}{4}}e^{-\frac{st}{2}}dt\right)^2\\ &\leq \int_0^\infty |f(t)|^2 e^{-st}t^{\frac{1}{2}}dt \int_0^\infty e^{-st}t^{-\frac{1}{2}}dt\end{aligned}$$

by the Cauchy-Schwarz Inequality. But

$$\begin{aligned}\int_0^\infty e^{-st}t^{-\frac{1}{2}}dt &= s^{-\frac{1}{2}}\int_0^\infty e^{-x}x^{-\frac{1}{2}}dx \text{ (via } st = x)\\ &= s^{-\frac{1}{2}}I \text{ (the "}I\text{" of Question 1)}\\ &= s^{-\frac{1}{2}}\sqrt{\pi}.\end{aligned}$$

Whence

$$|Lf(s)|^2 \leq s^{-\frac{1}{2}}\sqrt{\pi}\int_0^\infty |f(t)|^2 e^{-st}t^{\frac{1}{2}}dt.$$

Integrating the previous against s over $(0,\infty)$ yields

$$\|Lf\|_2^2 = \int_0^\infty |Lf(s)|^2 ds \leq \sqrt{\pi}\int_0^\infty\int_0^\infty |f(t)|^2 e^{-st}t^{\frac{1}{2}}s^{-\frac{1}{2}}dtds.$$

Applying the Fubini Theorem and making the following observation:

$$\int_0^\infty e^{-st}t^{\frac{1}{2}}s^{-\frac{1}{2}}ds = \int_0^\infty e^{-x}x^{-\frac{1}{2}}dx = \sqrt{\pi} \text{ (just as above)}$$

give us in the end

$$\|Lf\|_2^2 \leq \pi\|f\|_2^2, \text{ that is, } \|Lf\|_2 \leq \sqrt{\pi}\|f\|_2.$$

Therefore,

$$\|L\| \leq \sqrt{\pi}.$$

(3) Let $f, g \in L^2(\mathbb{R}_+^*)$. Then

$$< Lf, g >= \int_0^\infty Lf(s)\overline{g(s)}ds = \int_0^\infty\int_0^\infty f(t)e^{-st}\overline{g(s)}dtds.$$

Applying the Fubini Theorem gives

$$< Lf, g >= \int_0^\infty f(t)\overline{\left(\int_0^\infty e^{-st}g(s)ds\right)}dt =< f, Lg >,$$

establishing the self-adjointness of L.

(4) Let $f \in L^2(\mathbb{R}_+^*)$. Denote the Laplace transform by L. Then we can write (using the Fubini Theorem)

$$\begin{aligned}(L^2 f)(x) =& L(Lf(x))\\ =& \int_0^\infty Lf(s)e^{-sx}ds\\ =& \int_0^\infty \int_0^\infty f(t)e^{-st}e^{-sx}dsdt\\ =& \int_0^\infty f(t)\int_0^\infty e^{-st-sx}dsdt\\ =& \int_0^\infty \frac{f(t)}{x+t}dt \ (= Jf(x))\end{aligned}$$

(where J is the Hilbert Operator) for

$$\int_0^\infty e^{-st-sx}ds = \frac{1}{x+t}.$$

Since L is self-adjoint, by invoking Exercise 2.5.7 (in the case $p = 2$), we may write

$$\pi = \|J\| = \|L^2\| = \|L\|^2$$

and so

$$\|L\| = \sqrt{\pi},$$

as needed.

Solution 4.3.19. Let $f, g \in L^2(\mathbb{R})$. Then

$$\begin{aligned}< Uf, g > =& \int_{\mathbb{R}} Uf(x)\overline{g(x)}dx\\ =& \int_{\mathbb{R}} f(x+a)\overline{g(x)}dx\\ =& \int_{\mathbb{R}} f(y)\overline{g(y-a)}dy \text{ (by setting } x + a = y)\\ =& < f, U^*g >,\end{aligned}$$

where

$$(U^*f)(x) = f(x-a), \ x \in \mathbb{R}.$$

Now,

$$UU^*f(x) = U(U^*f(x)) = U(f(x-a)) = f(x-a+a) = f(x).$$

Similarly, we find that

$$U^*Uf(x) = f(x),$$

i.e. U is unitary.

Solution 4.3.20. First, we need to find the adjoint of A. Let $f, g \in L^2(0,1)$. Then

$$\begin{aligned}
< Af, g > &= \int_0^1 \sqrt{x} f(1-x^2)\overline{g(x)}dx \\
&= \int_0^1 f(y)\overline{\frac{1}{2(1-y)^{\frac{1}{4}}} g(\sqrt{1-y})}dy \text{ (we have set } 1-x^2 = y) \\
&=< f, A^*g >
\end{aligned}$$

where

$$A^*f(x) = \frac{1}{2(1-x)^{\frac{1}{4}}} f(\sqrt{1-x}).$$

Now, we have

$$\begin{aligned}
AA^*f(x) &= A(A^*f(x)) \\
&= A\left(\frac{1}{2(1-x)^{\frac{1}{4}}} f(\sqrt{1-x})\right) \\
&= \sqrt{x}\frac{1}{2(x^2)^{\frac{1}{4}}} f(x) \\
&= \frac{1}{2} f(x).
\end{aligned}$$

Similarly, we find that

$$\begin{aligned}
A^*Af(x) &= A^*(Af(x)) \\
&= A^*\left(\sqrt{x} f(1-x^2))\right) \\
&= \frac{1}{2(1-x)^{\frac{1}{4}}}(1-x)^{\frac{1}{4}} f(x) \\
&= \frac{1}{2} f(x).
\end{aligned}$$

Therefore, it becomes clear that

$$\left(\sqrt{2}A\right)\left(\sqrt{2}A^*\right) = \left(\sqrt{2}A^*\right)\left(\sqrt{2}A\right) = I,$$

i.e. $\sqrt{2}A$ is unitary.

SOLUTION 4.3.21. Since $Af(x) = xf(x)$ on $L^2(0,1)$, clearly its "square" $A^2 : L^2(0,1) \to L^2(0,1)$ is defined by $A^2f(x) = x^2f(x)$ with $x \in (0,1)$. We need to find a unitary $U \in B(L^2(0,1))$ such that

$$U^*A^2U = A.$$

After a few tries, we propose $Uf(x) = \sqrt{2x}f(x^2)$ defined from $L^2(0,1)$ into $L^2(0,1)$. Then as we did in the previous exercise we may readily show that

$$U^*f(x) = \frac{1}{\sqrt{2\sqrt{x}}}f(\sqrt{x}).$$

Hence as readers may easily check U is unitary. Finally, let $f \in L^2(0,1)$ (and $x \in (0,1)$). Then

$$U^*A^2Uf(x) = U^*A^2(\sqrt{2x}f(x^2)) = U^*(x^2\sqrt{2x}f(x^2)) = xf(x),$$

as required.

SOLUTION 4.3.22.

(1) It is easy to see that $M \in B(\ell^2(\mathbb{Z}))$ and that

$$M^*(\cdots, x_{-1}, \boldsymbol{x_0}, x_1, \cdots) = (\cdots, \overline{\alpha_{-1}}x_{-1}, \overline{\boldsymbol{\alpha_0}}\boldsymbol{x_0}, \overline{\alpha_1}x_1, \cdots).$$

(2) Let $x = (x_n)_{n\in\mathbb{Z}} \in \ell^2(\mathbb{Z})$. Then

$$\begin{aligned} A(\cdots, x_{-1}, \boldsymbol{x_0}, x_1, \cdots) &= MR(\cdots, x_{-1}, \boldsymbol{x_0}, x_1, \cdots) \\ &= M(\cdots, x_{-2}, \boldsymbol{x_{-1}}, x_0, \cdots) \\ &= (\cdots, \alpha_{-1}x_{-2}, \boldsymbol{\alpha_0 x_{-1}}, \alpha_1 x_0, \cdots). \end{aligned}$$

(3) To see when A is normal, we first have to find A^*. For $x \in \ell^2(\mathbb{Z})$, we have

$$A^*x = R^*M^*x = (\cdots, \alpha_0 x_0, \boldsymbol{\alpha_1 x_1}, \alpha_2 x_2, \cdots).$$

This implies that

$$AA^*(\cdots, x_{-1}, \boldsymbol{x_0}, x_1, \cdots) = (\cdots, |\alpha_{-1}|^2 x_{-1}, |\boldsymbol{\alpha_0}|^{\boldsymbol{2}}\boldsymbol{x_0}, |\alpha_1|^2 x_1, \cdots)$$

and

$$A^*A(\cdots, x_{-1}, \boldsymbol{x_0}, x_1, \cdots) = (\cdots, |\alpha_0|^2 x_{-1}, |\boldsymbol{\alpha_1}|^{\boldsymbol{2}}\boldsymbol{x_0}, |\alpha_2|^2 x_1, \cdots).$$

Then A is normal iff

$$|\alpha_n| = |\alpha_0|, \ \forall n \in \mathbb{Z}.$$

Finally, A is unitary iff

$$|\alpha_n| = 1, \ \forall n \in \mathbb{Z}.$$

SOLUTION 4.3.23.

(1) If $\dim H < \infty$, we may associate with A and B finite square matrices. Then we know that

$$AB \text{ is invertible } \Longleftrightarrow \det(AB) \neq 0 \Longleftrightarrow \det A \neq 0 \textit{ and } \det B \neq 0,$$

that is, iff A and B are both invertible.

(2) If S is the shift operator, then $S^*S = I$ is invertible whereas neither S nor S^* is invertible.

(3) This is a basic result from Algebra...

(4) Assume that $AB = BA$. If AB is invertible, then for some $C \in B(H)$, we have

$$(AB)C = C(AB) = I$$

and hence

$$A(BC) = (CB)A = I$$

which means that A has a left inverse CB and a right inverse BC. By Proposition 2.1.58, A is invertible. We may also show that B is invertible by applying an akin method.

(5) Since A is normal, $AA^* = A^*A$. Since AA^* is invertible, by the previous question A and A^* are invertible.

(6) If AB is invertible, then for some $C \in B(H)$, we have

$$(AB)C = C(AB) = I$$

and hence by passing to adjoints, we obtain

$$C^*BA = BAC^* = I^* = I.$$

A glance at the last two displayed equations allows us to say that A possesses a left inverse and a right inverse, and so does B. By Proposition 2.1.58, A and B are invertible. The proof is thus complete.

SOLUTION 4.3.24. Since $AB = I$, it means that A is right invertible. Passing to the adjoint (and as A is self-adjoint), we see that

$$B^*A = (AB)^* = I^* = I,$$

i.e. A is also left invertible. Proposition 2.1.58 tells us that A is invertible with an inverse given by $B = B^*$.

SOLUTION 4.3.25. It is easy to see that both A and B are self-adjoint. Hence, it only remains to verify that $A^2 = A$ and $B^2 = B$. Let $x = (x_n) \in \ell^2$. We have

$$A^2(x) = A(Ax) = A(x_1, 0, x_3, 0, x_5, \cdots) = (x_1, 0, x_3, 0, x_5, \cdots) = Ax.$$

Similarly, it can be proved that $B^2 = B$ which we leave to the interested reader.

SOLUTION 4.3.26. By hypothesis, $P^* = P = P^2$ and $Q^* = Q = Q^2$.

(1) If $PQ = QP$, then we have
$$(PQ)^2 = PQPQ = PPQQ = P^2Q^2 = PQ.$$
Also, $PQ = QP$ implies that PQ is self-adjoint. Hence PQ is an orthogonal projection.

Conversely, if PQ is an orthogonal projection, then it is self-adjoint and hence $PQ = (PQ)^* = QP$, and this finishes the proof.

(2) Let $x, y \in H$, then it is evident that
$$< Px, Qy >=< x, P^*Qy >=< x, PQy > .$$
Thus
$$PQ = 0 \Longleftrightarrow \text{ran}P \perp \text{ran}Q.$$

(3) Assume that $P + Q$ is an orthogonal projection, then necessarily $(P + Q)^2 = P + Q$, that is,
$$P^2 + QP + PQ + Q^2 = P + Q \text{ or } QP + PQ = 0$$
so that $QP = -PQ$. Hence
$$QP = QP^2 = -PQP = -P(-PQ) = P^2Q = PQ$$
and this leads to $PQ = 0$.

Conversely, if $PQ = 0$, then $QP = 0$. Hence $QP + PQ = 0$. Therefore, $(P + Q)^2 = P + Q$ and thus $P + Q$ is an orthogonal projection.

SOLUTION 4.3.27. First, consider a finite collection of mutually perpendicular (orthogonal) projections $P_1, \cdots, P_n$. Then as done in Exercise 4.3.26, we can show that $\sum_{k=1}^n P_k$ is again an orthogonal projection. Hence for all $x \in H$
$$\|P_1x\|^2 + \cdots + \|P_nx\|^2 = \left\|\sum_{k=1}^n P_kx\right\|^2 \leq \|x\|^2.$$

The (positive) real series $\sum_{n=1}^\infty \|P_nx\|^2$ therefore converges as its sequence of partial sums is bounded above. So, if we set $y_n = \sum_{k=1}^n P_kx$, then for all $n, m \in \mathbb{N}$ (with $n > m$)
$$\|y_n - y_m\|^2 = \left\|\sum_{k=m+1}^n P_kx\right\|^2 = \sum_{k=m+1}^n \|P_kx\|^2 \leq \sum_{k=m+1}^\infty \|P_kx\|^2.$$
Since the latter is a tail of a convergent series, it follows that (y_n) is a Cauchy sequence in the complete H. Therefore, (y_n) converges, that is, the series $\sum_{n=1}^\infty P_nx$ converges, completing the proof.

Solution 4.3.28.

(1) Let I be the identity on H. We then have

$$(A-I)(B-I) = \underbrace{AB - A - B}_{=0} + I = I.$$

Since $\dim H < \infty$, the previous means that $A - I$ and $B - I$ are both invertible and that also $(B - I)(A - I) = I$. Hence

$$(A-I)(B-I) = (B-I)(A-I),$$

i.e.

$$AB = BA.$$

(2) No! The counterexample is based (again!) on the shift operator S defined on ℓ^2. Recall that $S^*S = I$ and $SS^* \neq I$. Let

$$A = I - S^* \text{ and } B = I - S.$$

Then

$$A + B = I - S^* + I - S = 2I - (S^* + S) = 2(I - \text{Re}S)$$

and

$$AB = (I - S^*)(I - S) = I - S^* - S + S^*S = 2(I - \text{Re}S),$$

and so

$$AB = A + B.$$

If BA were equal to AB, then this would mean that

$$SS^* = I$$

and this the sought contradiction. Hence $AB \neq BA$.

Remark. We used $SS^* \neq I$ and $S^*S = I$ and nothing else. This tells us any *non-unitary* isometry V works as a counterexample ($A = I - V^*$ and $B = I - V$).

(3) (a) Let A be self-adjoint. Then $A - I$ is self-adjoint. Since

$$AB = A + B \iff (A - I)(B - I) = I,$$

it follows that the self-adjoint $A - I$ is right invertible. Exercise 4.3.24 then tells us that $A - I$ is invertible and so as above, we get $AB = BA$. Exercise 4.3.24 also says that $B - I$ is self-adjoint, from which we easily deduce that B is self-adjoint.

(b) By hypothesis, $A^*A = I$ and $BB^* = I$. Hence

$$\begin{aligned} A + B = AB &\Longrightarrow I + A^*B = B \\ &\Longrightarrow B^* + A^* = I \\ &\Longrightarrow B + A = I^* = I \\ &\Longrightarrow AB = I. \end{aligned}$$

Hence A is invertible as it is right and left invertible at the same time. Therefore, A commutes with B and thus both A and B^* are invertible isometries, i.e. A and B are unitary.

(c) Let $A + B = AB$. Left multiplying by A^{-1} and right multiplying by B^{-1} yield

$$B^{-1} + A^{-1} = I \text{ or } A^{-1} = I - B^{-1}.$$

Hence

$$B^{-1}A^{-1} = B^{-1}(I - B^{-1}) = B^{-1} - B^{-2} = (I - B^{-1})B^{-1} = A^{-1}B^{-1}.$$

This is equivalent to

$$AB = BA.$$

Solution 4.3.29.

(1) Let (P_n) be a sequence such that $\text{s-}\lim_{n\to\infty} P_n = P$ and $P_n^2 = P_n$ for all $n \in \mathbb{N}$.

We need to show that $P \in B(H)$ and $P^2 = P$. First, since (P_n) converges strongly to P, (P_nx) converges to Px in H (for all x) so that P may be defined as a bounded linear operator on H (mimic the arguments used in Exercise 2.3.5).

Next, as (P_n) converges to P strongly, then (P_n^2) converges to P^2 strongly too (Exercise 2.3.4). Now, let $x \in H$. We then have

$$0 \le \|P^2x - Px\| \le \|P^2x - P_n^2x\| + \|P_n^2x - P_nx\| + \|P_nx - Px\| \longrightarrow 0.$$

This leads to $P^2 = P$, i.e. to P being a projection.

(2) Let $x \in H$. Recalling that each P_n is self-adjoint, we can easily write

$$\|P_nx\|^2 = < P_nx, P_nx > = < P_n^2x, x > = < P_nx, x > \longrightarrow < Px, x > = \|Px\|^2.$$

The strong convergence now is implied by Theorem 3.1.16.

Solution 4.3.30. Let (A_n) be a sequence of self-adjoint operators in $B(H)$ (hence for all $n \in \mathbb{N}$, $A_n^* = A_n$) such that

$$\lim_{n\to\infty} \|A_n - A\| = 0.$$

We have to show that $A \in B(H)$ is self-adjoint. Observe first that $A \in B(H)$ as $B(H)$ is a Banach space. Now, since $\|A_n - A\| \to 0$, we equally have $\|A_n^* - A^*\| \to 0$ (why?). Hence

$$\begin{aligned}\|A - A^*\| &\leq \|A - A_n\| + \underbrace{\|A_n - A_n^*\|}_{=0} + \|A_n^* - A^*\| \\ &= 2\|A - A_n\| \\ &\longrightarrow 0,\end{aligned}$$

as $n \to \infty$. Thus, A is self-adjoint.

SOLUTION 4.3.31. Let (A_n) be a sequence of normal operators in $B(H)$ (i.e. $A_nA_n^* = A_n^*A_n$ for all n) such that $\|A_n - A\| \longrightarrow 0$ as n goes to ∞. Let us show that A is also bounded and normal. As in the previous exercise $A \in B(H)$. Next, we have

$$\begin{aligned}\|A^*A - AA^*\| \leq \|A^*A - A^*A_n\| + \|A^*A_n - A_n^*A_n\| + \underbrace{\|A_n^*A_n - A_nA_n^*\|}_{0} \\ + \|A_nA_n^* - A_nA^*\| + \|A_nA^* - AA^*\| \\ \leq \|A^*(A - A_n)\| + \|(A^* - A_n^*)A_n\| + \|A_n(A_n - A)^*\| \\ + \|(A_n - A)A^*\| \\ \leq 2\|A^*\|\|A_n - A\| + 2\|A_n\|\|(A - A_n)^*\| \longrightarrow 0,\end{aligned}$$

as n tends to infinity, because $(\|A_n\|)$ is bounded and $\|A_n - A\|$ goes to zero. Thus the limit A is normal, proving the closedness of the set of normal operators in $B(H)$.

SOLUTION 4.3.32.

(1) Let (V_n) be a sequence of isometries on H. This means that for all x and all n

$$\|V_nx\| = \|x\|.$$

Let (V_n) be strongly convergent to V, that is, for all x

$$\|V_nx - Vx\| \longrightarrow 0 \text{ when } n \longrightarrow \infty.$$

We must show that V is an isometry. Let $x \in H$. Since $\|V_nx - Vx\| \to 0$, we immediately see that

$$\lim_{n\to\infty} \|V_nx\| = \|Vx\|.$$

Thus, for all x

$$\|Vx\| = \|x\|,$$

i.e. V is an isometry.

(2) Consider the sequence of operators defined on ℓ^2 as

$$U_n(x_0, x_1, x_2, \cdots) = (x_n, x_0, x_1, \cdots, x_{n-1}, x_{n+1}, \cdots).$$

Then each U_n is unitary as

$$\|U_n x\| = \|x\|, \ \forall x \in \ell^2$$

and clearly each U_n is onto. Now, (U_n) converges strongly to the shift operator S (which is known to be non-unitary)...

Remark. In fact, the strong operator topology closure of unitary operators equals the set of isometries, i.e. unitary operators are dense in the set of isometries (w.r.t. S.O.T.). See [**49**]. See also [**66**] or [**93**] for more related results.

Solution 4.3.33.

(1) We know how to define e^A on $B(H)$. The way of defining e^{iA} is very similar, that is,

$$e^{iA} = \sum_{n=0}^{\infty} \frac{(iA)^n}{n!}.$$

To show that e^{iA} is unitary, we first need to find its adjoint. As A is self-adjoint, then so is A^n. Hence

$$\left(\sum_{n=0}^{k} \frac{(iA)^n}{n!}\right)^* = \sum_{n=0}^{k} \frac{(-iA)^n}{n!}$$

Now, since

$$\lim_{k\to\infty} \left\| \sum_{n=0}^{k} \frac{(iA)^n}{n!} - e^{iA} \right\| = 0,$$

by the continuity of the adjoint (w.r.t. the operator norm), we obtain

$$\lim_{k\to\infty} \left\| \left(\sum_{n=0}^{k} \frac{(iA)^n}{n!}\right)^* - (e^{iA})^* \right\| = 0.$$

Since clearly

$$\sum_{n=0}^{k} \frac{(-iA)^n}{n!} \longrightarrow e^{-iA},$$

we obtain :

$$(e^{iA})^* = e^{-iA}.$$

Thus

$$(e^{iA})^* e^{iA} = e^{-iA} e^{iA} = e^0 = I = e^{iA} e^{-iA} = e^{iA}(e^{iA})^*,$$

i.e. e^{iA} is unitary.

(2) By considering $(\sum_{n=0}^{k} \frac{A^n}{n!})^*$ and by applying a similar method to that of the previous question, we may show that (do it!)

$$(e^A)^* = e^A.$$

Solution 4.3.34. If $A = B = 0$, then clearly $T = 0$. Conversely, assume that $T = A + iB = 0$. Then since A and B are self-adjoint, we have

$$0 = T^* = A^* - iB^* = A - iB.$$

Hence

$$T + T^* = 0 \Longrightarrow A = 0 \text{ and } T - T^* = 0 \Longrightarrow B = 0.$$

Solution 4.3.35. Yes! We may rewrite the hypothesis as

$$(A - C) + i(B - D) = 0.$$

Then apply Exercise 4.3.34 to $(A-C)+i(B-D)$ by observing that $A - C$ and $B - D$ are self-adjoint!...

Solution 4.3.36. Since A and B are self-adjoint, it is clear that

$$T^* = (A + iB)^* = A - iB,$$

so that

$$TT^* = (A + iB)(A - iB) = A^2 + B^2 + i(BA - AB)$$

and

$$T^*T = (A - iB)(A + iB) = A^2 + B^2 + i(AB - BA).$$

Hence

$$[T, T^*] = TT^* - T^*T = 2i(BA - AB).$$

Thus, it is easily seen that T is normal iff $AB = BA$.

To show the other equivalence, observe first that for every $x \in H$

$$\begin{aligned} \|Tx\|^2 &= \|(A + iB)x\|^2 \\ &= < Ax + iBx, Ax + iBx > \\ &= \|Ax\|^2 + \|Bx\|^2 - 2\text{Re}(i < Ax, Bx >) \end{aligned}$$

and similarly,

$$\|T^*x\|^2 = \|Ax\|^2 + \|Bx\|^2 + 2\text{Re}(i < Ax, Bx >).$$

Therefore,

$$\|T^*x\|^2 = \|Tx\|^2 \Longleftrightarrow \text{Re}(i < Ax, Bx >) = 0,$$

i.e.

$$\|T^*x\|^2 = \|Tx\|^2 \iff \|Tx\|^2 = \|Ax\|^2 + \|Bx\|^2.$$

SOLUTION 4.3.37. By the foregoing exercise, we have

$$TT^* = T^*T = A^2 + B^2,$$

whenever T is normal.

(1) Suppose that T is invertible. Then T^* too is invertible. Hence TT^* is invertible, that is, $A^2 + B^2$ is invertible.

(2) Now, assume that $A^2 + B^2$ is invertible. Then

$$T^*T = A^2 + B^2 \Longrightarrow (A^2 + B^2)^{-1}T^*T = I.$$

Hence the self-adjoint T^*T is left invertible and so it is invertible. Since T is normal, Exercise 4.3.23 gives the invertibility of T, as suggested.

SOLUTION 4.3.38. Since N is normal, $N = A + iB$ where A and B are commuting self-adjoint operators (we denote A by $\operatorname{Re} N$). Hence

$$e^N = e^{A+iB} = e^A e^{iB}.$$

Since B is self-adjoint, e^{iB} is unitary. By Exercises 2.3.32 & 4.3.6, we may therefore write:

$$\|e^N\| = \|e^A e^{iB}\| = \|e^A\| \le e^{\|A\|} = e^{\|\operatorname{Re} N\|},$$

as desired.

SOLUTION 4.3.39. We may write

$$T^*T = \begin{pmatrix} 0 & B^* \\ A^* & 0 \end{pmatrix} \begin{pmatrix} 0 & A \\ B & 0 \end{pmatrix} = \begin{pmatrix} B^*B & 0 \\ 0 & A^*A \end{pmatrix}.$$

By Proposition 4.1.28, we have

$$\|T\|^2 = \|T^*T\| = \max(\|B^*B\|, \|A^*A\|) = \max(\|B\|^2, \|A\|^2)$$

and so

$$\|T\| = \max(\|B\|, \|A\|),$$

as needed.

SOLUTION 4.3.40. If A and B are invertible, we can easily check that

$$\begin{pmatrix} A & 0 \\ 0 & B \end{pmatrix} \begin{pmatrix} A^{-1} & 0 \\ 0 & B^{-1} \end{pmatrix} = \begin{pmatrix} A^{-1} & 0 \\ 0 & B^{-1} \end{pmatrix} \begin{pmatrix} A & 0 \\ 0 & B \end{pmatrix} = \begin{pmatrix} I_H & 0 \\ 0 & I_K \end{pmatrix}.$$

Conversely, suppose that $T = A \oplus B$ is invertible. We will show that A is bijective and the invertibility will then follow from the Banach Isomorphism Theorem.

Let $x \in H$ be such that $Ax = 0$. Then $T(x,0) = (Ax,0) = (0,0)$ and so $x = 0$ as T is injective. To see why A is surjective, let $z \in H$. If $t \in K$, then $(z,t) \in H \oplus K$, and since T is surjective, we know that

$$(z,t) = T(x,y) = (Ax, By)$$

for some (x,y) in $H \oplus K$, i.e. $z = Ax$ for some $x \in H$, i.e. A is surjective.

A similar method leads to the invertibility of B, and this marks the end of the proof.

Solution 4.3.41.

(1) The formal determinant of A is $S^*S - \mathbf{0}$, that is, it is just I, which is obviously invertible.
(2) To show that A is not invertible, it suffices (for example) to show that its kernel is non trivial. Let $e_1 = (1,0,0,\cdots)$ from the standard basis of ℓ^2 so that $(e_1, 0_{\ell^2}) \neq (0_{\ell^2}, 0_{\ell^2})$. But

$$A\begin{pmatrix} e_1 \\ 0_{\ell^2} \end{pmatrix} = \begin{pmatrix} S^* & \mathbf{0} \\ \mathbf{0} & S \end{pmatrix}\begin{pmatrix} e_1 \\ 0_{\ell^2} \end{pmatrix} = \begin{pmatrix} S^*e_1 \\ 0_{\ell^2} \end{pmatrix} = \begin{pmatrix} 0_{\ell^2} \\ 0_{\ell^2} \end{pmatrix}.$$

(3) It does not contradict Proposition 4.1.34 since even though $\mathbf{0}$ commutes with S, the invertibility of S is *not available.*

Solution 4.3.42.

(1) Let $z \in A^*(M^\perp)$, that is, $z = A^*y$ for some $y \in M^\perp$. By assumption, for any $x \in M$, Ax too remains in M. Therefore,

$$0 =< Ax, y >=< x, A^*y > .$$

Thus, $z = A^*y \in M^\perp$.
(2) The proof follows by applying the previous result twice and by remembering that $\overline{M} = M^{\perp\perp}$, $\overline{M'} = M'^{\perp\perp}$ and $A^{**} = A$.

Solution 4.3.43. Since $AB = 0$, plainly $\text{ran}B \subset \ker A$. Hence

$$B(\ker A) \subset \text{ran}B \subset \ker A.$$

So, if $B \neq 0$, then

$$\text{ran}B \neq \{0\} \text{ and } \ker A \neq \{0\}$$

and so $\overline{\text{ran}B} \neq \{0\}$ either. Also, if $A \neq 0$, then $\ker A \neq H$. But,

$$\overline{\text{ran}\, B} = H \Longrightarrow (\text{ran}B)^\perp = \{0\} \Longrightarrow (\ker A)^\perp = \{0\} \Longrightarrow \ker A = H$$

(as $\ker A$ is closed). Hence $\overline{\text{ran}B} \neq H$.

Consequently,

$$\{0\} \neq \ker A \neq H \text{ and } \{0\} \neq \overline{\text{ran}B} \neq H.$$

It is also clear that $A(\text{ran}B) = \{0\}$. So, by the continuity of A, we have

$$A(\overline{\text{ran}B}) \subset \overline{A(\text{ran}B)} \subset \overline{\text{ran}B}.$$

To summarize:

(1) We have shown that $\overline{\text{ran}B}$ is invariant for A and $\ker A$ is invariant for B.
(2) We have also shown that $\ker A$ and $\overline{\text{ran}B}$ are nontrivial.
(3) We already know that $\ker A$ and $\overline{\text{ran}B}$ are invariant for A and B respectively.

Therefore, we have shown $\ker A$ and $\overline{\text{ran}B}$ are nontrivial invariant subspaces for both A and B, as wished.

SOLUTION 4.3.44. We can equivalently work with the linear transformation associated with the matrix A and call it A as well. Then $A : \mathbb{R}^2 \to \mathbb{R}^2$ is given by $A(x, y) = (-y, x)$.

The aim is to show that Lat$A$ is trivial, that is, there is no invariant subspace besides $\{(0, 0)\}$ and $\mathbb{R}^2$. If there were another invariant subspace say M, then surely it would be unidimensional!

So, $M = \{a(x, y) : a \in \mathbb{R}\}$ where (x, y) is a non-zero vector. Now, $AM \subset M$ means that $A(x, y) = \alpha(x, y)$ for some real α. Hence $(-y, x) = \alpha(x, y)$ and so

$$(-x, -y) = A(-y, x) = \alpha A(x, y) = \alpha(-y, x) = \alpha^2(x, y)$$

which leads to $\alpha^2 = -1$ which is impossible over $\mathbb{R}$!

Consequently, Lat$A$ is trivial as required.

SOLUTION 4.3.45. Let H be a non-separable Hilbert space and let $A \in B(H)$. Let $x \in H$ with $x \neq 0$. Consider the **orbit** of x under A, that is, $\{x, Ax, A^2x, \cdots\} = \{A^n x\}_{n \geq 0}$. Set $M = \text{span}\{A^n x\}_{n \geq 0}$. Then $AM \subset M$ and so $A\overline{M} \subset \overline{M}$. Since M is a subspace, $\overline{M}$ too is a subspace (which is also closed). Thus, $\overline{M}$ is invariant under A.

Clearly, $\overline{M} \neq \{0\}$. Also, $\overline{M} \neq H$ as if it were, i.e. if $\overline{M} = H$, then H would be separable as M is countable! Therefore, $\overline{M}$ is not trivial and this completes the proof.

SOLUTION 4.3.46.

(1) A simple example is the following:

$$M = \text{ran}S.$$

It is simple to see that M is closed (e.g. using Proposition 2.1.63) and it is plainly an invariant subspace for S. It is non-trivial as $M \neq \{0\}$ (clear!) and $M \neq \ell^2$ as S is not onto.

Remark. In fact, we can have a collection of invariant subspaces under S. They are given by

$$M_n = \{x = (x_k) \in \ell^2 : \ x_k = 0, \ k = 1, \cdots, n\}.$$

Then M_n is an invariant subspace under S for each value of n. A question pops us: Do we have

$$\text{Lat}S = \{M_n\}?$$

Not really and this is a quite hard problem. See e.g. [**94**].

(2) Assume that S has a non-trivial reducing subspace M, that is, M is a closed subspace of ℓ^2 and

$$SM \subset M \text{ and } SM^\perp \subset M^\perp.$$

Denote the restriction of S to M and $M^\perp$ by $S|_M$ and $S|_{M^\perp}$ respectively. Hence we can write

$$S = \begin{pmatrix} S|_M & 0 \\ 0 & S|_{M^\perp} \end{pmatrix}$$

defined on $\ell^2 = M \oplus M^\perp$.

Since S is injective, so are $S|_M$ and $S|_{M^\perp}$. In fact, since S is an isometry, so are $S|_M$ and $S|_{M^\perp}$. Obviously, $S|_M$ and $S|_{M^\perp}$ cannot be invertible simultaneously! In fact, they cannot be non-invertible together either. Indeed, one of them only is surely surjective. To see this, suppose that they were both non-surjective, then $\text{ran}S|_M \neq \ell^2$ and $\text{ran}S|_{M^\perp} \neq \ell^2$. Since $\text{ran}S|_M$ and $\text{ran}S|_{M^\perp}$ are both closed, we get that

$$\ker S^*|_M \neq \{0\} \text{ and } \ker S^*|_{M^\perp} \neq \{0\}$$

which contradicts $\dim \ker S^* = 1$!

Assume that for instance $S|_M$ is onto. Then it becomes bijective and hence unitary. Therefore, for all $x \in M$ and all n, we have

$$\|(S^*)^n x\| = \|(S^*|_M)^n x\| = \|x\|$$

contradicting $\|(S^*)^n x\| \to 0$ (for all x) already known from Exercise 3.3.41. Consequently, S has non non-trivial reducing subspaces.

4.4. Hints/Answers to Tests

Solution 4.4.1. Try $f(t) = t\mathbb{1}_{[0,x]}(t)$...

Solution 4.4.2. For all $x = (x_n) \in \ell^2(\mathbb{Z})$, we have

$$RR^*(\cdots, x_{-1}, \boldsymbol{x_0}, x_1, \cdots) = R(\cdots, x_0, \boldsymbol{x_1}, x_2, \cdots) = (\cdots, x_{-1}, \boldsymbol{x_0}, x_1, \cdots),$$

i.e. $RR^* = I$. Similarly, we may show (do it!) that $R^*R = I$. Thus R is unitary.

Solution 4.4.3. Here is a (generous!) hint, write

$$ABA = ABA \text{ so that } A(BA)A^* = AB.$$

Conclude!...

Solution 4.4.4. The answer is yes (and for the equivalence!). There are different ways of seeing this. The simplest way perhaps is to use Exercise 4.3.13 from which we know that

$$\|AB\| = \|BA\|$$

provided that A and B are normal. Thereupon, $AB = 0$ iff $BA = 0$.

Solution 4.4.5. Set $S = A|_M$ and $T = A^*|_M$. Then S and T are both well-defined thanks to Proposition 4.1.40. For all $x, y \in M$, we have

$$< S^*x, y >=< x, Sy >=< x, Ay >=< A^*x, y >=< Tx, y > .$$

By Proposition 3.1.65, $S^* = T$.

Solution 4.4.6. The main problem is the implication

$$"A^2 + B^2 + i(AB - BA) = I \Longrightarrow A^2 + B^2 = I \text{ and } AB = BA!"$$

This would have been correct (according to Exercise 4.3.35) if $AB - BA$ were self-adjoint (given that $A^2 + B^2$ is already self-adjoint) and this is not the case...

Solution 4.4.7. We could reuse the method of Exercise 4.3.33 or: From Proposition 2.1.72, we know that

$$e^{iA} = \cos A + i \sin A.$$

Therefore,

$$\cos A = \frac{1}{2}(e^{iA} + e^{-iA}) \text{ and } \sin A = \frac{1}{2i}(e^{iA} - e^{-iA}).$$

Thus

$$(\cos A)^* = \left(\frac{1}{2}(e^{iA} + e^{-iA})\right)^* = \frac{1}{2}(e^{-iA} + e^{iA}) = \cos A.$$

The proof of self-adjointness of $\sin A$ is left to readers...

CHAPTER 5

Positive Operators. Square Root

5.2. True or False: Answers

ANSWERS.

(1) When we compare the operators A and B, we are in fact comparing the numbers $< Ax, x >$ and $< Bx, x >$. So we have to assume that A and B are self-adjoint so that $< Ax, x >$ and $< Bx, x >$ are *real numbers* and so they can be compared!

After all, if i is the usual complex number, then no mathematician would ever want to stare at something like "$1+i \geq i$" even though we do know that the difference $1 + i - i = 1$ is positive!

(2) False! If an operator A on H is not positive, then

$$\exists x \in H : < Ax, x >< 0$$

which is different from

$$\forall x \in H : < Ax, x >\leq 0,$$

which is the definition of a negative operator.

(3) The answer is no in general. Take $A = iI$ where $i^2 = -1$. Then $A^2x = A(Ax) = i(ix) = -x$ and

$$\forall x \in H : < A^2x, x >=< -x, x >= - < x, x >= -\|x\|^2 \leq 0,$$

i.e. A est negative.

This result becomes true if one assumes further that A is self-adjoint. The proof is simple. Let $x \in H$. Then

$$< A^2x, x >=< Ax, A^*x >=< Ax, Ax >= \|Ax\|^2 \geq 0.$$

(4) True! Since A is self-adjoint, so is its inverse A^{-1}. Hence by Proposition 5.1.6:

$$A \geq 0 \Longrightarrow A^{-1}AA^{-1} \geq 0 \Longrightarrow A^{-1} \geq 0.$$

(5) The answer is yes even for weak convergence. Indeed, by assumption

$$\forall x \in H : < A_nx, x >\geq 0.$$

Then passing to the limit as $n \to \infty$, we get

$$\forall x \in H : < Ax, x > \geq 0,$$

i.e. $A \geq 0$.

(6) False! Let

$$A = \begin{pmatrix} 0 & 1 \\ 0 & 0 \end{pmatrix}.$$

Then A does not have any square root. To see this, assume that A has a square root, B say, which is a 2×2 matrix of the form

$$B = \begin{pmatrix} a & b \\ c & d \end{pmatrix}.$$

Then

$$B^2 = A \iff \begin{pmatrix} a^2 + bc & ab + bd \\ ac + cd & bc + d^2 \end{pmatrix} = \begin{pmatrix} 0 & 1 \\ 0 & 0 \end{pmatrix}$$

and hence

$$\begin{cases} a^2 + bc = 0, \\ ab + bd = 1, \\ ac + cd = 0, \\ bc + d^2 = 0. \end{cases}$$

The previous system has no solution (as the reader may easily check) meaning that A has no square root.

We give another example to use a different approach (cf. [**95**]). Let

$$A = \begin{pmatrix} 0 & 1 & 0 \\ 0 & 0 & 1 \\ 0 & 0 & 0 \end{pmatrix}.$$

Then

$$A^2 = \begin{pmatrix} 0 & 0 & 1 \\ 0 & 0 & 0 \\ 0 & 0 & 0 \end{pmatrix} \text{ and } A^3 = \begin{pmatrix} 0 & 0 & 0 \\ 0 & 0 & 0 \\ 0 & 0 & 0 \end{pmatrix}.$$

Assume that A has a square root B, that is, $B^2 = A$. Hence $B^6 = A^3 = 0$, i.e. B is nilpotent. Since B is a 3×3 matrix, its index cannot exceed 3. Therefore, $B^3 = 0$, but this is just not consistent with $B^4 = A^2 \neq 0$. Thus A has no square root.

(7) The answer is no! Even when A is self-adjoint, then $A^2 = I$ must not imply that $A = I$ or $A = -I$. As a simple counterexample, consider:

$$\pm I \neq A = \begin{pmatrix} -1 & 0 \\ 0 & 1 \end{pmatrix}$$

and yet $A^2 = I$. These are not the only cases. In Example 5.1.27, the collection (A_x) (for $x \neq 0$) was not self-adjoint. Here is an infinite self-adjoint family of square roots of I:

$$B_x = \begin{pmatrix} x & \sqrt{1-x^2} \\ \sqrt{1-x^2} & -x \end{pmatrix}$$

where $x \in [-1, 1]$.

(8) True. First, A is invertible by Proposition 5.1.23. Since A^{-1} is positive and commutes with both A and I, Theorem 5.1.5 gives

$$A \geq I \Longrightarrow I = AA^{-1} \geq A^{-1}.$$

(9) False! See Exercise 9.3.17 for a counterexample.

(10) True. Since A is invertible, $AB = BA = I$ for some $B \in B(H)$. As seen above, because A is positive so is its inverse $B = A^{-1}$. Therefore, $\sqrt{A}$ and $\sqrt{B}$ both exist. Hence

$$\sqrt{A}\sqrt{B} = \sqrt{B}\sqrt{A} = \sqrt{I} = I.$$

Thus, $\sqrt{A}$ is invertible and

$$(\sqrt{A})^{-1} = \sqrt{B} = \sqrt{A^{-1}}$$

as suggested.

(11) True! We know that $\|B\|^2 = \|B^*B\|$ for any $B \in B(H)$. In particular for $B = \sqrt{A}$, we have

$$\|\sqrt{A}\|^2 = \|\sqrt{A}\sqrt{A}\| = \|A\| \text{ or } \|\sqrt{A}\| = \sqrt{\|A\|}.$$

Remark. See also Theorem 11.1.13 for a generalization. See Exercise 11.3.18 for yet another generalization.

5.3. Solutions to Exercises

Solution 5.3.1. Both A and B are positive. Let $x, y \in \mathbb{R}$. Then

$$< \begin{pmatrix} 1 & 1 \\ 1 & 1 \end{pmatrix} \begin{pmatrix} x \\ y \end{pmatrix}, \begin{pmatrix} x \\ y \end{pmatrix} > = < \begin{pmatrix} x+y \\ x+y \end{pmatrix}, \begin{pmatrix} x \\ y \end{pmatrix} >$$

$$= x^2 + yx + xy + y^2$$
$$= (x+y)^2 \geq 0.$$

As for B, despite the fact that

$$\begin{aligned}<\begin{pmatrix} 1 & 1 \\ 0 & 2 \end{pmatrix}\begin{pmatrix} x \\ y \end{pmatrix},\begin{pmatrix} x \\ y \end{pmatrix}> &=< \begin{pmatrix} x+y \\ 2y \end{pmatrix},\begin{pmatrix} x \\ y \end{pmatrix}> \\ &= x^2+yx+2y^2 \\ &= \left(x+\frac{y}{2}\right)^2+\frac{7}{4}y^2>0,\end{aligned}$$

we cannot consider it as a positive matrix as B is not symmetric!

In fine, C is not positive because

$$<\begin{pmatrix} 1 & 2 \\ 2 & 2 \end{pmatrix}\begin{pmatrix} x \\ y \end{pmatrix},\begin{pmatrix} x \\ y \end{pmatrix}>= x^2+4xy+2y^2$$

can be negative (e.g. if $x=1$ and $y=-1$).

Solution 5.3.2. The answer is yes. Let $x=(x_1,x_2,\cdots)\in\ell^2$. Then we already know that

$$S(S^*x)=S(x_2,x_3,\cdots)=(0,x_2,x_3,\cdots).$$

Hence

$$(I-SS^*)(x_1,x_2,\cdots)=(x_1,x_2,\cdots)-(0,x_2,\cdots)=(x_1,0,0,\cdots).$$

Thence

$$<(I-SS^*)x,x>=<(x_1,0,0,\cdots),(x_1,x_2,\cdots)>=x_1\overline{x_1}+0+\cdots=|x_1|^2.$$

Therefore, $I-SS^*$ is positive.

Remark. We know that $S^*S=I$. This means that we have just shown that $SS^*\leq S^*S$. In fact, any isometry A verifies $AA^*\leq A^*A$ (see Exercise 12.3.3). This seems to be an unnecessary observation but this shows that the shift operator belongs to an important class of operators (see Hyponormal Operators).

Solution 5.3.3. From Exercise 3.3.43, we know that A is self-adjoint iff α_n is real-valued for each n. If $\alpha_n\geq 0$ for all n, then clearly for any $x=(x_1,x_2,\cdots,x_n,\cdots)\in\ell^2$

$$<Ax,x>=\sum_{n=1}^{\infty}\alpha_n|x_n|^2\geq 0,$$

i.e. $A\geq 0$.

Conversely, if $A\geq 0$, then for *any* $x=(x_1,x_2,\cdots,x_n,\cdots)\in\ell^2$

$$<Ax,x>=\sum_{n=1}^{\infty}\alpha_n|x_n|^2\geq 0.$$

In particular, for $x = e_n$ (from the usual orthonormal basis), we have that $\alpha_n \geq 0$ for all n, as needed.

SOLUTION 5.3.4. Let $x \in H$. Since A is self-adjoint, $A/2$ too is self-adjoint so that $e^{\frac{A}{2}}$ is self-adjoint (by Exercise 4.3.33, say). We may then write for all $x \in H$

$$< e^A x, x >=< e^{\frac{A}{2}} e^{\frac{A}{2}} x, x >=< e^{\frac{A}{2}} x, e^{\frac{A}{2}} x >= \|e^{\frac{A}{2}} x\|^2 \geq 0.$$

SOLUTION 5.3.5. No! Consider the positive matrices

$$A = \begin{pmatrix} 1 & 1 \\ 1 & 1 \end{pmatrix} \text{ and } B = \begin{pmatrix} 0 & 0 \\ 0 & 1 \end{pmatrix}.$$

Then,

$$AB = \begin{pmatrix} 0 & 1 \\ 0 & 1 \end{pmatrix} \text{ and } BA = (AB)^* = \begin{pmatrix} 0 & 0 \\ 1 & 1 \end{pmatrix}.$$

But

$$AB + BA = \begin{pmatrix} 0 & 1 \\ 1 & 2 \end{pmatrix}$$

is not positive (why?).

SOLUTION 5.3.6. Let $x \in H$. We may write for all $x \in H$

$$0 =< (A + B)x, x >=< Ax, x > + < Bx, x > .$$

But $< Ax, x >$ and $< Bx, x >$ are two positive *real* numbers because A and B are positive operators. Therefore,

$$< Ax, x >= 0 \text{ and } < Bx, x >= 0 \text{ for all } x \in H,$$

i.e. $A = B = 0$.

SOLUTION 5.3.7. Since $A \geq 0$, A is self-adjoint. Hence it is diagonalizable (a well known fact or see e.g. [**10**]). Thus, for some invertible P,

$$P^{-1}AP = D,$$

where D is a diagonal matrix whose diagonal contains the eigenvalues of A which are all positive (why?). But, clearly

$$\text{tr}D = \text{tr}(P^{-1}AP) = \text{tr}(APP^{-1}) = \text{tr}A.$$

Since tr$A$=0, tr$D$=0, that is, the sum of the *positive* eigenvalues vanishes. This forces $D = 0$ or $A = 0$.

SOLUTION 5.3.8.

(1) Let $x \in H$. Then
$$< T^*ATx, x >=< ATx, T^{**}x >=< ATx, Tx > \geq 0$$
since A is positive. A similar argument applies to prove the other inequality.

(2) Since $A - B \geq 0$, we may just apply the previous results to have
$$T^*(A-B)T \geq 0 \text{ or } T^*AT \geq T^*BT$$
(since also T^*AT and T^*BT are self-adjoint) and
$$T(A-B)T^* \geq 0 \text{ or } TAT^* \geq TBT^*.$$

SOLUTION 5.3.9. Assume that $P - Q$ is an orthogonal projection. Then $(P-Q)^2 = P - Q$ so that for all $x \in H$, we have
$$< (P-Q)x, x >=< (P-Q)^2x, x >=< (P-Q)x, (P-Q)x >= \|(P-Q)x\|^2 \geq 0,$$
meaning that $P \geq Q$.

Conversely, assume that $P \geq Q$. Then we leave it to the reader to show that this is equivalent to saying that $PQ = Q$, and also equivalent to $QP = Q$. Hence
$$(P-Q)^2 = P^2 - PQ - QP + Q^2 = P - Q - Q + Q = P - Q.$$
Accordingly, $P - Q$ is an orthogonal projection (because $P - Q$ is also self-adjoint).

SOLUTION 5.3.10.

(1) Let $x, y \in H$. Define
$$[x, y] =< Ax, y > .$$
Then $[\cdot,\cdot]$ verifies all the properties of an inner product except perhaps that we may have $[x, x] = 0$ for some $x \neq 0$. So, to establish the required inequality, just proceed as in the first question of Exercise 3.3.7.

REMARK. ([**130**]) Another way of establishing the previous inequality is to set $< x, y >_r =< Ax, y > +r < x, y >$ where $r > 0$. Then show that $< \cdot,\cdot >_r$ is an inner product, apply the standard Cauchy-Schwarz Inequality to it, send $r \to 0$ and finally get the desired generalization!

(2) Setting $y = Ax$ in the previous result, we get
$$\|Ax\|^4 = |< Ax, Ax >|^2 \leq < Ax, x >< A^2x, Ax > \leq < Ax, x > \|A^2x\|\|Ax\|.$$
Whence
$$\|Ax\|^4 \leq < Ax, x > \|A\|\|Ax\|\|Ax\| \Longrightarrow \|Ax\|^4 \leq < Ax, x > \|A\|\|Ax\|^2.$$

Thus

$$\|Ax\|^2 \le \|A\| < Ax, x > .$$

Remark. Another way of proving the previous inequality is via the Reid Inequality (as observed in [**181**]). Indeed, setting $A = K$ in the Reid Inequality gives a shorter proof of this result.

Solution 5.3.11.

(1) Since A is self-adjoint, $< Ax, x >$ is real (for all $x \in H$). We may then write

$$\begin{aligned} < (\pm A - I)x, x > &= \pm < Ax, x > - \|x\|^2 \\ &= |< Ax, x >| - \|x\|^2 \\ &\le \|Ax\|\|x\| - \|x\|^2 \text{ (by the Cauchy-Schwarz Inequality)} \\ &= (\|Ax\| - \|x\|)\|x\|. \end{aligned}$$

If $\|A\| \le 1$, then clearly $\|Ax\| \le \|A\|\|x\| \le \|x\|$ for each $x \in H$. Hence

$$< (\pm A - I)x, x > \le 0 \text{ or merely } \pm A \le I,$$

i.e. $-I \le A \le I$.

To prove the other implication, notice that if $-I \le A \le I$, then

$$\forall x \in H: \pm < Ax, x > \le \|x\|^2 \text{ or } | < Ax, x > | \le \|x\|^2$$

for all $x \in H$. Passing to the supremum over $\|x\| = 1$ yields (by taking into account the self-adjointness of A)

$$\|A\| = \sup_{\|x\|=1} | < Ax, x > | \le 1$$

and this marks the end of the proof.

(2) If $\alpha = 0$, then the results is obvious. If $\alpha > 0$, then apply the previous question with $\frac{1}{\alpha}A$ instead of A.

Solution 5.3.12. By assumption, for all $x \in H$

$$- < Ax, x > \le < Bx, x > \le < Ax, x > \text{ or merely } |< Bx, x >| \le < Ax, x > .$$

Therefore,

$$\|B\| = \sup_{\|x\|=1} | < Bx, x > | \le \sup_{\|x\|=1} < Ax, x > = \|A\|,$$

as desired.

Solution 5.3.13. WLOG, we may assume that $\|A\| \geq \|B\|$. So we must show that

$$\|A - B\| \leq \|A\|.$$

Since $A, B \geq 0$, they are self-adjoint, and so is then $A - B$. Again, since $A, B \geq 0$, we have

$$-B \leq A - B \leq A.$$

Also for all $x \in H$, we have (by the Cauchy-Schwarz Inequality)

$$< Ax, x > \leq \|Ax\|\|x\| \leq \|A\| < Ix, x > = < \|A\|Ix, x >,$$

i.e. $A \leq \|A\|I$. Similarly, we find that $-B \geq -\|B\|I$. Thus,

$$-\|B\|I \leq A - B \leq \|A\|I.$$

Taking into account the choice $\|A\| \geq \|B\|$ yields

$$-\|A\|I \leq A - B \leq \|A\|I.$$

Finally, by Exercise 5.3.11, we then obtain

$$\|A - B\| \leq \|A\| = \max(\|A\|, \|B\|).$$

Solution 5.3.14. The proof presented here is mostly due to Reid in [**181**]. WLOG, we may assume that $\|K\| \leq 1$ (why?). Therefore, we need only show

$$| < AKx, x > | \leq < Ax, x >$$

for all $x \in H$.

Since AK is self-adjoint, it follows that $AK = K^*A$. Hence

$$AK^2 = K^*AK = (K^*)^2A = (AK^2)^*,$$
$$AK^3 = (K^*)^2AK = (K^*)^3A = (AK^3)^*, \cdots,$$

so by induction, for each n, AK^n is self-adjoint.

Since $AK \geq 0$, Corollary 5.1.14 yields for all $x \in H$:

$$\begin{aligned} | < AKx, x > | &\leq \frac{1}{2}[< Ax, x > + < AKx, Kx >] \\ &= \frac{1}{2}[< Ax, x > + < K^*AKx, x >] \\ &= \frac{1}{2}[< Ax, x > + < AK^2x, x >]. \end{aligned}$$

Thanks to the previous inequality and by doing a little induction, we get for all n (and all x)

$$| < AKx, x > | \leq (2^{-1} + \cdots + 2^{-n}) < Ax, x > + 2^{-n} < AK^{2^n}x, x > \text{...(1)}$$

Since $\|K\| \leq 1$, we have by the Cauchy-Schwarz Inequality

$$\begin{aligned} |< AK^{2^n}x, x >| &\leq \|AK^{2^n}x\|\|x\| \leq \|A\|\|K^{2^n}\|\|x\|^2 \\ &\leq \|A\|\|K\|^{2^n}\|x\|^2 \leq \|A\|\|x\|^2 \end{aligned}$$

and so passing to the limit as $n \to \infty$ in (1) gives clearly

$$|< AKx, x >| \leq < Ax, x >,$$

as suggested.

Solution 5.3.15. First, observe that

$$AK^* = KA \Longrightarrow A(K^*)^2 = KAK^* = K^2A.$$

Since A is positive, so is KAK^* or $A(K^*)^2$. Thereupon, using Theorem 5.1.15, we know that

$$< KAK^*x, x > = < A(K^*)^2x, x > = |< A(K^*)^2x, x >| \leq < Ax, x > .$$

So much for the proof.

Solution 5.3.16. WLOG, we may suppose that $0 \leq B \leq I$ (otherwise work with $\frac{B}{\|B\|}$). Hence $\|I - B\| \leq 1$. Since $A(I - B)$ is clearly self-adjoint and $A \geq 0$, it follows from Theorem 5.1.15 that

$$AB = A - A(I - B) \geq 0.$$

Solution 5.3.17. The proof follows by induction (using the fact that the product of two positive commuting operators remains positive). Alternatively, we can treat two cases: n being even and n being odd (remembering that a positive operator is self-adjoint). Details are left to the reader.

Solution 5.3.18.

(1) The answer is no! Anticipating a little bit, we know from Question 2 that we need to choose two non-commuting A and B. Consider

$$A = \begin{pmatrix} 2 & 1 \\ 1 & 1 \end{pmatrix} \text{ and } B = \begin{pmatrix} 1 & 0 \\ 0 & 0 \end{pmatrix}.$$

Observe that both A and B are positive. So it only remains to check that $A \geq B$ whereas $A^2 \not\geq B^2$, that is, we need to verify that $A - B \geq 0$ and that $A^2 - B^2 \not\geq 0$. We see that

$$A - B = \begin{pmatrix} 1 & 1 \\ 1 & 1 \end{pmatrix} \geq 0$$

whereas

$$A^2 - B^2 = \begin{pmatrix} 5 & 3 \\ 3 & 2 \end{pmatrix} - \begin{pmatrix} 1 & 0 \\ 0 & 0 \end{pmatrix} = \begin{pmatrix} 4 & 3 \\ 3 & 2 \end{pmatrix} \not\geq 0$$

(check it).

(2) Since $AB = BA$, we clearly have

$$A^2 - B^2 = (A+B)(A-B).$$

But, $A \geq B$ means that $A - B \geq 0$. Also, it is plain that $A + B \geq 0$.

The fact that $A - B$ commutes with $A + B$ (as $AB = BA$) with a look at Corollary 5.1.16 imply that

$$(A+B)(A-B) = A^2 - B^2 \geq 0,$$

and hence $A^2 \geq B^2$ (***remember that A^2 and B^2 are self-adjoint, a simple but a crucial point!***). This marks the end of the proof.

Solution 5.3.19. Since $AB = BA$, we have by Theorem 5.1.5

$$0 \leq A \leq B \Longrightarrow 0 \leq A^2 \leq AB$$

and

$$0 \leq A \leq B \Longrightarrow 0 \leq AB \leq B^2.$$

Theorem 5.1.5 (yet again) gives

$$A^2 \leq B^2$$

(which is another proof of the result of Exercise 5.3.18). Using a similar argument, and a proof by induction, we can easily prove the required inequality for $n \in \mathbb{N}$...

Solution 5.3.20.

(1) Let $x \in H$. By the Cauchy-Schwarz Inequality

$$c\|x\|^2 \leq < Ax, x > \leq \|Ax\|\|x\|.$$

Therefore $\|Ax\| \geq c\|x\|$. Since A is self-adjoint, the result follows by Corollary 5.1.24.

(2) By hypothesis $\triangle = a^2 - 4b < 0$. Then

$$p(A) = A^2 + aA + bI$$

is self-adjoint. We may write

$$A^2 + aA + bI = \left(A + \frac{a}{2}I\right)^2 + b - \frac{a^2}{4} = \left(A + \frac{a}{2}I\right)^2 - 4\triangle.$$

Since $A + a/2I$ is self-adjoint, $(A + \frac{a}{2}I)^2$ is positive. Hence for all $x \in H$

$$< p(A)x, x > \geq \underbrace{-4\triangle}_{>0} < x, x > .$$

Thus $p(A)$ is invertible by the foregoing question.

(3) Straightforward!

(4) Let

$$A = \begin{pmatrix} 0 & -1 \\ 1 & 0 \end{pmatrix}.$$

Then A is not self-adjoint. It is also easy to see that

$$A^2 = -I \text{ or } A^2 + I = 0.$$

With the above notation, $a = 0$ and $b = 1$ and so $a^2 - 4b < 0$. In the end, it is clear that $A^2 + I$ is not invertible.

SOLUTION 5.3.21.

(1) Let $x \in H$. By considering

$$\|(A + iI)x\|^2 = < (A + iI)x, (A + iI)x >,$$

one can easily see that

$$\forall x \in H: \ \|(A + iI)x\| \geq \|x\|.$$

Hence $A + iI$ is bounded below. Since A is self-adjoint, $A + iI$ is normal. Therefore, by Corollary 5.1.24, $A + iI$ is invertible.

(2) First we compute U^*. We have

$$\begin{aligned} U^* =& [(A - iI)(A + iI)^{-1}]^* \\ =& [(A + iI)^{-1}]^*(A - iI)^* \\ =& [(A + iI)^*]^{-1}(A^* + iI^*) \\ =& [(A^* - iI^*)]^{-1}(A^* + iI) \\ =& (A - iI)^{-1}(A + iI) \text{ (because } A \text{ is self-adjoint).} \end{aligned}$$

Since A commutes with multiples of the identity, we easily see that

$$\begin{aligned} U^*U =& [(A - iI)]^{-1}(A + iI)(A - iI)(A + iI)^{-1} \\ =& \underbrace{[(A - iI)]^{-1}(A - iI)}_{I} \underbrace{(A + iI)(A + iI)^{-1}}_{I} \\ =& I. \end{aligned}$$

In a similar vein, we find that $UU^* = I$, that is, U is unitary.

SOLUTION 5.3.22.

(1) "(1) ⇒ (2)": First, we set

$$A = \frac{1}{2}(U + I) \text{ and } B = \frac{1}{2}(V + I).$$

Then it is clear that

$$\|A - B\| = \frac{1}{2}\|U - V\| < 1.$$

Hence $\|A-B\|^2 < 1$ so that there exists some $\alpha > 0$ such that

$$\|(A-B)^*(A-B)\| = \|A-B\|^2 \leq 1-\alpha.$$

Whence

$$(A-B)^*(A-B) \leq (1-\alpha)I$$

or after simplification,

$$I - A^*A - B^*B + A^*B + B^*A \geq \alpha I.$$

It is clear that

$$A^*A = \frac{1}{2}(A+A^*) \text{ and } B^*B = \frac{1}{2}(B+B^*).$$

Since $U = 2A - I$ and $V = 2B - I$, we have

$$\begin{aligned}(U+V)^*(U+V) =& 4(A^* + B^* - I)(A+B-I)\\ =& 4(I - A^*A - B^*B + A^*B + B^*A)\\ \geq& 4\alpha I.\end{aligned}$$

Similarly, by considering

$$A^* = \frac{1}{2}(U^* + I) \text{ and } B^* = \frac{1}{2}(V^* + I),$$

we may show that

$$(U+V)(U+V)^* \geq 4\alpha I.$$

Thus $U + V$ is invertible.

(2) The other implication may be proved by going backwards in the previous proof (do the details!).

SOLUTION 5.3.23. It is clear that if $\alpha \in \mathbb{R}$, then

$$(A-\alpha I)^*(A-\alpha I) - \alpha^2 I = (A^* - \alpha I)(A-\alpha I) - \alpha^2 I = A^*A - \alpha(A^* + A) ...(1)$$

If the previous quantity is positive for all $\alpha < 0$, then we have

$$\alpha(A^* + A) \leq A^*A \text{ or } A^* + A \geq \frac{1}{\alpha}A^*A.$$

Taking the limit as $\alpha \to -\infty$ gives

$$A + A^* \geq 0, \text{ i.e. } \mathrm{Re}A \geq 0$$

and this proves "$\Leftarrow$".

Now assume that $\mathrm{Re}A \geq 0$ and let $\alpha < 0$. Since A^*A is positive, it is evident that

$$A + A^* \geq 0 \geq \frac{A^*A}{\alpha}.$$

This means that the quantities on each side of the equalities involved in Equation (1) are greater than or equal to zero, so that for any $\alpha < 0$,

$$(A - \alpha I)^*(A - \alpha I) \geq \alpha^2 I,$$

establishing "$\Rightarrow$".

SOLUTION 5.3.24.

(1) It is easy to see that A is positive (do the details!). It then follows that A has one and only one positive square root. As clearly $A^2 = A$, then $\sqrt{A} = A$ is the (unique) positive square root of A.

(2) The shift operator and its adjoint do not possess any square root whatsoever. Assume for the sake of contradiction that e.g. S^* does, i.e. $A^2 = S^*$, where $A \in B(H)$. Then, $A^2 S = S^* S = I$ and by the general theory A is right invertible and so it is surjective. Notice also that A cannot be injective (indeed, this would imply that $A^2 = S^*$ is injective and this is untrue).

Now, we show that $\ker A = \ker S^* = \mathbb{R}e_1$, where $e_1 = (1, 0, 0, \cdots)$. The equality $\ker S^* = \mathbb{R}e_1$ is known and clear. It also implies that $\dim \ker S^* = 1$. Now, we obviously have $\ker A \subset \ker S^*$ because $A^2 = S^*$. Since A is not injective, we are forced to have $\ker A = \ker S^*$ as $\ker A$ and $\ker S^*$ are vector spaces.

Since A is onto, for all $y \in \ell^2$, in particular for $e_1 \in \ell^2$, there is an $x \in \ell^2$ such that $Ax = e_1$ (and so $x \notin \ker A = \ker S^*$). Thus (as $e_1 \in \ker A$)

$$A^2 x = Ae_1 = 0 \neq S^* x.$$

This shows that S^* does not have any square root.

If S had a square root, then we would have $S = B^2$, where $B \in B(\ell^2)$. Therefore, $S^* = (B^2)^* = (B^*)^2$, i.e. S^* would possess a square root! This is a contradiction with what we have just seen. Accordingly, S cannot have a square root either!

SOLUTION 5.3.25. By assumption, we know that $A_1 \leq A_2 \leq \cdots \leq A_n \leq \cdots \leq A$ for some self-adjoint $A \in B(H)$. WLOG we may assume that $A_1 \leq A_2 \leq \cdots \leq A_n \leq \cdots \leq I$ (just divide each A_i by $\|A\|$ and relabel $\frac{A_i}{\|A\|}$ as A_i). There is also no loss of generality in assume that all $A_n \geq 0$ (e.g. we could use the sequence $(A_n - A_1)_n$, say). Therefore, we may work with $0 \leq A_1 \leq A_2 \leq \cdots \leq A_n \leq \cdots \leq I$.

The primary aim is to show that $(A_n x)$ converges for each x in H. By the completeness of H, this means that it suffices then to show that

$(A_n x)$ is Cauchy. Let $n > m$ and let $x \in H$. Then $A_n - A_m \geq 0$ and $A_n - A_m \leq I$. Hence $\|A_n - A_m\| \leq 1$. Now, we may write

$$\begin{aligned}\|A_n x - A_m x\|^4 &= < (A_n - A_m)x, (A_n - A_m)x >^2 \\ &\leq < (A_n - A_m)x, x >< (A_n - A_m)^2 x, (A_n - A_m)x > \\ &\leq < (A_n - A_m)x, x > \|(A_n - A_m)^2 x\| \|(A_n - A_m)x\| \\ &\leq < (A_n - A_m)x, x > \|A_n - A_m\| \|(A_n - A_m)x\|^2 \\ &\leq < (A_n - A_m)x, x > \|A_n x - A_m x\|^2\end{aligned}$$

where we have used Theorem 5.1.13 in the first inequality. Therefore,

$$\|A_n x - A_m x\|^2 \leq < (A_n - A_m)x, x > = < A_n x, x > - < A_m x, x > .$$

But $(< A_n x, x >)_n$ is an increasing *real* sequence which is bounded above by $(\|x\|^2)$. Whence, it converges and so it is Cauchy. Thereupon,

$$\lim_{n,m\to\infty} \|A_n x - A_m x\| = 0.$$

This means, as already observed above, that $\lim_{n\to\infty} A_n x$ exists for each $x \in H$.

Define now for each x

$$Ax = \lim_{n\to\infty} A_n x$$

(in the sense that $\|A_n x - Ax\| \to 0$ for all x). Then A is clearly linear. It only remains to see why A is bounded and self-adjoint. We prove these two requirements together: By the continuity of the inner product, we have for all $x, y \in H$

$$< Ax, y > = \lim_{n\to\infty} < A_n x, y > = \lim_{n\to\infty} < x, A_n y > = < x, Ay > .$$

Calling on the Hellinger-Toeplitz Theorem, i.e. Theorem 3.1.70 (or Exercise 2.3.52), we obtain that $A \in B(H)$, and clearly A is self-adjoint.

To summarize, the bounded monotone increasing sequence (A_n) converges strongly to the self-adjoint bounded operator A.

SOLUTION 5.3.26.

(1) Observe first that since A is positive and $\|A\| \leq 1$, we have $0 \leq A \leq I$. Another equally important observation is that the sequence (B_n) is a "polynomial" of A. This implies that all of B_n are pairwise commuting.

Next, $B_0 = 0$ is evidently self-adjoint. So, assuming that B_n is self-adjoint (and recalling that A is self-adjoint), we can easily check that B_{n+1} too is self-adjoint. Therefore, all B_n are self-adjoint.

Now, we claim that $B_n \leq I$ for all n. This is obviously true for $n = 0$. Assume that $B_n \leq I$. Observing that $(I - B_n)^2 \geq 0$ (why?), we then have

$$I - B_{n+1} = I - B_n - \frac{1}{2}(A - B_n^2) = \frac{1}{2}(I - B_n)^2 + \frac{1}{2}(I - A) \geq 0.$$

To prove that (B_n) is increasing, observe first that $B_0 \leq \frac{1}{2}A = B_1$. Assuming that $B_n \geq B_{n-1}$, we may write

$$B_{n+1} - B_n = \frac{1}{2}[(I - B_{n-1}) + (I - B_n)](B_n - B_{n-1})$$

which, being a product of commuting positive operators, itself is positive.

Consequently, we have shown that

$$0 = B_0 \leq B_1 \leq \cdots \leq B_n \leq \cdots \leq I,$$

as needed.

(2) Since (B_n) is bounded monotone increasing, by Theorem 5.1.38 we know that (B_n) converges strongly to some self-adjoint $B \in B(H)$. Since each B_n is positive, we have

$$< Bx, x >= \lim_{n \to \infty} < B_n x, x > \geq 0$$

as strong convergence implies weak one. Thus, $B \geq 0$.

It remains to show that $B^2 = A$. Let $x \in H$. We have by hypothesis

$$B_{n+1}x = B_n x + \frac{1}{2}(Ax - B_n^2 x).$$

Passing to the strong limit and using $\|B_n^2 x - B^2 x\| \to 0$ (why?), we finally get $B^2 = A$, as required.

Finally, assume that a $C \in B(H)$ commutes with A, i.e. $AC = CA$. We must show that $BC = CB$. Since C commutes with A, we may easily show that C commutes with B_n too, that is, $CB_n x = B_n Cx$ (for all n and all x). On the one hand, we clearly see that $B_n Cx \to BCx$. On the other hand, invoking the (sequential) continuity of C, we have that $CB_n x \to CBx$. By uniqueness of the strong limit, we get

$$BCx = CBx, \ \forall x \in H,$$

as desired.

(3) If $A = 0$, then $B = 0$ will do. So if $A \neq 0$, considering $T = \frac{A}{\|A\|}$ gives $0 \leq T \leq 1$. Then, apply what we have already done above.

(4) *The proof of uniqueness here, although not being complicated, is not as direct as one is used to with other theorems.*

We have already shown that $B^2 = A$. Assume that there is another positive $C \in B(H)$ such that $C^2 = A$. We must show that $Bx = Cx$ for all $x \in H$. Observe first that A plainly commutes with C. By Question (2), C commutes with B as well, i.e. $BC = CB$. This tells us that

$$(B + C)(B - C) = B^2 - C^2 = A - A = 0.$$

So, if we let $x \in H$ and set $y = (B - C)x$, then

$$< By, y > + < Cy, y >=< (B+C)y, y >=< (B+C)(B-C)x, y >= 0.$$

Because both B and C are positive, we obtain (cf. Exercise 5.3.6)

$$< By, y >=< Cy, y >= 0.$$

By Question (2) again, $B \geq 0$ has a square root which we denote by D, say. That is, $D^2 = B$. Therefore,

$$\|Dy\|^2 =< Dy, Dy >=< D^2y, y >=< By, y >= 0$$

and so $Dy = 0$. This implies that $By = D^2y = D(0) = 0$.

Using also a square root of C, we may similarly show that $Cy = 0$. Consequently,

$$\|Bx - Cx\|^2 =< (B - C)x, (B - C)x >=< (B - C)y, x >= 0.$$

Accordingly, $B = C$, i.e. we have proven that the positive A can *only have one* positive square root, marking the end of the proof.

Solution 5.3.27. Assume that $A \in B(H)$ is positive. Hence, there is a positive $B \in B(H)$ such that $B^2 = A$. Assume that there is another positive $C \in B(H)$ such that $A = C^2$ and so $B^2 = C^2$. We ought to show that $B = C$.

First, it is clear that

$$CA = C^3 = AC.$$

Hence C commutes with B as well (why?). This gives

$$(B - C)B(B - C) + (B - C)C(B - C) = (B^2 - C^2)(B - C) = 0.$$

As $B, C \geq 0$ and $B - C$ is *self-adjoint*, then $(B - C)B(B - C)$ and $(B - C)C(B - C)$ are both positive and so

$$(B - C)B(B - C) = (B - C)C(B - C) = 0.$$

Thereupon,

$$(B - C)B(B - C) - (B - C)C(B - C) = 0,$$

that is

$$(B-C)^3 = 0.$$

Whence

$$(B-C)^4 = 0.$$

Now, if $T \in B(H)$ is self-adjoint, then $\|T^2\| = \|T\|^2$. Since T^2 is self-adjoint, we get $\|T^4\| = \|T\|^4$ (more on this can be found in Exercise 8.3.2).

Consequently,

$$0 = \|(B-C)^4\| = \|B-C\|^4,$$

that is $B = C$, as required.

SOLUTION 5.3.28.

(1) Since A is positive, it admits a unique positive square root, which we denote by P (that is $P^2 = A$). Since B commutes with A, it commutes with P as well.

Let $x \in H$. We may write (remembering that positive operators are necessarily self-adjoint)

$$< ABx, x > = < P^2Bx, x > = < PBx, Px > = < BPx, Px > \geq 0$$

as B is positive. Therefore, $AB \geq 0$.

Since A and B are positive, both $A^{\frac{1}{2}}$ and $B^{\frac{1}{2}}$ exist and are well-defined. Since A and B also commute, AB is positive and it makes sense then to define $(AB)^{\frac{1}{2}}$. If we come to show that

$$(A^{\frac{1}{2}}B^{\frac{1}{2}})^2 = AB,$$

then by the uniqueness of the square root, the desired result follows.

Now since A and B commute, so do their square roots and we have

$$(A^{\frac{1}{2}}B^{\frac{1}{2}})^2 = A^{\frac{1}{2}}B^{\frac{1}{2}}A^{\frac{1}{2}}B^{\frac{1}{2}} = A^{\frac{1}{2}}A^{\frac{1}{2}}B^{\frac{1}{2}}B^{\frac{1}{2}} = AB.$$

The proof is complete.

(2) Let

$$A = \begin{pmatrix} 1 & 0 \\ 0 & 2 \end{pmatrix} \text{ and } B = \begin{pmatrix} 1 & 1 \\ 1 & 3 \end{pmatrix}.$$

Then both A and B are positive.

We may also check that

$$AB = \begin{pmatrix} 1 & 1 \\ 2 & 6 \end{pmatrix},$$

i.e. AB is not positive because it is not even self-adjoint and

$$AB = \begin{pmatrix} 1 & 1 \\ 2 & 6 \end{pmatrix} \neq \begin{pmatrix} 1 & 2 \\ 1 & 6 \end{pmatrix} = BA.$$

(3) Since A, B and AB are all positive operators, they are all self-adjoint. Accordingly,

$$BA = B^*A^* = (AB)^* = AB,$$

that is A and B commute.

Solution 5.3.29. Since $KA = AK$ and A is self-adjoint, it follows that $AK^* = K^*A$. Hence $AK^*K = K^*KA$. Theorem 5.1.28 then yields $A^{\frac{1}{2}}K^*K = K^*KA^{\frac{1}{2}}$ as $A \geq 0$.

Now, let $x \in H$. By the Generalized Cauchy-Schwarz Inequality, we may write

$$< K^*AKx, x >^2 = < AK^*Kx, x >^2 \leq < Ax, x >< AK^*Kx, K^*Kx > .$$

But,

$$\begin{aligned} < AK^*Kx, K^*Kx > &= < A^{\frac{1}{2}}K^*Kx, A^{\frac{1}{2}}K^*Kx > \\ &= \|A^{\frac{1}{2}}K^*Kx\|^2 = \|K^*KA^{\frac{1}{2}}x\|^2. \end{aligned}$$

Because $\|K^*K\| \leq 1$, we obtain

$$\|K^*KA^{\frac{1}{2}}x\|^2 \leq \|A^{\frac{1}{2}}x\|^2 = < A^{\frac{1}{2}}x, A^{\frac{1}{2}}x > = < Ax, x >$$

so that

$$< K^*AKx, x >^2 \leq < Ax, x >^2,$$

completing the proof.

Solution 5.3.30.

(1) Let $x \in H$. Since $0 \leq A \leq B$, we have for all $x \in H$

$$0 \leq < Ax, x > \leq < Bx, x > \Longleftrightarrow 0 \leq < \sqrt{A}x, \sqrt{A}x > \leq < \sqrt{B}x, \sqrt{B}x >$$

and so (for all x)

$$0 \leq \|\sqrt{A}x\|^2 \leq \|\sqrt{B}x\|^2.$$

So, by Theorem 3.1.69, we know that $\sqrt{A} = K\sqrt{B}$ for some contraction $K \in B(H)$. Since $\sqrt{A}$ is self-adjoint, it follows that $K\sqrt{B}$ too is self-adjoint, i.e. $K\sqrt{B} = \sqrt{B}K^*$. Since $\sqrt{B} \geq 0$, by the Reid Inequality (Theorem 5.1.15) we obtain:

$$< \sqrt{A}x, x > = < \sqrt{B}K^*x, x > \leq < \sqrt{B}x, x >,$$

that is,

$$\sqrt{A} \leq \sqrt{B},$$

as required.

(2) As before, we know that $\sqrt{A} = K\sqrt{B}$ for some contraction $K \in B(H)$. Since $\sqrt{A}$ is invertible (as A is), it follows that $I = (\sqrt{A})^{-1}K\sqrt{B}$, i.e. the self-adjoint $\sqrt{B}$ is left invertible. By taking adjoints, we see that $\sqrt{B}$ is also right invertible. Thus, B is invertible and

$$(\sqrt{B})^{-1} = (\sqrt{A})^{-1}K = K^*(\sqrt{A})^{-1}$$

by the self-adjointness of both $(\sqrt{B})^{-1}$ and $(\sqrt{A})^{-1}$.

Finally, let $x \in H$. Then (since K^* too is a contraction)

$$< B^{-1}x, x >= \|(\sqrt{B})^{-1}x\|^2 = \|K^*(\sqrt{A})^{-1}x\|^2 \leq \|(\sqrt{A})^{-1}x\|^2 =< A^{-1}x, x >,$$

as needed.

Solution 5.3.31. Since $AB = BA$ and $A, B \geq 0$, we have $\sqrt{A}\sqrt{B} = \sqrt{B}\sqrt{A}$. Hence $\sqrt{A}\sqrt{B} \geq 0$. Therefore,

$$A + B \leq A + 2\sqrt{A}\sqrt{B} + B = (\sqrt{A} + \sqrt{B})^2.$$

Since $\sqrt{A} + \sqrt{B} \geq 0$, by Theorem 5.1.39, we get

$$\sqrt{A+B} \leq \sqrt{A} + \sqrt{B},$$

establishing half of the result.

Finally, to prove the other inequality, reason similarly using $(\sqrt{A} - \sqrt{B})^2 \geq 0$...

Solution 5.3.32.

(1) We need only verify that $I - A^2$ is a positive operator. Let $x \in H$. We have

$$\begin{aligned} < (I - A^2)x, x > \geq 0 &\iff < x, x > - < A^2x, x > \geq 0 \\ &\iff < A^2x, x > \leq \|x\|^2 \\ &\iff < Ax, Ax > = \|Ax\|^2 \leq \|x\|^2. \end{aligned}$$

But by hypothesis, $\|A\| \leq 1$ which leads to

$$\|Ax\|^2 \leq \|A\|^2\|x\|^2 \leq \|x\|^2.$$

Therefore, $I - A^2 \geq 0$.

(2) We only prove U_+ is unitary (the proof for U_- is very akin). Since A is self-adjoint, one has

$$U_+^* = (A + i(I - A^2)^{\frac{1}{2}})^* = A - i(I - A^2)^{\frac{1}{2}}.$$

Since A and $I - A^2$ commute, so do A and $(I - A^2)^{\frac{1}{2}}$ and so

$$\begin{aligned} U_+U_+^* &= (A + i(I - A^2)^{\frac{1}{2}})(A - i(I - A^2)^{\frac{1}{2}}) \\ &= A^2 - iA(I - A^2)^{\frac{1}{2}} + i(I - A^2)^{\frac{1}{2}}A + I - A^2 \\ &= I. \end{aligned}$$

Similarly, one shows that $U_+^*U_+ = I$

Solution 5.3.33. We already know that any $A \in B(H)$ may be written as $A = \operatorname{Re} A + i \operatorname{Im} A$, that is, every $A \in B(H)$ may be expressed as a linear combination of *two self-adjoint* operators.

Now, suppose that $B \in B(H)$ is self-adjoint. WLOG, we may assume that $\|B\| \leq 1$ (otherwise, you know what you should do!). By Exercise 5.3.32, $B \pm i(I - B^2)^{\frac{1}{2}}$ are unitary operators and clearly

$$B = \frac{1}{2}[B + i(I - B^2)^{\frac{1}{2}}] + \frac{1}{2}[B - i(I - B^2)^{\frac{1}{2}}],$$

so that each self-adjoint operator may be expressed as a linear combination of *two unitary* operators, and this leads to the fact that any $A \in B(H)$ may be written as a linear combination of four unitary operators.

Solution 5.3.34.

(1) "$\Leftarrow$": Let $x \in H$. Then

$$0 \leq < KBx, Bx >=< Ax, Bx >=< BAx, x >,$$

that is, $BA \geq 0$.

(2) "$\Rightarrow$": Since $BA \geq 0$, it follows that BA is self-adjoint, i.e. $AB = BA$. As a consequence, $\ker A$ reduces A and B, and the restriction of A to $\ker A$ is the zero operator on $\ker A$. Hence, we can assume that A is injective. Therefore, because $\ker B \subset \ker A = \{0\}$, we see that B^{-1} is self-adjoint and **densely defined** (i.e. defined on a dense domain). Set $K_0 = AB^{-1}$. Then K_0 is densely defined and

$$||K_0(Bx)|| = ||AB^{-1}Bx|| = ||Ax|| \leq ||Bx||, \forall x \in H,$$

signifying that K_0 is a contraction with a unique contractive extension K to the whole H. Since

$$< K_0(Bx), Bx >=< Ax, Bx >=< BAx, x > \geq 0$$

for all $x \in H$, we see that K is positive as well. Clearly

$$KBx = K_0(Bx) = Ax$$

for all $x \in H$, and this completes the proof.

SOLUTION 5.3.35. Since $AB \geq 0$, we know by Lemma 5.1.40 that $\sqrt{A} = K\sqrt{B}$ for some positive contraction $K \in B(H)$ and $K\sqrt{B} = \sqrt{B}K$. Hence

$$A = K\sqrt{B}K\sqrt{B} = K^2 B.$$

So for all $x \in H$:

$$\|Ax\|^2 = \|K^2 Bx\|^2 \leq \|Bx\|^2$$

or merely

$$< A^2x, x >=< Ax, Ax >= \|Ax\|^2 \leq \|Bx\|^2 =< B^2x, x >,$$

as required.

SOLUTION 5.3.36.

(1) "$\Longrightarrow$": Assume that $T \geq 0$. By the Generalized Cauchy-Schwarz Inequality (applied to the vectors $(x, 0)$ and $(0, y)$), we have

$$\left| < T\begin{pmatrix} x \\ 0 \end{pmatrix}, \begin{pmatrix} 0 \\ y \end{pmatrix} > \right|^2 \leq < T\begin{pmatrix} x \\ 0 \end{pmatrix}, \begin{pmatrix} x \\ 0 \end{pmatrix} >< T\begin{pmatrix} 0 \\ y \end{pmatrix}, \begin{pmatrix} 0 \\ y \end{pmatrix} > .$$

But $T = \begin{pmatrix} A & C^* \\ C & B \end{pmatrix}$ and so the previous inequality becomes after simplifications:

$$| < Cx, y > |^2 \leq < Ax, x >< By, y >,$$

valid obviously for all $x, y \in H$.

(2) "$\Longleftarrow$": Now, suppose that

$$| < Cx, y > |^2 \leq < Ax, x >< By, y >, \ \forall x, y \in H.$$

To show that T is positive, let $x, y \in H$ and observe that

$$< T\begin{pmatrix} x \\ y \end{pmatrix}, \begin{pmatrix} x \\ y \end{pmatrix} >=< Ax, x > + < C^*y, x > + < Cx, y > + < By, y > .$$

Since

$$< C^*y, x > + < Cx, y >= \overline{< Cx, y >} + < Cx, y >= 2\text{Re} < Cx, y >,$$

it follows that

$$\begin{aligned} < T\begin{pmatrix} x \\ y \end{pmatrix}, \begin{pmatrix} x \\ y \end{pmatrix} > &=< Ax, x > +2\text{Re} < Cx, y > + < By, y > \\ &\geq 2 < Ax, x >^{\frac{1}{2}}< Bx, x >^{\frac{1}{2}} +2\text{Re} < Cx, y > \ \text{(why?)} \\ &\geq 2| < Cx, y > | + 2\text{Re} < Cx, y > \ \text{(by assumption)} \\ &\geq 2| < Cx, y > | - 2| < Cx, y > | \\ &= 0, \end{aligned}$$

marking the end of the proof.

Solution 5.3.37. Set

$$T = \begin{pmatrix} B & 0 \\ 0 & C \end{pmatrix} \text{ and } S = \begin{pmatrix} 0 & A \\ 0 & 0 \end{pmatrix}$$

both defined on $H \oplus H$. Since $B, C \geq 0$, it easily follows that $T \geq 0$ as

$$\begin{aligned} < \begin{pmatrix} B & 0 \\ 0 & C \end{pmatrix} \begin{pmatrix} x \\ y \end{pmatrix}, \begin{pmatrix} x \\ y \end{pmatrix} = & < \begin{pmatrix} Bx \\ Cy \end{pmatrix}, \begin{pmatrix} x \\ y \end{pmatrix} > \\ = & < Bx, x > + < Cy, y > \geq 0 \end{aligned}$$

for all $x, y \in H$. It is also clear that the square root of T is given by

$$\sqrt{T} = \begin{pmatrix} \sqrt{B} & 0 \\ 0 & \sqrt{C} \end{pmatrix}.$$

Since by assumption $BA = AC$, we get

$$TS = \begin{pmatrix} B & 0 \\ 0 & C \end{pmatrix} \begin{pmatrix} 0 & A \\ 0 & 0 \end{pmatrix} = \begin{pmatrix} 0 & BA \\ 0 & 0 \end{pmatrix} = \begin{pmatrix} 0 & AC \\ 0 & 0 \end{pmatrix} = ST.$$

Now, as $T \geq 0$, then by Corollary 5.1.29 we obtain $\sqrt{T}S = S\sqrt{T}$. This means that

$$\begin{pmatrix} \sqrt{B} & 0 \\ 0 & \sqrt{C} \end{pmatrix} \begin{pmatrix} 0 & A \\ 0 & 0 \end{pmatrix} = \begin{pmatrix} 0 & A \\ 0 & 0 \end{pmatrix} \begin{pmatrix} \sqrt{B} & 0 \\ 0 & \sqrt{C} \end{pmatrix}$$

or

$$\begin{pmatrix} 0 & \sqrt{B}A \\ 0 & 0 \end{pmatrix} = \begin{pmatrix} 0 & A\sqrt{C} \\ 0 & 0 \end{pmatrix},$$

i.e. $\sqrt{B}A = A\sqrt{C}$, as required.

Solution 5.3.38. First, recall that

$$[A, B] = AB - BA.$$

(1) Let B be a self-adjoint contraction. By Exercise 5.3.32, $U = B + i\sqrt{I - B^2}$ is unitary and $B = \text{Re } U = \frac{U+U^*}{2}$.

$$\begin{aligned}
\|AB - BA\| &= \left\| A\left(\frac{U+U^*}{2}\right) - \left(\frac{U+U^*}{2}\right)A \right\| \\
&= \frac{1}{2}\|AU - UA + AU^* - U^*A\| \\
&\leq \frac{1}{2}\|AU - UA\| + \frac{1}{2}\|AU^* - U^*A\| \\
&= \frac{1}{2}\|AU - UA\| + \frac{1}{2}\|U(AU^* - U^*A)U\| \text{ (Exercise 4.3.6)} \\
&= \frac{1}{2}\|AU - UA\| + \frac{1}{2}\|(UAU^* - A)U\| \\
&= \frac{1}{2}\|AU - UA\| + \frac{1}{2}\|UA - AU\| \\
&= \|AU - UA\| \\
&= \|(A - UAU^*)U\| \\
&= \|A - UAU^*\| \text{ (Exercise 4.3.6 again)} \\
&\leq \max(\|A\|, \|UAU^*\|) \text{ (Exercises 5.3.8 \& 5.3.13)} \\
&= \|A\| \text{ (Exercise 4.3.6 yet again!)},
\end{aligned}$$

establishing the result.

(2) Let B be self-adjoint. The inequality clearly holds for $B = 0$, so assume that $\|B\| > 0$. Hence $\frac{B}{\|B\|}$ remains self-adjoint and besides, it is a contraction. Therefore, the result of the previous question applies and yields

$$\left\| A\frac{B}{\|B\|} - \frac{B}{\|B\|}A \right\| \leq \|A\|,$$

that is,

$$\|AB - BA\| \leq \|A\|\|B\|,$$

as required.

(3) Let $B \in B(H)$. Define on $H \oplus H$

$$\tilde{A} = \begin{pmatrix} A & \mathbf{0} \\ \mathbf{0} & A \end{pmatrix} \text{ and } \tilde{B} = \begin{pmatrix} \mathbf{0} & B \\ B^* & \mathbf{0} \end{pmatrix},$$

where the $\mathbf{0}$ is the zero operator on H. Observe that $\tilde{B}$ is self-adjoint (even if B is not one), and that $\tilde{A}$ is self-adjoint because A is one! Hence, by the previous question we know that

$$\|\tilde{A}\tilde{B} - \tilde{B}\tilde{A}\| \leq \|\tilde{A}\|\|\tilde{B}\|.$$

But,

$$\tilde{A}\tilde{B}-\tilde{B}\tilde{A}=\begin{pmatrix} \mathbf{0} & AB-BA \\ AB^*-B^*A & \mathbf{0} \end{pmatrix}.$$

Also, from Proposition 4.1.28 and Corollary 4.1.30 (for all $C,D \in B(H)$), we have:

$$\left\|\begin{pmatrix} C & \mathbf{0} \\ \mathbf{0} & D \end{pmatrix}\right\| = \left\|\begin{pmatrix} \mathbf{0} & C \\ D & \mathbf{0} \end{pmatrix}\right\| = \max(\|C\|,\|D\|).$$

Hence (why?)

$$\|\tilde{A}\| = \|A\| \text{ and } \|\tilde{B}\| = \|B\|$$

With all these observations, we infer that

$$\begin{aligned}\|\tilde{A}\tilde{B}-\tilde{B}\tilde{A}\| &= \max(\|AB-BA\|,\|AB^*-B^*A\|)\\ &= \max(\|AB-BA\|,\|(AB^*-B^*A)^*\|)\\ &= \max(\|AB-BA\|,\|BA-AB\|)\\ &= \|AB-BA\|,\end{aligned}$$

so that finally we get

$$\|\tilde{A}\tilde{B}-\tilde{B}\tilde{A}\| \le \|\tilde{A}\|\|\tilde{B}\| \iff \|AB-BA\| \le \|A\|\|B\|,$$

and this completes the proof.

Solution 5.3.39. Write $T = A+iB$ where $A,B \in B(H)$ are self-adjoint with $A = \operatorname{Re} T$ and $B = \operatorname{Im} T$ as is known to readers. Then clearly

$$T^2 = A^2 - B^2 + i(AB+BA).$$

So, if $T^2 = 0$, then

$$A^2 - B^2 + i(AB+BA) = 0 \Longrightarrow \begin{cases} A^2 = B^2, \\ AB = -BA. \end{cases}$$

Hence, if $A \ge 0$ (a similar argument works when $B \ge 0$), then

$$AB = -BA \Longrightarrow A^2B = -ABA = BA^2 \Longrightarrow AB = BA.$$

Therefore, T is normal. Accordingly (by a look at Exercise 4.3.13)

$$\|T\|^2 = \|T^2\| = 0 \Longrightarrow T = 0,$$

as suggested.

5.4. Hints/Answers to Tests

SOLUTION 5.4.1. In fact, there is not much to do here! Indeed, the question would have been a little harder had we been asked to find $\sqrt{A}$!

First, A is positive... So is $\sqrt{A}$. Since the (positive) square root is unique, we are done if we show that $(\sqrt{A})^2 = A$. This is in effect the case!...

SOLUTION 5.4.2. We use a trace argument, then we invoke Exercise 5.3.7:

By assumption, $B - ABA^{-1} \geq 0$. Also,

$$\operatorname{tr}(B - ABA^{-1}) = \operatorname{tr}B - \operatorname{tr}[(AB)A^{-1}] = \operatorname{tr}B - \operatorname{tr}[A^{-1}(AB)] = ... = 0.$$

Accordingly, $B = ABA^{-1}$ or merely $AB = BA$, as needed.

SOLUTION 5.4.3. No! With our assumption $A^2 = I$ and $A \geq 0$. Then pass the square root!...

SOLUTION 5.4.4. Yes! Use Exercise 5.3.38...

SOLUTION 5.4.5. Since A and B are commuting and positive, we have $AB \geq 0$. Besides,

$$(AB)^n = \underbrace{(AB)(AB)\cdots(AB)}_{n \text{ times}} = A^n B^n.$$

Replacing A by $A^{\frac{1}{n}}$ and B by $B^{\frac{1}{n}}$, we get

$$(A^{\frac{1}{n}} B^{\frac{1}{n}})^n = AB.$$

By uniqueness of the positive nth root, we obtain

$$(AB)^{\frac{1}{n}} = A^{\frac{1}{n}} B^{\frac{1}{n}},$$

as desired.

SOLUTION 5.4.6. Here is a little push

$$A^2 B = A(AB) = ... = ...$$

Then using $A \geq 0$, obtain that $AB = 0$...

SOLUTION 5.4.7. Define

$$\tilde{A} = \begin{pmatrix} 0 & A \\ A^* & 0 \end{pmatrix} \text{ and } \tilde{B} = \begin{pmatrix} B & 0 \\ 0 & 0 \end{pmatrix}.$$

It is clear that $\tilde{B} \geq 0$ on $H \oplus H$. Besides,

$$\tilde{A}\tilde{B} = \tilde{B}\tilde{A} = \begin{pmatrix} 0 & 0 \\ 0 & 0 \end{pmatrix}$$

for $BA = A^*B = 0$ by assumption. Hence Theorem 5.1.28 yields

$$\tilde{A}\sqrt{\tilde{B}} = \sqrt{\tilde{B}}\tilde{A}$$

which then forces

$$\sqrt{B}A = 0 = A^*\sqrt{B},$$

as desired.

CHAPTER 6

Absolute Value. Polar Decomposition of an Operator

6.2. True or False: Answers

(1) True for the equivalence. If $A = 0$, then evidently
$$|A| = \sqrt{A^*A} = \sqrt{0} = 0.$$
If $|A| = 0$, then we have $|A|^2 = 0$ or $A^*A = 0$ and hence
$$\|A\|^2 = \|A^*A\| = 0,$$
that is, $A = 0$.

(2) True. If $A = |A|$, then clearly $A \geq 0$ because $|A| \geq 0$. Conversely, assume that $A \geq 0$. Then $A^2 \geq 0$ as A is self-adjoint. By uniqueness of the square root we get
$$A = \sqrt{A^2} = \sqrt{AA} = |A|,$$
as needed.

(3) False for both! If A is normal, then both inequalities hold. The proof will be seen in Exercise 6.3.6 or in Test 6.4.4 (Exercise 12.3.11 contains an even stronger result). For a counterexample, let
$$A = \begin{pmatrix} 0 & 2 \\ 0 & 0 \end{pmatrix}.$$
Then, we know from Examples 6.1.2 that
$$|A| = \begin{pmatrix} 0 & 0 \\ 0 & 2 \end{pmatrix} \text{ and } |\operatorname{Re} A| = |\operatorname{Im} A| = I.$$
Thence
$$|\operatorname{Re} A| = |\operatorname{Im} A| \not\leq |A|$$
as the matrix $\begin{pmatrix} -1 & 0 \\ 0 & 1 \end{pmatrix}$ is not positive.

(4) Only the implication "$\Leftarrow$" holds in general. The reason is simple: Since A is invertible, it follows that A^* and so A^*A are invertible. Hence $\sqrt{A^*A} = |A|$ is invertible.

The other implication is false in general. Just take S to be the shift operator on ℓ^2. Then $|S| = \sqrt{S^*S} = I$ is plainly invertible whereas we all know that S is not invertible. The reader may also check that if $|A|$ is invertible and A is normal, then A is invertible (use Exercise 4.3.23 or Exercise 6.3.26).

(5) The implication "$\Leftarrow$" is true. This is not so obvious. A proof may be found in Exercise 6.3.5. The reverse implication is false in general. A simple counterexample is to take $A \geq 0$ and invertible and set $B = -A$. Then $|A| + |B| = 2|A| = 2A$ is invertible whereas $A + B = 0$ is not invertible. Another yet simpler counterexample is to take $B = 0$ and call on the previous question.

(6) The answer is yes in the sense that if we have Proposition 6.1.7, then we can prove Theorem 6.1.8, and vice versa.

Indeed, we will see in Exercise 6.3.16, which contains the proof of Theorem 6.1.8, how we will use Proposition 6.1.7.

Theorem 6.1.8 also gives Proposition 6.1.7. To see this, let $A \in B(H)$ be normal. By the polar decomposition, $A = U|A| = |A|U$ where $U \in B(H)$ is unitary. Using Theorem 6.1.8 (with $A = |A|U$ normal, $\|U\| = 1$ and $|A| \geq 0$), we clearly have

$$|<Ax, x>= |<|A|Ux, x>| \leq <|A|x, x>,$$

which is the required inequality in Proposition 6.1.7.

(7) False! None of the three commutativity identities holds! Consider

$$A = \begin{pmatrix} 0 & 1 \\ 3 & 0 \end{pmatrix} = \begin{pmatrix} 0 & 1 \\ 1 & 0 \end{pmatrix} \begin{pmatrix} 3 & 0 \\ 0 & 1 \end{pmatrix} = U|A|$$

which is the *only polar decomposition*. Readers then can easily verify that

$$U|A| \neq |A|U,\ UA \neq AU \text{ and } A|A| \neq |A|A.$$

(8) The answer is no! The polar decomposition of a normal operator does not have to be unique. Indeed, let $A = 0$ be the zero operator on H. Then A is clearly normal yet for ***any unitary*** $U \in B(H)$:

$$0 = U|0| = |0|U.$$

6.3. Solutions to Exercises

Solution 6.3.1. Let $A, B \in B(H)$.

(1) Assume that $0 \leq A \leq B$. Then the Reid Inequality gives $\sqrt{A} \leq \sqrt{B}$ as in Exercise 5.3.30, that is, we are done with the first half of the proof.

(2) Now, assume that whenever $0 \leq A \leq B$, then $\sqrt{A} \leq \sqrt{B}$. Let $S, T \in B(H)$ be such that $S \geq 0$ and ST is self-adjoint. WLOG, assume that $T \neq 0$. Then $\frac{T}{\|T\|}$ is a contraction and so

$$TT^* \leq \|T\|^2 I.$$

Hence $STT^*S \leq \|T\|^2 S^2$. Therefore,

$$0 \leq \frac{|ST|^2}{\|T\|^2} = \frac{|(ST)^*|^2}{\|T\|^2} = \frac{STT^*S}{\|T\|^2} \leq S^2.$$

By hypothesis, we get $|ST| \leq \|T\|S$, i.e. for all $x \in H$ (by invoking Proposition 6.1.4) :

$$| < STx, x > | \leq < |ST|x, x > \leq \|T\| < Sx, x >,$$

which is nothing but the Reid Inequality. This completes the other half of the proof.

Solution 6.3.2.

(1) Let S be the shift operator. Set $A = S^*$ and $B = S$. Hence $AB = S^*S = I$ so that

$$|S^*S| = |I| = I$$

whereas

$$|S| = \sqrt{S^*S} = I \text{ and } |S^*| = \sqrt{SS^*}.$$

If $|AB| = |A|\ |B|$ held, we would then have

$$I = \sqrt{SS^*}I = \sqrt{SS^*}$$

which would lead to $SS^* = I$, i.e. S would be unitary and this is impossible.

(2) Let

$$A = \begin{pmatrix} -1 & 1 \\ 1 & -1 \end{pmatrix} \text{ and } B = \begin{pmatrix} 2 & 0 \\ 0 & 0 \end{pmatrix}.$$

To show that the required inequality does not hold for this couple of operators, we find A^*A, B^*B and $(A+B)^*(A+B)$ as well as their positive square roots. We have

$$A^*A = \begin{pmatrix} 2 & -2 \\ -2 & 2 \end{pmatrix},$$

$$B^*B = \begin{pmatrix} 4 & 0 \\ 0 & 0 \end{pmatrix}$$

and

$$(A+B)^*(A+B) = \begin{pmatrix} 2 & 0 \\ 0 & 2 \end{pmatrix}.$$

Hence

$$|A| = \begin{pmatrix} 1 & -1 \\ -1 & 1 \end{pmatrix}, \ |B| = \begin{pmatrix} 2 & 0 \\ 0 & 0 \end{pmatrix} \text{ and } |A+B| = \begin{pmatrix} \sqrt{2} & 0 \\ 0 & \sqrt{2} \end{pmatrix}.$$

If $|A+B| \leq |A| + |B|$ were true, then we would have for all vectors, in particular for $e_2 = \begin{pmatrix} 0 \\ 1 \end{pmatrix}$

$$< |A+B|e_2, e_2 > \leq < |A|e_2, e_2 > + < |B|e_2, e_2 > .$$

But

$$< |A+B|e_2, e_2 > = \sqrt{2} \not\leq 1 = < |A|e_2, e_2 > + < |B|e_2, e_2 > .$$

Solution 6.3.3.

(1) It is plain that as $|\alpha|^2 I$ commutes with A^*A. Hence

$$|\alpha A| = \sqrt{\overline{\alpha} A^* \alpha A} = \sqrt{\overline{\alpha}\alpha I}\sqrt{A^*A} = |\alpha|\ |A|.$$

(2) Since A is self-adjoint, we may write:

$$(A+\alpha I)^*(A+\alpha I) = A^2 + (\overline{\alpha}+\alpha)A + |\alpha|^2 I$$

and

$$(|A| + |\alpha| I)^2 = A^2 + |\alpha|^2 I + 2|\alpha||A|.$$

As $|\text{Re }\alpha| \leq |\alpha|$, Re α is real and the positive $|A|$ commutes with multiples of the identity, then Corollary 6.1.5 and the previous question give

$$|\alpha||A| \geq |\text{Re }\alpha||A| = |(\text{Re }\alpha)A| \geq (\text{Re }\alpha)A.$$

Therefore,

$$(A+\alpha I)^*(A+\alpha I) \leq (|A| + |\alpha| I)^2.$$

Since both sides are positive, passing to positive square roots gives

$$|A+\alpha I| \leq |A| + |\alpha| I,$$

as suggested.

Solution 6.3.4. The proof is straightforward. Recall that

$$|A+B|^2 = (A+B)^*(A+B), \ |A|^2 = A^*A \text{ and } |B|^2 = B^*B.$$

Thereupon,

$$\begin{aligned}|A+B|^2\leq 2(|A|^2+|B|^2) &\iff A^*A+A^*B+B^*A+B^*B\leq 2A^*A+2B^*B\\ &\iff 0\leq A^*A-A^*B-B^*A+B^*B\\ &\iff (A-B)^*(A-B)\geq 0.\end{aligned}$$

Since the last inequality always holds, the desired result follows.

Solution 6.3.5.

(1) It is clear that

$$\begin{aligned}A+B \text{ invertible } &\Longrightarrow |A+B| \text{ invertible}\\ &\Longrightarrow |A+B|^2 \text{ invertible}\\ &\Longrightarrow |A|^2+|B|^2 \text{ invertible (by Theorem 5.1.39 \&}\\ &\qquad\qquad\qquad\text{Exercise 6.3.4),}\end{aligned}$$

as required.

(2) (a) By the previous question, $|A|^2+|B|^2$ is invertible from which we readily get that $|A|^4+|B|^4$ is invertible and, by induction, we establish the invertibility of $|A|^{2^n}+|B|^{2^n}$.

(b) The invertibility of $|A|+|B|$ is not hard to prove. WLOG we may assume that $\|A\|\leq 1$ and $\|B\|\leq 1$. Hence

$$A^*A\leq I \text{ and } B^*B\leq I$$

and so

$$|A|\leq I \text{ and } |B|\leq I$$

by passing to the square root. Accordingly,

$$|A|^2\leq |A| \text{ and } |B|^2\leq |B|.$$

This implies that

$$|A|+|B|\geq |A|^2+|B|^2\geq 0.$$

A glance at Theorem 5.1.39 allows us to confirm the invertibility of $|A|+|B|$, as desired.

Finally, for the general case, assume that $\|A\|\geq\|B\|$ (why not?). As $A+B$ is invertible, then so is $\frac{A}{\|A\|}+\frac{B}{\|A\|}$. But, $\frac{|A|}{\|A\|}$ and $\frac{|B|}{\|A\|}$ have a norm smaller than one. Hence $\frac{|A|}{\|A\|}+\frac{|B|}{\|A\|}$ is invertible, that is, $|A|+|B|$ is invertible.

(3) Since $\alpha A+\beta B$ is invertible, by the previous question, so is $|\alpha||A|+|\beta||B|$ or merely $|\alpha|A+|\beta|B$ as $A,B\geq 0$. Since we can assume $|\alpha|\geq|\beta|>0$ (why?), we infer that

$$|\alpha|(A+B)\geq|\alpha|A+|\beta|B.$$

Consequently, $|\alpha|(A+B)$ or simply $A+B$ is invertible.

(4) Since C is invertible and $C = (C - A) + A$, it follows that $|C - A| + |A|$ too is invertible by the second question.

SOLUTION 6.3.6. It is clear that $A - A^*$ is anti-symmetric, and so $(A - A^*)^2 \leq 0$. Since $AA^* = A^*A$, we have

$$A^2 + A^{*^2} - 2AA^* \leq 0 \text{ or } A^2 + A^{*^2} + 2A^*A \leq 4A^*A,$$

and so $(A + A^*)^2 \leq 4A^*A$. Passing to (positive) square roots yields the desired inequality

$$|\mathrm{Re}A| \leq |A|.$$

A similar idea applies to prove the corresponding inequality for $\mathrm{Im}\, A$: Since $A + A^*$ is self-adjoint, it follows that $(A + A^*)^2 \geq 0$ and we leave details to readers...

SOLUTION 6.3.7.

(1) Let $x \in H$. Since A is self-adjoint, it possesses a polar decomposition, i.e. $A = |A|U = U|A|$, where U is unitary (as a bonus in this case, U is even self-adjoint but this is not needed in the proof). In fact, we can equally use the decomposition of Theorem 6.1.39. Since $|A|$ is positive and $|A|U = A$ is self-adjoint, by the Reid Inequality, we obtain

$$| < Ax, x > | = | < |A|Ux, x > | \leq \|U\| \, | < |A|x, x > | =< |A|x, x >,$$

as desired.

(2) Let $x \in H$. Since $< Ax, x >$ is real, we know that

$$| < Ax, x > | \geq < Ax, x > .$$

By the previous question, we get

$$< |A|x, x > \geq < Ax, x >,$$

that is, $|A| \geq A$. The inequality $|A| \geq -A$ is proved in a similar manner.

SOLUTION 6.3.8. Since $|A| \leq B$, Proposition 6.1.4 gives (for each $x \in H$):

$$| < Ax, x > | \leq < |A|x, x > \leq < Bx, x >$$

and so

$$- < Bx, x > \leq < Ax, x > \leq < Bx, x >,$$

that is, $-B \leq A \leq B$, as desired.

SOLUTION 6.3.9. The hardest part of the proof is the meticulousness! First C and D are self-adjoint. Besides, $CD = DC$ as N is normal. Then

$$|N|^2 = N^*N = (C - iD)(C + iD) = C^2 + D^2 \geq C^2$$

as $D^2 \geq 0$ because D is self-adjoint. By Theorem 5.1.39, we get $|N| \geq |C|$. Hence, by Proposition 6.1.6

$$-|N| \leq C \leq |N|$$

and so

$$|N| - C \geq 0 \text{ and } |N| + C \geq 0.$$

Therefore, it makes sense to define their *positive* square roots. Consider (the self-adjoint!)

$$A = \left(\frac{|N| + C}{2}\right)^{\frac{1}{2}} \text{ and } B = \left(\frac{|N| - C}{2}\right)^{\frac{1}{2}}.$$

Since $|N|$ commutes with C, it follows that A commutes with B. Consequently, the operator $M := A + iB$ is normal. Finally,

(1) If $D \geq 0$, then

$$\begin{aligned} M^2 &= (A+iB)(A+iB) \\ &= A^2 - B^2 + i(AB + BA) \\ &= \frac{|N| + C}{2} - \frac{|N| - C}{2} + 2i\left(\frac{|N|^2 - C^2}{4}\right)^{\frac{1}{2}} \\ &= C + i(D^2)^{\frac{1}{2}} \\ &= C + iD \\ &= N, \end{aligned}$$

that is, M is a normal square root of N.

(2) A similar argument applies when $D \leq 0$. In this case,

$$(M^*)^2 = N,$$

that is, M^* is a normal square root of N. The proof is therefore complete.

Solution 6.3.10. First, recall that A is normal iff $AA^* = A^*A$, that $\text{Re}A = \frac{A+A^*}{2}$ and that $\text{Im}A = \frac{A-A^*}{2i}$.

(1) "1) $\Rightarrow$ 2)": Assume that A is normal. Then $|A^*| = |A|$. So we must only show that $|A|$ commutes with ReA. We may write

$$\begin{aligned} A^*A(A + A^*) &= A^*AA + A^*AA^* \\ &= AA^*A + A^*A^*A \\ &= (A + A^*)A^*A \end{aligned}$$

or

$$A^*A(\text{Re}A) = (\text{Re}A)A^*A$$

and so

$$\sqrt{A^*A}(\text{Re}A) = (\text{Re}A)\sqrt{A^*A}, \text{ that is, } |A|(\text{Re}A) = (\text{Re}A)|A|.$$

(2) "2) $\Rightarrow$ 1)": We have

$$\begin{aligned}|A|\text{Re}A = \text{Re}A|A| &\Longleftrightarrow A^*A(A+A^*) = (A+A^*)A^*A\\ &\Longleftrightarrow A^*AA + A^*AA^* = AA^*A + A^*A^*A\\ &\Longleftrightarrow (A^*A - AA^*)A = A^*(A^*A - AA^*)\\ &\Longleftrightarrow (|A|^2 - |A^*|^2)A = A^*(|A|^2 - |A^*|^2)...(1)\end{aligned}$$

Similarly, we may find that

$$|A^*|\text{Re}A = \text{Re}A|A^*| \Longleftrightarrow (|A|^2 - |A^*|^2)A^* = A(|A|^2 - |A^*|^2)...(2)$$

Multiplying Equation (2) on the left by A^*, and using Equation (1), we get

$$|A|^2(|A|^2 - |A^*|^2) = (|A|^2 - |A^*|^2)|A^*|^2...(3)$$

In a similar way, multiplying Equation (1) on the left by A, and using Equation (2), we then obtain

$$|A^*|^2(|A|^2 - |A^*|^2) = (|A|^2 - |A^*|^2)|A|^2...(4)$$

Subtracting (3) from (4) yields

$$(|A|^2 - |A^*|^2)^2 = -(|A|^2 - |A^*|^2)^2.$$

Hence by Exercise 4.3.3 (and since $|A|^2 - |A^*|^2$ is self-adjoint!), we finally obtain that

$$|A|^2 = |A^*|^2 \text{ or } A^*A = AA^*,$$

i.e. A is normal.

(3) Let $A = S$ be the shift operator. Then

$$|A| = |S| = \sqrt{S^*S} = \sqrt{I} = I$$

and hence it commutes with ReS, but S is not normal.

For the other hypothesis, just consider the non-normal $A = S^*$, where S is still the shift operator.

Solution 6.3.11. Define an operator $S \in B(H \oplus H)$ by:

$$S = \begin{pmatrix} 0 & A^* \\ A & 0 \end{pmatrix}.$$

Then S is clearly self-adjoint. Hence $S^2 \geq 0$, that is,

$$S^2 = \begin{pmatrix} A^*A & 0 \\ 0 & AA^* \end{pmatrix} \geq 0.$$

Since the positive square root of a positive operator is unique, we obtain:

$$|S| = \begin{pmatrix} |A| & 0 \\ 0 & |A^*| \end{pmatrix}.$$

But S is self-adjoint and so Corollary 6.1.5 tells us that $S + |S| \geq 0$. Accordingly,

$$T = S + |S| = \begin{pmatrix} |A| & A^* \\ A & |A^*| \end{pmatrix} \geq 0,$$

as desired.

SOLUTION 6.3.12.

(1) Since $AB = BA$ and A is normal, invoking the Fuglede Theorem (Theorem 4.1.14) we have that $A^*B = BA^*$. We then clearly have from the previous two relations:

$$AB = BA \Longrightarrow A^*AB = A^*BA = BA^*A.$$

Hence

$$|A|B = B|A|.$$

Since $|A|$ is self-adjoint, the previous equality gives (by taking adjoints) $|A|B^* = B^*|A|$. Hence

$$B^*|A|B = |A|B^*B \Longrightarrow B^*B|A| = |A|B^*B \Longrightarrow |B||A| = |A||B|,$$

as required.

(2) Since $AB = BA$ and A is normal, we get $A^*B = BA^*$ or $AB^* = B^*A$. Hence

$$|AB|^2 = (AB)^*AB = B^*A^*AB = A^*B^*AB = A^*AB^*B.$$

By the preceding question, $A^*AB^*B = B^*BA^*A$. Consequently,

$$|AB| = \sqrt{A^*AB^*B} = \sqrt{A^*A}\sqrt{B^*B} = |A||B|,$$

as required.

(3) Since A and B are normal, $|A| = |A^*|$ and $|B| = |B^*|$. As $AB = BA$, then $|A||B| = |B||A|$ for A (or B!) is normal. Now, apply the foregoing question to each of the eight products.

(4) Since $AB = BA$ and B is invertible, we have $AB^{-1} = B^{-1}A$. Question (2) again does the remaining job.

(5) It is clear that

$$I = |AA^{-1}| = |A||A^{-1}|.$$

So, the self-adjoint $|A|$ is right invertible and so it is invertible (why?) and:

$$|A^{-1}| = |A|^{-1}.$$

SOLUTION 6.3.13. Clearly,

$$(A+B)^*(A+B) = A^*A + A^*B + B^*A + B^*B.$$

As $A^*B + B^*A \leq 0$, then

$$A^*A + A^*B + B^*A + B^*B \leq A^*A + B^*B.$$

By Theorem 5.1.39, we have

$$|A+B| = \sqrt{A^*A + A^*B + B^*A + B^*B} \leq \sqrt{A^*A + B^*B}.$$

Since $AB = BA$ and A is normal, Proposition 6.1.13 implies that $|A||B| = |B||A|$ or $|A|^2|B|^2 = |B|^2|A|^2$. Finally, Corollary 5.1.44 gives

$$|A+B| \leq |A| + |B|,$$

and this completes the proof.

SOLUTION 6.3.14. Since A is normal and $AB = BA$, we know from Proposition 6.1.13 that $|A||B| = |B||A|$. Hence

$$|A+B|^2 \leq (|A|+|B|)^2 \Longleftrightarrow (A+B)^*(A+B) \leq A^*A + B^*B + 2\sqrt{A^*A}\sqrt{B^*B}$$

$$\Longleftrightarrow A^*B + B^*A \leq 2\sqrt{A^*A}\sqrt{B^*B}.$$

We already know from above that $\sqrt{A^*A}\sqrt{B^*B} = \sqrt{A^*AB^*B}$. So, to prove the desired triangle inequality, we are only required to prove that

$$A^*B + B^*A \leq 2\sqrt{A^*AB^*B}.$$

But

$$A^*AB^*B = AA^*B^*B = AB^*A^*B = B^*AA^*B.$$

If we set $T = A^*B$, then are done with the proof if we come to show that the following holds:

$$T + T^* \leq 2\sqrt{T^*T}.$$

This is true (see e.g. Exercise 6.3.6) as soon as we show that A^*B is normal. This is in effect the case as A^*B is already known to be normal by Theorem 4.1.16.

Therefore, we have shown that

$$|A+B|^2 \leq (|A|+|B|)^2.$$

Hence, by Theorem 5.1.39 and given the positivity of both $|A+B|$ and $|A|+|B|$, we finally end up with

$$|A+B| \leq |A| + |B|,$$

and this is precisely what we were required to prove.

SOLUTION 6.3.15.

(1) Since A and B are commuting normal operators, we know that $A+B$ too is normal. Since $A+B$ commutes with B, by Corollary 6.1.24 we have

$$|A| = |A+B-B| \leq |A+B| + |B| \Longrightarrow |A| - |B| \leq |A+B|.$$

Similarly, as $A+B$ commutes with A, we get

$$|B| - |A| \leq |A+B|.$$

Whence

$$-|A+B| \leq |A| - |B| \leq |A+B|.$$

By Exercise 5.3.12 (and remembering that $\|T\| = \|\ |T|\ \|$ for $T \in B(H)$), we obtain

$$\||A| - |B|\| \leq \|\ |A+B|\ \| = \|A+B\|,$$

as required.

(2) This inequality follows from the previous one by observing that $A(-B) = (-B)A$ and the fact that $(-B)$ remains normal...

(3) We easily see as $|A||B| = |B||A|$ that

$$\begin{aligned}||A|-|B||^2 \leq |A-B|^2 &\Longleftrightarrow |A|^2+|B|^2-2|A||B| \leq |A|^2+|B|^2-A^*B-B^*A\\ &\Longleftrightarrow A^*B+B^*A \leq 2\sqrt{A^*A}\sqrt{B^*B}\\ &\Longleftrightarrow A^*B+B^*A \leq 2\sqrt{B^*AA^*B}.\end{aligned}$$

But, this is always true (just as in Exercise 6.3.14!). Therefore, we have shown

$$||A| - |B||^2 \leq |A-B|^2.$$

A glance at Theorem 5.1.39 finally gives

$$||A| - |B|| \leq |A-B|.$$

SOLUTION 6.3.16. The inequality is evident when $K = 0$. So, assume that $K \neq 0$. It is then clear that $\frac{K}{\|K\|}$ satisfies

$$KK^* \leq \|K\|^2 I.$$

Ergo, $AKK^*A \leq \|K\|^2A^2$. But AK is normal, that is, $|AK| = |(AK)^*|$. Thereupon,

$$0 \leq \frac{|AK|^2}{\|K\|^2} = \frac{|(AK)^*|^2}{\|K\|^2} = \frac{AKK^*A}{\|K\|^2} \leq A^2.$$

Passing to (positive) square roots yields

$$|AK| \leq \|K\|A.$$

The previous, with a glance at Proposition 6.1.7, allows us to write for each $x \in H$

$$|<AKx, x>| \leq <|AK|x, x> \leq \|K\| <Ax, x>,$$

which is nothing but the required result.

Solution 6.3.17. Since A is normal, A and A^* commute. Hence so do $\text{Re}A$ and $|A|$. Now, the self-adjointness of $\text{Re}A$ implies that $|A|$ commutes with $|\text{Re}A|$, i.e.

$$|A||\text{Re}A| = |\text{Re}A||A|.$$

Since $|A| \leq |\text{Re}A|$ (and $|A| \geq 0$!), the previous equation implies that

$$|A|^2 \leq |A||\text{Re}A| \text{ and } |A||\text{Re}A| \leq |\text{Re}A|^2 = (\text{Re}A)^2$$

or just

$$|A|^2 \leq (\text{Re}A)^2.$$

The self-adjointness of A finally follows from Theorem 6.1.29, and this completes the proof.

Solution 6.3.18. First, recall that

$$J(A) = \begin{pmatrix} (I - |A^*|^2)^{\frac{1}{2}} & A \\ -A^* & (I - |A|^2)^{\frac{1}{2}} \end{pmatrix}$$

where

$$(I - |A^*|^2)^{\frac{1}{2}} = (I - AA^*)^{\frac{1}{2}} \text{ and } (I - |A|^2)^{\frac{1}{2}} = (I - A^*A)^{\frac{1}{2}}$$

which are self-adjoint (in fact positive!).

Hence

$$J(A)^*J(A) = \begin{pmatrix} (I-|A^*|^2)^{\frac{1}{2}} & -A \\ A^* & (I-|A|^2)^{\frac{1}{2}} \end{pmatrix}\begin{pmatrix} (I-|A^*|^2)^{\frac{1}{2}} & A \\ -A^* & (I-|A|^2)^{\frac{1}{2}} \end{pmatrix}$$
$$= \begin{pmatrix} I & (I-|A^*|^2)^{\frac{1}{2}}A - A(I-|A|^2)^{\frac{1}{2}} \\ A^*(I-|A^*|^2)^{\frac{1}{2}} - (I-|A|^2)^{\frac{1}{2}}A^* & I \end{pmatrix}.$$

We then must show that the two off-diagonal operators are zero. It is clear that

$$(I - AA^*)A = A(I - A^*A).$$

Since $(I - AA^*)$ and $(I - A^*A)$ are positive, Corollary 5.1.31 yields

$$(I - AA^*)^{\frac{1}{2}}A = A(I - A^*A)^{\frac{1}{2}}.$$

By taking adjoints

$$A^*(I - AA^*)^{\frac{1}{2}} = (I - A^*A)^{\frac{1}{2}}A^*.$$

Therefore,

$$J(A)^*J(A) = \begin{pmatrix} I & 0 \\ 0 & I \end{pmatrix}.$$

Similarly, we may prove that

$$J(A)J(A)^* = \begin{pmatrix} I & 0 \\ 0 & I \end{pmatrix},$$

which therefore means that $J(A)$ is unitary.

Solution 6.3.19. Let S be the shift operator. Assume that $S = UP$, with U unitary and P positive. Then

$$S^* = (UP)^* = P^*U^* = PU^*.$$

From here

$$S^*S = PU^*UP = P^2.$$

Now, remember that $S^*S = I$, whence $P^2 = I$. But P is positive and so is I. Thus $P = I$ which would imply that $S = U$, i.e. S would be unitary and this is the contradiction we have been after.

Solution 6.3.20. Since A is invertible, so are A^* and A^*A. Since A^*A is positive, it has a unique positive square root (this is just $|A|$), which is again invertible. Put $U = A|A|^{-1}$. Hence U is invertible and (since $|A|$ is self-adjoint)

$$U^*U = (A|A|^{-1})^*A|A|^{-1} = |A|^{-1}A^*A|A|^{-1} = |A|^{-1}|A|^2|A|^{-1} = I,$$

i.e. U is also an isometry. Thus U is unitary (why?).

Solution 6.3.21. Basically, we are just asked to find the polar decomposition of A. It is clear that A is invertible. By the constructive proof of Theorem 6.1.32 (which may be found in Exercise 6.3.20), that is, by the solution of Exercise 6.3.20, we know how to do it. Recall that we want to write A as $U|A|$.

First, we need to find $|A|$. We have

$$A^* = \begin{pmatrix} -1 & 2 \\ -2 & 1 \end{pmatrix}$$

Hence

$$A^*A = \begin{pmatrix} 5 & 4 \\ 4 & 5 \end{pmatrix}.$$

Then, we observe (otherwise diagonalize A^*A!) that

$$|A| = \sqrt{A^*A} = \begin{pmatrix} 2 & 1 \\ 1 & 2 \end{pmatrix}.$$

Now, U is given by $A|A|^{-1}$. Since

$$|A|^{-1} = \frac{1}{3}\begin{pmatrix} 2 & -1 \\ -1 & 2 \end{pmatrix},$$

we obtain

$$U = A|A|^{-1} = \begin{pmatrix} 0 & -1 \\ 1 & 0 \end{pmatrix}.$$

Thus the polar decomposition of A is given by:

$$A = U|A| = \begin{pmatrix} 0 & -1 \\ 1 & 0 \end{pmatrix}\begin{pmatrix} 2 & 1 \\ 1 & 2 \end{pmatrix}.$$

SOLUTION 6.3.22.

(1) If S is the shift operator, then $S^*S = I$. Hence clearly $SS^*S = S$ and $S^*SS^* = S^*$, that is, S has a generalized inverse.
    We only needed the fact that $S^*S = I$. Accordingly, if A is *any* isometry, then it always has a generalized inverse.

(2) If A is invertible with inverse B, then $BA = I$ and $AB = I$. Left multiplying the first equation by A and the second one by B yields the desired result.

(3) (a) $(a) \Rightarrow (b)$: If A is a partial isometry, then it is a contraction. Hence by Proposition 6.1.38, $AA^*A = A$ and $A^*AA^* = A^*$, that is, A has a generalized inverse, which is just A^* and it is a contraction.

(b) $(b) \Rightarrow (a)$: First, observe that if $A = 0$, then the result trivially holds. Let us then assume that $A \neq 0$, and also $B \neq 0$ (why?).
Since A has a generalized inverse, there exists a $B \in B(H)$ such that $ABA = A$ and $BAB = B$. Since $\|A\| \leq 1$, we have

$$\|A\| = \|ABA\| \leq \|AB\| \leq \|A\|\|B\|$$

which gives $\|A\| \geq 1$, and so $\|A\| = 1$. Similarly, we obtain $\|B\| = 1$. By assumption again, we have for all $x \in H$,

$$\|Bx\| = \|BABx\| \leq \|ABx\| \leq \|Bx\|$$

leading to

$$\|Bx\| = \|ABx\|$$

for any $x \in H$. Hence for all $x \in H$,

$$
\begin{aligned}
< (I - A^*A)Bx, Bx > &=< Bx, Bx > - < A^*ABx, Bx > \\
&= \|Bx\|^2 - < ABx, ABx > \\
&= \|Bx\|^2 - \|ABx\|^2 \\
&= 0.
\end{aligned}
$$

Since A is a contraction, $I - A^*A$ is positive and so it possesses a unique positive square root, denoted by $(I - A^*A)^{\frac{1}{2}}$. We then have, for every $x \in H$,

$$\|(I - A^*A)^{\frac{1}{2}}Bx\|^2 =< (I - A^*A)^{\frac{1}{2}}Bx, (I - A^*A)^{\frac{1}{2}}Bx >$$

and hence

$$\|(I - A^*A)^{\frac{1}{2}}Bx\|^2 =< (I - A^*A)Bx, Bx >= 0.$$

Accordingly, we have $(I - A^*A)^{\frac{1}{2}}Bx = 0$ for all x, that is, $(I - A^*A)^{\frac{1}{2}}B = 0$ or simply $(I - A^*A)B = 0$. Thus $B = A^*AB$ so that finally we have

$$A = ABA = AA^*ABA = AA^*A,$$

as required (using Proposition 6.1.38).

Solution 6.3.23. By definition V is an isometry means that $V^*V = I$. Hence clearly, V^* is surjective. Now, recall that V^* is a partial isometry if

$$\|V^*x\| = \|x\|, \ \forall x \in (\ker V^*)^\perp.$$

We know that $(\ker V^*)^\perp = \overline{\text{ran}V}$, and since V is an isometry, $\text{ran}V$ is closed (cf. Exercise 2.3.12). Hence $(\ker V^*)^\perp = \text{ran}V$.

To show that V^* is a partial isometry, let $x \in (\ker V^*)^\perp = \text{ran}V$. Then $x = Vy$ for some $y \in H$ so that

$$\|V^*x\| = \|V^*Vy\| = \|Iy\| = \|y\| = \|Vy\| = \|x\|,$$

establishing the result.

Solution 6.3.24. First, observe that for all $x \in H$

$$\||A|x\|^2 =< |A|^2x, x >=< A^*Ax, x >=< Ax, Ax >= \|Ax\|^2 ...(1).$$

Hence $\ker A = \ker |A|$. Now, define $V : \text{ran}|A| \subset H \to \text{ran}A$ by

$$V(|A|x) = Ax$$

for all $x \in H$. First, V is well-defined as whenever $|A|x = |A|y$, then clearly $Ax = Ay$ by (1). The linearity of V is left to readers.

Now, for all x, we have

$$\|V(|A|x)\| = \|Ax\| = \||A|x\|,$$

that is, V is an isometry from $\operatorname{ran}|A|$ onto $\operatorname{ran}A$. Therefore, we can extend V to an isometry from $\overline{\operatorname{ran}|A|}$ onto $\overline{\operatorname{ran}A}$ (by Corollary 2.1.105). We can in the end extend V to all of H by letting $V = 0$ on $(\operatorname{ran}|A|)^\perp$.

Let us show now that $\ker V = (\operatorname{ran}|A|)^\perp$.

(1) Let $x \in (\operatorname{ran}|A|)^\perp$, then $Vx = 0$ (by definition). This shows that $x \in \ker V$ and so $(\operatorname{ran}|A|)^\perp \subset \ker V$.

(2) Conversely, let $x \in \ker V$. Hence $x \in H$ and $Vx = 0$. We can also uniquely express $x = y + z$ where $y \in \overline{\operatorname{ran}|A|}$ and $z \in (\operatorname{ran}|A|)^\perp$. Therefore,

$$0 = \|Vx\| = \|Vy\| = \|y\|$$

leading to $y = 0$, i.e. $x = z \in (\operatorname{ran}|A|)^\perp$. Thus, $\ker V \subset (\operatorname{ran}|A|)^\perp$.

Therefore, V is a partial isometry because it is an isometry on $(\ker V)^\perp = \overline{\operatorname{ran}|A|}$.

As for the other identity, remember that as V is an isometry on $\operatorname{ran}|A|$, then $V^*Vy = y$ for any $y \in \operatorname{ran}|A|$. Hence

$$V^*Ax = V^*V|A|x = |A|x$$

for every $x \in H$, that is, $V^*A = |A|$ as needed.

Finally, as $|A|$ is self-adjoint, then

$$\ker V = (\operatorname{ran}|A|)^\perp = \ker |A| = \ker A.$$

Solution 6.3.25. By Proposition 6.1.41, we know that if A is normal, then $A = U|A|$ where U is unitary.

By the normality of A and the self-adjointness of $|A|$, we have

$$AA^* = A^*A \Longleftrightarrow U|A|(U|A|)^* = (U|A|)^*U|A| \Longleftrightarrow U|A|^2U^* = |A|^2$$

or $U|A|^2 = |A|^2U$. Since $|A| \geq 0$, Theorem 5.1.28 yields

$$(A =)\ U|A| = |A|U,$$

as needed.

Solution 6.3.26. Consider a right invertible normal operator A (the proof is very much alike in the case of left invertibility), i.e. $AB = I$. Since A is normal, its polar decomposition is

$$A = UP = PU,$$

where U is unitary and P is positive. Hence $A^* = U^*P = PU^*$.

Now, clearly

$$AB = I \Longrightarrow B^*A^* = I$$

so that

$$B^*PU^* = I \Longrightarrow B^*P = U \Longrightarrow U^*B^*P = I.$$

This means that (*the self-adjoint*) P is left invertible. Exercise 4.3.24 then tells us that P is invertible. Since U is unitary, $A = UP$ is invertible.

6.4. Hints/Answers to Tests

SOLUTION 6.4.1. There are at least two ways of seeing this:

(1) Use Corollary 5.1.24 by remembering that $A + A^*$ is self-adjoint...
(2) Alternatively, since A is normal, we know that $|\text{Re}A| \leq |A|$. In view of the invertibility of $A+A^*$ or $\text{Re}A$, it follows that $|\text{Re}A|$ is invertible. On that account, $|A|$ too is invertible (why?) and calling on again the normality of A, we finally infer that A is invertible.

SOLUTION 6.4.2.

(1) If $A_1, A_2, \cdots, A_{n-1}$ are the normal factors, then using a proof by induction we can show that their product is normal as they are already commuting (cf. Theorem 4.1.16)... Then use Theorem 6.1.16... Otherwise, just use commutativity to push the non normal factor to the right as many times as needed until it will be the last factor on the right of the product "$\prod_{i=1}^{n} A_i$". Then proceed as just indicated three lines above...
(2) The case $n \geq 0$ follows from the previous question. The case $n < 0$ follows also from the previous question and Corollary 6.1.19 as well...

SOLUTION 6.4.3. The answer is no! If such A existed, then we would also have $A^{*2} = 0$. Hence we would get

$$A^2 + A^{*2} + AA^* + A^*A \geq 4A^*A$$

or

$$(\text{Re}A)^2 \geq |A|^2.$$

Accordingly, A must be self-adjoint (have a look at Theorem 6.1.29!)...

SOLUTION 6.4.4. Write $A = |A|U = U|A|$ where $U \in B(H)$ is unitary. Hence $A^* = |A|U^* = U^*|A|$ and so

$$\text{Re}A = |A|(\text{Re}U) = (\text{Re}U)|A|.$$

Now, since $(U - U^*)^2 \leq 0$, clearly we have

$$(\text{Re}U)^2 = \frac{U^2 + (U^*)^2 + 2I}{4} \leq I.$$

Ergo,

$$|A|(\text{Re}U)^2|A| \leq |A|^2.$$

Thereupon (using Theorem 5.1.39)

$$|\text{Re}A| = \sqrt{|A|(\text{Re}U)^2|A|} \leq |A|.$$

The proof for the second inequality is left to readers...

Solution 6.4.5. Let $x \in H$ be such that $\|x\| = 1$. By the Cauchy-Schwarz Inequality,

$$\sup_{\|x\|=1} < |A|x, x > \leq \||A|\| = \|A\|...$$

By Proposition 6.1.3, we obtain

$$\sup_{\|x\|=1} \|Ax\|^2 \leq \|A\| \sup_{\|x\|=1} < |A|x, x >$$

and so

$$\|A\| \leq \sup_{\|x\|=1} < |A|x, x >,$$

as needed.

Solution 6.4.6. We have

$$A + B = AB \iff (A - I)(B - I) = I.$$

The normal $A - I$ is right invertible and so it is invertible by Exercise 6.3.26. Therefore,

$$(A - I)(B - I) = (B - I)(A - I)$$

from which we get $AB = BA$.

Solution 6.4.7. Let $x \in H$. Then (using the Cauchy-Schwarz Inequality and simple properties)

$$\begin{aligned} < ABx, x > &= | < ABx, x > | \\ &= | < Bx, A^*x > | \\ &\leq \|A^*x\|\|Bx\| \\ &\leq \frac{1}{2}(\|A^*x\|^2 + \|Bx\|^2) \\ &= \frac{1}{2}(< AA^*x, x > + < B^*Bx, x >) \\ &=< \frac{1}{2}(AA^* + B^*B)x, x >, \end{aligned}$$

that is,

$$AB \leq \frac{1}{2}(AA^* + B^*B) = \frac{1}{2}(|A^*|^2 + |B|^2).$$

A quick look at Theorem 5.1.39 allows us to establish the invertibility of $|A^*|^2 + |B|^2$, as required.

Solution 6.4.8. By Theorem 5.1.39, it suffices to show

$$\left|\sum_{k=1}^{n} a_k A_k\right|_{B(H)}^2 \leq \sum_{k=1}^{n} |a_k|_{\mathbb{C}}^2 \sum_{k=1}^{n} |A_k|_{B(H)}^2$$

or equivalently

$$\left(\sum_{k=1}^{n} a_k A_k\right)^* \left(\sum_{k=1}^{n} a_k A_k\right) \leq \left(\sum_{k=1}^{n} |a_k|^2\right) \left(\sum_{k=1}^{n} A_k^* A_k\right).$$

...In the end, we find that the previous inequality is also equivalent to

$$\left\|\sum_{k=1}^{n} a_k A_k x\right\|^2 \leq \left(\sum_{k=1}^{n} |a_k|^2\right) \left(\sum_{k=1}^{n} \|A_k x\|^2\right)$$

which clearly holds for all $x \in H$ (why?)...

CHAPTER 7

Spectrum of an Operator

7.2. True or False: Answers

ANSWERS.

(1) Let A be such that $\sigma(A) = [0,1]$ (for instance A can be the multiplication operator by x on $L^2[0,1]$). Taking $B = -A$ yields

$$\sigma(A+B) = \sigma(A-A) = \{0\} \neq \sigma(A) + \sigma(B)$$

since $1 \in \sigma(A) + \sigma(B)$.

(2) True. Indeed, since $\sigma(A)$ and $\sigma(B)$ are both compact, then $\sigma(A) + \sigma(B)$ is closed. In fact, less than that suffices by Exercise 1.3.12.

REMARK. Exercise 1.3.12 says that $\sigma(A) + \sigma(B)$ is closed even when one of the spectra is only closed as in the case of a closed operator (see Chapter 10).

(3) The implication "$\Rightarrow$" is correct. A proof is given in Exercise 7.3.24.

As for "$\Leftarrow$", it is untrue. For a counterexample, take

$$A = \begin{pmatrix} 0 & 2018 \\ 0 & 1 \end{pmatrix}.$$

Then A is not an orthogonal projection as $A \neq A^*$. On the other hand, it is immediate that

$$\sigma(A) = \sigma_p(A) = \{0, 1\}.$$

But, if we assume further that A is self-adjoint, then the result is true. See Exercise 8.3.6. In fact, just a normal A will do! This will be proved in Exercise 11.3.28.

(4) False in general! Needless to repeat that the result is true in the finite dimensional setting since in such case $\sigma_p(A) = \sigma(A)$ and hence just apply Proposition 7.1.6.

If, however $\dim H = \infty$, then the result is not necessarily true anymore. Indeed, let S be the shift operator on ℓ^2. Then we will see in Exercise 7.3.14 that $\sigma_p(S) = \varnothing$ and $\sigma_p(S^*) = \{\lambda \in \mathbb{C} : |\lambda| < 1\}$. So

$$\sigma_p(S^*) \neq \{\overline{\lambda} : \lambda \in \sigma_p(S)\} = \varnothing.$$

Remark. It is easy to see that if $A \in B(H)$ is normal, then $\lambda \in \sigma_p(A) \Leftrightarrow \overline{\lambda} \in \sigma_p(A^*)$.

(5) Exercise 7.3.8 shows that "$\Rightarrow$" holds. The implication "$\Leftarrow$" does not hold in general. For example, if $\dim H < \infty$, then $\sigma_r(A) = \varnothing$ ***for any*** $A \in B(H)$.
(6) The proof is false (*but the result is correct*!). The reason is that $(\sigma(P))^2 = \sigma(P)$ does not necessarily imply that $\sigma(P) = \{0, 1\}$. Indeed, $[0, 1]$ also satisfies $[0, 1]^2 = [0, 1]$. A correct proof may be found in Exercise 7.3.24.
(7) The answer is no! In $AX - XB = C$, set $A = S^*$ (the shift's adjoint) and $B = 0$. The Sylvester Equation then becomes: $S^*X = C$. This equation has always a solution X whichever $C \in B(\ell^2)$. It is given by $X = SC$ as

$$S^*X = S^*SC = C,$$

yet

$$\sigma(B) = \{0\} \subset \sigma(S^*) = \{\lambda \in \mathbb{C} : |\lambda| \leq 1\}$$

(as it will be shown in Exercise 7.3.14).

7.3. Solutions to Exercises

Solution 7.3.1. We show that by assuming $\sigma(A) = \varnothing$ (hence $\rho(A) = \mathbb{C}$), we will get contradictory results.

We already know that $R(\lambda, A)$ is analytic at each point of $\rho(A)$ (Proposition 7.1.18). Since $\rho(A) = \mathbb{C}$, $R(\lambda, A)$ becomes entire. A glance at Proposition 7.1.15 yields

$$\lim_{|\lambda| \to \infty} R(\lambda, A) = 0.$$

This means that $R(\cdot, A)$ is bounded at infinity. Away from infinity, $R(\lambda, A)$ is bounded too. Hence $\lambda \mapsto R(\lambda, A)$ is a bounded entire function. The Liouville Theorem (Theorem 2.1.98) then implies that $\lambda \mapsto R(\lambda, A)$ must be a constant function. Since $\lim_{|\lambda| \to \infty} R(\lambda, A) = 0$, we infer that $R(\lambda, A) = 0$ something impossible to happen unless $H = \{0\}$! Thus $\sigma(A) \neq \varnothing$.

Solution 7.3.2.

(1) If $|\lambda| > \|A\|$, then it is clear that $\lambda \neq 0$ so that $\|\lambda^{-1}A\| < 1$. By Theorem 2.1.88, we know that $I - \lambda^{-1}A$ is invertible. Hence $\lambda I - A$ too is invertible, that is, $\lambda \notin \sigma(A)$.

(2) Define a function $f : \mathbb{C} \to B(H)$ by

$$f(\lambda) = \lambda I - A.$$

Then for all $\lambda, \mu \in \mathbb{C}$, we have

$$\|f(\lambda) - f(\mu)\| = \|\lambda I - A - (\mu I - A)\| = \|(\lambda - \mu)I\| = |\lambda - \mu|$$

and hence f is an isometry, implying that it is continuous. Since the set of invertible operators, denoted by $B_i(H)$, is open in $B(H)$ (see Exercise 2.3.39), we clearly have

$$\begin{aligned}\sigma(A) &= \{\lambda \in \mathbb{C} : \ f(\lambda) \text{ is not invertible}\} \\ &= \{\lambda \in \mathbb{C} : \ f(\lambda) \notin B_i(H)\} \\ &= \{\lambda \in \mathbb{C} : \ f(\lambda) \in [B_i(H)]^c\} \\ &= f^{-1}([B_i(H)]^c),\end{aligned}$$

so that $\sigma(A)$ is closed in $\mathbb{C}$ because f is continuous and $[B_i(H)]^c$ is closed in $B(H)$.

By Question 1), $\sigma(A) \subset B(0, \|A\|)$ (the closed unit ball in usual $\mathbb{C}$), i.e. $\sigma(A)$ is bounded. By Question 2), it is closed so it is a compact set in $\mathbb{C}$.

SOLUTION 7.3.3.

(1) Let $U \in B(H)$ be unitary. It is clear that

$$||\lambda| - |1|| \, \|x\| = |\|\lambda x\| - \|x\|| = |\|\lambda x\| - \|Ux\|| \leq \|(\lambda I - U)x\|.$$

So if $|\lambda| \neq 1$, then $\lambda I - U$ is bounded below. Since U is unitary, it is normal and so $\lambda I - U$ too is normal. Thus, by Corollary 5.1.24, $\lambda I - U$ is invertible, i.e. $\lambda \in \rho(U)$. Therefore, we have just proved that

$$\sigma(U) \subset \{\lambda \in \mathbb{C} : \ |\lambda| = 1\}.$$

(2) Since $A \in B(H)$ is self-adjoint, $< Ax, x >=< x, Ax > \in \mathbb{R}$ for all $x \in H$. Hence if $\alpha, \beta \in \mathbb{R}$, then

$$< (\alpha I - A)x, \beta x >=< \beta x, (\alpha I - A)x > \in \mathbb{R}$$

so that

$$\text{Re} < i\beta x, (\alpha I - A)x >= 0.$$

Now, let $\lambda = \alpha + i\beta$ and $x \in H$. Then

$$\begin{aligned}
\|(\lambda I - A)x\|^2 &= \|i\beta x + (\alpha I - A)x\|^2 \\
&= |\beta|^2\|x\|^2 + 2\underbrace{\operatorname{Re} < i\beta x, (\alpha I - A)x >}_{0} + \|(\alpha I - A)x\|^2 \\
&= |\beta|^2\|x\|^2 + \|(\alpha I - A)x\|^2 \\
&\geq |\beta|^2\|x\|^2.
\end{aligned}$$

So if $\lambda \notin \mathbb{R}$ (that is, $\beta \neq 0$), then $\lambda I - A$ is clearly bounded below. Since A is self-adjoint and λ is complex, $\lambda I - A$ is normal. So by calling on Corollary 5.1.24, we know that $\lambda I - A$ is invertible, i.e. $\lambda \in \rho(A)$. Thus, we have just proved that

$$\sigma(A) \subset \mathbb{R}.$$

Remark. ([**156**]) Another way of proving that the spectrum of a self-adjoint is real is as follows: By Exercises 6.3.6 & 6.3.14, it is easy to see that if $T = A + iB$ is normal, then T is invertible iff $|A| + |B|$ is invertible. Now, let $\lambda \notin \mathbb{R}$, i.e. $\lambda = \alpha + i\beta$ ($\alpha \in \mathbb{R}$, $\beta \in \mathbb{R}^*$). Since $A - \alpha I$ is self-adjoint, it follows that $A - \alpha I - i\beta I$ is normal. By the invertibility of $|\beta| I$, it follows that of $|A - \alpha I| + |\beta I|$ (by Theorem 5.1.39). This just means that $A - \lambda I$ is invertible, that is, $\lambda \notin \sigma(A)$.

(3) Let $\lambda > 0$. Since $A \geq 0$, so is $A + \lambda I$. Let $x \in H$. Then

$$\lambda\|x\|^2 =< \lambda x, x > \leq < (A + \lambda I)x, x > \leq \|(A + \lambda I)x\|\|x\|$$

by the Cauchy-Schwarz Inequality. If $x \neq 0$, then

$$\|(A + \lambda I)x\| \geq \lambda\|x\|.$$

Since the previous inequality holds for $x = 0$ as well, we obtain for *all* $x \in H$

$$\|(A + \lambda I)x\| \geq \lambda\|x\|.$$

Hence $A + \lambda I$ is bounded below. Since it is also self-adjoint, Corollary 5.1.24 implies that $A + \lambda I$ is invertible. Whence $-\lambda \notin \sigma(A)$.

Therefore, we have proved that if $-\lambda \in \sigma(A)$, then $\lambda \leq 0$, or equivalently, we have just proved that $\lambda \in \sigma(A)$ implies $\lambda \geq 0$, which is the desired result.

Remark. A much simpler way is as follows: If $\lambda < 0$, then $-\lambda I > 0$ and so $A - \lambda I \geq -\lambda I$ because $A \geq 0$. Hence, by Theorem 5.1.39 $A - \lambda I$ is invertible as $-\lambda I$ is, i.e. $\lambda \notin \sigma(A)$.

SOLUTION 7.3.4. Since (λ_n) converges, it is Cauchy. By the resolvent equation (see Proposition 7.1.16), we have

$$R(\lambda_n, A) - R(\lambda_m, A) = (\lambda_m - \lambda_n)R(\lambda_n, A)R(\lambda_m, A).$$

By assumption, $(R(\lambda_n, A))_n$ is bounded, i.e.

$$\exists M > 0, \ \|R(\lambda_n, A)\| \le M, \ \forall n \in \mathbb{N}.$$

Therefore,

$$\|R(\lambda_n, A) - R(\lambda_m, A)\| \le M^2|\lambda_m - \lambda_n|$$

so that $(R(\lambda_n, A))_n$ is Cauchy because (λ_n) is Cauchy.

Since $B(H)$ is complete, $(R(\lambda_n, A))_n$ converges (with respect to the operator norm) to some $B \in B(H)$. But, for all n we have

$$R(\lambda_n, A)(\lambda_n I - A) = (\lambda_n I - A)R(\lambda_n, A) = I.$$

Passing to the limit when n goes to infinity yields (and using Exercise 2.3.4)

$$B(\lambda I - A) = (\lambda I - A)B = I,$$

i.e. $\lambda \in \rho(A)$, quod erat demonstrandum.

SOLUTION 7.3.5.

(1) We claim that the eigenvalues are the λ_n themselves. If λ is an eigenvalue for A, then for some non zero vector $x = (x_n)$ in ℓ^2 we have

$$\forall n \in \mathbb{N} : \ (\lambda_n x_n) = (\lambda x_n).$$

Since x is not zero, we automatically have $\lambda = \lambda_n$. Thus the only possible eigenvalues are the λ_n. On the other hand, a vector e_n (from the orthonormal basis as above) is an eigenvector for λ_n as

$$A(e_n) = (\lambda_n e_n).$$

Therefore,

$$\sigma_p(A) = \{\lambda_n : \ n \in \mathbb{N}\}.$$

We now pass to $\sigma(A)$. We know that $\sigma_p(A) \subset \sigma(A)$ so that

$$\overline{\sigma_p(A)} \subset \overline{\sigma(A)} = \sigma(A)$$

because $\sigma(A)$ is closed. Let us show the reverse inclusion. Let $\lambda \notin \overline{\sigma_p(A)}$. Hence

$$d := d(\lambda, \sigma_p(A)) = \inf_{n \in \mathbb{N}} |\lambda_n - \lambda| > 0$$

(remember that: $x \in \overline{Y} \Leftrightarrow \inf_{y\in Y} d(x,y) = d(x,Y) = 0$, in some metric space (X,d) where $Y \subset X$). Therefore, it becomes clear that $A - \lambda I$ has a bounded inverse, defined on $\ell^2(\mathbb{N})$, and given by

$$(A-\lambda I)^{-1}(x_1, x_2, \cdots, x_n, \cdots) = \left(\frac{1}{\lambda_1-\lambda}x_1, \frac{1}{\lambda_2-\lambda}x_2, \cdots, \frac{1}{\lambda_n-\lambda}x_n, \cdots\right),$$

which just means that $\lambda \notin \sigma(A)$, i.e. we have proved $\overline{\sigma_p(A)} \supset \sigma(A)$. Accordingly

$$\sigma(A) = \overline{\sigma_p(A)} = \overline{\{\lambda_n : n \in \mathbb{N}\}}.$$

(2) Since A is normal, by Proposition 7.1.23 we have $\sigma_r(A) = \varnothing$. Consequently, $\sigma_c(A) = \overline{\sigma_p(A)} - \sigma_p(A)$.

Solution 7.3.6. Since K is separable, it contains a countable dense subset, which we denote by $\{\lambda_n\}_n$, i.e. $K = \overline{\{\lambda_n\}_n}$. Define an A on ℓ^2 (as $((\lambda_n)_n)$ is bounded) by

$$A(x_1, x_2, \cdots) = (\lambda_1 x_1, \lambda_2 x_2, \cdots).$$

Then from Exercise 7.3.5, A is bounded and that $\{\lambda_n\}_n$ is the set of eigenvalues. Besides, it was shown that $\sigma(A) = \overline{\sigma_p(A)}$. Thus

$$\sigma(A) = \overline{\sigma_p(A)} = \overline{\{\lambda_n\}_n} = K.$$

Solution 7.3.7.

(1) The statement follows from the fact that we always have

$$\|A + I\| \le \|A\| + \|I\| = \|A\| + 1.$$

(2) It is clear that $A = I$ does satisfy D.E.

Let $B = -I$. Then $\|B\| = \|-I\| = 1$. Hence

$$\|B + I\| = \|-I + I\| = \|0\| = 0 \neq \|B\| + 1 = 2,$$

i.e. B does not satisfy D.E.

(3) We have

$$\begin{aligned} A \text{ satisfies D.E. } &\Longleftrightarrow \|A\| + 1 = \|A + I\| \\ &\Longleftrightarrow \|A^*\| + 1 = \|(A + I)^*\| \\ &\Longleftrightarrow \|A^*\| + 1 = \|A^* + I\| \\ &\Longleftrightarrow A^* \text{ satisfies D.E.} \end{aligned}$$

(4) Let $\|A\| \in \sigma(A)$. Since $\sigma(A) \subset B_c(0, \|A\|)$, it follows that $\|A\| \in \partial\sigma(A)$ so that $\|A\| \in \sigma_a(A)$ (cf. Exercise 7.3.10), meaning that

$$\|Ax_n - \|A\|x_n\| \longrightarrow 0$$

(as $n \to \infty$) for some unit vector (x_n) in H. Hence

$$\begin{aligned}\|A+I\| &= \sup_{\|x\|=1} \|(A+I)x\| \\ &\geq \|(A+I)x_n\| \\ &= \|Ax_n + x_n + \|A\|x_n - \|A\|x_n\| \\ &= \|x_n + \|A\|x_n - (\|A\|x_n - Ax_n)\| \\ &\geq \|x_n + \|A\|x_n\| - \|\|A\|x_n - Ax_n\| \text{ (cf. Exercise 1.3.2).}\end{aligned}$$

But

$$\|x_n+\|A\|x_n\|^2 =< x_n+\|A\|x_n, x_n+\|A\|x_n >= (1+2\|A\|+\|A\|^2)\|x_n\|^2$$

and hence

$$\|x_n + \|A\|x_n\| = 1 + \|A\|.$$

Therefore,

$$\|A+I\| \geq 1 + \|A\| - \|\|A\|x_n - Ax_n\|,$$

which, upon passing to the limit as n approaches infinity, gives

$$\|A+I\| \geq 1 + \|A\|,$$

that is, A satisfies D.E.

(5) Let A be such that $\|A+I\| = \|A\| + 1$. By Exercise 1.3.6 we may then write

$$\|(\alpha A) + I\| = \alpha\|A\| + 1 = \|\alpha A\| + 1,$$

i.e. αA also satisfies D.E.

Solution 7.3.8. Let A be a normal operator. We need to show that $\sigma_r(A) = \varnothing$. Let $\lambda \in \sigma(A)$. Since A is normal, so is $A - \lambda I$ for any λ (not only for those in $\sigma(A)$!). Two cases are to be discussed:

(1) If $A - \lambda I$ is not injective, then λ is an eigenvalue for A, i.e. $\lambda \in \sigma_p(A)$.
(2) If $A - \lambda I$ is injective, then since $A - \lambda I$ is normal,

$$\{0\} = \ker(A - \lambda I) = \ker(A - \lambda I)^* = (\operatorname{ran}(A - \lambda I))^\perp,$$

i.e. $A - \lambda I$ has a dense image and this precisely means that $\lambda \in \sigma_c(A)$. Therefore, there is no room for λ to be somewhere else.

Thus in both cases:

$$\sigma_r(A) = \varnothing.$$

SOLUTION 7.3.9. Let $\lambda \in \sigma_c(A)$. Then $\lambda I - A$ is injective and $\overline{\operatorname{ran}(\lambda I - A)}$ is dense without being closed. Hence, by Theorem 3.1.66, $\operatorname{ran}(\overline{\lambda} I - A^*) = H$ and $\ker(\overline{\lambda} I - A^*) = \{0\}$.

Now, as $\operatorname{ran}(\lambda I - A)$ is not closed, then we know, by Exercise 3.3.48, that $\operatorname{ran}[(\lambda I - A)^*] = \operatorname{ran}(\overline{\lambda} I - A^*)$ is not closed either. Thus, we easily see that $\overline{\lambda} \in \sigma_c(A^*)$, i.e. we have proven that $\lambda \in \sigma_c(A) \Rightarrow \overline{\lambda} \in \sigma_c(A^*)$. Hence

$$\overline{\lambda} \in \sigma_c(A^*) \Longrightarrow \overline{\overline{\lambda}} = \lambda \in \sigma_c(A^{**}) = \sigma_c(A)$$

that is, we have shown that

$$\sigma_c(A^*) = \{\overline{\lambda} : \lambda \in \sigma_c(A)\}.$$

To prove the second equality, we notice first (using Theorem 3.1.66 again) that

$$\begin{aligned}\overline{\operatorname{ran}(\lambda I - A)} \neq H &\Longleftrightarrow \ker(\overline{\lambda} I - A^*) \neq \{0\} \\ &\Longleftrightarrow \overline{\lambda} \in \sigma_p(A^*) \\ &\Longleftrightarrow \lambda \in \{\overline{\lambda} : \lambda \in \sigma_p(A^*)\}.\end{aligned}$$

Having these relations on our hands, let $\lambda \in \sigma_r(A)$. Then, by definition, $\lambda I - A$ is injective and $\overline{\operatorname{ran}(\lambda I - A)} \neq H$. Then λ cannot be in $\sigma_p(A)$, and $\overline{\operatorname{ran}(\lambda I - A)} \neq H$ gives $\lambda \in \{\overline{\lambda} : \lambda \in \sigma_p(A^*)\}$. We have therefore shown that

$$\sigma_r(A) \subset \{\overline{\lambda} : \lambda \in \sigma_p(A^*)\} \setminus \sigma_p(A).$$

To prove the converse, let λ be in $\{\overline{\lambda} : \lambda \in \sigma_p(A^*)\} \setminus \sigma_p(A)$. Then $\overline{\operatorname{ran}(\lambda I - A)} \neq H$ and $\lambda \notin \sigma_p(A)$ (i.e. $\lambda I - A$ is injective), which precisely means that $\lambda \in \sigma_r(A)$, marking the end of the proof.

SOLUTION 7.3.10.

(1) (a) (i) "(i) $\Rightarrow$ (ii)": It is clear that for all $n \in \mathbb{N}$

$$0 \leq \inf_{\|x\|=1} \|\lambda x - Ax\| \leq \|\lambda x_n - Ax_n\|.$$

Taking the limit $n \to \infty$ yields $f(\lambda) = 0$.

(ii) The implication "(ii) $\Rightarrow$ (i)" follows form the definition of the "infimum".

(iii) Here we show that "(iii) $\Leftrightarrow$ (i)":

If $\lambda \in \sigma_a(A)$, then for any $\varepsilon > 0$, there exists $x \in H$, with $\|x\| = 1$, such that $\|(\lambda - A)x\| < \varepsilon$. In particular, if we let $\varepsilon = \frac{1}{n}$, then the implication "$\Rightarrow$" becomes evident. Conversely, if

$$\exists x_n \in H,\ \|x_n\| = 1 \text{ such that } \lim_{n\to\infty} \|(\lambda - A)x_n\| = 0,$$

then we readily see that $\lambda - A$ cannot be bounded below.

(b) Let $\lambda, \mu \in \mathbb{C}$ and let $x \in H$ be such that $\|x\| = 1$. We then have

$$\begin{aligned}\|\lambda x - Ax\| =& \|\lambda x - \mu x + \mu x - Ax\| \\ \leq& \|\lambda x - \mu x\| + \|\mu x - Ax\| \\ =& |\lambda - \mu|\|x\| + \|\mu x - Ax\| \\ =& |\lambda - \mu| + \|\mu x - Ax\|\end{aligned}$$

Passing to "inf" on $\{x \in H : \|x\| = 1\}$ gives

$$f(\lambda) \leq |\lambda - \mu| + f(\mu).$$

Inverting the roles of λ and μ yields

$$f(\mu) \leq |\mu - \lambda| + f(\lambda) = |\lambda - \mu| + f(\lambda).$$

Therefore,

$$|f(\lambda) - f(\mu)| \leq |\lambda - \mu|.$$

Thus f is 1-Lipschitz and so it is uniformly continuous or simply continuous.

(2) Let $\lambda \in \rho(A)$. Whence $\lambda - A$ is invertible, and let $R(\lambda, A)$ be the associated resolvent. If $\|x\| = 1$, then

$$1 = \|x\| = \|R(\lambda, A)(\lambda - A)x\| \leq \|R(\lambda, A)\|\|\lambda x - Ax\|$$

which implies, after taking the "inf" (on $\|x\| = 1$) in both sides, that

$$1 \leq \|R(\lambda, A)\| f(\lambda) \text{ or } f(\lambda) \geq \frac{1}{\|R(\lambda, A)\|}.$$

To prove the reverse inequality, first notice that we have

$$\left\| \frac{R(\lambda, A)x}{\|R(\lambda, A)x\|} \right\| = 1.$$

Now, for $\|x\| = 1$:

$$\left\| (\lambda - A) \frac{R(\lambda, A)x}{\|R(\lambda, A)x\|} \right\| = \frac{\|x\|}{\|R(\lambda, A)x\|} = \frac{1}{\|R(\lambda, A)x\|}.$$

Therefore,

$$f(\lambda) \leq \left\| (\lambda - A) \frac{R(\lambda, A)x}{\|R(\lambda, A)x\|} \right\| = \frac{1}{\|R(\lambda, A)x\|}.$$

Thus

$$\|R(\lambda, A)\| \leq \frac{1}{f(\lambda)},$$

so that finally

$$f(\lambda) = \frac{1}{\|R(\lambda, A)\|}.$$

(3) (a) Since $\lambda \in \sigma_a(A)$ iff $f(\lambda) = 0$, we can write $\sigma_a(A)$ as

$$\sigma_a(A) = f^{-1}(\{0\}).$$

As f is continuous and $\{0\}$ is closed in $\mathbb{C}$, then $f^{-1}(\{0\})$ is closed, that is, $\sigma_a(A)$ is closed.

(b) Let $\lambda \in \sigma_p(A)$. Then λ is an eigenvalue associated with a non-null eigenvector x_λ. Hence $Ax_\lambda = \lambda x_\lambda$. As usual, we may take that $\|x_\lambda\| = 1$. Thus

$$f(\lambda) \le \|\lambda x_\lambda - Ax_\lambda\| = 0.$$

Therefore,

$$\sigma_p(A) \subset \sigma_a(A) \text{ or } \overline{\sigma_p(A)} \subset \sigma_a(A)$$

for $\sigma_a(A)$ is closed.
It remains to prove that $\sigma_a(A) \subset \sigma(A)$, or $\rho(A) \subset \sigma_a(A)^c$. Let $\lambda \in \rho(A)$. Then, by Question (2),

$$f(\lambda) = \frac{1}{\|R(\lambda, A)\|}.$$

But, this clearly implies that $f(\lambda) \neq 0$, so that by Question 1-a), $\lambda \notin \sigma_a(A)$, i.e. $\lambda \in \sigma_a(A)^c$.

(c) Since $\sigma(A)$ is bounded and $\sigma_a(A) \subset \sigma(A)$, $\sigma_a(A)$ too is bounded. Since we have already seen that it is closed, $\sigma_a(A)$ is compact (as a closed and bounded subset in *usual* $\mathbb{C}$).

Remark. Alternatively, we could have said that $\sigma_a(A)$, being a *closed* subset in the *compact* $\sigma(A)$, is also compact.

(4) (a) It may easily be seen that (and since $\sigma(A)$ is closed!)

$$\partial\sigma(A) = \overline{\sigma(A)} \setminus \overset{\circ}{\sigma(A)} = \sigma(A) \setminus \overset{\circ}{\sigma(A)} = \sigma(A) \cap \overline{\rho(A)}.$$

Let $\lambda \in \partial\sigma(A) = \sigma(A) \cap \overline{\rho(A)}$. Then $\lambda \in \sigma(A)$ and $\lambda \in \overline{\rho(A)}$. Hence, there is $\lambda_n \in \rho(A)$ such that $\lambda_n \to \lambda$. Since $\lambda \in \sigma(A)$, $\lambda - A$ is not invertible. Calling on Exercise 7.3.4, we see that the sequence $(R(\lambda_n, A))_n$ cannot be bounded, that is, the sequence $(\|(\lambda_n - A)^{-1}\|)_n$ (of positive reals!), cannot be bounded.

WLOG, we may then assume that $\|(\lambda_n - A)^{-1}\| \to \infty$ as n tends to infinity (otherwise, this sequence surely has a subsequence that goes to infinity, in such case, just work with this new subsequence). By Question 2),

$$f(\lambda_n) = \frac{1}{\|R(\lambda_n, A)\|} \longrightarrow 0$$

when $n \to \infty$.
Finally, since $\lambda_n \to \lambda$, by the sequential continuity of f we have

$$0 \longleftarrow f(\lambda_n) \longrightarrow f(\lambda),$$

so that by uniqueness of the limit, $f(\lambda) = 0$, i.e. $\lambda \in \sigma_a(A)$.

(b) There are different ways of seeing that $\sigma_a(A)$ is not empty. For instance, $\sigma_a(A)$ does contain $\partial\sigma(A)$ so that it suffices to check that $\partial\sigma(A)$ is not void. Assume $\partial\sigma(A) = \varnothing$. Therefore,

$$\varnothing = \partial\sigma(A) = \overline{\sigma(A)} \setminus \overset{\circ}{\widehat{\sigma(A)}} \Longrightarrow \sigma(A) \subset \overset{\circ}{\widehat{\sigma(A)}}.$$

But, it is well-known that $\sigma(A) \supset \overset{\circ}{\widehat{\sigma(A)}}$. Hence, $\sigma(A) = \overset{\circ}{\widehat{\sigma(A)}}$, i.e. $\sigma(A)$ is open. Since it is already closed, $\sigma(A)$ is clopen[1] in $\mathbb{C}$. But surely, this cannot happen as $\mathbb{C}$ is connected! Thus, $\sigma_a(A)$ is not empty.

(5) If $\lambda \notin \sigma(A)$, then $\lambda - A$ is invertible. Ergo, for some $\alpha > 0$ and all $x \in H$, we have

$$\|(\lambda - A)x\| \geq \alpha\|x\|.$$

Hence it is clear that $\lambda \notin \sigma_a(A)$. Besides, by Exercise 2.3.12, $\operatorname{ran}(\lambda - A)$ is closed, and since $\lambda - A$ is bijective, $\operatorname{ran}(\lambda - A)$ is actually dense too. This gives $\lambda \notin \sigma_r(A)$, and proves $\sigma_a(A) \cup \sigma_r(A) \subset \sigma(A)$.

Conversely, assume that $\lambda \notin \sigma_r(A)$ *and* $\lambda \notin \sigma_a(A)$. Then $\operatorname{ran}(\lambda - A)$ is dense and $\lambda \notin \overline{\sigma_p(A)}$ (by Question 3)-b)). From here $\lambda \notin \sigma_p(A)$, which in turn implies that $\lambda - A$ is injective.

Therefore, $\lambda - A : H \to \operatorname{ran}(\lambda - A)$ is bijective. Hence it makes sense to define $(\lambda - A)^{-1} : \operatorname{ran}(\lambda - A) \to H$ only

[1]Recall that a clopen set is a simultaneously open and closed subset in a given topological space.

"formally" (to be called the inverse of $\lambda - A$ if it is continuous). But since $\lambda \notin \sigma_a(A)$, we know that

$$\exists \alpha > 0,\ \forall x \in H: \ \|(\lambda - A)x\| \geq \alpha \|x\|$$

so that

$$\forall y \in \operatorname{ran}(\lambda - A): \ \|(\lambda - A)^{-1}y\| \leq \frac{1}{\alpha}\|y\|,$$

i.e. $(\lambda - A)^{-1}$, which is defined on the dense $\operatorname{ran}(\lambda - A)$, is bounded (or continuous!). By Theorem 2.1.104, we may extend it to a linear bounded operator defined on the whole of H. This means that $\lambda - A$ is invertible in $B(H)$, that is, $\lambda \notin \sigma(A)$. This proves $\sigma(A) \subset \sigma_a(A) \cup \sigma_r(A)$ and finishes the proof.

(6) The equality is obvious as the classes of normal and self-adjoint operators are known to have empty residual spectra.

Solution 7.3.11. If $A = 0$, then $\|A\| = 0$ and $\sigma(A) = \{0\}$, hence there is nothing to prove. Hence let $A \neq 0$. WLOG, we may assume that $\|A\| = 1$ (otherwise, just work with $\|A\|^{-1}A$). Consequently, we need to show that 1 or -1 belongs $\sigma(A)$. By the definition of the norm of A, we know that there exists a sequence of *unit* vectors (x_n) such that $\lim_{n\to\infty} \|Ax_n\| = 1$. We may then write

$$\begin{aligned}
\|(I - A^2)x_n\|^2 &= < (I - A^2)x_n, (I - A^2)x_n > \\
&= \|x_n\|^2 + \|A^2x_n\|^2 - 2 < A^2x_n, x_n > \ (\text{as } < A^2x_n, x_n > \in \mathbb{R}) \\
&\leq 1 + 1 - 2 < Ax_n, Ax_n > \\
&= 2 - 2\|Ax_n\|^2 \\
&\longrightarrow 0 \text{ as } n \text{ tends to infinity.}
\end{aligned}$$

Thus,

$$\lim_{n\to\infty} \|(I - A^2)x_n\| = 0$$

leading to $1 \in \sigma_a(A^2)$. But, since A^2 is self-adjoint, its spectrum coincides with its approximate point spectrum. Hence $1 \in \sigma(A^2)$ or $1 \in (\sigma(A))^2$. Finally, $1 \in \sigma(A)$ or $-1 \in \sigma(A)$.

Solution 7.3.12. Let $P \in B(H)$ be invertible and such that $P^{-1}AP = B$.

(1) Let I be the identity operator. Let $\lambda \notin \sigma(A)$. Then $A - \lambda I$ is invertible and hence $P^{-1}(A - \lambda I)P$ too is invertible. But

$$P^{-1}(A - \lambda I)P = P^{-1}AP - \lambda I = B - \lambda I.$$

Thus $\lambda \notin \sigma(B)$. So we have proved $\sigma(B) \subset \sigma(A)$. The other inclusion follows by a similar method.

(2) True. Let λ be an eigenvalue for A. Hence

$$\exists x \neq 0: \; Ax = \lambda x.$$

So

$$BP^{-1}x = P^{-1}AP(P^{-1}x) = P^{-1}\lambda x = \lambda P^{-1}x.$$

If $P^{-1}x \neq 0$, then we are done, i.e. λ is an eigenvalue for B. If $P^{-1}x$ were 0, then we would have $x = P(P^{-1}x) = P(0) = 0$, a contradiction. Thus $\lambda \in \sigma_p(B)$. The other inclusion follows similarly. Accordingly,

$$\sigma_p(B) = \sigma_p(P^{-1}AP) = \sigma_p(A).$$

(3) Yes for both. Let's start with the residual spectrum. Then (utilizing Proposition 7.1.26)

$$\begin{aligned}\sigma_r(B) &= \sigma_r(P^{-1}AP)\\ &= \{\overline{\lambda}: \; \lambda \in \sigma_p[(P^{-1}AP)^*]\} \setminus \sigma_p(P^{-1}AP)\\ &= \{\overline{\lambda}: \; \lambda \in \sigma_p(P^*A^*(P^*)^{-1})\} \setminus \sigma_p(P^{-1}AP)\\ &= \{\overline{\lambda}: \; \lambda \in \sigma_p(A^*)\} \setminus \sigma_p(A) \text{ (by the previous question)}\\ &= \sigma_r(A).\end{aligned}$$

Let's prove now the corresponding result for the continuous spectrum. By Corollary 7.1.27, we know that

$$\sigma_c(A) = \sigma(A) \setminus (\sigma_p(A) \cup \sigma_r(A)).$$

Hence

$$\begin{aligned}\sigma_c(B) &= \sigma_c(P^{-1}AP)\\ &= \sigma(P^{-1}AP) \setminus (\sigma_p(P^{-1}AP) \cup \sigma_r(P^{-1}AP))\\ &= \sigma(A) \setminus (\sigma_p(A) \cup \sigma_r(A)) \text{ (by the previous results)}\\ &= \sigma_c(A).\end{aligned}$$

Solution 7.3.13.

(1) Since $A, B \in B(H)$ are similar, we know that there exists an invertible $T \in B(H)$ such that $A = T^{-1}BT$. If B is bounded below, then

$$\exists \alpha > 0: \; B^*B \geq \alpha I.$$

Since T is invertible, we have

$$\exists \beta > 0: \; T^*T \geq \beta I.$$

Also, T^* or $(T^*)^{-1}$ is invertible so that

$$\exists\gamma > 0: \ (TT^*)^{-1} = (T^*)^{-1}T^{-1} \geq \gamma I.$$

With these observations, we easily see that

$$\begin{aligned}A^*A &= T^*B^*(T^*)^{-1}T^{-1}BT\\ &= T^*B^*(TT^*)^{-1}BT\\ &= (BT)^*(TT^*)^{-1}BT\\ &\geq (BT)^*\gamma IBT \text{ (by Exercise 5.3.8)}\\ &= \gamma T^*B^*BT\\ &\geq \gamma T^*\alpha T\\ &= \gamma\alpha T^*T\\ &\geq \alpha\beta\gamma I,\end{aligned}$$

i.e. A is bounded below.

Interchanging the roles of A and B allows us to prove the reverse implication, meaning that we have proved the whole equivalence.

(2) To establish the desired result, observe first that if A is similar to B, then $A - \lambda I$ is similar to $B - \lambda I$ (for any complex λ). On that account,

$$\begin{aligned}\lambda \notin \sigma_a(A) &\Longleftrightarrow A - \lambda I \text{ is bounded below}\\ &\Longleftrightarrow B - \lambda I \text{ is bounded below}\\ &\Longleftrightarrow \lambda \notin \sigma_a(B).\end{aligned}$$

Solution 7.3.14. First recall that

$$S^*(x_1, x_2, \cdots, x_n, \cdots) = (x_2, \cdots, x_n, \cdots).$$

(1) Let λ be complex and such that

$$S(x_1, x_2, \cdots, x_n, \cdots) = \lambda(x_1, x_2, \cdots, x_n, \cdots).$$

Then

$$(0, x_1, x_2, \cdots, x_n, \cdots) = (\lambda x_1, \lambda x_2, \cdots, \lambda x_n, \cdots).$$

If $\lambda = 0$, then $x_1 = x_2 = \cdots = x_n = \cdots = 0$, i.e. $x = 0$ and hence it cannot be an eigenvector. If $\lambda \neq 0$, then $\lambda x_1 = 0$ gives $x_1 = 0$; $\lambda x_2 = 0$ gives $x_2 = 0$, etc... In this case, we also obtain $x = 0$. Thus, no $\lambda \in \mathbb{C}$ is an eigenvalue for S, that is

$$\sigma_p(S) = \varnothing.$$

(2) Let $\lambda \in \mathbb{C}$ be such that

$$S^*(x_1, x_2, \cdots, x_n, \cdots) = \lambda(x_1, x_2, \cdots, x_n, \cdots).$$

Then

$$(x_2, x_3, \cdots, x_n, \cdots) = (\lambda x_1, \lambda x_2, \cdots, \lambda x_n, \cdots).$$

This obviously implies that $x_n = \lambda^{n-1} x_1$ (we may take any $x_1 \neq 0$). Then λ is an eigenvalue for S^* iff $(x_n) \subset \ell^2$, that is, iff $|\lambda| < 1$. Therefore,

$$\sigma_p(S^*) = \{\lambda \in \mathbb{C} : |\lambda| < 1\}.$$

(3) We have

$$\overline{\sigma_p(S^*)} = \overline{\{\lambda \in \mathbb{C} : |\lambda| < 1\}} = \{\lambda \in \mathbb{C} : |\lambda| \leq 1\} \subset \overline{\sigma(S^*)} = \sigma(S^*).$$

Since $\|S^*\| = 1$, we get (why?)

$$\sigma(S^*) \subset \{\lambda \in \mathbb{C} : |\lambda| \leq \|S^*\| = 1\}.$$

Therefore,

$$\sigma(S^*) = \{\lambda \in \mathbb{C} : |\lambda| \leq 1\}.$$

(4) It is known that

$$\sigma(S^*) = \{\overline{\lambda} : \lambda \in \sigma(S)\} \text{ and so } \sigma(S) = \{\overline{\lambda} : \lambda \in \sigma(S^*)\}.$$

Hence

$$\sigma(S) = \{\overline{\lambda} \in \mathbb{C} : |\lambda| \leq 1\} = \{\lambda \in \mathbb{C} : |\overline{\lambda}| \leq 1\} = \{\lambda \in \mathbb{C} : |\lambda| \leq 1\}.$$

(5) We know from Exercise 7.3.10, that $\partial\sigma(S) \subset \sigma_a(S)$. Since $\sigma(S) = \{\lambda \in \mathbb{C} : |\lambda| \leq 1\}$, we have

$$\partial\sigma(S) = \{\lambda \in \mathbb{C} : |\lambda| = 1\} \subset \sigma_a(S).$$

To prove the other inclusion, let $\lambda \in \sigma_a(S)$. Then $\lambda \in \sigma(S)$ (by Exercise 7.3.10 again) so that $|\lambda| \leq 1$. Now, if $x \in H$ with $\|x\| = 1$, then

$$1 = \|x\| = \|Sx\| = \|Sx - \lambda x + \lambda x\| \leq \|Sx - \lambda x\| + |\lambda|\|x\| = \|Sx - \lambda x\| + |\lambda|.$$

Passing to "inf" over x such that $\|x\| = 1$ gives

$$1 \leq \underbrace{\inf_{\|x\|=1} \|Sx - \lambda x\|}_{=0 \text{ as } \lambda \in \sigma_a(S)} + |\lambda| = |\lambda|.$$

Hence $|\lambda| = 1$ proving that

$$\{\lambda \in \mathbb{C} : |\lambda| = 1\} \supset \sigma_a(S)$$

and thus

$$\sigma_a(S) = \{\lambda \in \mathbb{C} : |\lambda| = 1\},$$

as suggested.

(6) (a) **First Method (the longest):** First remember that σ_p, σ_r and σ_c are mutually disjoint and that their union recovers the whole spectrum σ. According to the definitions of σ_r and σ_c, we need to see when $S-\lambda$ and $S^*-\lambda$ are/are not one-to-one/onto. But, having already dealt with the point spectrum of both S and S^*, we need only study the surjectivity of $S-\lambda$ and $S^*-\lambda$. In fact, we only have to see when $S-\lambda$ is not onto (why?). Let $\lambda \in \mathbb{C}$. We may write

$$(S-\lambda)(x_1, x_2, \cdots, x_n, \cdots) = (-\lambda x_1, x_1 - \lambda x_2, \cdots, x_{n-1} - \lambda x_n, \cdots).$$

(i) If $\lambda = 0$, then clearly e_1 (from the orthonormal basis of ℓ^2) is not in $\operatorname{ran}(S-\lambda)$ (=ran$S$).

(ii) If $\lambda \neq 0$, then e_1 again has no antecedent for

$$-\lambda x_1 = 1,\ x_1 - \lambda x_2 = 0, \cdots, x_{n-1} - \lambda x_n = 0, \cdots,$$

and since $\lambda \neq 0$, we would have

$$x_n = -\frac{1}{\lambda^n},\ n \in \mathbb{N}.$$

But with the previous coordinates,

$$x \in \ell^2 \Longleftrightarrow \sum_{n=1}^{\infty} \frac{1}{|\lambda|^{2n}} < \infty \Longleftrightarrow |\lambda| > 1,$$

but those values are outside $\sigma(S)$! Thus, $S-\lambda$ is not onto for $|\lambda| \leq 1$.

Now, we ask when $S-\lambda$ has a dense range in ℓ^2? We have already computed that $\sigma_p(S^*) = \{\lambda \in \mathbb{C} : \ |\lambda| < 1\}$ so that $S^*-\lambda$ is injective iff $|\lambda| = 1$. Hence

$$\overline{\operatorname{ran}(S-\lambda)} = \ell^2 \Longleftrightarrow \ker[(S-\lambda)^*] = \ker(S^* - \overline{\lambda}) = \{0\} \Longleftrightarrow |\lambda| = |\overline{\lambda}| = 1.$$

Finally, the intersection of the values of λ for which $S-\lambda$ is one-to-one, not onto and having a dense range simultaneously reduces to the unit circle of $\mathbb{C}$. Ergo

$$\sigma_c(S) = \{\lambda \in \mathbb{C} : \ |\lambda| = 1\}$$

so that

$$\sigma_r(S) = \{\lambda \in \mathbb{C} : \ |\lambda| < 1\}.$$

Things are a little less tricky for S^*. We know by Corollary 7.1.28 that if $\lambda \in \sigma_r(S^*)$, then $\overline{\lambda} \in \sigma_p(S^*)$. But

$\sigma_p(S^*) = \varnothing$ and hence $\sigma_r(S^*) = \varnothing$. Therefore, this clearly leaves us with

$$\sigma_c(S^*) = \{\lambda \in \mathbb{C} : |\lambda| = 1\}.$$

(b) **Second Method (the shortest):** We know from Proposition 7.1.26 that

$$\sigma_c(S^*) = \{\overline{\lambda} : \lambda \in \sigma_c(S)\}$$

and

$$\sigma_r(S) = \{\overline{\lambda} : \lambda \in \sigma_p(S^*)\} \setminus \sigma_p(S).$$

Hence

$$\sigma_r(S) = \{\lambda \in \mathbb{C} : |\lambda| < 1\} \setminus \varnothing = \{\lambda \in \mathbb{C} : |\lambda| < 1\}.$$

It is easy to see that

$$\sigma_r(S^*) = \varnothing.$$

Thus

$$\sigma_c(S) = \{\lambda \in \mathbb{C} : |\lambda| = 1\}$$

and

$$\sigma_c(S^*) = \{\lambda \in \mathbb{C} : |\lambda| = 1\}.$$

(7) We know that $S^*S = I$ and so

$$\sigma(S^*S) = \sigma(I) = \{1\}.$$

(8) Remember that

$$SS^*(x_1, x_2, \cdots, x_n, \cdots) = (0, x_2, \cdots, x_n, \cdots).$$

It is therefore apparent that 0 is an eigenvalue for SS^* as

$$SS^*(1, 0, \cdots, 0, \cdots) = (0, 0, \cdots, 0, \cdots) = 0(1, 0, \cdots, 0, \cdots).$$

This implies that $0 \in \sigma_p(SS^*) \subset \sigma(SS^*)$. Since SS^* and S^*S have the same non-zero elements (see Theorem 7.1.33), we immediately conclude that

$$\sigma(SS^*) = \{0, 1\}.$$

Solution 7.3.15.

(1) Let $\lambda \in \sigma_p(R)$. Hence $Rx = \lambda x$ for some $x \in \ell^2(\mathbb{Z})$ such that $x \neq 0$.

(a) If $\lambda = 0$, then $Rx = 0$ and this implies that $x_n = 0$ for all $n \in \mathbb{Z}$, that is, $x = 0$. So $0 \notin \sigma_p(R)$.

(b) Let $\lambda \neq 0$. Then $Rx = \lambda x$ implies

$$\begin{cases} \vdots \\ x_{-2} = \lambda x_{-1} \\ x_{-1} = \lambda x_0 \\ x_0 = \lambda x_1 \\ x_1 = \lambda x_2 \\ \vdots \end{cases}$$

Hence for all $n \geq 0$

$$x_n = \frac{1}{\lambda^n} x_0 \text{ and } x_{-n} = \lambda^n x_0.$$

As $x \in \ell^2(\mathbb{Z})$, then $\sum_{n=-\infty}^{\infty} |x_n|^2 < \infty$. But, we have (for $k \in \mathbb{N}$)

$$\sum_{n=-k}^{k} |x_n|^2 = |x_0|^2 \sum_{n=1}^{k} \left(|\lambda|^{2n} + \frac{1}{|\lambda|^{2n}} \right) \to \infty$$

when $k \to \infty$. This is the sought contradiction as $\sum_{n=1}^{\infty} |\lambda|^{2n}$ is finite if $|\lambda| < 1$, and $\sum_{n=1}^{\infty} \frac{1}{|\lambda|^{2n}}$ is finite if $|\lambda| > 1$!

Thus

$$\sigma_p(R) = \varnothing.$$

(2) The equality $\sigma(R) = \sigma_c(R)$ is trivial since we already know that $\sigma_p(R) = \varnothing$, and that $\sigma_r(R) = \varnothing$ (merely because R is unitary). Also, we know by Exercise 7.3.10, that σ_a and σ coincide for normal operators, in particular, for unitary ones. Hence

$$\sigma(R) = \sigma_a(R).$$

To prove that $\sigma(R) = \{\lambda \in \mathbb{C} : |\lambda| = 1\}$, first notice that since R is unitary, by Exercise 7.3.3, we have that

$$\sigma(R) \subset \{\lambda \in \mathbb{C} : |\lambda| = 1\}.$$

So, it only remains to show that $\{\lambda \in \mathbb{C} : |\lambda| = 1\} \subset \sigma(R)$. Let λ be such that $|\lambda| = 1$. We must show that for these values, $\lambda I - R$ is not invertible. It then suffices to show that (for example!) $\lambda I - R$ is not surjective. Let

$$e_0 = (\cdots, 0, \mathbf{1}, 0, \cdots) \; (\in \ell^2(\mathbb{Z})).$$

If $\lambda I - R$ were surjective, then there would exist some $x \in \ell^2(\mathbb{Z})$ such that $(\lambda I - R)x = e_0$, that is, we would have

$$(\cdots, \lambda x_{-1} - x_{-2}, \boldsymbol{\lambda x_0 - x_{-1}}, \lambda x_1 - x_0, \cdots) = (\cdots, 0, \mathbf{1}, 0, \cdots).$$

Hence

$$\lambda x_0 - x_{-1} = 1 \text{ and } \lambda x_n - x_{n-1} = 0 \text{ for } n = \pm 1, \pm 2, \pm 3, \cdots.$$

Now, we want to show that $x_0 = 0$. It is clear that for $n \geq 1$, we would have $x_n = \lambda^{-n} x_0$, so remembering that $|\lambda| = 1$ and $x \in \ell^2(\mathbb{Z})$, we would then have

$$\|x\|_{\ell^2(\mathbb{Z})}^2 = \sum_{n=-\infty}^{\infty} |x_n|^2 \geq \sum_{n=0}^{\infty} |x_0|^2 \frac{1}{|\lambda|^{2n}} = \sum_{n=0}^{\infty} |x_0|^2$$

which is finite iff $x_0 = 0$. Now, we turn to negative n. Since $x_0 = 0$, we would also obtain

$$x_{-1} = -1 \text{ so that } x_{-2} = -\lambda,\ x_{-3} = -\lambda^2, \cdots x_{-n} = -\lambda^{n-1},\ n \geq 1.$$

This would imply that

$$\|x\|_{\ell^2(\mathbb{Z})}^2 = \sum_{n=-\infty}^{n=\infty} |x_n|^2 \geq \sum_{n=-\infty}^{-1} |x_n|^2 = \sum_{n=1}^{\infty} |x_{-n}|^2 = \sum_{n=1}^{\infty} |\lambda|^{2n-2} = \sum_{n=1}^{\infty} 1 = \infty$$

and so there cannot exist an $x \in \ell^2(\mathbb{Z})$ such that $(\lambda I - R)x = e_0$, proving that $\lambda I - R$ is not onto. Thus

$$\{\lambda \in \mathbb{C} : |\lambda| = 1\} \subset \sigma(R)$$

and so

$$\sigma(R) = \{\lambda \in \mathbb{C} : |\lambda| = 1\},$$

as needed.

(3) (a) **First method:** We use the given operator U. We find that

$$U^*(\cdots, x_{-1}, \boldsymbol{x_0}, x_1, \cdots) = (\cdots, x_1, \boldsymbol{x_0}, x_{-1}, \cdots).$$

Then we may easily check that for all $x \in \ell^2(\mathbb{Z})$

$$\begin{aligned} UU^*(\cdots, x_{-1}, \boldsymbol{x_0}, x_1, \cdots) &= U^*U(\cdots, x_{-1}, \boldsymbol{x_0}, x_1, \cdots) \\ &= I(\cdots, x_{-1}, \boldsymbol{x_0}, x_1, \cdots), \end{aligned}$$

i.e. U is unitary. Now, we have for all $x \in \ell^2(\mathbb{Z})$

$$\begin{aligned} U^*RU(\cdots, x_{-1}, \boldsymbol{x_0}, x_1, \cdots) &= U^*R(\cdots, x_2, x_1, \boldsymbol{x_0}, x_{-1}, x_{-2}, \cdots) \\ &= U^*(\cdots, x_2, \boldsymbol{x_1}, x_0, x_{-1}, \cdots) \\ &= (\cdots, x_{-1}, x_0, \boldsymbol{x_1}, x_2, \cdots) \\ &= R^*(\cdots, x_{-1}, \boldsymbol{x_0}, x_1, \cdots), \end{aligned}$$

i.e.

$$U^*RU = R^*,$$

that is, R is unitarily equivalent to R^*. Thus, by Exercise 7.3.12 we recover $\sigma(R^*)$ and its subparts just from those of R. Accordingly,

$$\sigma_r(R^*) = \sigma_p(R^*) = \varnothing$$

and

$$\sigma(R^*) = \sigma_c(R^*) = \{\lambda \in \mathbb{C} : \ |\lambda| = 1\}.$$

As for $\sigma_a(R^*)$ we may either use Exercise 7.3.10 (since R^* is normal!) or Exercise 7.3.13 to obtain

$$\sigma_a(R^*) = \sigma(R^*) = \{\lambda \in \mathbb{C} : \ |\lambda| = 1\}.$$

(b) **Second method:** Since R^* is unitary, $\sigma_r(R^*) = \varnothing$ and $\sigma_a(R^*) = \sigma(R^*)$. Also, a similar method to the one used for R may be applied to show that $\sigma_p(R^*) = \varnothing$. By Proposition 7.1.26, we have

$$\sigma_c(R^*) = \{\overline{\lambda} : \ \lambda \in \sigma_c(R)\} = \{\lambda \in \mathbb{C} : \ |\lambda| = 1\}.$$

Thus

$$\sigma(R^*) = \sigma_c(R^*) = \{\lambda \in \mathbb{C} : \ |\lambda| = 1\}.$$

SOLUTION 7.3.16.

(1) First, remember that in a separable space, every orthonormal set is at most countable. Let λ be an eigenvalue for A. We know that the associated eigenvectors for each λ, denoted by x_λ, are pairwise orthogonal (Proposition 7.1.25). Thus $\{x_\lambda : \lambda \in \sigma_p(A)\}$ is an orthonormal set (if they are not normalized, then use $\frac{x_\lambda}{\|x_\lambda\|}$...etc) in H which is separable. Therefore, $\{x_\lambda : \ \lambda \in \sigma_p(A)\}$ is countable and hence so is the set of the corresponding λ.

(2) Let S^* be the adjoint of the unilateral shift operator on ℓ^2 (which is not normal as readers know). Then from Exercise 7.3.14, we know that

$$\sigma_p(S^*) = \{\lambda \in \mathbb{C} : \ |\lambda| < 1\}$$

which is clearly uncountable, and yet ℓ^2 is separable.

SOLUTION 7.3.17.

(1) The proof is known from a Linear Algebra Course, yet we include a proof. We treat two separate cases: $\lambda = 0$ and $\lambda \neq 0$.

(a) Let $\lambda = 0$. Then

$$\begin{aligned}\lambda \in \sigma_p(AB) &\Longleftrightarrow \det(AB) = 0\\ &\Longleftrightarrow \det(A)\det(B) = 0\\ &\Longleftrightarrow \det(BA) = 0\\ &\Longleftrightarrow \lambda \in \sigma_p(BA).\end{aligned}$$

(b) Let $\lambda \neq 0$. If $\lambda \in \sigma_p(AB)$, then for some $x \neq 0$, $ABx = \lambda x \neq 0$. Set $y = Bx$. If $y = 0$, then $A(Bx) = 0$ a contradiction. So $y \neq 0$. Hence

$$BAy = B(ABx) = B(\lambda x) = \lambda Bx = \lambda y,$$

i.e. λ is an eigenvalue for BA. The other inclusion follows by interchanging the roles of A and B.

(2) We can prove instead that the non-zero elements of $\sigma(ST)$ and $\sigma(TS)$ are the same. Let $\lambda \neq 0$. Let us show that $ST - \lambda I$ is invertible iff $TS - \lambda I$ is invertible. To this end, it suffices to show that $ST - I$ is invertible iff $TS - I$ is invertible. For

$$ST - \lambda I = \lambda I\left(S\frac{T}{\lambda} - I\right)$$

and λI is invertible and by renaming $\frac{T}{\lambda}$ by T. We claim that

$$(I - TS)^{-1} = I + T(I - ST)^{-1}S.$$

If $I - ST$ is invertible, call C its inverse. Hence

$$(I - ST)C = C(I - ST) = I$$

or

$$CST = STC = C - I.$$

Whence

$$\begin{aligned}(I + TCS)(I - TS) =& I - TS + TCS - TCSTS\\ =& I - TS + TCS - T(C - I)S\\ =& I.\end{aligned}$$

Similarly, we can prove that

$$(I - TS)(I + TCS) = I,$$

establishing the result.

(3) One example, which appeared in Exercise 7.3.14, consists of taking S to be the shift operator and T its adjoint. Then we have

$$\sigma(SS^*) = \{0, 1\} \neq \sigma(S^*S) = \{1\}.$$

(4) Easy! If $\sigma(ST) = \{0\}$, then $\sigma(TS) \cup \{0\} = \{0\}$, forcing (as $\sigma(TS)$ is never empty!) $\sigma(TS) = \{0\}$. Conversely, if $\sigma(TS) = \{0\}$, then similarly, we obtain that $\sigma(ST) = \{0\}$.

(5) If S is invertible, then observe that

$$TS = S^{-1}(ST)S,$$

that is, TS and ST are similar and so they have the same spectrum in virtue of Exercise 7.3.12.

Solution 7.3.18.

(1) We know that

$$\ker(B) = (\text{ran}B^*)^\perp \text{ so that } \ker(B) = (\overline{\text{ran}B^*})^\perp = H^\perp = \{0\}.$$

Hence B is injective, and since by hypothesis it is also surjective, it is bijective. Since H is a Hilbert space, the Banach Isomorphism Theorem yields the invertibility of B.

(2) Since BA is invertible, it is surjective and hence so is B (why?).

Since $0 \notin \sigma_r(B^*)$, either 0 is an eigenvalue for B^* or $0 \in \sigma_c(B^*)$. So two cases must be investigated.

(a) If $0 \in \sigma_c(B^*)$, then by definition $\overline{\text{ran}(B^*)} = H$. Thus B is invertible by the first question.

(b) If $0 \in \sigma_p(B^*)$, then $\ker(B^*) \neq \{0\}$. But, since B is surjective, $\text{ran}(B) = H$ so that

$$\ker(B^*) = (\text{ran}B)^\perp = H^\perp = \{0\}$$

which is not consistent with $0 \in \sigma_p(B^*)$.

Thus, B is invertible.

Solution 7.3.19. We already know that the non-zero elements of $\sigma(AB)$ and $\sigma(BA)$ coincide. It only remains to show that $0 \in \sigma(AB)$ iff $0 \in \sigma(BA)$, or equivalently $0 \in \rho(AB)$ iff $0 \in \rho(BA)$, i.e. AB is invertible iff BA is invertible.

Assume that BA is invertible. We claim that $0 \notin \sigma_r(B^*) = \sigma_r(B)$ (remember that by assumption $B = B^*$). If $0 \in \sigma_r(B)$ held, then we would have

$$\ker B = \{0\} \text{ and } \overline{\text{ran}(B)} \neq H.$$

But since B is self-adjoint, the injectivity of B would yield $\overline{\text{ran}(B)} = H$! This is the sought contradiction. Therefore, $0 \notin \sigma_r(B)$. By Exercise 7.3.18, B is invertible. Hence B^{-1} too is invertible. Thus $B^{-1}BA = A$ is invertible. In the end, AB is invertible.

Conversely, if AB is invertible, then $(AB)^*$ is invertible. Hence BA^* is invertible. The reasoning of the first part of the proof may then be applied to give the invertibility of A^*B. Whence $(A^*B)^* = BA^{**} = BA$ is invertible. Therefore, we have proved the first equality.

To prove the second equality (the one with "P"), observe that

$$AB = AP^2 = APP.$$

Since $P \geq 0$, by the first equality we have

$$\sigma(AB) = \sigma[(AP)P] = \sigma(PAP),$$

as required.

SOLUTION 7.3.20. As in Exercise 7.3.19, we only have to show that BA is invertible iff AB is invertible.

WLOG, assume that B is normal. Then

$$\begin{aligned} BA \textbf{ is invertible } &\Longrightarrow B \text{ is right invertible} \\ &\Longrightarrow B \text{ is invertible (by Exercise 6.3.26)} \\ &\Longrightarrow B^{-1} \text{ is invertible} \\ &\Longrightarrow B^{-1}BA = A \text{ is invertible} \\ &\Longrightarrow AB \textbf{ is invertible} \\ &\Longrightarrow B \text{ is left invertible} \\ &\Longrightarrow B \text{ is invertible (by Exercise 6.3.26)} \\ &\Longrightarrow B^{-1} \text{ is invertible} \\ &\Longrightarrow A = ABB^{-1} \text{ is invertible} \\ &\Longrightarrow BA \textbf{ is invertible}, \end{aligned}$$

and this completes the proof.

REMARK. The previous proof is a slight modification of the original one which appeared in [**15**].

The other proof (much quicker), relies on the results of Exercises 7.3.17 & 7.3.19 (it appeared in the first version of [**54**]): Since B is normal, write (why?) $B = UP = PU$, where P is positive and U is unitary. Then (since $P \geq 0$ and U is invertible)

$$\sigma(BA) = \sigma(PUA) = \sigma(UAP) = \sigma(APU) = \sigma(AB).$$

SOLUTION 7.3.21. We have been asked to provide a new proof of the result of Exercise 2.3.11. Since

$$AB = BA + I,$$

we have

$$\sigma(AB) = \sigma(BA + I) = \sigma(BA) + \{1\}$$

which is clearly not consistent (why?) with

$$\sigma(AB) \cup \{0\} = \sigma(BA) \cup \{0\}.$$

SOLUTION 7.3.22.

(1) Let $\lambda \in \mathbb{C}$. Set $r(x) = p(x) - p(\lambda)$. Then r is a polynomial and λ is obviously a root for r. Hence, there exists a polynomial q (of degree smaller than that of r) such that

$$p(x) - p(\lambda) = (x - \lambda)q(x)$$

or symbolically

$$p(A) - p(\lambda)I = (A - \lambda I)q(A).$$

Now if $p(\lambda) \notin \sigma(p(A))$, then $p(A) - p(\lambda)I$ is invertible, call B its inverse. This means that

$$I = B[p(A) - p(\lambda)I] = [p(A) - p(\lambda)I]B = B[(A - \lambda I)q(A)] = [(A - \lambda I)q(A)]B,$$

and since $A - \lambda I$ commutes with $q(A)$, we end up with

$$I = (A - \lambda I)q(A)B = Bq(A)(A - \lambda I).$$

This signifies that $A - \lambda I$ has a left inverse and a right inverse. By Proposition 2.1.58, $A - \lambda I$ is invertible, i.e. $\lambda \notin \sigma(A)$. Therefore, we have proved that $p(\sigma(A)) \subset \sigma(p(A))$.

Now, let $\mu \in \sigma(p(A))$. If p is of degree n, say, then by the Fundamental Theorem of Algebra $p - \mu$ must have n roots (distinct or not) on $\mathbb{C}$. Hence, we may write (for $a \neq 0$)

$$p(x) - \mu = a(x - \lambda_1) \cdots (x - \lambda_n)$$

and hence

$$p(A) - \mu I = a(A - \lambda_1 I) \cdots (A - \lambda_n I).$$

Since $\mu \in \sigma(p(A))$, $p(A) - \mu I$ is not invertible. If all of the factors $A - \lambda_i I$, $i = 1, \cdots, n$, were invertible, then $p(A) - \mu I$ would be invertible. Therefore, at least one $A - \lambda_j I$ is not invertible, i.e. $\lambda_j \in \sigma(A)$. But, $p(\lambda_j) = \mu$ which implies that $\mu \in p(\sigma(A))$. Thus $\sigma(p(A)) \subset p(\sigma(A))$ and this completes the proof.

(2) For the counterexample, let

$$A = \begin{pmatrix} 0 & 1 \\ -1 & 0 \end{pmatrix}$$

considered on $\mathbb{R}^2$. Then A has no real eigenvalues, i.e.

$$\sigma(A) = \varnothing.$$

Now, let $p(x) = x^2$. Then

$$p(A) = A^2 = \begin{pmatrix} -1 & 0 \\ 0 & -1 \end{pmatrix} = -I$$

whose spectrum is reduced to

$$\sigma(p(A)) = \{-1\}.$$

Thus

$$\sigma(p(A)) = \{-1\} \neq p(\sigma(A)) = p(\varnothing) = \varnothing.$$

SOLUTION 7.3.23. Since we are in a finite-dimensional setting, the spectrum is reduced to eigenvalues only. The eigenvalues of A are easily seen to be -1 and 3, that is, $\sigma(A) = \{-1, 3\}$.

(1) We know that

$$\sigma(A^2) = [\sigma(A)]^2 = \{1, 9\}.$$

(2) Similarly,

$$\sigma(A^n) = [\sigma(A)]^n = \{(-1)^n, 3^n\}$$

for $n = 1, 2, \cdots$.

(3) It is clear that A is invertible and hence we have

$$\sigma(A^{-1}) = \left\{\frac{1}{\lambda} : \lambda \in \sigma(A)\right\} = \left\{-1, \frac{1}{3}\right\}.$$

SOLUTION 7.3.24.

(1) We have

$$A^2 = (2P - I)(2P - I) = 4P^2 - 2P - 2P + I = I$$

(as $P^2 = P$).

Hence by the Spectral Mapping Theorem, we obtain

$$(\sigma(A))^2 = \sigma(A^2) = \sigma(I) = \{1\}.$$

Hence if $\lambda \in \sigma(A)$, then

$$\lambda^2 \in (\sigma(A))^2 = \{1\}.$$

Therefore,

$$\lambda = \pm 1 \text{ and so } \sigma(A) \subset \{1, -1\}.$$

On the other hand, it is clear that we have

$$\sigma(2P) = \sigma(A + I) = \sigma(A) + \{1\} \subset \{0, 2\}.$$

Thus

$$\sigma(P) \subset \{0, 1\}.$$

(2) We claim that 0 and 1 are two eigenvalues for P. Since

$$0 \neq P = P^2 \neq I,$$

we deduce from Proposition 4.1.7 that

$$\operatorname{ran} P = \ker(I - P) \neq \{0\} \text{ and } \ker P = \operatorname{ran}(I - P) \neq \{0\}.$$

Hence $I - P$ is not injective, that is,

$$\exists x \in H,\ x \neq 0 : Px = x,$$

proving that 1 is an eigenvalue for P. Similarly, P is not injective and hence

$$\exists x \in H,\ x \neq 0 : Px = 0,$$

which amounts to say that that 0 is an eigenvalue for P.

(3) From the preceding two answers we may conclude that

$$\sigma_p(P) \supset \{0, 1\} \text{ so that } \sigma(P) = \sigma_p(P) = \{0, 1\}.$$

Remark. As a bonus, we have that

$$\sigma_r(P) = \sigma_c(P) = \varnothing.$$

Solution 7.3.25.

(1) It is plain that

$$A\underbrace{(1, 0, 0, 0, \cdots)}_{\neq 0} = (-1, 0, 0, \cdots) = -1 \times (1, 0, 0, 0, \cdots)$$

and

$$A\underbrace{(0, 1, 0, 0, \cdots)}_{\neq 0} = (0, 1, 0, \cdots) = 1 \times (0, 1, 0, 0, \cdots).$$

This proves that 1 and -1 are two eigenvalues for A.

(2) Let $x = (x_1, x_2, \cdots) \in \ell^2$. We may write

$$A^2(x) = A(Ax) = A(-x_1, x_2, -x_3, x_4, \cdots) = (x_1, x_2, x_3, x_4, \cdots) = x,$$

that is, we have shown that $A^2 = I$.

(3) By the Spectral Mapping Theorem

$$[\sigma(A)]^2 = \sigma(A^2) = \sigma(I) = \{1\}.$$

Hence if $\lambda \in \sigma(A)$, then $\lambda^2 = 1$. Hence $\sigma(A) \subset \{1, -1\}$. But

$$\{1, -1\} \subset \sigma_p(A) \subset \sigma(A) \text{ so that } \sigma(A) = \{1, -1\}.$$

Solution 7.3.26. First, since $\mathcal{F}$ is unitary, its spectrum is contained in the unit cercle of center $(0, 0)$.

(1) Let f in $L^2(\mathbb{R})$. We may easily prove that

$$\mathcal{F}^2 f = \mathcal{F}\mathcal{F}f = \check{f}$$

where $\check{f}(t) = f(-t)$. Hence

$$\mathcal{F}^4 f = \mathcal{F}^2(\mathcal{F}^2 f) = \mathcal{F}^2(\check{f}) = \check{\check{f}} = f.$$

Therefore, we have shown that $\mathcal{F}^4 = I_{L^2(\mathbb{R})}$ with $\mathcal{F}^2 \neq I_{L^2(\mathbb{R})}$.

(2) Since $\mathcal{F}^4 = I_{L^2(\mathbb{R})}$ (while $\mathcal{F}^2 \neq I_{L^2(\mathbb{R})}$), the Spectral Mapping Theorem yields

$$(\sigma(\mathcal{F}))^4 = \sigma(\mathcal{F}^4) = \sigma(I) = \{1\} \text{ (whereas } (\sigma(\mathcal{F})^2 = \sigma(\mathcal{F}^2) \neq \{1\}).$$

Let $\lambda \in \sigma(\mathcal{F})$. Then $\lambda^4 = 1$ (and $\lambda^2 \neq 1$) so that the solutions are $1, -1, i$ and $-i$. This proves that $\sigma(\mathcal{F}) \subset \{1, -1, i, -i\}$.

(3) The aim of showing that $1, -1, i$ and $-i$ are eigenvalues for $\mathcal{F}$ is to prove the reverse inclusion $\sigma(\mathcal{F}) \supset \{1, -1, i, -i\}$. Let $f \in L^2(\mathbb{R})$. We then have by the linearity of $\mathcal{F}$

$$\mathcal{F}(pf) = \frac{1}{4}\mathcal{F}(f+\mathcal{F}f+\mathcal{F}^2 f+\mathcal{F}^3 f) = \frac{1}{4}(\mathcal{F}f+\mathcal{F}^2 f+\mathcal{F}^3 f+\underbrace{\mathcal{F}^4 f}_{f}) = 1\times pf.$$

Hence pf is an eigenvector for $\lambda = 1$ making the latter an eigenvalue for $\mathcal{F}$. Similarly, we easily check that

$$\mathcal{F}(qf)=\frac{1}{4}\mathcal{F}(f-i\mathcal{F}f-\mathcal{F}^2 f+i\mathcal{F}^3 f)=\frac{1}{4}(\mathcal{F}f-i\mathcal{F}^2 f-\mathcal{F}^3 f+i\mathcal{F}^4 f)=i\times qf,$$

so that qf is an eigenvector associated with the eigenvalue i. We leave it to the interested reader to check that

$$\mathcal{F}(rf) = (-1) \times rf \text{ and } \mathcal{F}(sf) = (-i) \times rf.$$

Finally, since $\{1, -1, i, -i\} \subset \sigma_p(\mathcal{F})$, we infer that

$$\sigma(\mathcal{F}) = \{1, -1, i, -i\}.$$

Remark. Explicit eigenvectors of the Fourier transform may be obtained via the so-called **Hermite functions**.

Solution 7.3.27. Consider the following operator A defined on ℓ^2 by

$$A(x_1, x_2, \cdots, x_n, \cdots) = (x_2, 0, x_4, 0, \cdots).$$

That A is bounded and $A \neq 0$ are clear.

Let us show that $\sigma(A) = \{0\}$. We have for any $(x_n) \in \ell^2$

$$A^2(x_1, x_2, \cdots, x_n, \cdots) = A(x_2, 0, x_4, 0, \cdots) = (0, 0, \cdots) = 0_{\ell^2}.$$

Now if $\lambda \in \sigma(A)$, then $\lambda^2 \in \sigma(A^2) = \{0\}$. That is $\sigma(A) \subset \{0\}$. To establish the equality, ***either*** observe that

$$A\underbrace{(1,0,0,\cdots,0,\cdots)}_{\neq 0} = (0,0,\cdots,0,\cdots) = 0(0,0,\cdots,0,\cdots)$$

so that 0 is an eigenvalue for A and thus

$$\{0\} \subset \sigma_p(A) \subset \sigma(A) \subset \{0\},$$

i.e. $\sigma(A) = \{0\}$ (or this simply follows from the fact that $\sigma(A) \neq \varnothing$ over $\mathbb{C}$!).

Remark. We will meet another (more sophisticated) example in Exercise 9.3.21.

Solution 7.3.28.

(1) Let A be a nilpotent operator. Then $A^n = 0$ for some $n \in \mathbb{N}$. So by the Spectral Mapping Theorem

$$[\sigma(A)]^n = \sigma(A^n) = \sigma(0) = \{0\}.$$

So if $\lambda \in \sigma(A)$, then $\lambda^n = 0$ or $\lambda = 0$, that is, we have shown that $\sigma(A) \subset \{0\}$.

Now, either by showing that $0 \in \sigma_p(A)$ or simply by observing that $\sigma(A) \neq \varnothing$, we necessarily have

$$\sigma(A) = \{0\}.$$

(2) Since the spectrum of A is reduced to the singleton $\{0\}$, it is clear that nilpotent operators are not invertible.

Solution 7.3.29. Assume, for instance, that $N^n = 0$ for some $n \in \mathbb{N}$. We first show that $\sigma(A+N) \subset \sigma(A)$. Contra positively, we may instead show that $\rho(A) \subset \rho(A+N)$. Let $\lambda \in \rho(A)$. Then $R(\lambda, A) = (\lambda - A)^{-1}$ exists and is bounded.

Now, since A commutes with N, so does $R(\lambda, A)$ with N. It is then apparent by the nilpotence of N that

$$(NR(\lambda, A))^n = N^n R(\lambda, A)^n = 0,$$

that is, $NR(\lambda, A)$ too is nilpotent.

By Exercise 2.3.31, $I - NR(\lambda, A)$ (and also $I - R(\lambda, A)N$) is invertible and

$$\begin{aligned}(I - R(\lambda, A)N)^{-1} &= (I - NR(\lambda, A))^{-1} \\ &= \sum_{k=0}^{n-1} N^k [R(\lambda, A)]^k \\ &= \sum_{k=0}^{n-1} [R(\lambda, A)]^k N^k.\end{aligned}$$

Now, we have

$$\begin{aligned}R(\lambda, A + N) &= (\lambda - A - N)^{-1} \\ &= [(\lambda - A)(I - (\lambda - A)^{-1}N)]^{-1} \\ &= [(\lambda - A)(I - N(\lambda - A)^{-1})]^{-1} \\ &= [I - N(\lambda - A)^{-1}]^{-1}(\lambda - A)^{-1} \\ &= (I - NR(\lambda, A))^{-1}R(\lambda, A).\end{aligned}$$

Hence $R(\lambda, A + N)$ exists and is bounded since both $(I - NR(\lambda, A))^{-1}$ and $R(\lambda, A)$ exist and are bounded. Therefore, $\lambda \in \rho(A + N)$, i.e. we have proved that $\sigma(A + N) \subset \sigma(A)$.

To prove the reverse inclusion, remember first that since N is nilpotent, $\sigma(N) = \{0\}$ (Exercise 7.3.28). We have already proved that $\sigma(A + N) \subset \sigma(A)$ whenever $AN = NA$. Hence as

$$(A + N)(-N) = (-N)(A + N),$$

then

$$\sigma(A) = \sigma[A + N + (-N)] \subset \sigma(A + N) \text{ (since } -N \text{ is nilpotent).}$$

Consequently,

$$\sigma(A + N) = \sigma(A).$$

Solution 7.3.30.

(1) Linearity is very easy... Boundedness is also easy. Let $X \in B(H)$. Then

$$\|TX\| = \|AX - XB\| \le \|AX\| + \|XB\| \le (\|A\| + \|B\|)\|X\|,$$

and so $T \in B(B(H))$.

(2) We shall prove one inclusion (e.g. the one involving A), the other one is left to readers.

To show $\sigma(L_A) \subset \sigma(A)$, we can equivalently show that $\rho(A) \subset \rho(L_A)$. So, let $\lambda \in \mathbb{C}$ be such that $A - \lambda I$ is invertible. It is easy to see that for any $X \in B(H)$

$$(A-\lambda I)^{-1}(A-\lambda I)X = (A-\lambda I)(A-\lambda I)^{-1}X = X.$$

This shows that $L_A - \lambda I$ is invertible in $B(B(H))$ and that

$$(L_A - \lambda I)^{-1}X = (A-\lambda I)^{-1}X = L_{(A-\lambda I)^{-1}}X,$$

i.e. $\lambda \in \rho(L_A)$.

(3) We claim that $L_A R_B = R_B L_A$: Let $X \in B(H)$. Then

$$L_A R_B(X) = L_A(XB) = AXB = R_B(AX) = R_B L_A(X)$$

as needed. Thus, by Proposition 7.1.40 and the preceding question, the following

$$\sigma(T) = \sigma(L_A - R_B) \subset \sigma(L_A) - \sigma(R_B) \subset \sigma(A) - \sigma(B)$$

holds.

Solution 7.3.31. By the remark just below Theorem 7.1.44, we are only required to show that $T : B(H) \to B(H)$ defined by $T(X) = AX - XB$ is invertible, that is $0 \notin \sigma(T)$. By Theorem 7.1.43 (the previous exercise!),

$$\sigma(T) \subset \sigma(A) - \sigma(B).$$

The assumption $\sigma(A) \cap \sigma(B) = \varnothing$ then yields $0 \notin \sigma(A) - \sigma(B)$ and so $0 \notin \sigma(T)$ as desired.

Solution 7.3.32. Define $T : B(H) \to B(H)$ by

$$T = L_A R_B - L_C R_D$$

where

$$T(X) = L_A R_B X - L_C R_D X = AXB - CXD.$$

If we can show that T is invertible, then $AXB - CXD = E$ has a unique solution $X \in B(H)$ for each $E \in B(H)$.

As in Exercise 7.3.30, we can show that (proofs are left to readers)

$$(L_A R_B)(L_C R_D) = (L_C R_D)(L_A R_B),\ L_A R_B = R_B L_A \text{ and } L_C R_D = R_D L_C.$$

Hence (by invoking Proposition 7.1.40)

$$\sigma(T) = \sigma(L_A R_B - L_C R_D) \subset \sigma(L_A R_B) - \sigma(L_C R_D).$$

Also

$$\sigma(L_A R_B) \subset \sigma(L_A)\sigma(R_B) \text{ and } \sigma(L_C R_D) \subset \sigma(L_C)\sigma(R_D).$$

Finally, as in Exercise 7.3.30:

$$\sigma(L_A) \subset \sigma(A),\ \sigma(R_B) \subset \sigma(B),\ \sigma(L_C) \subset \sigma(C) \text{ and } \sigma(R_D) \subset \sigma(D).$$

All these inclusions give us

$$\sigma(T) \subset \sigma(A)\sigma(B) - \sigma(C)\sigma(D).$$

By the assumption on the separation of the sets $\sigma(A)\sigma(B)$ and $\sigma(C)\sigma(D)$, $0 \notin \sigma(A)\sigma(B)-\sigma(C)\sigma(D)$. Therefore, $0 \notin \sigma(T)$, that is, T is invertible, and this completes the proof.

Solution 7.3.33. By hypothesis, $C(A+B) = (A+B)C$ and $C(AB) = (AB)C$. Hence

$$ACA + ACB = AAC + ABC = AAC + CAB$$

and so

$$A(AC-CA) = (AC-CA)B \text{ or } A(AC-CA) - (AC-CA)B = 0.$$

As $\sigma(A)\cap\sigma(B) = \varnothing$, Theorem 7.1.44 tells us that the previous equation has a unique solution. The apparent solution $AC-CA = 0$ is therefore the only solution and so $AC = CA$, as required.

By the same token C commutes with B (another way to see this is to write $B = B+A-A$ and use the assumption), and this completes the proof.

Solution 7.3.34. By Example 4.1.32, the matrix operator $\begin{pmatrix} I & X \\ 0 & I \end{pmatrix}$ is invertible. Hence, a similarity of the two matrix operators $\begin{pmatrix} A & C \\ 0 & B \end{pmatrix}$ and $\begin{pmatrix} A & 0 \\ 0 & B \end{pmatrix}$ is realized if for example

$$\begin{pmatrix} A & C \\ 0 & B \end{pmatrix}\begin{pmatrix} I & X \\ 0 & I \end{pmatrix} = \begin{pmatrix} I & X \\ 0 & I \end{pmatrix}\begin{pmatrix} A & 0 \\ 0 & B \end{pmatrix}$$

holds.

By examining the entries of the two matrices resulting from the two products on each side, we see that the previous holds when exactly $AX + C = XB$ or $AX - XB = -C$. So the similarity of two matrices of operators is possible if such an X exists. A glance at the assumption $\sigma(A)\cap\sigma(B) = \varnothing$ as well as Theorem 7.1.44 give the desired result.

Solution 7.3.35. Assume that $\dim H = \infty$ and let A be a non-unitary isometry in $B(H)$, i.e. $A^*A = I$ and $C := I - AA^* \neq 0$ (an explicit example would be the shift operator on ℓ^2). The matrix of operators

$$\begin{pmatrix} A & C \\ 0 & A^* \end{pmatrix}$$

defined on $B(H \oplus H)$ is unitary (as one can check) and so

$$\begin{pmatrix} A & C \\ 0 & A^* \end{pmatrix}^{-1} = \begin{pmatrix} A & C \\ 0 & A^* \end{pmatrix}^* = \begin{pmatrix} A^* & 0 \\ C^* & A \end{pmatrix} = \begin{pmatrix} A^* & 0 \\ C & A \end{pmatrix}.$$

Now, it is easy to see that

$$\begin{pmatrix} A^* & 0 \\ C & A \end{pmatrix}\begin{pmatrix} A & 0 \\ 0 & 0 \end{pmatrix}\begin{pmatrix} A & C \\ 0 & A^* \end{pmatrix} = \begin{pmatrix} A & C \\ 0 & 0 \end{pmatrix}.$$

This signifies that the matrix operators $\begin{pmatrix} A & 0 \\ 0 & 0 \end{pmatrix}$ and $\begin{pmatrix} A & C \\ 0 & 0 \end{pmatrix}$ are *unitarily equivalent* on $B(H \oplus H)$. Consider now the Sylvester Equation:

$$AX - XB = C$$

in case $B = 0$, i.e. $AX = C$. This equation has no solution in $B(H)$. If such an X existed, then we would get

$$AX = C \Longrightarrow X = A^*C = A^*(I - AA^*) = A^* - A^* = 0$$

and hence $C = 0$, and this is the contradiction we have been after!

Solution 7.3.36.

(1) We know that $\ker A$ is a hyperinvariant subspace for A. Since A is nilpotent, $A^n = 0$ for some $n \in \mathbb{N}$ (chosen to be the smallest). Then $A^{n-1} \neq 0$. Since A is also non zero, and $A^{n-1}A = A^n = 0$, a glance at Exercise 4.3.43 allows us that $\ker A$ is a nontrivial hyperinvariant subspace for A.

(2) Let A be algebraic and assume that $p(A) = 0$ with p having the smallest degree. By the Fundamental Theorem of Algebra, we may factorize p as

$$p(A) = (\lambda I - A)q(A)$$

for some polynomial q of degree deg q=deg $p - 1$.

But, by assumption A is non scalar and this means that $\lambda I - A \neq 0$. Moreover, $q(A) \neq 0$ as we have taken p to be minimal. Accordingly, calling on again Exercise 4.3.43, we may state with ease that $\ker(\lambda I - A)$ is a nontrivial hyperinvariant subspace for A.

7.4. Hints/Answers to Tests

Solution 7.4.1. The answer is no! Assume that a self-adjoint $A \in B(\mathbb{R}^3)$ enjoys:

$$A(1,1,1) = (0,0,0) = 0(1,1,1) \text{ and } A(1,2,2) = (2,4,4) = 2(1,2,2),$$

i.e. 0 and 2 are eigenvalues for A. Since they are distinct, the corresponding eigenvectors must therefore be perpendicular! Are $(1,1,1)$ and $(1,2,2)$ so?...

Solution 7.4.2. Consider

$$\varphi(x)=\begin{cases} 0, & 0<x<\frac{1}{3},\\ 3(x-\frac{1}{3}), & \frac{1}{3}\le x\le\frac{2}{3},\\ 1, & \frac{2}{3}<x<1.\end{cases}$$

Now, do the details...

Solution 7.4.3. Let S be the shift. Then

$$\overline{\sigma_p(S)}=\varnothing\subsetneq\sigma_a(S)=\{\lambda\in\mathbb{C}:\ |\lambda|=1\}\subsetneq\sigma(S)=\{\lambda\in\mathbb{C}:\ |\lambda|\le 1\}.$$

Solution 7.4.4. Yes! We do know that $\|S\|=1$. What is $\sigma(S)$?...

Solution 7.4.5. First, notice that the condition $\operatorname{Re}\sigma(A)<0$ yields $\operatorname{Re}\sigma(A^*)<0$ too. Hence

$$\sigma(A)\cap\sigma(-A^*)=\varnothing.$$

Theorem 7.1.44 is applicable and guarantees the existence of a solution X. By taking adjoints,

$$AX+XA^*=-I\Longleftrightarrow AX^*+X^*A^*=-I.$$

By uniqueness of the solution, we must have $X=X^*$, as required.

Solution 7.4.6. First, show that $(A-\lambda_1)(A-\lambda_2)\cdots(A-\lambda_n)$ is bounded below iff $(A-\lambda_i)$ is bounded below for each $i=1,\cdots,n$...

Solution 7.4.7.

(1) Clearly, $\ker(\lambda I-A)$ is invariant for $\lambda I-A$, and other very simple details are left to readers...
(2) Since A is non scalar, $\ker(\lambda I-A)\neq H$. Also, we have that $\ker(\lambda I-A)\neq\{0\}$ (why?)...

CHAPTER 8

Spectral Radius. Numerical Range

8.2. True or False: Answers

ANSWERS.

(1) True. To see why, let $A \in B(H)$. Clearly for any $n \in \mathbb{N}$

$$(A^n)^* = (A^*)^n \text{ and so } \|A^n\| = \|(A^n)^*\| = \|(A^*)^n\|.$$

Therefore,

$$r(A) = \lim_{n\to\infty} \|A^n\|^{\frac{1}{n}} = \lim_{n\to\infty} \|(A^*)^n\|^{\frac{1}{n}} = r(A^*),$$

as needed.

REMARK. Consequently, A is normaloid iff A^* is normaloid

(2) Only "$\Leftarrow$" is true. Let us prove it (before giving a counterexample to the backward implication). Since $\sigma(A) \subset \sigma(B)$ and since A is self-adjoint, we have by the Spectral Radius Theorem

$$\|A\| = \sup\{|\lambda| : \lambda \in \sigma(A)\} \leq \sup\{|\lambda| : \lambda \in \sigma(B)\} = \|B\|$$

as B too is self-adjoint.

For the counterexample, just consider

$$A = \begin{pmatrix} 1 & 0 \\ 0 & 2 \end{pmatrix} \text{ and } B = \begin{pmatrix} 1 & 0 \\ 0 & 3 \end{pmatrix}.$$

Then plainly $\|A\| \leq \|B\|$. But

$$\sigma(A) = \sigma_p(A) = \{1,2\} \not\subset \sigma(B) = \sigma_p(B) = \{1,3\}.$$

(3) True! By hypothesis, $A = U^*BU$ for some unitary $U \in B(H)$. By Exercise 7.3.12, we know that $\sigma(A) = \sigma(B)$ so that

$$r(A) = \sup\{|\lambda| : \lambda \in \sigma(A)\} = \sup\{|\lambda| : \lambda \in \sigma(B)\} = r(B).$$

(4) According to Proposition 8.1.12, A and B must not be chosen to commute. Let

$$A = \begin{pmatrix} 0 & 0 \\ 1 & 0 \end{pmatrix} \text{ and } B = A^* = \begin{pmatrix} 0 & 1 \\ 0 & 0 \end{pmatrix}.$$

It is then clear that $\sigma(A) = \sigma(B) = \{0\}$ so that

$$r(A) = r(B) = 0.$$

Now,

$$A + B = \begin{pmatrix} 0 & 1 \\ 1 & 0 \end{pmatrix}$$

so that $\sigma(A + B) = \{1, -1\}$, it follows that $r(A + B) = 1$. Thus

$$r(A + B) > r(A) + r(B).$$

REMARK. In [**219**], other hypotheses, different from commutativity, also lead to $r(A+B) \leq r(A)+r(B)$. See also [**119**] for further related results.

(5) True! See Exercise 8.3.3.
(6) If you look at the proof of $W(A) = \{\alpha\}$ iff $A = \alpha I$ (Exercise 8.3.10), you will see that we do need the fact that H is a $\mathbb{C}$-Hilbert space (cf. Exercise 3.3.45).
(7) False! $W(A)$ is *never* empty. To see this, assume that $W(A) = \varnothing$. Then $\overline{W(A)} = \overline{\varnothing} = \varnothing$. Surely, this cannot happen as we know by Theorem 8.1.21 that $\overline{W(A)}$ *contains* $\sigma(A)$ which is never empty (as H is over $\mathbb{C}$). Thus, $W(A)$ is never empty.
(8) False! On $B(\ell^2)$, define

$$A(x_1, x_2, x_3, \cdots) = (x_1, \frac{1}{2}x_2, \frac{1}{3}x_3, \cdots)$$

and

$$B(x_1, x_2, x_3, \cdots) = (0, x_2, \frac{1}{2}x_3, \frac{1}{3}x_4, \cdots).$$

Then A and B are clearly self-adjoint,

$$\sigma(A) = \overline{\left\{\frac{1}{n} : n \in \mathbb{N}\right\}} = \left\{\frac{1}{n} : n \in \mathbb{N}\right\} \cup \{0\} = \overline{\left\{\frac{1}{n} : n \in \mathbb{N}\right\} \cup \{0\}} = \sigma(B)$$

whereas

$$W(A) = (0, 1] \neq W(B) = [0, 1].$$

REMARK. As observed in [**128**], if two normal operators A and B have equal spectra, then $W(A)$ and $W(B)$ can differ only by their boundaries $\partial W(A)$ and $\partial W(B)$. This comes from the following observation

$$\overline{W(A)} = \operatorname{conv}(\sigma(A)) = \operatorname{conv}(\sigma(B)) = \overline{W(B)}.$$

(9) False! We know by Proposition 8.1.18 that in finite dimensions, $W(A)$ is compact in $\mathbb{C}$ and hence it is closed. But, as soon as we leave the finite dimensional setting the closedness of $W(A)$ is not guaranteed anymore. For a counterexample, consider the shift operator S. Then (see Exercise 8.3.14)

$$W(S) = B(0,1) = \{\lambda \in \mathbb{C} : |\lambda| < 1\}.$$

Thus $W(S)$ is not closed.

(10) None of the inclusions needs to hold. By the remark below Proposition 8.1.19, we may have a self-adjoint A with spectrum $[0,1]$, and with numerical range e.g. $(0,1]$. This shows that

$$\sigma(A) \not\subset W(A).$$

A similar argument applies to the other inclusion. Indeed, if A is a self-adjoint multiplication operator on $B(\ell^2)$ (multiplication by $(\frac{1}{n})_n$, say), then as noticed above

$$\sigma(A) = \overline{\left\{\frac{1}{n} : n \in \mathbb{N}\right\}} = \{0\} \cup \left\{\frac{1}{n} : n \in \mathbb{N}\right\} \not\supset W(A) = (0,1].$$

8.3. Solutions to Exercises

SOLUTION 8.3.1.

(1) First, we find S^2. Let $(x_1, x_2, \cdots) \in \ell^2$. Then

$$S^2(x_1,x_2,\cdots) = S(S(x_1,x_2,\cdots)) = S(0,x_1,x_2,\cdots) = (0,0,x_1,x_2,\cdots).$$

Similarly,

$$S^3(x_1,x_2,\cdots) = (0,0,0,x_1,x_2,\cdots).$$

Hence, if $n \in \mathbb{N}$, we obtain by induction,

$$S^n(x_1,x_2,\cdots) = (\underbrace{0,\cdots,0}_{n \text{ first terms}}, x_1, x_2, \cdots).$$

(2) The previous work implies that for all $x = (x_1, x_2, \cdots) \in \ell^2$

$$\|S^n x\|_2^2 = |x_1|^2 + |x_2|^2 + \cdots = \|x\|_2^2$$

so that

$$\|S^n\| = 1, \ \forall n \in \mathbb{N}.$$

Accordingly,

$$\|S\| = 1 = \lim_{n\to\infty} \|S^n\|^{\frac{1}{n}},$$

as required.

SOLUTION 8.3.2.

(1) (a) Since $\|A^*A\| = \|A\|^2$ and A is self-adjoint, we obtain

$$\|A^2\| = \|A\|^2.$$

(b) We use a proof by induction. The statement holds true for $n = 1$. Assume that $\|A^{2^n}\| = \|A\|^{2^n}$ and let us show that $\|A^{2^{n+1}}\| = \|A\|^{2^{n+1}}$. We have

$$\begin{aligned}
\|A^{2^{n+1}}\| &= \|A^{2^n \times 2}\| \\
&= \|A^{2^n} A^{2^n}\| \\
&= \|A^{2^n}\| \|A^{2^n}\| \text{ (Question 1 (a) as } A^{2^n} \text{ is self-adjoint)} \\
&= \|A\|^{2^n} \|A\|^{2^n} \text{ (Induction hypothesis)} \\
&= \left(\|A\|^{2^n}\right)^2 \\
&= \|A\|^{2^{n+1}},
\end{aligned}$$

as needed.

(c) By Proposition 8.1.2, the limit $\lim_{n\to\infty} \|A^n\|^{\frac{1}{n}}$ always exists and equals to $r(A)$. Hence

$$r(A) = \lim_{n\to\infty} \|A^n\|^{\frac{1}{n}} = \lim_{n\to\infty} \|A^{2^n}\|^{\frac{1}{2^n}} = \lim_{n\to\infty} \left(\|A\|^{2^n}\right)^{\frac{1}{2^n}} = \|A\|.$$

(2) We have

$$\begin{aligned}
\|A^n\| &= r(A^n) \\
&= \sup\{|\lambda| : \lambda \in \sigma(A^n)\} \\
&= \sup\{|\lambda| : \lambda \in (\sigma(A))^n\} \\
&= \sup\{|\mu|^n : \mu \in \sigma(A)\} \\
&= [\sup\{|\mu| : \mu \in \sigma(A)\}]^n \\
&= (r(A))^n \\
&= \|A\|^n.
\end{aligned}$$

Solution 8.3.3.

(1) Since A is normal, we have

$$\|A^*Ax\| = \|AAx\| = \|A^2x\|, \ \forall x \in H.$$

By taking the supremum over $x \in H$ with $\|x\| = 1$, we then obtain

$$\|A^*A\| = \|A^2\|.$$

Therefore,

$$\|A^2\| = \|A\|^2.$$

(2) We use a proof by induction together with the fact the powers of normal operators stay normal (see Exercise 4.5.8). The statement clearly holds for $n = 1$ (Question 1!). Assume $\|A^{2^n}\| = \|A\|^{2^n}$. We have

$$\begin{aligned}\|A^{2^{n+1}}\| =&\|A^{2^n}A^{2^n}\|\\ =&\|A^{2^n}\|^2 \text{ (since } A^{2^n} \text{ is normal)}\\ =&\left(\|A\|^{2^n}\right)^2 \text{ (induction hypothesis)}\\ =&\|A\|^{2^{n+1}},\end{aligned}$$

as needed.

(3) Then as in Exercise 8.3.2, we may show that

$$r(A) = \|A\|.$$

REMARK. We also obtain $\|A^n\| = \|A\|^n$ for all n.

(4) Since A is normal, so is $A - \lambda I$. But $\sigma(A - \lambda I) = \{0\}$ and hence

$$\|A - \lambda I\| = r(A - \lambda I) = \sup\{|\mu| :\ \mu \in \sigma(A - \lambda I)\} = 0 \Longrightarrow A = \lambda I,$$

as suggested.

(5) If $\sigma(A)$ contained one point, λ say, then by the previous question A would be a scalar operator, namely $A = \lambda I$. But, this violates the assumption A being non-scalar! Therefore, $\sigma(A)$ contains at least two different numbers.

SOLUTION 8.3.4. Recall that

$$p(A) = a_0 I + a_1 A + \cdots + a_n A^n$$

where $a_0, a_1, \cdots, a_n \in \mathbb{C}$. A glance at Exercise 4.5.8 allows us to easily establish the normality of $p(A)$.

Since $p(A) \in B(H)$ is normal, the Spectral Radius Theorem yields

$$\|p(A)\| = r(p(A)) = \sup\{|\lambda| :\ \lambda \in \sigma(p(A))\}.$$

Invoking Theorem 7.1.56 implies

$$\sup\{|\lambda| :\ \lambda \in \sigma(p(A))\} = \sup\{|p(\lambda)| :\ \lambda \in \sigma(A)\}.$$

Consequently,

$$\|p(A)\| = \max_{\lambda \in \sigma(A)} |p(\lambda)|$$

as needed.

SOLUTION 8.3.5. We have

$$\begin{aligned} r(AB) &= \sup_{\lambda\in\sigma(AB)} |\lambda| \\ &= \sup_{\lambda\in\sigma(AB)\cup\{0\}} |\lambda| \text{ (why?)} \\ &= \sup_{\lambda\in\sigma(BA)\cup\{0\}} |\lambda| \text{ (from Exercise 7.3.17)} \\ &= \sup_{\lambda\in\sigma(BA)} |\lambda| \\ &= r(BA). \end{aligned}$$

SOLUTION 8.3.6. Since by hypothesis A is already self-adjoint, it only remains to show that A is a projection, i.e. $A^2 = A$. Define the following polynomial $p(x) = x^2 - x$ on $\mathbb{R}$. Then p restricted to $\sigma(A)$ vanishes, i.e. $p(\sigma(A)) = \{0\}$. Let $r(A^2 - A)$ be the spectral radius of $A^2 - A$. Then

$$\begin{aligned} r(A^2-A) =& r(p(A)) \\ =& \sup\{|\lambda| : \lambda\in\sigma(p(A))\} \\ =& \sup\{|\lambda| : \lambda\in p(\sigma(A))\} \text{ (by the Spectral Mapping Theorem)} \\ =& \sup\{|\lambda| : \lambda\in\{0\}\} \\ =& 0. \end{aligned}$$

Since $A^2 - A$ is self-adjoint, we immediately deduce that

$$\|A^2 - A\| = r(A^2 - A) = 0 \text{ and hence } A^2 = A.$$

SOLUTION 8.3.7. First, the conditions $\sigma(A) \subset \{z \in \mathbb{C} : |z| > r\}$ and $\sigma(B) \subset \{z \in \mathbb{C} : |z| < r\}$ imply the separation of spectra and so Theorem 7.1.44 insures the *existence* of a *unique* solution for each C.

The "harder part" is to prove the convergence of the series appearing in the expression of X. It suffices to show that the series is absolutely convergent.

Observe that we can find $r_1, r_2 > 0$ such that $r_1 < r < r_2$ (so $r_1/r_2 < 1$) with

$$\sigma(A) \subset \{z \in \mathbb{C} : |z| > r_2\} \text{ and } \sigma(B) \subset \{z \in \mathbb{C} : |z| < r_1\}.$$

The condition $\sigma(A) \subset \{z \in \mathbb{C} : |z| > r_2\}$ implies that A is invertible and that

$$\sigma(A^{-1}) \subset \{z \in \mathbb{C} : |z| < r_2^{-1}\}.$$

Passing to "sup" gives

$$r(B) < r_1 \text{ and } r(A^{-1}) < r_2^{-1}.$$

Recall that $r(B) = \lim_{n\to\infty} \|B^n\|^{\frac{1}{n}}$. Hence,

$$\exists N \in \mathbb{N},\ \forall n \geq N:\ \|B^n\| \leq r_1^n.$$

Similarly, $\forall n \geq N':\ \|A^{-n}\| \leq r_2^{-n}$ for some $N' \in \mathbb{N}$. Therefore, for all $n \geq \max(N, N')$

$$\|A^{-n-1}CB^n\| \leq \|A^{-n}\|\|A^{-1}C\|\|B^n\| \leq \|A^{-1}C\| \left(\frac{r_1}{r_2}\right)^n$$

from which we easily infer the convergence of the series $\sum_{n=0}^{\infty}\|A^{-n-1}CB^n\|$.

Finally, we ask readers to check that the given series does satisfy $AX - XB = C$ for each $C \in B(H)$ (cf. Exercise 2.3.37).

Solution 8.3.8.

(1) We may write:

$$BAB + A = 0 \Longrightarrow BABA + A^2 = 0 \Longrightarrow (BA)^2 = -A^2.$$

Hence

$$\sigma[(BA)^2] = \sigma(-A^2) = -\sigma(A^2) \subset \mathbb{R}^-$$

as A^2 is positive because A is self-adjoint.

(2) Set $C = BA$. We know that $\sigma(C^2) = (\sigma(C))^2$. Let $\lambda \in \mathbb{C}$. If λ is in $\sigma(C)$, then $\lambda^2 \leq 0$. Thus λ is purely imaginary.

(3) Since B is positive, it possesses a (unique) square root, which we denote by $\sqrt{B}$ (remember that it is also positive). Hence we can do the following (cf. Theorem 7.1.36)

$$\sigma(C) = \sigma(BA) = \sigma(\sqrt{B}\sqrt{B}A) = \sigma(\sqrt{B}A\sqrt{B}) \subset \mathbb{R}$$

as $\sqrt{B}A\sqrt{B}$ is self-adjoint. Combining this result with the previous question, we may conclude by saying that $\sigma(BA) = \{0\}$. This leads to $\sigma[(BA)^2] = \{0\}$ so that by the first answer above, we get

$$-\sigma(A^2) = \{0\} \text{ or } [\sigma(A)]^2 = \{0\} \text{ or } \sigma(A) = \{0\}.$$

(4) Since A is self-adjoint, we know that

$$\|A\| = r(A) = \sup\{|\lambda| :\ \lambda \in \sigma(A)\}.$$

So by $\sigma(A) = \{0\}$, we finally obtain $\|A\| = 0$ or $A = 0$, as required.

Solution 8.3.9. Recall that A and B are self-adjoint. The normality of T is equivalent to $AB = BA$. Hence $TA = AT$. Since $T - A = iB$, we have by Proposition 7.1.40

$$i\sigma(B) = \sigma(iB) = \sigma(T - A) \subset \sigma(T) + \sigma(-A) \subset \mathbb{R}.$$

Thus, necessarily $\sigma(B) = \{0\}$. Accordingly, the Spectral Radius Theorem gives us $B = 0$ and so $T = A$, i.e. T is self-adjoint.

Solution 8.3.10.

(1) Let λ be an eigenvalue for A with an (non-zero) eigenvector x_λ. Then the *unit vector* $y_\lambda = \frac{x_\lambda}{\|x_\lambda\|}$ is also an eigenvector associated with λ. Hence

$$\lambda = \lambda\|y_\lambda\|^2 =< \lambda y_\lambda, y_\lambda >=< Ay_\lambda, y_\lambda >\in W(A).$$

(2) Let $\lambda \in W(A)$. We need to show that $|\lambda - 0| = |\lambda| \leq \|A\|$. Since $\lambda \in W(A)$, $\lambda =< Ax, x >$ for some unit vector x. But then

$$|\lambda| = |< Ax, x >| \leq \|Ax\|\|x\| \leq \|A\|\|x\|\|x\| = \|A\|,$$

as required.

(3) If $A = \alpha I$, then for any unit vector $x \in H$ we have

$$< Ax, x >=< \alpha Ix, x >= \alpha < x, x >= \alpha,$$

i.e. $W(A) = \{\alpha\}$.

Now, suppose that $W(A) = \{\alpha\}$. If x is a unit vector, then

$$< Ax, x >= \alpha = \alpha < x, x >=< \alpha x, x > .$$

Hence by Exercise 3.3.45, we have $A = \alpha I$.

The other property is easily verifiable and hence it is left to the interested reader.

(4) Easy, just use the definition of the numerical range and the sum of two sets.

(5) We have

$$\begin{aligned} W(A^*) &= \{< A^*x, x >: \ \|x\| = 1\} \\ &= \{< x, Ax >: \ \|x\| = 1\} \\ &= \{\overline{< Ax, x >}: \ \|x\| = 1\} \\ &= \{\overline{\lambda}: \ \lambda \in W(A)\}. \end{aligned}$$

(6) It is known that

$$A \text{ is self-adjoint } \Longleftrightarrow < Ax, x >\in \mathbb{R}, \ \forall x \in H.$$

Hence $< Ax, x >\in \mathbb{R}$, $\forall x \in H$ such that $\|x\| = 1$.

Conversely, if $x \in H$ is such that $\|x\| = 1$ and $< Ax, x >\in \mathbb{R}$, then for any $y \in H$ $(y \neq 0)$,

$$< A\frac{y}{\|y\|}, \frac{y}{\|y\|} >\in \mathbb{R}$$

or simply $< Ay, y > \in \mathbb{R}$ for any y (including $y = 0$) as $\frac{1}{\|y\|^2} \in \mathbb{R}$. Therefore, A is self-adjoint.

(7) We have for any $x \in H$:

$$< UAU^*x, x > = < AU^*x, U^*x >$$

Since U is unitary, $\|Ux\| = \|U^*x\| = \|x\|$ for any $x \in H$. Therefore,

$$W(A) = W(UAU^*).$$

Solution 8.3.11.

(1) We show that $W(A) = [0, 1]$. Let $x = (x_1, x_2) \in \mathbb{C}^2$ of norm equal to one. Then $Ax = (x_1, 0)$. Hence

$$< Ax, x > = < (x_1, 0), (x_1, x_2) > = x_1\overline{x_1} = |x_1|^2.$$

Since $\|x\|^2 = 1$, it becomes clear that

$$0 \leq < Ax, x > \leq |x_1|^2 + |x_2|^2 = \|x\|^2 = 1,$$

i.e. $W(A) \subset [0, 1]$.

Now, let $t \in [0, 1]$. Consider the *unit* (check it!) vector $(\sqrt{t}e^{i\theta}, 0)$, $\theta \in [0, 2\pi)$. Then

$$< Ax, x > = |\sqrt{t}e^{i\theta}|^2 = t|e^{i\theta}|^2 = t,$$

that is, $t \in W(A)$, and this proves the remaining half of the equality.

(2) We show that $W(B) = \{z \in \mathbb{C} : |z| \leq \frac{1}{2}\}$. Let $x = (x_1, x_2) \in \mathbb{C}^2$. Then $Bx = (x_2, 0)$. Hence

$$< Bx, x > = x_2\overline{x_1}.$$

So

$$| < Bx, x > | = |x_2\overline{x_1}| \leq \frac{1}{2}(|x_1|^2 + |x_2|^2) = \frac{1}{2}\|x\|^2 = \frac{1}{2}.$$

This demonstrates that $W(B) \subset \{z \in \mathbb{C} : |z| \leq \frac{1}{2}\}$.

Now, let $z = re^{i\theta}$, with $0 \leq r \leq \frac{1}{2}$. Take $x = (\cos\alpha, e^{i\theta}\sin\alpha)$ (remember x must be a unit vector and it is in this case!) with $\sin 2\alpha = 2r \leq 1$ and $0 \leq \alpha \leq \frac{\pi}{4}$. With this choice, we clearly have

$$< Bx, x > = e^{i\theta}\sin\alpha\overline{\cos\alpha} = e^{i\theta}\sin\alpha\cos\alpha = \frac{1}{2}e^{i\theta}\sin(2\alpha) = re^{i\theta} = z,$$

so that $W(B) \supset \{z \in \mathbb{C} : |z| \leq \frac{1}{2}\}$. The proof is over.

(3) We could compute $W(C)$ as we have done before, or much more quickly we may do the following: It is clear that we have

$$C = \begin{pmatrix} 3 & 0 \\ 0 & 1 \end{pmatrix} = \begin{pmatrix} 1 & 0 \\ 0 & 1 \end{pmatrix} + 2\begin{pmatrix} 1 & 0 \\ 0 & 0 \end{pmatrix} = I + 2A,$$

where A is the matrix of Question 1). Hence

$$W(C) = W(I + 2A) = W(I) + 2W(A) = \{1\} + [0, 2] = [1, 3].$$

Another way of seeing this is to use Proposition 8.1.19. Since C is a self-adjoint diagonal operator, we have (since $W(C)$ is closed in a finite dimensional setting)

$$W(C) = \overline{W(C)} = \operatorname{conv}(\{1, 3\}) = [1, 3]$$

(4) Obviously

$$D = \begin{pmatrix} 0 & 0 \\ 1 & 0 \end{pmatrix} = \begin{pmatrix} 0 & 1 \\ 0 & 0 \end{pmatrix}^* = B^*$$

(the B of Question 2). Hence

$$W(D) = W(B^*) = \{\overline{\lambda} : \lambda \in W(B)\} = \{z \in \mathbb{C} : |z| \leq \frac{1}{2}\}.$$

SOLUTION 8.3.12.

(1) Let $P \in B(H)$ be such that $P^* = P$, $P^2 = P$ and $P \neq 0, I$. Recall that $\|P\| = 1$. Hence if $x \in H$ is such that $\|x\| = 1$, then

$$\|Px\| \leq \|P\|\|x\| = \|x\| = 1.$$

We also have

$$< Px, x >=< P^2x, x >=< Px, P^*x >=< Px, Px >= \|Px\|^2$$

which shows that

$$0 \leq< Px, x >\leq 1.$$

Accordingly,

$$W(P) \subset [0, 1].$$

Let us now prove the reverse inclusion. Let $\lambda \in [0, 1]$.

By the solution of Exercise 7.3.24, we know that both P and $I-P$ are not injective. Recalling that $\operatorname{ran}P = \ker(I-P) \neq \{0\}$ we may then write

$$\exists y \in \ker P,\ y \neq 0 \text{ and } \exists z \in \operatorname{ran}P,\ z \neq 0.$$

WLOG we may assume that $\|y\| = \|z\| = 1$.

Now let

$$x = \sqrt{1-\lambda}y + \sqrt{\lambda}z$$

($\sqrt{1-\lambda}$ and $\sqrt{\lambda}$ have been introduced for scaling reasons). Since $\ker P$ and $\operatorname{ran}P$ are vector spaces, $\sqrt{1-\lambda}y \in \ker P$ and $\sqrt{\lambda}z \in \operatorname{ran}P$. Hence

$$x = \sqrt{1-\lambda}y + \sqrt{\lambda}z \in \ker P \oplus \operatorname{ran}P = \ker P \oplus (\ker P)^{\perp} = H.$$

Since $y \perp z$, we have

$$\begin{aligned} \|x\|^2 &= \|\sqrt{1-\lambda}y + \sqrt{\lambda}z\|^2 \\ &= \|\sqrt{1-\lambda}y\|^2 + \|\sqrt{\lambda}z\|^2 \\ &= (\sqrt{1-\lambda})^2 \underbrace{\|y\|^2}_{=1} + (\sqrt{\lambda})^2 \underbrace{\|z\|^2}_{=1} \\ &= 1 - \lambda + \lambda \\ &= 1. \end{aligned}$$

In the end, since $Px = \sqrt{\lambda}z$, we obtain

$$< Px, x > = \|Px\|^2 = \|\sqrt{\lambda}z\|^2 = \lambda,$$

i.e. $\lambda \in W(P)$, that is, $[0,1] \subset W(P)$.

(2) Let us re-consider the example of A which appeared in Exercise 8.3.11, that is,

$$A = \begin{pmatrix} 1 & 0 \\ 0 & 0 \end{pmatrix}.$$

Then it is evident that

(a) $A \neq 0, I$;
(b) A is self-adjoint;
(c) $A^2 = A$;

that is, A is a non-trivial orthogonal projection. Hence the result of Question (1) is applicable and gives

$$W(A) = [0,1].$$

Solution 8.3.13. Let

$$A = \begin{pmatrix} 1 & 0 \\ 0 & 0 \end{pmatrix} \text{ and } B = \begin{pmatrix} 0 & 0 \\ 0 & 1 \end{pmatrix}.$$

Hence both A and B are non-trivial orthogonal projections. Hence, by Exercise 8.3.12,

$$W(A) = W(B) = [0,1].$$

But

$$A + B = I \Longrightarrow W(A+B) = \{1\}.$$

Therefore,

$$W(A+B) \not\supset W(A) + W(B)$$

(for instance $2 = 1 + 1 \in W(A) + W(B)$ and $2 \notin W(A+B)$).

Solution 8.3.14. Let S be the shift operator on ℓ^2. We claim that

$$W(S) = B(0,1) = \{\lambda \in \mathbb{C} : \ |\lambda| < 1\}.$$

Recall that for any $x = (x_1, x_2, \cdots) \in \ell^2$,

$$S(x_1, x_2, \cdots) = (0, x_1, x_2, \cdots).$$

Then, if $\|x\| = 1$, it is clear by the Cauchy-Schwarz Inequality that we have

$$| < Sx, x > | \leq \|Sx\| \|x\| = \|x\| \|x\| = 1.$$

We claim that $| < Sx, x > | < 1$ so let us prove that $| < Sx, x > | \neq 1$. Assume for the sake of contradiction that $| < Sx, x > | = 1$. We know that the equality holds in the Cauchy-Schwarz Inequality iff each vector (non-zero) is a multiple of the other (and here also $\|x\| = 1$). So

$$| < Sx, x > | = 1 \Longleftrightarrow \|Sx\| = 1 \text{ and } Sx = \lambda x \text{ for some } \lambda \in \mathbb{C},$$

with $|\lambda| = 1$ necessarily (so that e.g. $\lambda = 0$ is excluded!). By a similar method to that applied to find the spectrum of the shift (see Exercise 7.3.14), we know that $Sx = \lambda x$ implies that $x = 0$. This is the desired contradiction as x is taken to be a unit vector. Thus we are half way through as we have only proved that

$$W(S) \subset B(0,1) = \{\lambda \in \mathbb{C} : \ |\lambda| < 1\}.$$

To prove the reverse inclusion, let $\lambda \in \mathbb{C}$ be such that $|\lambda| < 1$. By Exercise 7.3.14, any complex number inside the open unit disk is an eignevalue for S^*. Since $|\overline{\lambda}| = |\lambda| < 1$, $\overline{\lambda}$ is an eigenvalue for S^*. Let x be one corresponding eigenvector for $\overline{\lambda}$. If it is not normalized, just normalize it to make it of norm one. Now we have

$$< Sx, x > = < x, S^*x > = < x, \overline{\lambda} x > = \overline{\overline{\lambda}} \underbrace{< x, x >}_{=1} = \lambda,$$

that is, we have just proved that $\lambda \in W(S)$. Hence

$$B(0,1) = \{\lambda \in \mathbb{C} : \ |\lambda| < 1\} \subset W(S),$$

and the proof is therefore complete.

Solution 8.3.15.

(1) Since H is *finite* dimensional (with the usual metric on $\mathbb{C}$!), $W(A)$ is compact iff it is closed and bounded. But by Exercise 8.3.10, we already know that $W(A)$ is bounded. To prove its compactness, it only remains to show its closedness. To this end, let $\lambda_n \in W(A)$ be convergent to some $\lambda \in \mathbb{C}$. We must show that $\lambda \in W(A)$. Since $\lambda_n \in W(A)$,

$$\exists x_n \in H, \ \|x_n\| = 1 \text{ and } \lambda_n =< Ax_n, x_n >$$

Since $\dim H$ is finite, the closed unit ball is compact, equivalently, it is sequentially compact, so that (x_n) contains a convergent subsequence $(x_{nk})_k$ to some x, with $\|x\| = 1$. Thus

$$< Ax, x >= \lim_{k\to\infty} < Ax_{nk}, x_{nk} >= \lim_{n\to\infty} < Ax_n, x_n >= \lim_{n\to\infty} \lambda_n = \lambda,$$

i.e. $\lambda \in W(A)$.

(2) The hypothesis "finite dimensional" is primordial. Indeed, we have just seen in Exercise 8.3.14 that the shift operator S has a numerical range given by

$$W(S) = \{\lambda \in \mathbb{C} : \ |\lambda| < 1\}.$$

Hence $W(S)$ is not compact because it is not closed.

8.4. Hints/Answers to Tests

Solution 8.4.1.

(1) The matrix A has two eigenvalues 0 and 4. As A is self-adjoint, then $\|A\| = 4$.

(2) Observe that $BB^* = A$. Hence

$$\|B\|^2 = \|BB^*\| = \|A\| = 4,$$

i.e. $\|B\| = 2$.

Solution 8.4.2. Let $A \in B(H)$ be self-adjoint. We clearly have for all n

$$\|A^n\| \le \|A\|^n \Longrightarrow \|A^n\|^{\frac{1}{n}} \le \|A\|$$

so that

$$r(A) \le \|A\|.$$

Since either $\|A\|$ or $-\|A\|$ is in $\sigma(A)$, it follows that $\|A\|$ is in $\{|\lambda| : \ \lambda \in \sigma(A)\}$ so that $\|A\| = r(A)$, i.e. A is normaloid.

Solution 8.4.3. If P is a non-trivial orthogonal projection, then $\sigma(P) = \{0, 1\}$ (see Exercise 7.3.24). Since P is self-adjoint, we may then obtain by invoking the Spectral Radius Theorem:

$$\|P\| = \sup\{|\lambda| : \ \lambda \in \sigma(P)\} = \sup\{|\lambda| : \ \lambda \in \{0, 1\}\} = 1,$$

as required.

SOLUTION 8.4.4. Using one (or more) of the exercises above, there is nothing to prove! Only a good observation suffices... The answer is $W(A) = [0, 1]$.

SOLUTION 8.4.5. By Theorem 8.1.21, $0 \notin \overline{W(A)}$ yields $0 \notin \sigma(A)$, that is, A is invertible.

Assume for the sake of contradiction that $0 \in \overline{W(A^{-1})}$. Hence, for some sequence of unit vectors (x_n) we have

$$\lim_{n\to\infty} < A^{-1}x_n, x_n >= 0.$$

Hence

$$\lim_{n\to\infty} < \frac{A^{-1}x_n}{\|A^{-1}x_n\|}, A\left(\frac{A^{-1}x_n}{\|A^{-1}x_n\|}\right) >= 0$$

because the sequence $(\|A^{-1}x_n\|^{-1})$ is non-zero and bounded because A^{-1} is bounded below (and so $\|A^{-1}x_n\| \geq \alpha$ for some $\alpha > 0$).

Since $\frac{A^{-1}x_n}{\|A^{-1}x_n\|}$ is a unit vector, the previous will violate $0 \notin \overline{W(A)}$! Therefore, $0 \notin \overline{W(A^{-1})}$.

CHAPTER 9

Compact Operators

9.2. True or False: Answers

ANSWERS.

(1) True and this is in fact a simple application of Theorem 9.1.8. Indeed, by this theorem, we know that $K(H)$ is *closed* with respect to the topology of the operator norm. Since $K(H) \subset B(H)$ and $(B(H), \|\cdot\|_{B(H)})$ is a Banach space, so is $(K(H), \|\cdot\|_{B(H)})$.

(2) If A is compact, then so is A^2 (by Proposition 9.1.5). But if A^2 is compact, then A need not be compact, even if $A^2 = 0$. A counterexample may be found in Exercise 9.3.7. See the remark below Exercise 9.3.7 for further discussion.

(3) True if $\dim H < \infty$ but false if $\dim H = \infty$. Indeed, if A were compact, so would be A^n. Hence so would be I, which is a contradiction when $\dim H = \infty$.

(4) False! We will see that Volterra Operator is compact but it is not normaloid (see Exercise 9.3.21).

(5) True. We have that $U^*AU = B$ for some unitary $U \in B(H)$ (in fact, this equivalence is true for two *similar* compact operators). If A is compact, then so is B (by Proposition 9.1.5). If B is compact, reason similarly after writing $A = UBU^*$.

 REMARK. The question of when two compact operators are unitarily equivalent is an interesting problem in Operator Theory. See **[48]** for a necessary and sufficient condition that two compact *and normal* operators be unitarily equivalent. See also **[45]** for further reading.

(6) True. A proof is supplied in Exercise 9.3.16.

(7) If B is compact, then so is A. This is shown in Exercise 9.3.16. But if A is compact, then B need not be compact. As a counterexample, consider (on ℓ^2)

$$A(x_1, x_2, \cdots, x_n, \cdots) = \left(x_1, \frac{1}{2}x_2, \cdots, \frac{1}{n}x_n, \cdots\right).$$

Then A is compact (see Exercise 9.3.5). Let $B = I \in B(\ell^2)$. Then I is not compact. But, $0 \leq A \leq I$. Indeed, for any $x = (x_1, x_2, \cdots, x_n, \cdots)$,

$$
\begin{aligned}
< Ax, x > &=< (x_1, \frac{1}{2}x_2, \cdots, \frac{1}{n}x_n, \cdots), (x_1, x_2, \cdots, x_n, \cdots) > \\
&= |x_1|^2 + \frac{1}{2}|x_2|^2 + \cdots + \frac{1}{n}|x_n|^2 + \cdots \\
&\leq |x_1|^2 + |x_2|^2 + \cdots + |x_n|^2 + \cdots \\
&=< Ix, x > .
\end{aligned}
$$

9.3. Solutions to Exercises

SOLUTION 9.3.1.

(1) Since A is of finite rank, it is clear that $\operatorname{ran}(A)$ is a finite dimensional normed vector space. Let (x_n) be a sequence in H such that $\|x_n\| = 1$ for all n. Hence the sequence (Ax_n) is bounded in $\operatorname{ran}(A)$. Since $\dim \operatorname{ran}(A) < \infty$, we immediately infer that (Ax_n) has a convergence subsequence (why?), that is, A is compact.
(2) Since $\dim H < \infty$, we know that A is bounded. Also, $\operatorname{ran}A = A(H)$ is finite dimensional. By the previous question, A is therefore compact.
(3) Let (x_n) be an orthonormal sequence in H. Recall that this means that $\|x_n\| = 1$ for all n and that $< x_n, x_m >= 0$ for all $n \neq m$. Then

$$
\begin{aligned}
\|Ix_n - Ix_m\| &= \|x_n - x_m\| \\
&=< x_n - x_m, x_n - x_m > \\
&= \|x_n\|^2 - < x_n, x_m > - < x_m, x_n > + \|x_m\|^2 \\
&= 1 - 0 - 0 + 1 \\
&= 2.
\end{aligned}
$$

Thus (Ix_n) contains no Cauchy subsequence and a fortiori it contains no convergent subsequence.

SOLUTION 9.3.2. Assume that A is not bounded. So for all $n \in \mathbb{N}$, there exists a unit vector x_n in H such that $\|Ax_n\| \geq n$. But, A is compact and hence (Ax_n) has a convergent subsequence, call this subsequence (Ax_{n_k}). However, this is not consistent with $\|Ax_{n_k}\| \geq n_k$. Therefore, A must be bounded.

SOLUTION 9.3.3. Let

$$f_n(x) = \sqrt{n}\mathbb{1}_{[0,\frac{1}{n}]}(x).$$

Now, by computing $\|Af_n - Af_m\|$, show that (Af_n) cannot have a Cauchy subsequence...

SOLUTION 9.3.4.

(1) The linearity of A is clear since $< \cdot, \cdot >$ is an inner product. Let $x \in H$. We have

$$\|Ax\| = \| < x, t > t\| = | < x, t > | \; \|t\| \le \|x\| \; \|t\| \; \|t\| = \|x\| \; \|t\|^2.$$

This proves that $A \in B(H)$. It also gives that

$$\|A\| \le \|t\|^2.$$

If $t = 0$, then $Ax = 0$ so that $\|A\| = 0$. If $t \ne 0$, let $x_0 = \frac{t}{\|t\|}$. Hence

$$\|A\| = \sup_{\|x\|=1} \|Ax\| \ge \|Ax_0\| = \| < \frac{t}{\|t\|}, t > t\| = \|\|t\| \; t\| = \|t\| \; \|t\| = \|t\|^2,$$

which proves $\|A\| = \|t\|^2$.

(2) Let $x, y \in H$. We have

$$\begin{aligned} < Ax, y > &= << x, t > t, y > \\ &= < x, t >< t, y > \\ &= < t, y >< x, t > \\ &= \overline{< y, t >} < x, t > \\ &= < x, < y, t > t > \\ &= < x, Ay > . \end{aligned}$$

This proves that A is self-adjoint.

It is easy to see that A is positive since for any $x \in H$, we have

$$\begin{aligned} < Ax, x > &= << x, t > t, x > \\ &= < x, t >< t, x > \\ &= < x, t > \overline{< x, t >} \\ &= | < x, t > |^2 \ge 0. \end{aligned}$$

(3) Let $x \in H$. We then have

$$\begin{aligned} A^2x = A(Ax) = A(< x, t > t) &= << x, t > t, t > t \\ &= < x, t > \|t\|^2 t = \|t\|^2 Ax, \end{aligned}$$

that is, $A^2 = \lambda A$ where $\lambda = \|t\|^2 \ge 0$.

To find $\sqrt{A}$, notice that as $A^2 = \lambda A$ and A is positive, then it is clear (is it not?) that

$$A = \sqrt{A^2} = \sqrt{\lambda A} = \sqrt{\lambda}\sqrt{A}$$

so that

$$\sqrt{A} = \frac{1}{\sqrt{\lambda}} A \text{ if } \lambda > 0,$$

while if $\lambda = 0$, then $A^2 = 0$ or $A = 0$. To summarize

$$\sqrt{A} = \begin{cases} \frac{1}{\sqrt{\lambda}} A, & \lambda > 0, \\ 0, & \lambda = 0. \end{cases}$$

(4) Since $A^2 = \lambda A$, we may prove by induction that $A^n = \lambda^{n-1} A$ for all $n \in \mathbb{N}$. Hence

$$\|A^n\| = |\lambda|^{n-1} \|A\| \text{ or } \|A^n\|^{\frac{1}{n}} = |\lambda|^{\frac{n-1}{n}} \|A\|^{\frac{1}{n}}.$$

Passing to the limit as n tends to infinity leaves us with

$$|\lambda|^{\frac{n-1}{n}} \|A\|^{\frac{1}{n}} \longrightarrow |\lambda| = \lambda.$$

Thus

$$r(A) = \lim_{n \to \infty} \|A^n\|^{\frac{1}{n}} = \lambda = \|t\|^2.$$

Finally, since A is self-adjoint, we have

$$\|A\| = r(A) = \|t\|^2.$$

(5) A is compact because it is a finite rank operator, and it is of a finite rank because its range is included in $\operatorname{span}\{t\}$.

Solution 9.3.5.

(1) Assume that $\lim_{n \to \infty} \lambda_n = 0$. Define for all $x = (x_1, \cdots, x_n, \cdots)$ in ℓ^2,

$$A_k(x_1, x_2, \cdots, x_n, \cdots) = (\lambda_1 x_1, \lambda_2 x_2, \cdots, \lambda_k x_k, 0, \cdots).$$

Then obviously A_k is bounded and is of a finite rank (since $\operatorname{ran}(A_k)$ is spanned by the first k vectors of $\{e_1, e_2, \cdots, e_k, \cdots\}$

of the standard basis of ℓ^2). Now, we may write for all $x \in \ell^2$

$$\begin{aligned}\|(A-A_k)x\|_2^2 &= \|(\lambda_1 x_1, \lambda_2 x_2, \cdots, \lambda_n x_n, \cdots) \\ &\quad - (\lambda_1 x_1, \lambda_2 x_2, \cdots, \lambda_k x_k, 0, \cdots)\|_2^2 \\ &= \|(0, \cdots, 0, \lambda_{k+1} x_{k+1}, \lambda_{k+2} x_{k+2}, \cdots)\|_2^2 \\ &= \sum_{n=k+1}^{\infty} |\lambda_n|^2 |x_n|^2 \\ &\leq \sup_{n \geq k+1} |\lambda_n|^2 \sum_{n=k+1}^{\infty} |x_n|^2 \\ &\leq \sup_{n \geq k+1} |\lambda_n|^2 \sum_{n=1}^{\infty} |x_n|^2 \\ &= \sup_{n \geq k+1} |\lambda_n|^2 \|x\|^2.\end{aligned}$$

Hence

$$\|A - A_k\| \leq \sup_{n \geq k+1} |\lambda_n|$$

and letting $k \to \infty$ yields (remembering that $\lim_{n\to\infty} \lambda_n = 0$)

$$\lim_{k\to\infty} \|A - A_k\| = 0.$$

Thus A is the limit (w.r.t. the operator norm) of a sequence of finite rank operators signifying that A is compact.

Conversely, assume that A is compact. We must show that $\lim_{n\to\infty} \lambda_n = 0$. Assume, for the sake of contradiction, that $\lim_{n\to\infty} \lambda_n \neq 0$. Hence there exists $\varepsilon > 0$, and a subsequence (λ_{n_k}) such that for all $k \in \mathbb{N}$, we have $|\lambda_{n_k}| \geq \varepsilon$. Now, for any $k \neq l$, we have (if (e_n) is the standard basis on ℓ^2)

$$\begin{aligned}\|Ae_{n_k} - Ae_{n_l}\|^2 &= \|(0, \cdots, 0, \underset{k}{\lambda_{n_k}}, 0, \cdots) - (0, \cdots, 0, \underset{l}{\lambda_{n_l}}, 0, \cdots)\|^2 \\ &= \|(0, \cdots, 0, \underset{k}{\lambda_{n_k}}, 0, \cdots, \underset{l}{-\lambda_{n_l}}, 0, \cdots)\|^2 \\ &= |\lambda_{n_k}|^2 + |\lambda_{n_l}|^2 \\ &\geq 2\varepsilon^2,\end{aligned}$$

and this yields the non-compactness of A, completing the proof.

(2) We have

$$I(x_1, x_2, \cdots, x_n, \cdots) = (x_1, x_2, \cdots, x_n, \cdots),$$

i.e. it corresponds to A in the case $\lambda_n = 1$ for all $n \in \mathbb{N}$. Hence I is not compact for this λ_n does not tend to zero as n tends to infinity.

(3) Let (e_n) be the standard basis of ℓ^2. Then obviously (for each n)

$$\|Ae_n\|^2 = |\lambda_n|^2$$

and so

$$\sum_{n=1}^{\infty} \|Ae_n\|^2 = \sum_{n=1}^{\infty} |\lambda_n|^2.$$

It then becomes apparent that A is a Hilbert-Schmidt operator iff $\sum_{n=1}^{\infty} |\lambda_n|^2$ converges.

Solution 9.3.6. First A is clearly non-invertible: Indeed, since $\lim_{n\to\infty} \frac{1}{n} = 0$, A is compact; and as $\dim \ell^2 = \infty$, A cannot be invertible.

Now, let $x = (x_1, x_2, \cdots, x_n, \cdots) \neq 0_{\ell^2}$. Then

$$< Ax, x >= \sum_{n=1}^{\infty} \frac{|x_n|^2}{n} > 0,$$

i.e. A is strictly positive.

Solution 9.3.7. First, A^2 is compact as $A^2 = 0$. Indeed, for all $x \in \ell^2$ we have

$$A^2(x) = A^2(x_1, x_2, x_3, \cdots) = A(A(x_1, x_2, x_3, \cdots)) = A(0, x_1, 0, x_3, \cdots)$$

and hence

$$A^2(x_1, x_2, x_3, \cdots) = (0, 0, 0, , \cdots).$$

Let us now turn to proving that A is not compact. We may show that (Ax_n) has no convergent subsequence whichever bounded (x_n) in H. Alternatively, we may proceed as follows:

It is easy to see that

$$A^*(x_1, x_2, x_3, \cdots) = (x_2, 0, x_4, 0, \cdots)$$

so that

$$A^*A(x_1, x_2, x_3, \cdots) = (x_1, 0, x_3, 0, \cdots)$$

and

$$AA^*(x_1, x_2, x_3, \cdots) = (0, x_2, 0, x_4, \cdots).$$

So if A were compact, by Proposition 9.1.5, so would be A^*A and AA^* and hence so would be $A^*A + AA^*$ too, that is,

$$(A^*A + AA^*)(x_1, x_2, x_3, \cdots) = (x_1, x_2, x_3, \cdots) = I(x_1, x_2, x_3, \cdots)$$

would be compact, a contradiction! Thus A cannot be compact.

Solution 9.3.8.

(1) Let (x_n) be a bounded sequence in H. As A is compact, then (Ax_n) has a convergent subsequence, denoted by (Ax_{n_k}). Since (x_{n_k}) is bounded (do not reason directly with (x_n) and B!), and B is compact, we have that (Bx_{n_k}) too has a convergent subsequence, which we denote by $(Bx_{n_{k_l}})$. Hence clearly the sequence $(\alpha Ax_{n_{k_l}} + \beta Bx_{n_{k_l}})$ converges, i.e. $\alpha A + \beta B$ is compact. Accordingly, $K(H)$ is a linear subspace of $B(H)$.

(2) Let (x_n) be a bounded sequence.

  (a) Since A is compact, (Ax_n) has a convergent subsequence denoted by (Ax_{n_k}). Since B is bounded, (BAx_{n_k}) converges, i.e. BA is compact.

  (b) The proof for the compactness of AB is slightly different. Since (x_n) is bounded (by M say), so is (Bx_n) as

$$\|Bx_n\| \leq \|B\|\|x_n\| \leq M\|B\|.$$

  Since (Bx_n) is a bounded sequence, and A is compact, (ABx_n) has a convergent subsequence showing that AB is compact.

(3) If A is compact, then the previous question tells us that A^*A is compact.

Conversely, suppose that A^*A is compact. To show that A is compact, let (x_n) be a bounded sequence in H. Hence, (A^*Ax_n) admits a convergent subsequence, denoted by $(A^*Ax_{n_k})$. Now, we may write (for all $n, l \in \mathbb{N}$)

$$\begin{aligned}
\|Ax_{n_k} - Ax_{n_l}\|^2 &= \|A(x_{n_k} - x_{n_l})\|^2 \\
&=< A(x_{n_k} - x_{n_l}), A(x_{n_k} - x_{n_l}) > \\
&=< A^*A(x_{n_k} - x_{n_l}), (x_{n_k} - x_{n_l}) > \\
&\leq \|A^*A(x_{n_k} - x_{n_l})\|\|x_{n_k} - x_{n_l}\| \\
&\leq M\|A^*A(x_{n_k} - x_{n_l})\| \text{ (for some } M > 0) \\
&= M\|A^*Ax_{n_k} - A^*Ax_{n_l}\|.
\end{aligned}$$

Since $(A^*Ax_{n_k})$ converges, it is Cauchy. Hence, it is clear from the previous inequality that (Ax_{n_k}) too is a Cauchy sequence in H which is complete. Thus, (Ax_{n_k}) must converge. This proves that $A \in K(H)$.

The proof is very similar if we replace A^*A by AA^*.

(4) If A is compact, then AA^* is compact because A^* is bounded. By the previous question, A^* is compact.

Now, if A^* is compact, then by other half of this proof, A^{**} is compact, that is, A is compact.

SOLUTION 9.3.9. The answer is no. Since $A \in B(H)$ is bounded below, we have for some $\alpha > 0$ and all $x \in H$,

$$\|Ax\| \geq \alpha\|x\|.$$

Let (x_n) be an *orthonormal* sequence in H. Hence for all $n \neq m$,

$$\|Ax_n - Ax_m\| = \|A(x_n - x_m)\| \geq \alpha\|x_n - x_m\| = \alpha\sqrt{2}.$$

Therefore, (Ax_n) cannot have a convergent subsequence. Thus A is not compact.

SOLUTION 9.3.10. Let $A \in B(H)$ be a Hilbert-Schmidt operator. We show that A is expressible as a limit, w.r.t. the operator norm, of finite rank operators. Let (e_n) be an orthonormal basis of H. Hence any x may be expressed as $x = \sum_{n=1}^{\infty} < x, e_n > e_n$.

Now, let $x \in H$ and define for each $k \in \mathbb{N}$, a sequence of operators $A_k \in B(H)$ by

$$A_k x = A\left(\sum_{n=1}^{\infty} < x, e_n > e_n\right) = \sum_{n=1}^{k} < x, e_n > Ae_n.$$

Before carrying on the proof, we observe that the rank of A_k is at most k, i.e. (A_k) is indeed a sequence of finite rank operators.

Now, let $x \in H$. Then

$$\begin{aligned}
\|(A_k - A)x\| &= \left\|\sum_{n=1}^{k} < x, e_n > Ae_n - \sum_{n=1}^{\infty} < x, e_n > Ae_n\right\| \\
&= \left\|\sum_{n=k+1}^{\infty} < x, e_n > Ae_n\right\| \\
&\leq \sum_{n=k+1}^{\infty} | < x, e_n > | \|Ae_n\| \\
&\leq \left(\sum_{n=k+1}^{\infty} | < x, e_n > |^2\right)^{\frac{1}{2}} \left(\sum_{n=k+1}^{\infty} \|Ae_n\|^2\right)^{\frac{1}{2}} \\
&\leq \left(\sum_{n=1}^{\infty} | < x, e_n > |^2\right)^{\frac{1}{2}} \left(\sum_{n=k+1}^{\infty} \|Ae_n\|^2\right)^{\frac{1}{2}} \\
&\leq \|x\| \left(\sum_{n=k+1}^{\infty} \|Ae_n\|^2\right)^{\frac{1}{2}} \text{ (by the Bessel Inequality).}
\end{aligned}$$

Therefore, for all $k \in \mathbb{N}$, we have

$$0 \leq \|A_k - A\| \leq \left(\sum_{n=k+1}^{\infty} \|Ae_n\|^2 \right)^{\frac{1}{2}}.$$

Passing to the limit as $k \to \infty$, and observing that $\sum_{n=k+1}^{\infty} \|Ae_n\|^2$ is the remainder of a convergent series, we finally obtain

$$\lim_{k \to \infty} \|A_k - A\| = 0,$$

establishing the result.

Solution 9.3.11. First, recall in the case of infinite intervals, and using Tonelli Theorem (cf. [**211**]) that as $K \in L^2((a,b) \times (c,d))$, then $|K(x,\cdot)|^2 \in L^1(a,b)$

Having said this, let's now proceed with the proof. Since $L^2(c,d)$ is separable, let $\{e_n\}$ be an orthonormal basis in $L^2(c,d)$. Then for any $x \in (a,b)$, we have

$$Ae_n(x) = \int_c^d K(x,y)e_n(y)dy = < K(x,\cdot), \overline{e_n} > .$$

Hence

$$\|Ae_n\|^2 = \int_a^b |Ae_n(x)|^2 dx = \int_a^b | < K(x,\cdot), \overline{e_n} > |^2 dx.$$

To finish the proof, we have

$$\begin{aligned}
\sum_{n=1}^{\infty} \|Ae_n\|^2 &= \sum_{n=1}^{\infty} \int_a^b | < K(x,\cdot), \overline{e_n} > |^2 dx \\
&= \int_a^b \sum_{n=1}^{\infty} | < K(x,\cdot), \overline{e_n} > |^2 dx \text{ (why?)} \\
&= \int_a^b \|K(x,\cdot)\|^2 dx \text{ (by the Parseval Formula)} \\
&= \int_a^b \int_c^d |K(x,y)|^2 dx dy \\
&< \infty \text{ (by assumption).}
\end{aligned}$$

Solution 9.3.12. It is clear by Exercise 9.3.5 that A is compact merely because $\lim_{n\to\infty} \frac{1}{\sqrt{n}} = 0$.

To show that A is not Hilbert-Schmidt, let (e_n) be the standard orthonormal basis of ℓ^2. Then we have

$$\sum_{n=1}^{\infty} \|Ae_n\|^2 = \sum_{n=1}^{\infty} \|n^{-\frac{1}{2}} e_n\|^2 = \sum_{n=1}^{\infty} \frac{1}{n} = \infty.$$

Accordingly, A is not Hilbert-Schmidt by Theorem 9.1.20.

SOLUTION 9.3.13. Let $S \in B(\ell^2)$ be the usual shift operator. Then S is not compact because:

(1) $\sigma(S)$ is uncountable (Exercise 7.3.14), *or because*
(2) If S were compact, then so would be S^*S (why?). But $S^*S = I$, and hence I would be compact, a contradiction as $\dim \ell^2 = \infty$! *or because*
(3) If S were compact, every non-zero element of $\sigma(S)$ would be an eigenvalue, but this cannot happen as $\sigma_p(S) = \varnothing$ *or because*
(4) If S were compact, then we would have (cf. the remark just below Proposition 7.1.35)

$$\sigma(SS^*) = \sigma(S^*S)$$

but this is not the case (see Exercise 7.3.14) *or because*
(5) ...

SOLUTION 9.3.14.

(1) Assume that $0 \notin \sigma(A)$, that is, $0 \in \rho(A)$. Hence A is invertible and so AA^{-1} exists and would be compact, i.e. I would be compact which is not consistent with $\dim H = \infty$. Thus $0 \in \sigma(A)$.
(2) We know that the spectrum of a compact operator is at most countable. If $\overbrace{\sigma(A)}^{\circ} \neq \varnothing$, then it must contain at least one complex number, λ say. That is, by definition

$$\exists r > 0,\ B(\lambda, r) \subset \sigma(A),$$

where $B(\lambda, r)$ is the open ball in $\mathbb{C}$ with center λ and radius r. Obviously, this cannot happen as $B(\lambda, r)$ is uncountable while $\sigma(A)$ is countable. Thus $\overbrace{\sigma(A)}^{\circ} = \varnothing$.
(3) Since H is infinite dimensional, we know that $0 \in \sigma(A)$. Also, since $\sigma(A)$ is closed and has empty interior, we may write in the end

$$\partial\sigma(A) = \overline{\sigma(A)} \setminus \overbrace{\sigma(A)}^{\circ} = \sigma(A).$$

Since $0 \in \sigma(A)$, we also have $0 \in \partial\sigma(A)$. Now, by Exercise 7.3.10, we know that $\partial\sigma(A) \subset \sigma_a(A)$. Hence $0 \in \sigma_a(A)$.
(4) Since we know that (for any $A \in B(H)$),

$$\sigma_p(A) \subset \overline{\sigma_p(A)} \subset \sigma_a(A).$$

It only remains to prove that $\sigma_a(A)-\{0\} \subset \sigma_p(A)-\{0\}$. Let λ be a *non zero* approximate eigenvalue for A. This means that for some $x_n \in H$ such that $\|x_n\| = 1$, we have $\|\lambda x_n - Ax_n\| \to 0$ as n tends to ∞. Since A is compact, we do know that (Ax_n) has a convergent subsequence, call it $(Ax_{n(k)})_k$ and let y be its limit (in H!).

Now, since $\lambda x_n - Ax_n$ converges to zero, so does $\lambda x_{n(k)} - Ax_{n(k)}$ (in H). Hence we may write

$$\lambda x_{n(k)} = (\lambda x_{n(k)} - Ax_{n(k)}) + Ax_{n(k)} \longrightarrow 0 + y \text{ or } x_{n(k)} \longrightarrow \frac{y}{\lambda}.$$

Since A is continuous (and linear!), we obtain

$$Ax_{n(k)} \longrightarrow A\left(\frac{y}{\lambda}\right) = \frac{1}{\lambda}Ay.$$

Since also $Ax_{n(k)} \to y$, we get by the uniqueness of the limit that $Ay = \lambda y$.

Now, we proceed to show that y is an eigenvector associated with λ (i.e. we have to show that $y \neq 0$). Since $\lambda x_{n(k)} \to y$ (in H) and $\|x_{n(k)}\| = 1$, we may write

$$0 \leq ||\lambda| - \|y\|| = |\|\lambda x_{n(k)}\| - \|y\|| \leq \|\lambda x_{n(k)} - y\| \longrightarrow 0$$

as k goes to infinity. Hence $|\lambda| = \|y\|$ so that $y \neq 0$ for $\lambda \neq 0$. Thus λ is an eigenvalue, that is, $\lambda \in \sigma_p(A) - \{0\}$.

(5) Recall that $\sigma(A) = \sigma_p(A) \cup \{0\}$. By the previous question, we have $\sigma_p(A) - \{0\} = \sigma_a(A) - \{0\}$ or $\sigma_p(A) \cup \{0\} = \sigma_a(A) \cup \{0\}$ (why?). Therefore

$$\sigma(A) = \sigma_a(A) \cup \{0\}.$$

But, we have just seen that $0 \in \sigma_a(A)$. Thus

$$\sigma(A) = \sigma_a(A),$$

as needed.

SOLUTION 9.3.15. It is clear that AS is compact. Since $\dim \ell^2 = \infty$, $0 \in \sigma(AS)$. All other eventual elements of $\sigma(AS)$ are just eigenvalues.

Let $\lambda \neq 0$ be an eigenvalue of AS and let $0 \neq y \in \ell^2$ be its associated eigenvector. Then $ASy = \lambda y$, i.e.

$$ASy = (0, \alpha_2 y_1, \alpha_3 y_2, \cdots) = (\lambda y_1, \lambda y_2, \cdots).$$

Hence $\lambda y_1 = 0$ which, since $\lambda \neq 0$, yields $y_1 = 0$. Similarly $y_2 = 0$, etc... Hence $y_n = 0$ for all n, that is, $y = 0$. Therefore, there cannot be non-zero eigenvalues.

It only remains to treat the case whether 0 is an eigenvalue for AS. Two cases must be investigated (remembering that $\alpha_n \to 0$):

(1) If $\alpha_n \neq 0$ for all $n \in \mathbb{N}$, then $ASy = 0$ (still with $y \neq 0$) will certainly lead to $y = 0$. In such case, 0 is not an eigenvalue for AS and hence $\sigma_p(AS) = \varnothing$ and $\sigma(AS) = \{0\}$.

(2) If $\alpha_n = 0$ for some $n \in \mathbb{N}$, and if (e_n) is the standard basis of ℓ^2, then

$$ASe_{n-1} = (0, \alpha_2 \cdot 0, \alpha_3 \cdot 0, \cdots, \alpha_n \cdot 1, \cdots) = 0e_{n-1},$$

i.e. the non-zero (e_{n-1}) is associated with the eigenvalue 0. Thus

$$\sigma_p(AS) = \sigma(AS) = \{0\}.$$

SOLUTION 9.3.16.

(1) If $\sqrt{A}$ is compact, then so is $A = (\sqrt{A})^2$ by Proposition 9.1.5.

Conversely, assume that A is compact. To show that $\sqrt{A}$ is compact, let (x_n) be a sequence in H such that $\|x_n\| = 1$. Since $\sqrt{A}$ is self-adjoint, we may write for all $x \in H$

$$\|\sqrt{A}x\|^2 =< \sqrt{A}x, \sqrt{A}x >=< (\sqrt{A})^2 x, x >=< Ax, x >\leq \|Ax\|\|x\|.$$

Since $\|x_n\| = 1$ and A is compact, we know that (Ax_n) has a convergent subsequence, which we denote by (Ax_{n_k}). Hence it is Cauchy. By the previous inequality, we can assert that $(\sqrt{A}x_{n_k})$ too is Cauchy (in the complete H). Hence it converges, which settles the question of the compactness of $\sqrt{A}$.

(2) If A is compact, so is A^*A. But this latter is positive, and hence $\sqrt{A^*A}$ is compact by the previous question, that is, $|A|$ is compact.

Now, if $|A|$ is compact, then so is $|A|^2$ or A^*A. By Exercise 9.3.8, A is compact.

REMARK. The previous implication may have also been established by writing $A = U|A|$, where U is a partial isometry. Then if $|A|$ is compact, then $U|A|$ is also compact, that is, A is compact.

(3) Let $x \in H$. Since $A \geq 0$, $\sqrt{A}$ is self-adjoint, and hence we may write

$$< Ax, x >=< (\sqrt{A})^2 x, x >=< \sqrt{A}x, \sqrt{A}x >= \|\sqrt{A}x\|^2.$$

Similarly,

$$< Bx, x >= \|\sqrt{B}x\|^2.$$

As $A \leq B$, then

$$\|\sqrt{A}x\|^2 \leq \|\sqrt{B}x\|^2$$

for all $x \in H$. Now, let (x_n) be in H such that $\|x_n\| = 1$ (for all n). Since B is compact, $\sqrt{B}$ is compact and hence $(\sqrt{B}x_n)$ admits a convergent subsequence $(\sqrt{B}x_{n_k})$. Since this latter is Cauchy, the last displayed inequality gives us that $(\sqrt{A}x_{n_k})$ is equally Cauchy and hence it converges. Thus $\sqrt{A}$ is compact or A is compact. So much for the longer proof.

There is another much simpler way to prove this result: Indeed, since $A \leq B$, we have

$$\|\sqrt{A}x\|^2 \leq \|\sqrt{B}x\|^2$$

for all $x \in H$. Hence, Theorem 3.1.69 tells us that $\sqrt{A} = K\sqrt{B}$ for some contraction $K \in B(H)$. Because B is compact, so is $\sqrt{B}$. Hence so is $\sqrt{A}$ and finally so is A.

SOLUTION 9.3.17.

(1) It is clear that $(x, y) \mapsto K(x, y) \in L^2([0, 1]^2)$. By Theorem 9.1.23, A is compact. Since

$$K(x, y) = \overline{K(y, x)},$$

A is self-adjoint.

(2) Let $Af = \lambda f$, with $\lambda \neq 0$. Rewriting Af (for all x and all f) as

$$Af(x) = (x + 1) \int_x^1 yf(y)dy + x \int_0^x (y + 1)f(y)dy,$$

we clearly see that Af is continuous. Now, since $\lambda \neq 0$, we can write

$$Af = \lambda f \Longrightarrow f = \lambda^{-1}Af,$$

so that f too is continuous.

(3) Let $Af = \lambda f$, with $\lambda \in \mathbb{R}$. To find the eigenvalues we have to solve

$$(x + 1) \int_x^1 yf(y)dy + x \int_0^x (y + 1)f(y)dy = \lambda f(x).$$

Differentiating twice with respect to x, we obtain

$$\lambda f''(x) - f(x) = 0, \; f(0) = f'(0) \text{ and } f(1) = f'(1).$$

We then see that 0 cannot be an eigenvalue for A as otherwise we would have $f = 0$! Now, let $Af = \lambda f$, with $\lambda \neq 0$. We need to investigate the roots of $\lambda r^2 - 1 = 0$, where $\lambda \in \mathbb{R}^*$. Hence $r^2 = \frac{1}{\lambda}$.

(a) $\lambda > 0$: The solution is then

$$f(x) = \alpha e^{\left(\frac{x}{\sqrt{\lambda}}\right)} + \beta e^{-\left(\frac{x}{\sqrt{\lambda}}\right)}.$$

As $f(0) = f'(0)$ and $f(1) = f'(1)$, then

$$\begin{cases} (\sqrt{\lambda} - 1)\alpha + (\sqrt{\lambda} + 1)\beta = 0, \\ (\sqrt{\lambda} - 1)e^{\frac{1}{\sqrt{\lambda}}}\alpha + (\sqrt{\lambda} + 1)e^{-\frac{1}{\sqrt{\lambda}}}\beta = 0. \end{cases}$$

If the determinant of the previous system is non zero, that is, in case $\lambda \neq 1$, then $(\alpha, \beta) = (0, 0)$ is the unique solution. Hence $f = 0$ and we do not want that. Now, if $\lambda = 1$, then

$$f(x) = \alpha e^x, \ \alpha \in \mathbb{R}.$$

Therefore, $1 \in \sigma_p(A)$.

(b) $\lambda < 0$: Then $r = \pm \frac{i}{\sqrt{-\lambda}}$. Whence, the solution is

$$f(x) = \alpha \cos\left(\frac{x}{\sqrt{-\lambda}}\right) + \beta \sin\left(\frac{x}{\sqrt{-\lambda}}\right).$$

As $f(0) = f'(0)$ and $f(1) = f'(1)$, then

$$\begin{cases} \alpha = \frac{\beta}{\sqrt{-\lambda}} \\ \beta(1 - \frac{1}{\lambda}) \sin\left(\frac{1}{\sqrt{-\lambda}}\right) = 0. \end{cases}$$

As above, to have a non zero eigenvector, we need to have $\sin \frac{1}{\sqrt{-\lambda}} = 0$, i.e.

$$\lambda = -(n^2\pi^2)^{-1}, \ n \in \mathbb{N}.$$

The corresponding eigenvectors are then

$$n\pi \cos(n\pi x) + \sin(n\pi x), \ n \in \mathbb{N}.$$

(4) No! If A were positive, the fact that A is self-adjoint would then mean that necessarily $\sigma(A) \subset \mathbb{R}^+$. But, we have just seen that $\sigma(A)$ contains negative values! Therefore, A is not positive (and yet its kernel $K \geq 0$).

SOLUTION 9.3.18. As before, A is self-adjoint and compact.

(1) We find that the eigenvalues are $\frac{1}{(n\pi)^2}$, $n \in \mathbb{N}$. Hence

$$\sigma(A) = \left\{\frac{1}{(n\pi)^2} : \ n \in \mathbb{N}\right\} \cup \{0\}.$$

(2) Yes, A is positive since $\sigma(A) \subset \mathbb{R}^+$ and A is self-adjoint.

(3) Since A is self-adjoint, the Spectral Radius Theorem yields

$$\|A\| = \frac{1}{\pi^2}.$$

Solution 9.3.19. Let A be such that $\dim \operatorname{ran}(A) < \infty$. Then A is compact. Assume for the sake of contradiction that $\sigma(A)$ is infinite (it has to be countable!). Hence A has an infinite number of (non-zero) eigenvalues, which we denote by $\{\lambda_n : n \in \mathbb{N}\}$. Let $\{x_n : n \in \mathbb{N}\}$ be the associated eigenvectors. This means that

$$\forall n \in \mathbb{N} : \; Ax_n = \lambda_n x_n.$$

Setting $X = \{x_1, \cdots, x_n, \cdots\}$, we immediately see that $X \subset \operatorname{ran}(A)$ as (for any n)

$$Ax_n = \lambda_n x_n \Longrightarrow x_n = \lambda_n^{-1} A x_n = A(\lambda_n^{-1} x_n) \in \operatorname{ran}(A)$$

which is not consistent with $\dim \operatorname{ran}(A) < \infty$! Therefore, $\sigma(A)$ is a finite set.

Solution 9.3.20. The result follows from the fact that if A is compact, then $\sigma(A) = \sigma_p(A) \cup \{0\}$...

Solution 9.3.21.

(1) Let f be in $L^2[0,1]$. We may write

$$(Vf)(x) = \int_0^x f(t)dt = \int_0^1 \mathbb{1}_{[0,x]}(t) f(t)dt,$$

so that

$$|(Vf)(x)| \leq \int_0^1 \mathbb{1}_{[0,x]}(t)|f(t)|dt \leq \sqrt{x}\|f\|_2,$$

by the Cauchy-Schwarz Inequality. Hence

$$\|Vf\|_2^2 = \int_0^1 |(Vf)(x)|^2 dx \leq \|f\|_2^2 \int_0^1 x dx = \frac{1}{2}\|f\|_2^2,$$

so that $\|V\| \leq \frac{1}{\sqrt{2}}$, proving the boundedness of V on $L^2[0,1]$.

(2) Let $f \in L^2[0,1]$. We have

$$\begin{aligned}
\|Vf\|_2^2 &= \int_0^1 |Vf(x)|^2 dx \\
&= \int_0^1 \left| \int_0^x f(t)dt \right|^2 dx \\
&= \int_0^1 \left| \int_0^x \sqrt{\cos(\frac{\pi}{2}t)} \frac{f(t)}{\sqrt{\cos(\frac{\pi}{2}t)}} dt \right|^2 dx \\
&\leq \int_0^1 \left(\int_0^x \cos(\frac{\pi}{2}t)dt \int_0^x \frac{|f(t)|^2}{\cos(\frac{\pi}{2}t)} dt \right) dx
\end{aligned}$$

by the Cauchy-Schwarz Inequality. But it is clear that

$$\int_0^x \cos(\frac{\pi}{2}t)dt = \frac{2}{\pi}\sin(\frac{\pi}{2}x).$$

Hence

$$\begin{aligned}
\|Vf\|_2^2 &\leq \frac{2}{\pi}\int_0^1 \int_0^x \sin(\frac{\pi}{2}x) \frac{|f(t)|^2}{\cos(\frac{\pi}{2}t)} dtdx \\
&= \frac{2}{\pi}\int_0^1 \int_0^1 \mathbb{1}_{[0,x]}(t) \sin(\frac{\pi}{2}x) \frac{|f(t)|^2}{\cos(\frac{\pi}{2}t)} dtdx \\
&= \frac{2}{\pi}\int_0^1 \int_0^1 \mathbb{1}_{[t,1]}(x) \sin(\frac{\pi}{2}x) \frac{|f(t)|^2}{\cos(\frac{\pi}{2}t)} dtdx \\
&= \frac{2}{\pi}\int_0^1 \left(\int_t^1 \sin(\frac{\pi}{2}x)dx \right) \frac{|f(t)|^2}{\cos(\frac{\pi}{2}t)} dt \\
&= \frac{2}{\pi}\int_0^1 \frac{2}{\pi}\cos(\frac{\pi}{2}t) \frac{|f(t)|^2}{\cos(\frac{\pi}{2}t)} dt \\
&= \frac{4}{\pi^2}\int_0^1 \cos(\frac{\pi}{2}t) \frac{|f(t)|^2}{\cos(\frac{\pi}{2}t)} dt \\
&= \frac{4}{\pi^2}\int_0^1 |f(t)|^2 dt \\
&= \frac{4}{\pi^2}\|f\|_2^2.
\end{aligned}$$

Therefore,

$$\|Vf\|_2 \leq \frac{2}{\pi}\|f\|_2 \text{ so that } \|V\| \leq \frac{2}{\pi}.$$

The norm is attained at $f_0(t) = \cos(\frac{\pi}{2}t)$. We have

$$\|f_0\|_2^2 = \int_0^1 \cos^2(\frac{\pi}{2}t)dt = \frac{1}{2}[t + \frac{1}{\pi}\sin(\pi t)]_0^1 = \frac{1}{2}$$

or $\|f_0\|_2 = \frac{1}{\sqrt{2}}$. Now, we have

$$Vf_0(x) = \int_0^x \cos(\frac{\pi}{2}t)dt = \frac{2}{\pi}\sin(\frac{\pi}{2}x).$$

Hence

$$\|Vf_0\|_2^2 = \int_0^1 \frac{4}{\pi^2}\sin^2(\frac{\pi}{2}x)dx = \frac{4}{\pi^2} \times \frac{1}{2}[x - \frac{1}{\pi}\sin(\pi x)]_0^1 = \frac{2}{\pi^2}$$

or

$$\|Vf_0\|_2 = \frac{\sqrt{2}}{\pi}.$$

Finally,

$$\|V\| = \sup_{f\neq 0}\frac{\|Vf\|_2}{\|f\|_2} \geq \frac{\|Vf_0\|_2}{\|f_0\|_2} = \frac{\sqrt{2}}{\pi}\frac{\sqrt{2}}{1} = \frac{2}{\pi},$$

which, combined with what we saw above, implies that

$$\|V\| = \frac{2}{\pi}.$$

Remarks.

(a) The preceding elegant proof is due to some guy from Poland (apparently a physicist!). See [**170**]. I came across it by chance. I was not aware of it and apparently neither was "The Halmos"! (who said in Problem 188 in [**94**] that: "A direct attack on the problem seems not to lead anywhere.").

(b) When defined on $L^p(0,1)$, where $1 < p < \infty$, the norm of the Volterra Operator is given by

$$\|V\|_{B(L^p(0,1))} = \frac{p\sin(\frac{\pi}{p})}{\pi(p-1)^{\frac{1}{p}}}.$$

For a proof, see e.g. [**139**]. The sharp cases $p = 1$ and $p = \infty$ are easy to handle, and the reader may check that the norm in this case is equal to one.

(3) Let us start by finding V^2. Let $f \in L^2[0,1]$. We have

$$V^2f(x) = V(Vf(x)) = \int_0^x Vf(t)dt = \int_0^x\int_0^t f(s)dsdt$$

and hence

$$V^2f(x) = \int_0^x \int_0^1 \mathbb{1}_{[0,t]}(s)f(s)dsdt.$$

In order to avoid complicated formulae, we will transform this double integral into a simple one via an easy trick. It is clear that

$$\mathbb{1}_{[0,t]}(s) = \begin{cases} 1, & 0 \leq s \leq t \\ 0, & t < s \leq 1 \end{cases}$$

Taking into account that $0 \leq s \leq t \leq x$ we see that

$$\mathbb{1}_{[0,t]}(s) = \begin{cases} 1, & 0 \leq s \leq t \leq x \\ 0, & x < t \leq 1 \end{cases} = \mathbb{1}_{[s,x]}(t).$$

Hence

$$\begin{aligned} V^2f(x) &= \int_0^x \int_0^1 \mathbb{1}_{[s,x]}(t)f(s)dsdt \\ &= \int_0^x \left(\int_0^1 \mathbb{1}_{[s,x]}(t)dt\right) f(s)ds \\ &= \int_0^x \left(\int_s^x dt\right) f(s)ds \\ &= \int_0^x (x-s)f(s)ds \\ &= \int_0^x (x-t)f(t)dt. \end{aligned}$$

Similarly, we find that

$$V^3f(x) = \frac{1}{2!}\int_0^x (x-t)^2 f(t)dt,$$

so that we may conjecture that

$$V^n f(x) = \frac{1}{(n-1)!}\int_0^x (x-t)^{n-1}f(t)dt, \ n \in \mathbb{N}.$$

(and we leave the proof by induction to the interested reader).

Remark. By considering $f(t) = 1$, say, readers may check that the Volterra Operator is not nilpotent.

(4) Let f be in $L^2[0,1]$ with $\|f\|_2 \leq 1$. We could use the formulae of the previous question or we may just use a proof by induction. The statement obviously holds for $n = 1$. Assume that

$$|(V^{n+1}f)(x)| \leq \frac{x^n}{n!}$$

and let us show that

$$|(V^{n+2}f)(x)| \leq \frac{x^{n+1}}{(n+1)!}.$$

We have

$$\begin{aligned}|(V^{n+2}f)(x)| &= |V(V^{n+1}f)(x)| \\ &= \left|\int_0^x (V^{n+1}f)(t)dt\right| \\ &\leq \int_0^x \left|\frac{t^n}{n!}\right| dt \text{ (induction hypothesis)} \\ &= \left[\frac{t^{n+1}}{(n+1)!}\right]_0^x \\ &= \frac{x^{n+1}}{(n+1)!},\end{aligned}$$

establishing the required inequality.

(5) From the previous question, we may do the following

$$\|V^{n+1}f\|_2^2 \leq \frac{1}{(n!)^2}\int_0^1 x^{2n}dx \leq \frac{1}{(n!)^2}$$

so that

$$0 \leq \|V^{n+1}\| \leq \frac{1}{n!}$$

for all n. This easily implies (e.g. via the Stirling formula) that

$$\lim_{n\to\infty} \|V^{n+1}\|^{\frac{1}{n+1}} = 0.$$

If $r(V)$ is the spectral radius of V, then

$$0 = \lim_{n\to\infty} \|V^{n+1}\|^{\frac{1}{n+1}} = r(V) = \sup_{\lambda\in\sigma(V)} |\lambda|.$$

Thus $\sigma(V) = \{0\}$.

In the end, V is not self-adjoint. Indeed, if V were self-adjoint, then we would get by the Spectral Radius Theorem:

$$\|V\| = \sup\{|\lambda| : \lambda \in \underbrace{\sigma(V)}_{\{0\}}\} = \sup\{|\lambda| : \lambda = 0\} = 0,$$

contradicting what we saw above that $\|V\| = \frac{2}{\pi}$. Therefore, V is not self-adjoint.

(6) We may compute V^* in a usual way or we may just use Exercise 3.3.40, from which we know that the adjoint of the integral operator $\int_0^1 K(x,y)f(y)dy$ is given by

$$\int_0^1 \overline{K(y,x)}f(y)dy.$$

In our case, $K(t,x) = \mathbb{1}_{[0,x]}(t)$ so that

$$\begin{aligned} K^*(t,x) = K(x,t) &= \mathbb{1}_{[0,t]}(x) \\ &= \begin{cases} 1, & 0 \le x \le t \\ 0, & t < x \le 1 \end{cases} \\ &= \begin{cases} 1, & x \le t \le 1 \\ 0, & 0 \le t < x \end{cases} \\ &= \mathbb{1}_{[x,1]}(t). \end{aligned}$$

Thus

$$V^*f(x) = \int_x^1 f(t)dt.$$

No V is not normal! If V were normal, then we would have $VV^* = V^*V$, i.e. $VV^*f = V^*Vf$ for any $f \in L^2[0,1]$. This means that $VV^*f(x) = V^*Vf(x)$ almost everywhere. But, Vf and V^*f are both absolutely continuous (aren't they?), hence they are both continuous. Thus we would have $VV^*f(x) = V^*Vf(x)$ *everywhere*. Hence, in particular, we would have $VV^*1 = V^*V1$, where 1 is the constant function $x \mapsto 1$ on $[0,1]$ (which in $L^2[0,1]$ of course). However,

$$VV^*1 = V\left(\int_x^1 1dt\right) = V(1-x) = \int_0^x (1-t)dt = x - \frac{x^2}{2},$$

and

$$V^*V1 = V^*\left(\int_0^x 1dt\right) = V^*x = \int_x^1 tdt = \frac{1}{2}(1-x^2).$$

Thus we would finally end up with (by the "assumed normality" of V)

$$\frac{1}{2}(1-x^2) = x - \frac{x^2}{2},$$

(for any x) which is clearly not the case. Thus V is not normal.

Remark. Another way of seeing that V is not normal is to use the method given above to prove that V is not self-adjoint: If V were normal, then we would have by the Spectral

Radius Theorem

$$\frac{2}{\pi} = \|V\| = \sup\{|\lambda| : \lambda \in \sigma(V)\} = \sup\{|\lambda| : \lambda = 0\} = 0$$

which is absurd! Consequently, V is not normal.

(7) Let $f \in L^2[0,1]$ (and let $x \in [0,1]$). We have

$$(V+V^*)f(x) = Vf(x) + V^*f(x) = \int_0^x f(t)dt + \int_x^1 f(t)dt = \int_0^1 f(t)dt$$

so that

$$\begin{aligned} < (V+V^*)f, f >&= \int_0^1 \left(\int_0^1 f(t)dt \right) \overline{f(x)}dx \\ &= \int_0^1 \int_0^1 f(t)\overline{f(x)}dtdx \\ &= \int_0^1 f(t)dt \int_0^1 \overline{f(x)}dx \\ &= \int_0^1 f(t)dt \overline{\int_0^1 f(x)dx} \\ &= \left| \int_0^1 f(t)dt \right|^2 \geq 0. \end{aligned}$$

(8) We have just seen that $\sigma(V) = \{0\}$. Hence if λ is an eigenvalue for V, then it is necessarily equal to 0 (since $\sigma_p(V) \subset \sigma(V)$). We claim that V is one-to-one. Let $f \in \ker V$. Then for all $x \in [0,1]$

$$\int_0^x f(t)dt = 0 \text{ or } \int_0^1 \mathbb{1}_{[0,x]}(t)f(t)dt = 0.$$

Hence f is perpendicular to all simple functions as (for all x, t)

$$< \mathbb{1}_{[x,t]}, f >= \int_0^t f(\lambda)d\lambda - \int_0^x f(\mu)d\mu = 0.$$

But the set of simple functions (which we denote by $\mathcal{S}$) is dense in $L^2[0,1]$. Hence

$$f \in \mathcal{S}^\perp = \overline{\mathcal{S}}^\perp = (L^2[0,1])^\perp = \{0\}$$

so that $f = 0$, that is, $\ker V = \{0\}$. Thus V does not admit an eigenvalue. Therefore

$$\sigma_p(V) = \varnothing.$$

Remark. Alternatively, we could have seen that V was injective by first observing that V is absolutely continuous and hence its derivative exists almost everywhere. Then $Vf = 0$ would lead to $f = 0$ on L^2, i.e. V is one-to-one.

We claim that $\sigma_c(V) = \{0\}$. To establish it, we first show that $\overline{\text{ran}(V)} = L^2[0,1]$. By mimicking the case of $\ker V = \{0\}$, we may easily show that $\ker V^* = \{0\}$. Hence

$$\overline{\text{ran}(V)} = \text{ran}(V)^{\perp\perp} = (\ker V^*)^{\perp} = \{0\}^{\perp} = L^2[0,1]$$

so that $0 \in \sigma_c(V)$. Accordingly,

$$\sigma_c(V) = \sigma(V) = \{0\}.$$

Consequently,

$$\sigma_r(V) = \varnothing.$$

(9) We know from Theorem 9.1.23 that an integral operator of the form (since also clearly $(x,y) \mapsto \mathbb{1}_{[0,x]}(y)$ is in $L^2([0,1]^2)$)

$$Vf(x) = \int_0^x f(y)dy = \int_0^1 \mathbb{1}_{[0,x]}(y)f(y)dy,$$

is Hilbert-Schmidt and hence it is compact.

(10) First, recall that for all $f \in L^2[0,1]$, we have

$$V^*Vf(x) = \int_x^1 \int_0^t f(s)dsdt.$$

Since V^*V is self-adjoint and positive, all its eigenvalues are real and positive. Let $\lambda \geq 0$ (in fact $\lambda > 0$ as we will shortly see) be such that

$$V^*Vf(x) = \lambda f(x), \text{ i.e. } \int_x^1 \int_0^t f(s)dsdt = \lambda f(x)$$

for *all* x. Then $f(1) = 0$ and f is differentiable. By differentiating (against x) the functions in the previous equality, we obtain

$$-\lambda f'(x) = \int_0^x f(s)ds$$

(which gives us $f'(0) = 0$ and that f' is differentiable).

Differentiating the previous equation once more (still with respect to x) yields

$$\lambda f''(x) = -f(x)$$

(which also means that $\lambda \neq 0$, otherwise $f = 0!$). The solution of the previous equation is given by

$$f(x) = A\cos\frac{x}{\sqrt{\lambda}} + B\sin\frac{x}{\sqrt{\lambda}}.$$

Let us find the constants A and B using the initial conditions $f(1) = 0$ and $f'(0) = 0$. We have

$$\begin{cases} A\cos\frac{1}{\sqrt{\lambda}} + B\sin\frac{1}{\sqrt{\lambda}} = 0, \\ -\frac{A}{\sqrt{\lambda}}\sin\frac{0}{\sqrt{\lambda}} + \frac{B}{\sqrt{\lambda}}\cos\frac{0}{\sqrt{\lambda}} = 0 \end{cases}$$

or

$$\begin{cases} A\cos\frac{1}{\sqrt{\lambda}} = 0, \\ B = 0. \end{cases}$$

Hence

$$\begin{cases} \frac{1}{\sqrt{\lambda}} = (2n+1)\frac{\pi}{2},\ n \geq 0, \\ B = 0, \end{cases}$$

Therefore

$$B = 0 \text{ and } \lambda = \frac{4}{(2n+1)^2\pi^2},\ n \geq 0.$$

Thus, the eigenvalues are given by $\lambda_n = \frac{4}{(2n+1)^2\pi^2}$, $n \geq 0$ with corresponding eigenvectors $y_n = A\cos((2n+1)\frac{\pi}{2}x)$, $n \geq 0$.

(11) Since V^*V is self-adjoint, the eigenvectors are mutually orthogonal as the eigenvalues are distinct. By Theorem 9.1.26, there is an orthonormal basis constituted of eigenvectors of V^*V. But, since

$$\int_0^1 |\sqrt{2}\cos((2n+1)\frac{\pi}{2}x)|^2 dx = 1,$$

$\{\sqrt{2}\cos((2n+1)\frac{\pi}{2}x)\}$ works as the orthonormal basis. Thus

$$V^*Vf = \sum_{n=1}^{\infty} \lambda_n < f, f_n > f_n,$$

where f_n are the eigenvectors in the previous basis.

(12) Since V is compact, V^*V too is compact. Hence

$$\sigma(V^*V) = \{0\} \cup \sigma_p(V^*V)$$

so that

$$\sigma(V^*V) = \{0\} \cup \left\{\frac{4}{(2n+1)^2\pi^2} : n \geq 0\right\}.$$

Since V^*V is self-adjoint, the Spectral Radius Theorem gives

$$\begin{aligned}\|V^*V\| &= \sup\{|\lambda| : \lambda \in \sigma(V^*V)\}\\ &= \sup\{|\lambda| : \lambda \in \{0\} \cup \sigma_p(V^*V)\}\\ &= \sup\left\{|\lambda| : \lambda \in \left\{\frac{4}{(2n+1)^2\pi^2} : n \geq 0\right\} \cup \{0\}\right\}\\ &= \sup\left\{0, \frac{4}{\pi^2}, \frac{4}{9\pi^2}, \cdots\right\}\\ &= \frac{4}{\pi^2}.\end{aligned}$$

Hence

$$\frac{4}{\pi^2} = \|V^*V\| = \|V\|^2, \text{ i.e. } \|V\| = \frac{2}{\pi}.$$

(13) Since

$$\|-V\| = \|V\| = \frac{2}{\pi} < 1,$$

by Theorem 2.1.88, both $I - V$ and $I + V$ are invertible. Besides, using the Neumann Series we have

$$(I - V)^{-1} = \sum_{n=0}^{\infty} V^n.$$

Let $f \in L^2[0,1]$. We may then write

$$\begin{aligned}(I - V)^{-1}f(x) &= \sum_{n=0}^{\infty} V^n f(x)\\ &= f(x) + \sum_{n=1}^{\infty} V^n f(x)\\ &= f(x) + \sum_{n=1}^{\infty} \frac{1}{(n-1)!} \int_0^x (x-t)^{n-1} f(t)dt\\ &= f(x) + \sum_{n=1}^{\infty} \int_0^x \frac{(x-t)^{n-1}}{(n-1)!} f(t)dt\\ &= f(x) + \int_0^x \sum_{n=1}^{\infty} \frac{(x-t)^{n-1}}{(n-1)!} f(t)dt \text{ (why?)}\\ &= f(x) + \int_0^x e^{x-t} f(t)dt\\ &= f(x) + e^x \int_0^x e^{-t} f(t)dt.\end{aligned}$$

In a very similar manner (do it!), we obtain

$$(I+V)^{-1}f(x) = \sum_{n=0}^{\infty}(-1)^n V^n f(x) = f(x) - e^{-x}\int_0^x e^t f(t)dt.$$

(14) We need to find an *invertible* $A \in B(L^2[0,1])$ such that

$$A^{-1}(I-V)A = (I+V)^{-1} \text{ or } (I-V)A = A(I+V)^{-1}.$$

According to the answer of the previous question, the adequate choice of A seems to be clear. Let $A \in B(L^2[0,1])$ [1] by

$$Af(x) = e^x f(x).$$

Then A is invertible. Its inverse $A^{-1} \in B(L^2[0,1])$ is given by

$$A^{-1}f(x) = e^{-x}f(x).$$

Let $f \in L^2[0,1]$. We have on the one hand

$$\begin{aligned} A(I+V)^{-1}f(x) &= A[(I+V)^{-1}f(x)] \\ &= A\left(f(x) - e^{-x}\int_0^x e^t f(t)dt\right) \\ &= e^x f(x) - e^x e^{-x}\int_0^x e^t f(t)dt \\ &= e^x f(x) - \int_0^x e^t f(t)dt. \end{aligned}$$

On the other hand, we have

$$\begin{aligned} (I-V)Af(x) &= (I-V)(e^x f(x)) \\ &= e^x f(x) - V(e^x f(x)) \\ &= e^x f(x) - \int_0^x e^t f(t)dt. \end{aligned}$$

Therefore, for all $f \in L^2[0,1]$, we have

$$A(I+V)^{-1}f(x) = (I-V)Af(x), \text{ that is, } A(I+V)^{-1} = (I-V)A,$$

establishing the similarity between $(I+V)^{-1}$ and $I-V$.

(15) It is clear that we may write

$$V^*Vf(x) = \int_0^1 [1 - \max(x,t)]f(t)dt.$$

[1]For the originality of the work, credit is due to Pedersen. This appeared in [**5**]. We gave a slightly different approach.

Now, using Proposition 9.1.27, we then have

$$\int_0^1 \int_0^1 [1 - \max(x,t)]^2 dx dt = \sum_{n=0}^{\infty} \frac{16}{(2n+1)^4 \pi^4}.$$

Let us then find the double integral of the left hand side. Let

$$\Delta = \{(x,t) \in [0,1] \times [0,1] : \ x \leq t\}$$

(a triangle!). Hence

$$\int_0^1 \int_0^1 [1 - \max(x,t)]^2 dx dt = 2 \iint_{\Delta} (1-t)^2 dx dt = \frac{1}{6}.$$

Therefore

$$\sum_{n=0}^{\infty} \frac{1}{(2n+1)^4} = \frac{\pi^4}{96}.$$

In the end, it is clear that

$$\sum_{n=0}^{\infty} \frac{1}{n^4} = \sum_{k=0}^{\infty} \frac{1}{(2k+1)^4} + \sum_{k=1}^{\infty} \frac{1}{(2k)^4} = \sum_{k=0}^{\infty} \frac{1}{(2k+1)^4} + \frac{1}{16} \sum_{k=1}^{\infty} \frac{1}{k^4}$$

so that

$$\sum_{n=0}^{\infty} \frac{1}{n^4} = \frac{\pi^4}{96} \times \frac{16}{15} = \frac{\pi^4}{90}.$$

SOLUTION 9.3.22. Let V be the Volterra Operator as defined in Exercise 9.3.21. Set

$$A = (I + V)^{-1}.$$

We already know from Exercise 9.3.21 why $I + V$ is invertible. Now, it is plain that $A \neq I$ (otherwise $V = 0$!). Since $\sigma(V + I) = \{1\}$, we also have

$$\sigma(A) = \left\{\lambda^{-1} : \ \lambda \in \sigma(V + I)\right\} = \{1\}.$$

It only remains to show that $\|A\| = 1$. One inequality is immediate as

$$r(A) = \sup\{|\lambda| : \ \lambda \in \sigma(A)\} = 1 \leq \|A\|.$$

To show the reverse inequality, we may show that $\|Af\| \leq \|f\|$ for all f, or equivalently,

$$\|A^{-1}f\| \geq \|f\|, \ \forall f \in L^2[0,1].$$

Let $f \in L^2[0,1]$. Then

$$\begin{aligned}\|A^{-1}f\|^2 =& \|(V+I)f\|^2\\ =& <(V+I)f,(V+I)f>\\ =& \|Vf\|^2+\|f\|^2+<Vf,f>+<V^*f,f>\\ =& \|Vf\|^2+\|f\|^2+<(V+V^*)f,f>\\ \geq& 0\end{aligned}$$

for we already proved in Exercise 9.3.21 that $V+V^*$ is positive.

9.4. Hints/Answers to Tests

Solution 9.4.1. No! Assume that $\dim H=\infty$. If A is compact, then AI is also compact but I is not compact...

Solution 9.4.2. All of them! Why?...

Solution 9.4.3. Obvious if $\dim H<\infty$ (without compactness!)... If A or B is compact, then AB and BA are both compact. So if $\dim H=\infty$, then both $0\in\sigma(AB)$ and $0\in\sigma(BA)$...

Solution 9.4.4. Theorem 3.1.69 provides another way of showing the non-compactness of A. Indeed,

$$\|Ax\|\geq\alpha\|x\|=\|\alpha x\|$$

for some $\alpha>0$ and for all x means that $\alpha I=KA$ for some contraction $K\in B(H)$. If A were compact, so would be KA, that is, so would be I! Contradiction as $\dim H=\infty$!

Solution 9.4.5. If $p(0)=0$, then $p(A)$ is compact... If $p(A)$ is compact and $p(0)\neq 0$, then I would be compact!

Solution 9.4.6. WLOG assume that $A\neq 0$. Since A is compact and self-adjoint, we know that at least $-\|A\|$ or $\|A\|$ is an eigenvalue. In either case, A has at least an eigenvalue, λ say. Indeed, $\ker(A-\lambda I)$ is a non-trivial invariant subspace for A.

(1) $\ker(A-\lambda I)$ is clearly closed...
(2) It is non-trivial: $\ker(A-\lambda I)\neq\{0\}$ for λ is an eigenvalue; $\ker(A-\lambda I)\neq H$ as otherwise $A=\lambda I$ which is impossible...
(3) $\ker(A-\lambda I)$ is an invariant subspace for $A-\lambda I$. But

$$A(A-\lambda I)=(A-\lambda I)A$$

and so $\ker(A-\lambda I)$ is an invariant subspace for A.

CHAPTER 10

Closed Operators

10.2. True or False: Answers

Answers.

(1) False! In fact the correct conclusion should be $Ax = 0$ for all $x \in D(A)$, that is, $A = 0$ on $D(A)$. The reader is asked to show that in Exercise 10.5.1. For the sake of exhaustiveness, we include a counterexample: It suffices to take A to be zero on some dense but not closed domain (this can come from e.g. $A = B - B$ where $D(A) = D(B)$ is non-closed and dense). Then

$$\forall x \in D(A) : < Ax, x >= 0 \text{ but only } A \subset 0$$

where the last 0 is the everywhere defined zero operator!

(2) False! If A is unbounded, then A^2 may only be defined at zero (as already observed). Also, in Exercise 10.5.4, we have an example of an unbounded A (closed even!) such that $A^2 = 0$, i.e. A^2 is bounded.

(3) Only "$\Rightarrow$" is true. We know that if $A \in B(H)$, then $\sigma(A)$ is bounded (in fact compact). So contra positively, if $\sigma(A)$ is not bounded, then A is an unbounded operator.

As for "$\Leftarrow$", it is not true in general. In fact, not only unbounded operators may have a bounded spectrum, their spectrum may well be empty! For two examples, see Exercises 10.3.23 & 10.3.24. The spectrum of an unbounded operator may even equal the entire $\mathbb{C}$. Indeed, according to Corollary 10.1.17, any unclosed operator A necessarily verifies $\sigma(A) = \mathbb{C}$.

The only thing which we are sure of is that if A is a *closed* operator, then $\sigma(A)$ is (only) closed. For a proof, see Exercise 10.3.20.

Remark. The result may, however, be true if some conditions are imposed on A. In [**126**], the authors proved that if A is an "unbounded self-adjoint operator", then $\sigma(A)$ is bounded forces A to be bounded. Their proof was elementary and did not use any form of the spectral theorems.

In fact, the authors in **[165]** tell us that we may go at least as weak as "subnormality" (a notion not discussed in this book but it is weaker than normality). They proved that if A is **subnormal**, then A is bounded iff $\sigma(A)$ is bounded.

(4) False! Let A be an unbounded operator with domain $D(A) \subset H$ and let $B = 0$ everywhere ($B \in B(H)$!). Hence

$$AB = 0 \text{ on } H \text{ and } BA = 0 \text{ on } D(BA) = D(A) \text{ (cf. Exercise 10.3.13).}$$

Then BA is not closed on $D(BA)$ and hence $\sigma(BA) = \mathbb{C}$ (by Proposition 10.1.16). It is also clear that $\sigma(AB) = \{0\}$. Therefore,

$$\sigma(AB) \cup \{0\} \neq \sigma(BA) \cup \{0\}.$$

In **[100]**, it is shown that if A and B are two non-necessarily bounded operators such that $\sigma(AB) \neq \mathbb{C}$ and $\sigma(BA) \neq \mathbb{C}$ (implying that both AB and BA are closed), then

$$\sigma(AB) \cup \{0\} = \sigma(BA) \cup \{0\}.$$

(5) False! (cf. Exercise 10.5.2) For a counterexample, just consider $A = B|D$ (B restricted to D) where D is dense in H but not closed. Since D is not closed, A, which is bounded on D, cannot be closed. Observe in the end that $A \neq B$ merely because $D \neq H$!

(6) False! (cf. Exercise 10.5.2) Just consider $A = 0$ (the zero operator) on the trivial domain $D(A) = \{0\}$. Take B to be any non-zero bounded operator. Since $A(0) = 0 = B(0)$, we see plainly that $A \subset B$. Finally, it is clear that A is closed on $D(A)$, that $D(A)$ is not dense in H and that $A \neq B$.

10.3. Solutions to Exercises

Solution 10.3.1. Take $A = B = -C$. Then we are required to find an A such that

$$A(A - A) \not\subset A^2 - A^2.$$

But

$$D[A(A - A)] = \{x \in D(A) : (A - A)x = 0 \in D(A)\} = D(A)$$

and

$$D(A^2 - A^2) = D(A^2).$$

Therefore, we need to exhibit an A such that $D(A) \not\subset D(A^2)$. Example 10.1.3 does the remaining job as obviously $C_0^\infty(\mathbb{R}) \not\subset \{0\}$!

REMARK. We may also call on Chernoff's example for a stronger counterexample.

SOLUTION 10.3.2. We have to show that $B \subset A$. Let $y \in D(B) \subset H$. Since A is surjective (and $By \in H$), there exists an $x \in D(A)$ such that $Ax = By$. But, $Ax = Bx$ for all $x \in D(A)$. Therefore $Bx = By$ so that $x = y$ by the injectivity of B, i.e. $x = y \in D(A)$, proving that $B \subset A$.

SOLUTION 10.3.3. Suppose that $A \subset B$, i.e. $Ax = Bx$ for all $x \in D(A) \subset D(B)$. We then have

$$\begin{aligned} D(CA) &= \{x \in D(A) : \ Ax \in D(C)\} \\ &= \{x \in D(A) : \ Bx \in D(C)\} \\ &\subset \{x \in D(B) : \ Bx \in D(C)\} \\ &= D(CB). \end{aligned}$$

Moreover, if $x \in D(CA)$, then clearly $CAx = CBx$.

Let us prove the second "inclusion". We have

$$D(AC)=\{x\in D(C) : Cx\in D(A)\}\subset\{x\in D(C) : Cx\in D(B)\}=D(BC)$$

and for all $x \in D(AC)$, $ACx = BCx$, as required.

SOLUTION 10.3.4. By hypothesis $BA \subset AB$. So, by Exercise 10.3.3, we obtain

$$BA \subset AB \Longrightarrow B^2A = B(BA) \subset B(AB) = (BA)B \subset ABB = AB^2.$$

Similarly, we can show for all $n \in \mathbb{N}$ that

$$B^nA \subset AB^n.$$

Hence if $p(z) = a_0 + a_1z + \cdots + a_nz^n$ (with complex coefficients), then clearly (as $p(B) \in B(H)$)

$$p(B)A \subset Ap(B),$$

as required.

SOLUTION 10.3.5. We only show that $\overline{\operatorname{ran}(A)} \subset \operatorname{ran}(A)$. Let $y \in \overline{\operatorname{ran}(A)}$. Then there exists (y_n) in $\operatorname{ran}(A)$ such that $y_n \to y$, where $y_n = Ax_n$ for some x_n in $D(A)$. Hence from our hypothesis

$$\|Ax_n\| \geq \alpha\|x_n\| \text{ for all } n \in \mathbb{N}.$$

Since (Ax_n) converges, it is Cauchy. Whence for any $\varepsilon > 0$, there exists $N \in \mathbb{N}$ such that for all $n, m \geq N$ we have

$$\varepsilon > \|Ax_n - Ax_m\| \geq \alpha\|x_n - x_m\|$$

which certainly means that (x_n) too is Cauchy. Therefore, (x_n) converges to some $x \in H$.

Recapitulating, we see that we have on our hands

$$\begin{cases} x_n \longrightarrow x, & x_n \in D(A), \\ Ax_n \longrightarrow y. \end{cases}$$

Since A is closed, $x \in D(A)$ and $y = Ax$ yielding $y \in \operatorname{ran}(A)$.

Solution 10.3.6.

(1) (a) Assume that A is closed. Let (x_n) be a Cauchy sequence in $D(A)$ with respect to the graph norm. Then (x_n) and (Ax_n) are Cauchy sequences in H. Hence they converge by the completeness of H. That is, there exist $x \in H$ and $y \in H$ such that $x_n \to x$ and $Ax_n \to y$. But A is closed, so $x \in D(A)$ and $y = Ax$. It remains to check that $x_n \to x$ in the graph norm of A. We have

$$\|x_n - x\|_{G(A)}^2 = \|x_n - x\|_H^2 + \|Ax_n - Ax\|_H^2 \longrightarrow 0$$

as $n \to \infty$. Thus $(D(A), \|\cdot\|_{G(A)})$ is a Banach space.

(b) Assume that $(D(A), \|\cdot\|_{G(A)})$ is a Banach space. Let (x_n) be in $D(A)$ and such that $x_n \to x$ and $Ax_n \to y$ in the norm associated with H. Then both (x_n) and (Ax_n) are Cauchy so that (x_n) is Cauchy with respect to the norm $\|\cdot\|_{G(A)}$ too.

Since $(D(A), \|\cdot\|_{G(A)})$ is a Banach space, $\|x_n - x'\|_{G(A)} \to 0$ as $n \to \infty$ for some $x' \in D(A)$. Whence

$$\|x_n - x'\|_H \longrightarrow 0 \text{ and } \|Ax_n - Ax'\|_H \longrightarrow 0$$

as $n \to \infty$. By uniqueness of the limit,

$$x = x' \in D(A) \text{ and } Ax = Ax' = y,$$

i.e. A is closed, establishing the result.

(2) We first show that A is unbounded. Let $f_n(x) = e^{-n|x|}$ for $n = 1, 2, \cdots$ and $x \in \mathbb{R}$. Then $f_n \in H^1(\mathbb{R})$ for all n. Indeed, let $n \in \mathbb{N}$. Then $f_n \in L^2(\mathbb{R})$ because

$$\|f_n\|_2^2 = \int_{\mathbb{R}} e^{-2n|x|} dx = \frac{1}{2n} + \frac{1}{2n} = \frac{1}{n}$$

and $f_n' \in L^2(\mathbb{R})$ (the distributional derivative) for

$$\|f_n'\|_2^2 = \int_{\mathbb{R}} n^2 e^{-2n|x|} dx = n.$$

Hence
$$\frac{\|f_n'\|_2}{\|f_n\|_2} = n \longrightarrow \infty$$
as n tends to ∞. This proves that A is unbounded on $H^1(\mathbb{R})$.

Now, we prove that A is closed. By the previous question, A is closed iff $H^1(\mathbb{R})$ is a Banach space with respect to the graph norm of A. Let us thus show that $H^1(\mathbb{R})$ is complete with respect to the norm
$$\|f\|_{H^1(\mathbb{R})} = (\|f\|_2 + \|f'\|_2)^{\frac{1}{2}}.$$

Let (f_n) be a Cauchy sequence in $H^1(\mathbb{R})$. This means that
$$\|f_n - f_m\|_{H^1(\mathbb{R})} \longrightarrow 0 \text{ as } n, m \longrightarrow \infty,$$
which in turn means that (f_n) and (f_n') are both Cauchy sequences in $L^2(\mathbb{R})$. Since $L^2(\mathbb{R})$ is complete, we know that there exist $f, g \in L^2(\mathbb{R})$ such that
$$\lim_{n\to\infty} \|f_n - f\|_2 = 0 \text{ and } \lim_{n\to\infty} \|f_n' - g\|_2 = 0.$$
It becomes clear that $H^1(\mathbb{R})$ is complete once we show that $g = f'$.

Let $\varphi \in \mathcal{D}(\mathbb{R})$. Then clearly $\varphi \in L^2(\mathbb{R})$, that is, $\|\varphi\|_2$ is finite. Besides, it is independent of n so that by the Cauchy-Schwarz Inequality
$$0 \le \left| \int_{\mathbb{R}} \varphi(x)(f(x) - f_n(x))dx \right| \le \|\varphi\|_2 \|f - f_n\|_2 \longrightarrow 0$$
as $n \to \infty$. Therefore,
$$\lim_{n\to\infty} \int_{\mathbb{R}} \varphi(x) f_n(x) dx = \int_{\mathbb{R}} \varphi(x) f(x) dx$$
which precisely means that (f_n) converges to f in $\mathcal{D}'(\mathbb{R})$.

A similar approach leads to the convergence of (f_n') to g in $\mathcal{D}'(\mathbb{R})$. Finally, we infer that (by the continuity of $\frac{d}{dx}$ in $\mathcal{D}'(\mathbb{R})$)
$$f' = \lim_{n\to\infty} f_n' = g$$
by uniqueness of the limit in $\mathcal{D}'(\mathbb{R})$. Since $g \in L^2(\mathbb{R})$, we finally obtain that $f' \in L^2(\mathbb{R})$, and
$$\|f_n - f\|_{H^1(\mathbb{R})} = (\|f_n - f\|_2 + \|f_n' - f'\|_2)^{\frac{1}{2}} \longrightarrow 0$$
when $n \to \infty$. Accordingly, $(H^1(\mathbb{R}), \|\cdot\|_{H^1(\mathbb{R})})$ is a Banach space.

REMARKS.

(1) Clearly, we may replace "Banach" by "Hilbert" in the statements of the previous exercise. The associated inner product is defined by
$$< x,y >_{G(A)} = < x,y >_H + < Ax, Ay >_H .$$

(2) A slightly deeper Fourier Analysis (cf. [**129**]) tells us that $H^1(\mathbb{R})$ is just an $L^2(\mathbb{R})$ space with a measure which differs from Lebesgue's. Indeed, with our notation (of the Fourier transform, cf. Exercise 2.1.49),
$$\widehat{f'(x)} = -it\hat{f}(t).$$
Hence, by the Plancherel Theorem
$$\|f\|^2_{H^1(\mathbb{R})} = \|\hat{f}\|_2^2 + \|\widehat{f'}\|_2^2 = \int_{\mathbb{R}} |\hat{f}(t)|^2(1+|t|^2)dt$$
and so we may regard $(H^1(\mathbb{R}), dx)$ as $(L^2(\mathbb{R}), (1+|t|^2)dt)$.

SOLUTION 10.3.7. Let $x \in D(A)$. Then
$$\|x\| \le \|x\|_{G(A)} \le (1 + \|A\|^2)^{\frac{1}{2}}\|x\|.$$
This means that the norms $\|\cdot\|$ and $\|\cdot\|_{G(A)}$ are *equivalent*. This implies that $(D(A), \|\cdot\|_{G(A)})$ is complete *iff* $(D(A), \|\cdot\|)$ is complete, that is, A is closed *iff* $D(A)$ is closed, as required.

SOLUTION 10.3.8.

(1) It is fairly easy to see that A is closed w.r.t. D_1. The "hard" part is notational. Let (x^k) be a sequence in ℓ^2 which converges to x and such that (Ax^k) converges to some y (both limits are taken as k tends to infinity). This means that
$$\|x^k - x\|_2 \longrightarrow 0 \text{ and } \|Ax^k - y\|_2 \longrightarrow 0$$
(remember that (x^k), x, y all depend on n, but we keep it fixed). We need to show that $Ax = y$, i.e. $nx_n = y_n$ (for each n). Now, we have (for all k)
$$|y_n - nx_n| \le |y_n - nx_n^k| + |nx_n^k - nx_n| = |y_n - nx_n^k| + n|x_n^k - x_n|.$$
Let $\varepsilon > 0$. Since (x^k) converges to x, we have (don't we?) for some $K \in \mathbb{N}$
$$|x_n^k - x_n| \le \|x^k - x\|_2 < \frac{\varepsilon}{n}$$
when $k \ge K$. By the same token
$$|y_n - nx_n^k| \le \|y - Ax^k\|_2 < \varepsilon.$$

Therefore, for $k \geq K$, we have in the end

$$|y_n - nx_n| \leq \varepsilon + n \times \frac{\varepsilon}{n} = 2\varepsilon.$$

Thus, $y_n - nx_n = 0$ for all n, i.e. $Ax = y$ (where as a consequence we also have $Ax \in \ell^2$). So much for the proof of the closedness of A w.r.t. D_1.

(2) To show that A is not closed in this case, we must exhibit a sequence (x^k) in c_{00} satisfying $x^k \to x$ and $Ax^k \to y$ (both limits in ℓ^2) and such that

$$x \notin c_{00}, \text{ or } x \in c_{00} \text{ but } Ax \neq y.$$

This time we show that $x \notin c_{00}$. Let (x^k)

$$x^k = \left(1, \frac{1}{2^2}, \cdots, \frac{1}{k^2}, 0, 0, \cdots\right)$$

and it is clear that (x^k) is in c_{00} (we will see below why this is a good choice and that replacing $\frac{1}{k^2}$ by $\frac{1}{k}$ does not help). We know, from previous exercises (cf. Exercises 1.3.10 & 1.3.18) that (x^k) converges in ℓ^2 to

$$x = \left(1, \frac{1}{2^2}, \cdots, \frac{1}{k^2}, \frac{1}{(k+1)^2}, \cdots\right) \notin c_{00}.$$

To finish the proof, we have to show that (Ax^k) converges to some $y \in \ell^2$. We may write

$$Ax^k = \left(1 \times 1, 2 \times \frac{1}{2^2}, \cdots, k \times \frac{1}{k^2}, 0, 0, \cdots\right) = \left(1, \frac{1}{2}, \cdots, \frac{1}{k}, 0, 0, \cdots\right).$$

Arguing as above, we may easily prove that (Ax^k) converges (in ℓ^2) to

$$y = \left(1, \frac{1}{2}, \cdots, \frac{1}{k}, \frac{1}{k+1}, \cdots\right).$$

This proves that A is not closed when defined on c_{00}.

Finally, both domains are dense in ℓ^2. In fact, c_{00} is dense in ℓ^2 by Exercise 1.3.18. To show that D_1 is dense in ℓ^2, observe first that it is plain that $c_{00} \subset D_1$ (why?), i.e.

$$c_{00} \subset D_1 \subset \ell^2.$$

Passing to the closure w.r.t. the topology of ℓ^2, and remembering that c_{00} is dense in ℓ^2, we obtain

$$\ell^2 = \overline{c_{00}} \subset \overline{D_1} \subset \overline{\ell^2} = \ell^2$$

so that $\overline{D_1} = \ell^2$.

SOLUTION 10.3.9.

(1) No! (This is why we are introducing $D(M_\varphi)$ in the next question). Indeed, unless e.g. $\varphi \in L^\infty(\mathbb{R})$ there seems to be no reason why φf should be in $L^2(\mathbb{R})$ (e.g. take $f(x) = \frac{1}{x}\mathbb{1}_{[1,\infty)}(x)$ and $\varphi(x) = x$).

(2) The space $D(M_\varphi)$ is not void (by looking at the second part of the question, it is far from being empty!). Let us anyway show that $D(M_\varphi) \neq \varnothing$.

Let $f(x) = \frac{1}{1+|\varphi(x)|}\mathbb{1}_{[0,1]}(x)$. Then $f \in L^2(\mathbb{R})$ as $\frac{1}{1+|\varphi(x)|} \leq 1$ a.e., so that we may write

$$\|f\|_2^2 = \int_{\mathbb{R}} \frac{1}{(1+|\varphi(x)|)^2}\mathbb{1}_{[0,1]}(x)dx \leq 1\int_0^1 dx = 1 < \infty.$$

Also, $\varphi f \in L^2(\mathbb{R})$. Indeed, as $\frac{|\varphi(x)|}{1+|\varphi(x)|} \leq 1$ a.e. (also!), then

$$\|\varphi f\|_2^2 = \int_{\mathbb{R}} \frac{|\varphi(x)|^2}{(1+|\varphi(x)|)^2}\mathbb{1}_{[0,1]}(x)dx \leq 1.$$

Now, let us turn to showing that $D(M_\varphi)$ is dense in $L^2(\mathbb{R})$, consider the sequence of sets $X_n = \{x \in \mathbb{R} : |\varphi(x)| \leq n\}$. Then it is not hard to see that

$$X_n \subset X_{n+1} \text{ and } \bigcup_{n=1}^{\infty} X_n = \mathbb{R}.$$

Let $f \in L^2(\mathbb{R})$. We need to find a sequence (f_n) in $D(M_\varphi)$ such that

$$\lim_{n\to\infty} \|f_n - f\|_{L^2(\mathbb{R})} = 0.$$

Let $f_n = \mathbb{1}_{X_n} f$. Then $f_n \in D(M_\varphi)$ for $f_n \in L^2(\mathbb{R})$ as $f \in L^2(\mathbb{R})$ and $\mathbb{1}_{X_n} \in L^\infty(\mathbb{R})$, and $\varphi f_n \in L^2(\mathbb{R})$ because

$$\|\varphi f_n\|_2^2 = \int_{|\varphi(x)|\leq n} |\varphi(x) f(x)|^2 dx \leq n^2 \|f_n\|_2^2.$$

Lastly, we have

$$\|f_n - f\|_2^2 = \int_{\mathbb{R}} |f_n(x) - f(x)|^2 dx$$
$$= \int_{\mathbb{R}} |(\mathbb{1}_{X_n} - 1)f(x)|^2 dx.$$

Since $|(\mathbb{1}_{X_n} - 1)f(x)|^2$ tends to zero a.e., since for all n

$$|(\mathbb{1}_{X_n} - 1)f(x)|^2 \leq 2|f(x)|^2$$

and $2|f|^2 \in L^1(\mathbb{R})$, by the Dominated Convergence Theorem it is guaranteed that

$$\lim_{n\to\infty} \|f_n - f\|_2 = 0$$

holds.

Remarks.

(a) A little digression, we could have defined M_φ on other dense domains. For example, we could have defined it on $C_0^\infty(\mathbb{R})$ (or $\mathcal{S}(\mathbb{R})$: the Schwartz space).
Hence, we could have used this last property to prove the density of $D(M_\varphi)$ in $L^2(\mathbb{R})$. Indeed, it is apparent that

$$C_0^\infty(\mathbb{R}) \subset D(M_\varphi) \subset L^2(\mathbb{R}).$$

Passing to the $L^2(\mathbb{R})$-*closure* yields

$$L^2(\mathbb{R}) = \overline{C_0^\infty(\mathbb{R})} \subset \overline{D(M_\varphi)} \subset \overline{L^2(\mathbb{R})} = L^2(\mathbb{R}),$$

proving the density of $D(M_\varphi)$ in $L^2(\mathbb{R})$.

(b) One may wonder why we are caring about the density of the domain anyway? Remember that we already mentioned that the density is imperative if we want to compute the adjoint of an unbounded operator (a notion not considered in this book). If the domain is dense, then the adjoint is unique exactly like in the bounded case. Otherwise, the unbounded operator may have more than one adjoint (how bad!).

(3) Let $f_n \in D(M_\varphi)$ be such that

$$f_n \to f \text{ and } M_\varphi f_n = \varphi f_n \to g$$

(both limits being in the L^2-norm).

By Theorem 1.1.73, $L^2(\mathbb{R})$ is complete. Since (f_n) and (φf_n) are Cauchy, their respective limits f, g are both in $L^2(\mathbb{R})$. By the same theorem, there exists a subsequence (f_{n_k}) of (f_n) such that

$$\lim_{k\to\infty} f_{n_k}(x) = f(x) \text{ for almost every } x \text{ in } \mathbb{R}.$$

Hence $\varphi(x)f_{n_k}(x) \to \varphi(x)f(x)$ for a.e. x in $\mathbb{R}$.

On the other hand, since $\varphi f_n \to g$ in $L^2(\mathbb{R})$, every subsequence (φf_{n_p}) of (φf_n) converges to g, still in $L^2(\mathbb{R})$. But (φf_{n_k}) is one of them (isn't?) so that (φf_{n_k}) too has an a.e.

convergent subsequence to g. Denote this last subsequence by $(\varphi f_{n_{k_j}})$. Recapitulating, we have a.e. in $\mathbb{R}$

$$\begin{cases} \lim_{j\to\infty} \varphi f_{n_{k_j}} = g, \\ \lim_{k\to\infty} \varphi f_{n_k} = \varphi f. \end{cases}$$

Therefore $g = \varphi f$ a.e. in $\mathbb{R}$, that is, we have a full equality in $L^2(\mathbb{R})$. Since $g \in L^2(\mathbb{R})$, we get that $\varphi f \in L^2(\mathbb{R})$, so that $f \in D(M_\varphi)$. This finishes the proof.

(4) Assume that φ is essentially bounded. Denote by $\|\varphi\|_\infty$ its norm. Then for all $f \in D(M_\varphi) = L^2(\mathbb{R})$ we have

$$\|M_\varphi f\|_2^2 = \int_{\mathbb{R}} |\varphi(x)f(x)|^2 dx \le \|\varphi\|_\infty^2 \int_{\mathbb{R}} |f(x)|^2 dx = \|\varphi\|_\infty^2 \|f\|_2^2.$$

This implies that $\|M_\varphi\| \le \|\varphi\|_\infty$. If $\|\varphi\|_\infty = 0$, then $\|M_\varphi\| = 0$. Now if $\|\varphi\|_\infty > 0$, then for all $\varepsilon \in (0, \|\varphi\|_\infty)$, the set

$$X_\varepsilon = \{x \in \mathbb{R} : \ |\varphi(x)| \ge \|\varphi\|_\infty - \varepsilon\}$$

has a strictly positive measure (otherwise $\|\varphi\|_\infty$ would not be the essential supremum of φ anymore!). Now, let f_0 be an element of $L^2(\mathbb{R})$ which *vanishes outside* X_ε. Then

$$\begin{aligned} \|M_\varphi f_0\|_2^2 &= \int_{\mathbb{R}} |\varphi(x)f_0(x)|^2 dx \\ &= \int_{X_\varepsilon} |\varphi(x)f_0(x)|^2 dx + \underbrace{\int_{X_\varepsilon^c} |\varphi(x)f_0(x)|^2 dx}_{0} \\ &= \int_{X_\varepsilon} |\varphi(x)f_0(x)|^2 dx \\ &\ge (\|\varphi\|_\infty - \varepsilon)^2 \int_{X_\varepsilon} |f_0(x)|^2 dx \\ &= (\|\varphi\|_\infty - \varepsilon)^2 \|f_0\|_2^2. \end{aligned}$$

Hence

$$\|M_\varphi\| = \sup_{0\ne f\in L^2(\mathbb{R})} \frac{\|M_\varphi f\|_2}{\|f\|_2} \ge \frac{\|M_\varphi f_0\|_2}{\|f_0\|_2} \ge \|\varphi\|_\infty - \varepsilon, \ \forall \varepsilon \in (0, \|\varphi\|_\infty)$$

so that $\|M_\varphi\| \ge \|\varphi\|_\infty$. Therefore,

$$\|M_\varphi\| = \|\varphi\|_\infty.$$

This proves half of the first equivalence and the second assertion. To prove the other implication of the first assertion, assume that φ is not essentially bounded, then for all $n \in \mathbb{N}$, the following set

$$X_n = \{x \in \mathbb{R} : |\varphi(x)| \geq n\}$$

has a strictly positive measure. For (any) $f \in D(M_\varphi)$ which *vanishes outside* X_n, we have (as before) for any n

$$\|M_\varphi f\|_2^2 \geq n^2 \int_{X_n} |f(x)|^2 dx = n^2 \int_{\mathbb{R}} |f(x)|^2 dx = n^2 \|f\|_2^2$$

so that

$$\|M_\varphi f\|_2 \geq n\|f\|_2,$$

i.e. M_φ is not bounded.

(5) If φ is essentially bounded, then it is clear that for any $f \in L^2(\mathbb{R})$, we have $\varphi f \in L^2(\mathbb{R})$. Hence $D(M_\varphi) = L^2(\mathbb{R})$, which proves half of the equivalence.

Now suppose that $D(M_\varphi) = L^2(\mathbb{R})$. Then as $L^2(\mathbb{R})$ is a Banach space, by calling on the Closed Graph Theorem and the closedness of M_φ, we immediately deduce that M_φ is bounded so that by the previous question φ is essentially bounded.

(6) Let $f, g \in L^2(\mathbb{R})$. Then we have

$$\begin{aligned} < M_\varphi f, g > &= \int_{\mathbb{R}} M_\varphi f(x)\overline{g(x)}dx \\ &= \int_{\mathbb{R}} \varphi(x)f(x)\overline{g(x)}dx \\ &= \int_{\mathbb{R}} f(x)\overline{\overline{\varphi(x)}g(x)}dx \\ &=< f, M_\varphi^* g >, \end{aligned}$$

where

$$M_\varphi^* f = \overline{\varphi} f = M_{\overline{\varphi}} f, \ \forall f \in L^2(\mathbb{R}).$$

Thus it is clear that:

(a) *M_φ is self-adjoint iff φ is real valued almost everywhere:* One way of seeing this is that $M_\varphi - M_\varphi^* = M_\varphi - M_{\overline{\varphi}}$ is the multiplication operator by $\varphi - \overline{\varphi}$. Hence

$$\|M_\varphi - M_\varphi^*\|_{B(L^2(\mathbb{R}))} = \|\varphi - \overline{\varphi}\|_\infty = \operatorname*{ess\,sup}_{x\in\mathbb{R}} |\varphi(x) - \overline{\varphi(x)}|$$

so that it becomes clear that M_φ is self-adjoint iff φ is real-valued a.e.

(b) *M_φ is positive iff φ is positive almost everywhere:* It is plain that $\varphi \geq 0$ a.e. implies that

$$\varphi(x)|f(x)|^2 \geq 0$$

for any $f \in L^2(\mathbb{R})$. Thus, for all f

$$< M_\varphi f, f >= \int_{\mathbb{R}} \varphi(x)|f(x)|^2 dx \geq 0,$$

i.e. $M_\varphi \geq 0$.

Now, if $M_\varphi \geq 0$, then $\varphi \geq 0$ a.e. This result, as intuitive as it may seem, must be made clear to the reader. It is seldom that authors give a detailed proof of it (it can be fairly long if we work out details, as in **[50]** on Pages 345, but at least they have given it). So I propose the following proof: By assumption,

$$0 \leq < M_\varphi f, f >= \int_{\mathbb{R}} \varphi(x)|f(x)|^2 dx, \ \forall f \in L^2(\mathbb{R}).$$

In particular, for the $L^2(\mathbb{R})$-function $f(x) = e^{-x^2}\mathbb{1}_I(x)$ (where I is any measurable subset of $\mathbb{R}$). Then we still have

$$\int_I \varphi(x)e^{-x^2} dx \geq 0, \ \forall I \subset \mathbb{R}.$$

Since the previous holds for all I, it follows from a result on Integration that $\varphi(x)e^{-x^2} \geq 0$ for almost every x and so $\varphi \geq 0$ a.e., as required.

Now, to find the positive square root of M_φ, assume that $\varphi \geq 0$ a.e. in $\mathbb{R}$, then M_φ is positive. Observe that $M_{\sqrt{\varphi}}$ satisfies (for all $f \in L^2(\mathbb{R})$)

$$\begin{aligned} M^2_{\sqrt{\varphi}} f(x) &= M_{\sqrt{\varphi}}(M_{\sqrt{\varphi}} f(x)) = M_{\sqrt{\varphi}}(\sqrt{\varphi(x)} f(x)) \\ &= (\sqrt{\varphi(x)})^2 f(x) = M_\varphi f(x), \end{aligned}$$

that is, $M_{\sqrt{\varphi}}$ is a positive square root of M_φ. Since, the square root of a positive operator is unique, $M_{\sqrt{\varphi}}$ is *the* square root of M_φ.

(c) *M_φ is always normal*: For all $f \in L^2(\mathbb{R})$, we have

$$M_\varphi M_\varphi^* f = M_\varphi(\overline{\varphi} f) = \varphi\overline{\varphi} f = |\varphi|^2 f$$

and

$$M_\varphi^* M_\varphi f = M_\varphi^*(\varphi f) = \overline{\varphi}\varphi f = |\varphi|^2 f.$$

Remark. It may also be easily shown that if $\varphi, \psi \in L^\infty(\mathbb{R})$, then

$$M_{\varphi\psi} = M_\varphi M_\psi.$$

Hence, *any* two multiplication operators (defined on the same space) always commute.

(d) M_φ *is unitary iff* $|\varphi| = 1$ *almost everywhere* (just apply a similar idea to Part (a)).

(7) First, we prove that (a) $\Leftrightarrow$ (c). Recall that:

$$\ker M_\varphi^* = (\text{ran} M_\varphi)^\perp.$$

Since M_φ is (always) normal, we know that

$$\ker M_\varphi^* = \ker M_\varphi$$

so that

$$\ker M_\varphi = (\text{ran} M_\varphi)^\perp.$$

Hence

$$\begin{aligned} M_\varphi \text{ is injective } &\Longleftrightarrow \ker M_\varphi = \{0\} \\ &\Longleftrightarrow (\text{ran} M_\varphi)^\perp = \{0\} \\ &\Longleftrightarrow (\text{ran} M_\varphi)^{\perp\perp} = L^2(\mathbb{R}) \\ &\Longleftrightarrow \overline{\text{ran} M_\varphi} = L^2(\mathbb{R}). \end{aligned}$$

Now, we prove (b) $\Rightarrow$ (c). Let $f \in L^2(\mathbb{R})$ be such that $M_\varphi f = 0$, i.e. $\varphi(x)f(x) = 0$ a.e. in $\mathbb{R}$. Hence, and since $\varphi(x) \neq 0$ for almost every $x \in \mathbb{R}$, we obtain $f(x) = 0$ a.e. in $\mathbb{R}$, that is, $f = 0$ in $L^2(\mathbb{R})$. Therefore, M_φ is one-to-one.

Finally, we prove that (c) $\Rightarrow$ (b). Set

$$X_0 = \{x \in \mathbb{R} : \varphi(x) = 0\}.$$

Then for any $f \in L^2(\mathbb{R})$ which *vanishes outside* X_0, we have a.e. in $\mathbb{R}$

$$M_\varphi f(x) = \varphi(x)f(x) = 0$$

(because if $x \in X_0$, then $\varphi(x) = 0$; and if $x \notin X_0$, then $f(x) = 0$). This implies that $f \in \ker M_\varphi$. Since M_φ is injective, we have $f = 0$. This forces X_0 to have measure zero (otherwise, f would not have vanished).

(8) We claim that

$$\text{ran}M_\varphi = L^2(\mathbb{R}) \Longleftrightarrow \exists \alpha > 0 : \ |\varphi(x)| \geq \alpha \text{ a.e. in } \mathbb{R}.$$

Then by the previous question

$$M_\varphi \text{ is invertible} \Longleftrightarrow \text{ran}M_\varphi = L^2(\mathbb{R}).$$

Since M_φ is normal, we do know that

$$M_\varphi \text{ is invertible} \Longleftrightarrow \exists \alpha > 0 : \ \|M_\varphi(f)\| \geq \alpha\|f\|, \ \forall f \in L^2(\mathbb{R}).$$

Thus we need only show that

$$\exists \alpha > 0 : \|M_\varphi(f)\|_2 \geq \alpha\|f\|_2, \ \forall f \in L^2(\mathbb{R}) \Longleftrightarrow \exists \alpha > 0 : \ |\varphi(x)| \geq \alpha \text{ a.e. in } \mathbb{R}.$$

(a) "$\Longleftarrow$": Assume that for some $\alpha > 0 : \ |\varphi(x)| \geq \alpha$ a.e. in $\mathbb{R}$. Let $f \in L^2(\mathbb{R})$. Then

$$\|M_\varphi f\|_2^2 = \int_{\mathbb{R}} |\varphi(x)f(x)|^2 dx \geq \alpha^2 \int_{\mathbb{R}} |f(x)|^2 dx = \alpha^2\|f\|_2^2,$$

proving one implication.

(b) "$\Longrightarrow$": Assume that for some $\alpha > 0$ and all $f \in L^2(\mathbb{R})$, $\|M_\varphi f\|_2 \geq \alpha\|f\|_2$ holds. Then $\|M_\varphi f\|_2^2 \geq \alpha^2\|f\|_2^2$ so that

$$< M_\varphi f, M_\varphi f > \geq \alpha^2 < f, f > \text{ or } < M_\varphi^* M_\varphi f, f > \geq < \alpha^2 f, f > .$$

Hence

$$< (M_{|\varphi|^2} - \alpha^2)f, f > \geq 0, \ \forall f \in L^2(\mathbb{R}),$$

so that by Question (6), we obtain

$$|\varphi(x)| \geq \alpha \text{ for a.e. } x,$$

completing the proof.

Remark. In fact, we have shown that M_φ is invertible iff $\varphi \neq 0$ a.e.

If $\varphi\psi = 1$ a.e., then M_ψ acts as the inverse of M_φ as

$$M_\varphi M_\psi = M_{\varphi\psi} = M_1 = I_{L^2(\mathbb{R})},$$

where 1 in the last equality is just the function $x \mapsto 1$.

(9) Recall that $\lambda \in \sigma_p(M_\varphi)$ iff $\lambda - M_\varphi$ is *not* injective. Now, clearly $\lambda - M_\varphi$ is the multiplication operator by the function $\lambda - \varphi$, so by Question (7), $\lambda - M_\varphi$ is not injective iff $\lambda - \varphi(x) = 0$ almost everywhere. Therefore,

$$\sigma_p(M_\varphi) = \{\lambda \in \mathbb{C} : \ \varphi^{-1}(\lambda) \text{ has a strictly positive measure}\},$$

where as usual $\varphi^{-1}(\lambda) = \{x \in \mathbb{R} : \lambda = \varphi(x)\}$.

Now, we exploit Question (8) to find $\sigma(M_\varphi)$. We know that $\lambda \in \rho(M_\varphi)$ iff $\lambda - M_\varphi$ is invertible, that is, iff $\lambda - M_\varphi$ is bijective. By Question (8), this happens exactly when there exists some $\alpha > 0$ such that $|\lambda - \varphi(x)| \geq \alpha$ a.e. in $\mathbb{R}$.

Therefore, by passing to the complement, we obtain

$$\begin{aligned}\sigma(M_\varphi) &= [\rho(M_\varphi)]^c \\ &= \{\lambda \in \mathbb{C} : \forall \varepsilon > 0, \{x \in \mathbb{R} : |\lambda - \varphi(x)| < \varepsilon\} \text{ has a positive measure}\} \\ &= \text{essran}\varphi.\end{aligned}$$

Remark. We already knew that the essential range of an essentially bounded function is compact (see Exercise 1.5.9). Since the spectrum of a bounded operator is always compact (see Theorem 7.1.4), we have just obtained another proof that the essential range of a function is compact.

(10) Assume that φ is not zero a.e. in $\mathbb{R}$. Let $\alpha > 0$ be such that the set $\{x \in \mathbb{R} : |\varphi(x)| > \alpha\}$ has a strictly positive measure. Then by Question (8), M_φ is invertible (i.e. M_φ^{-1} is bounded), so that if M_φ were compact, then $M_\varphi M_\varphi^{-1}$ would be compact too. But, since $M_\varphi M_\varphi^{-1} = I$, this would mean that the identity operator I (on an infinite dimensional space!) would be compact, a contradiction! Thus, we have proved that $\varphi = 0$ a.e. in $\mathbb{R}$ if M_φ compact.

To prove the backward implication, just recall that the zero operator is always compact. After a "trek", the solution of the exercise is over!

Solution 10.3.10. We claim that

$$\|A\|_{B(L^2(\mathbb{R}))} = \|\varphi\|_\infty.$$

To show this, we may do it by the "brutal" way, or, we may use some finesse instead. Write $A = MU$, where

$$Mf(x) = \varphi(x+a)f(x) \text{ and } Uf(x) = f(x-a),$$

with $f \in L^2(\mathbb{R})$.

Now, by Exercise 4.3.19, U is unitary. By Exercise 4.3.6, we have

$$\|A\| = \|MU\| = \|M\|.$$

By Exercise 10.3.9, $\|M\| = \|\varphi\|_\infty$. Hence

$$\|A\| = \|M\| = \operatorname*{esssup}_{x\in\mathbb{R}}|\varphi(x+a)| = \operatorname*{esssup}_{x\in\mathbb{R}}|\varphi(x)|.$$

Solution 10.3.11.

(1) Let $x \in D(A)$. Then

$$\|Ax + ix\|^2 = \|x\|^2 + \|Ax\|^2 + i < x, Ax > -i < Ax, x > .$$

Since A is symmetric, it follows that

$$\|Ax + ix\|^2 = \|x\|^2 + \|Ax\|^2.$$

(2) By the previous result, we have for all $x \in D(A)$

$$\|Ax + ix\| \geq \|x\|.$$

The equivalence then follows from Exercises 10.3.5 & 10.3.6.

Solution 10.3.12. Let

$$Bf(x) = \frac{1}{1+x^2} f(x).$$

Then B is bounded on $L^2(\mathbb{R})$. Then B is the inverse of A if

$$AB = I \text{ and } BA \subset I.$$

This is indeed the case as

$$D(AB) = \{f \in D(B) : Bf \in D(A)\} = \{f \in L^2(\mathbb{R}) : \frac{1}{1+x^2} f \in D(A)\}.$$

Since $\frac{1}{1+x^2} \in L^\infty(\mathbb{R})$, we obtain

$$D(AB) = L^2(\mathbb{R}).$$

Now, as B is bounded, then

$$D(BA) = D(A) \subset L^2(\mathbb{R}).$$

Let $f \in D(AB)$ and let $g \in D(BA) = D(A)$. Then

$$ABf = f \text{ and } BAg = g.$$

Accordingly,

$$AB = I \text{ and } BA \subset I.$$

Solution 10.3.13.

(1) Not always! Let A be an unbounded and invertible operator (with a non-closed domain $D(A)$) and let B be its bounded inverse. Then $BA \subset I$ because

$$BA = I_{D(BA)} \subset I = I_H.$$

This shows that BA cannot be closed since this would lead to the closedness of $D(BA) = D(A)$ (see Proposition 10.1.11).

(2) Let $x_n \in D(AB)$ be such that $x_n \to x$ and $ABx_n \to y$. Since B is bounded, $Bx_n \to Bx$. But A is closed, so $ABx = y$ and $Bx \in D(A)$, i.e. $x \in D(AB)$. Therefore, AB is closed.

(3) Let $x_n \in D(CA)$ such that $x_n \to x$ and $CAx_n \to y$ (remember that we must show that $x \in D(CA)$ and $CAx = y$). Since C^{-1} is bounded, we have $Ax_n \to C^{-1}y$. But A is closed and so

$$C^{-1}y = Ax \text{ and } x \in D(A).$$

Whence $CC^{-1}y = y = CAx$ and $Ax \in \operatorname{ran}(C^{-1}) = D(C)$. So $x \in D(CA)$ and $CAx = y$, which completes the proof.

(4) The answer is no! Let A be any closed operator with a non-closed domain $D(A)$. Then $-A$ too is closed but $A - A$ is not closed as

$$A - A = 0_{D(A)},$$

i.e. it is the zero operator defined on $D(A)$ (cf. Proposition 10.1.11).

(5) Recall that $A+B$ is defined on $D(A) \cap D(B)$. But $D(B) = H$ and $D(A) \subset H$. Hence the domain of $A+B$ is reduced to $D(A)$.

To show that $A+B$ is closed in $D(A)$, let $x_n \in D(A)$ be such that $x_n \to x$ and $(A+B)x_n \to y$. Since B is bounded, $Bx_n \to Bx$. This implies that

$$Ax_n = (A+B)x_n - Bx_n \longrightarrow y - Bx.$$

But A is closed, hence

$$x \in D(A) \text{ and } y - Bx = Ax \text{ or } (A+B)x = y,$$

i.e. $A+B$ is closed.

Solution 10.3.14. We may proceed as in Exercise 10.3.8. Alternatively, we may use the results of Exercise 10.3.13. We try to write A as a product of two operators, the one on the left is closed and the one on the right is bounded.

Let S^* be the shift's adjoint. Then if M is the multiplication operator by (α_n), i.e.

$$M(x_1, x_2, \cdots) = (\alpha_1 x_1, \alpha_2 x_2, \cdots)$$

defined on

$$D(M) = \{(x_n) \in \ell^2 : (\alpha_n x_n) \in \ell^2\},$$

then proceeding just as in Exercise 10.3.8, we may show that M is closed (do it!). Now, it is plain that on $D(MS^*)$:

$$A(x_1, x_2, \cdots) = MS^*(x_1, x_2, \cdots) = (\alpha_1 x_2, \alpha_2 x_3, \cdots).$$

Since S^* is bounded, Exercise 10.3.13 gives the closedness of MS^*, i.e. that of A.

Solution 10.3.15.

(1) No! We give an explicit counterexample. Consider the following two operators defined by

$$Af(x) = e^{2x}f(x) \text{ and } Cf(x) = (e^{-x}+1)f(x)$$

on their respective domains

$$D(A) = \{f \in L^2(\mathbb{R}) : e^{2x}f \in L^2(\mathbb{R})\}$$

and

$$D(C) = \{f \in L^2(\mathbb{R}) : e^{-2x}f, e^{-x}f \in L^2(\mathbb{R})\}$$

(think of C being a sum of two operators $S = e^{-2x} + e^{-x} + 1$ and $T = -e^{-2x}$, say, on their "natural" domains). Then C is not closed.

Now the operator CA is defined by

$$CAf(x) = (e^{2x} + e^{x})f(x)$$

on

$$D(CA) = \{f \in D(A) : Af \in D(C)\} = \{f \in L^2(\mathbb{R}) : e^{2x}f, e^{x}f \in L^2(\mathbb{R})\}$$

and one can see that CA is closed on this domain.

To treat the other case, just consider

$$Af(x) = e^{2x}f(x) \text{ and } Cf(x) = (e^{-x}+1)f(x)$$

on their respective domains

$$D(A) = \{f \in L^2(\mathbb{R}) : e^{x}f, e^{2x}f \in L^2(\mathbb{R})\}$$

and

$$D(C) = \{f \in L^2(\mathbb{R}) : e^{-x}f \in L^2(\mathbb{R})\}.$$

Details are left to the reader...

(2) Let (x_n) be in $D(BA) = D(A)$ such that

$$BAx_n \longrightarrow y \text{ and } x_n \longrightarrow x.$$

Then (as B is continuous!)

$$B^2Ax_n \longrightarrow By \text{ and } x_n \longrightarrow x.$$

Since B^2A is closed, we obtain

$$B^2Ax = By \text{ and } x \in D(B^2A) = D(A).$$

Since $B(BAx - y) = 0$ and B is one-to-one, we finally infer that

$$BAx - y = 0 \text{ and } x \in D(A) = D(BA),$$

that is, BA is closed.

(3) Let (x_n) be in $D(BA) = D(A)$ such that

$$BAx_n \longrightarrow y \text{ and } x_n \longrightarrow x.$$

Then it is clear that $y \in \overline{\operatorname{ran}(B)}$ for $BAx_n \in \operatorname{ran}(B)$. Since B is continuous, we obtain

$$B^2Ax_n \longrightarrow By \text{ and } x_n \longrightarrow x.$$

As B^2A is closed, we then obtain

$$B^2Ax = By \text{ and } x \in D(B^2A) = D(A).$$

Hence $B(y - BAx) = 0$, that is, $y - BAx \in \ker(B)$. Since also $y - BAx \in \overline{\operatorname{ran}(B)}$ and B is self-adjoint, we get $y - BAx \in [\ker(B)]^\perp$. Thus $y - BAx = 0$ or $BAx = y$. Since we already know that $x \in D(A) = D(BA)$, the proof of the closedness of BA is complete.

If B is normal, then $\ker(B) = \ker(B^*)$. Then, use the same proof as before.

Solution 10.3.16. Let $C \in B(H)$ be the inverse of BA. This implies that $BAC = I$.

Since $BA : D(BA) = D(A) \to \operatorname{ran}(BA)$ is invertible, $\operatorname{ran}C = D(A)$ so that

$$D(AC) = \{x \in H : Cx \in D(A)\} = H.$$

Finally, since AC is closed, by the Closed Graph Theorem we infer that AC is bounded. This leads to B being right invertible.

Solution 10.3.17.

(1) Let $x \in D(A) = D(A + B)$. We have

$$\|(A + B)x\| \le \|Ax\| + \|Bx\| \le (1 + a)\|Ax\| + b\|x\| ...(1)$$

and

$$\|(A + B)x\| \ge \|Ax\| - \|Bx\| \ge (1 - a)\|Ax\| - b\|x\| ...(2)$$

Hence

$$\|Ax\| \le \frac{1}{1 - a}\left(\|(A + B)x\| + b\|x\|\right) ...(3)$$

Next, let $x_n \in D(A)$ be such that $x_n \to x$ and $(A+B)x_n \to y$. Whence they are both Cauchy. By Inequality (3), we easily see that (Ax_n) is a Cauchy sequence. So there exists a vector $z \in H$ such that $Ax_n \to z$. However, A is closed, and so $x \in D(A)$ and $z = Ax$.

Going back to Inequality (1), we may then write

$$0 \le \|(A + B)(x_n - x)\| \le (1 + a)\|Ax_n - Ax\| + b\|x_n - x\| \longrightarrow 0$$

as n tends to infinity. Thus $(A+B)x = y$, proving the closedness of $A+B$.

(2) No! Let A be closed with a non-closed domain. Set $B = -A$ which remains closed. Then

$$\|Bx\| = \|Ax\| \leq \|Ax\| + b\|x\|$$

for all $x \in D(A) = D(B)$ (and any positive b). So B is A-bounded and yet $A+B$ is not closed.

SOLUTION 10.3.18.

(1) Left to the interested reader...

(2) We give two methods:

(a) First method: We will construct the counterexample by using a linear interpolation. We define $(x,y) \mapsto \varphi(x,y)$ on $\mathbb{R} \times (y_n, y_{n+1})$ by

$$\varphi(x,y) = \frac{1}{y_{n+1} - y_n}[(y - y_n)f_{n+1}(x) - (y - y_{n+1})f_n(x)]$$

where $f_n(x) = \varphi(x, y_n)$, and the f_n and y_n are yet to be determined.

Hence on $\mathbb{R} \times (y_1, \infty)$ we have

$$\|\varphi\|_2^2 = \iint\limits_{\mathbb{R}\times(y_1,\infty)} |\varphi(x,y)|^2 dxdy = \sum_1^\infty \iint\limits_{\mathbb{R}\times(y_n,y_{n+1})} |\varphi(x,y)|^2 dxdy.$$

Doing some arithmetic, we find that the condition that makes φ in L^2 is

$$\sum_1^\infty (y_{n+1} - y_n)(\|f_n\|_2^2 + \|f_{n+1}\|_2^2) < \infty.$$

We also have

$$\frac{\partial^2 \varphi}{\partial x \partial y} = \frac{1}{y_{n+1} - y_n}(f'_{n+1} - f'_n) \in L^2(\mathbb{R} \times (y_1, \infty))$$

if

$$\sum_1^\infty \frac{1}{y_{n+1} - y_n}\|\psi'_n\|_2^2 < \infty$$

where

$$\psi_n(x) = f_{n+1}(x) - f_n(x).$$

Since $\psi_n(x) = f_{n+1}(x) - f_n(x)$, we obtain

$$f_n(x) = -\sum_n^\infty \psi_k(x).$$

We also want $f_n \notin L^\infty(\mathbb{R})$ so that $\varphi \notin L^\infty(\mathbb{R}^2)$.
Consider

$$\psi_n(x) = \begin{cases} \frac{e^n}{n}x + \frac{1}{n} & \text{if } -e^{-n}, \leq x \leq 0 \\ -\frac{e^n}{n}x + \frac{1}{n} & \text{if } 0 \leq x \leq e^{-n}, \\ 0 & \text{if } |x| \geq e^{-n}. \end{cases}$$

Thus

$$\|\psi_n'\|_2^2 \sim \frac{e^n}{n^2} \text{ and } \|\psi_n\|_2^2 \sim \frac{e^{-n}}{n^2}.$$

We also have

$$\|f_n\|_2 \leq \sum_{k=n}^{\infty} \|\psi_k\|_2 = a\sum_{k=n}^{\infty} \frac{e^{-\frac{n}{2}}}{n} \simeq \int_n^\infty \frac{e^{-\frac{x}{2}}}{x}dx \leq \frac{1}{n}\int_n^\infty e^{-\frac{x}{2}}dx \sim \frac{e^{-\frac{n}{2}}}{n}.$$

Now if we choose $y_{n+1} - y_n = e^n$, then the series

$$\sum_1^\infty \frac{1}{y_{n+1} - y_n}\|\psi_n'\|_2^2 = \sum_1^\infty \frac{1}{e^n} \times \frac{e^n}{n^2} = \sum_1^\infty \frac{1}{n^2}$$

converges and so does

$$\sum_1^\infty (y_{n+1} - y_n)(\|f_n\|_2^2 + \|f_{n+1}\|_2^2) \leq \sum_1^\infty e^n \times [\frac{e^{-(n+1)}}{(n+1)^2} + \frac{e^{-n}}{n^2}] \sim \sum_1^\infty \frac{1}{n^2}.$$

Now the φ defined on $\mathbb{R} \times (y_n, y_{n+1})$ is given by

$$\varphi(x, y) = e^{-n}[(y - y_n)(-\sum_{n+1}^\infty \psi_k(x)) - (y - y_{n+1})(-\sum_n^\infty \psi_k(x))].$$

This φ is actually defined only for $x \in \mathbb{R}$ and $y \geq y_1$. To extend it to the case $y < y_1$ we define φ for $x \in \mathbb{R}$ and $y_1 - y_{n+1} < y < y_1 - y_n$ as follows:

$$\varphi(x, y) = \frac{1}{y_{n+1} - y_n}[(y - y_1 + y_n)f_{n+1}(x) - (y - y_1 + y_{n+1})f_n(x)].$$

This φ is in $D(A)$ for sure. Now we need to show that φ is not in $L^\infty(\mathbb{R}^2)$. Let $x > 0$ and $x \leq e^{-k}$ then $\ln x \leq -k$ or $\ln \frac{1}{x} \geq k$. So

$$f_n(x) = -x\sum_n^{[\ln \frac{1}{x}]} \frac{e^{-k}}{k} + \sum_n^{[\ln \frac{1}{x}]} \frac{1}{k} \geq (-x^2 + 1)\sum_n^{[\ln \frac{1}{x}]} \frac{1}{k}.$$

Now as $x \to 0$, then $[\ln \frac{1}{x}] \to \infty$ hence $\ln[\ln \frac{1}{x}] \to \infty$. Thus $f_n(x) \to \infty$ which implies that $\varphi(x, y) \to \infty$. So that finally $\varphi \notin L^\infty(\mathbb{R}^2)$.

(b) Second method: ([**141**]) First, consider $f : \mathbb{R}^2 \to \mathbb{R}$ such that

$$f(u,v) = \frac{1}{1+|uv|}.$$

The function f is obviously positive. Besides, it does not belong to $L^2(\mathbb{R}^2)$ since

$$\iint_{\mathbb{R}^2} \frac{1}{(1+|uv|)^2}\,dudv \geq \int_0^\infty \int_0^\infty \frac{1}{(1+uv)^2}\,dudv = \lim_{R\to\infty} \int_0^R \int_0^R \frac{1}{(1+uv)^2}\,dudv$$

But

$$\int_0^R \frac{1}{(1+uv)^2}du = [-\frac{1}{v}(1+uv)^{-1}]_0^R = \frac{1}{v} - \frac{1}{v(1+Rv)} = \frac{R}{1+Rv}.$$

So

$$\|f\|_2^2 \geq \lim_{R\to\infty} \int_0^R \frac{R}{1+Rv}dv = \lim_{R\to\infty} [\ln(1+Rv)]_0^R = \lim_{R\to\infty} \ln(1+R^2) = \infty.$$

Now by Proposition 1.1.66, we know that there exists a positive $\psi \in L^2$ such that $\psi f \notin L^1$. Since $f \in L^\infty$, ψf belongs to L^2 and it legitimate to define $\varphi = \mathcal{F}^{-1}(\psi f)$ where $\mathcal{F}$ is the L^2-Fourier transform. By the Plancherel Theorem φ is in L^2. Also

$$\mathcal{F}\left(\frac{\partial^2 \varphi}{\partial x \partial y}\right) = uv\mathcal{F}\varphi = uv\psi(u,v)f(u,v).$$

Since $(u,v) \mapsto \frac{uv}{1+|uv|} \in L^\infty(\mathbb{R}^2)$ and since $\psi \in L^2(\mathbb{R}^2)$ it follows that $\frac{uv}{1+|uv|}\psi \in L^2(\mathbb{R}^2)$ and hence, as a consequence of the Plancherel Theorem, one gets that $\frac{\partial^2 \varphi}{\partial x \partial y} \in L^2(\mathbb{R}^2)$. Since $\mathcal{F}(\varphi) \in L^2$, since it is positive and since $\mathcal{F}(\varphi) \notin L^1$, Proposition 2.1.53 allows us to say that $\varphi \notin L^\infty(\mathbb{R}^2)$.

Solution 10.3.19.

(1) The way of showing that A is unbounded and closed on the domain $D(A) = H^2(\mathbb{R}^n)$ is very similar to that of Exercise 10.3.6. We leave details to readers...

(2) We only give the proof in the case $n = 1$ (the reader may do the cases $n = 2$ and $n = 3$ similarly). Let $f \in H^2(\mathbb{R})$. We first show that $\hat{f} \in L^1(\mathbb{R})$. Since $f \in H^2(\mathbb{R})$, by the Plancherel Theorem we have

$$(1+t^2)\hat{f} \in L^2(\mathbb{R}).$$

Besides $(1+t^2)^{-1} \in L^2(\mathbb{R})$ and so by the Cauchy-Schwarz Inequality we have

$$\hat{f} = \frac{1}{1+t^2}(1+t^2)\hat{f} \in L^1(\mathbb{R}),$$

and that

$$\|\hat{f}\|_1 \le \alpha\|(1+t^2)\hat{f}\|_2 \le \alpha(\|t^2\hat{f}\|_2 + \|\hat{f}\|_2)$$

where

$$\alpha^2 = \int_{\mathbb{R}} \frac{dt}{(1+t^2)^2} < \infty.$$

Thus, by the inversion formula and the Plancherel Theorem, we obtain

$$\|f\|_\infty \le \alpha\|Af\|_2 + \alpha\|f\|_2 ...(1)$$

To obtain the desired condition on the constant, we use a scaling argument. Set

$$f_r(x) = f(rx),\ r > 0.$$

Then we easily obtain that

$$\|f_r\|_\infty = \|f\|_\infty,\ \|f_r\|_2 = \frac{1}{\sqrt{r}}\|f\|_2 \text{ and } \|Af_r\|_2 = r^{\frac{3}{2}}\|Af\|_2.$$

Hence by substituting f by f_r in Inequality (1) we get

$$\|f\|_\infty \le \alpha r^{\frac{3}{2}}\|Af\|_2 + \alpha\frac{1}{\sqrt{r}}\|f\|_2,$$

and by choosing r small enough we obtain the desired result.

(3) Let $n \ge 4$. We apply a similar method as the second method of Question 2 of Exercise 10.3.18. Therefore, we only have to show that the function $f : \mathbb{R}^n \to \mathbb{R}$ defined by

$$g(x_1, x_2, \cdots, x_n) = \frac{1}{1 + x_1^2 + x_2^2 + \cdots + x_n^2}$$

is not in $L^2(\mathbb{R}^n)$. This is easily seen, using the generalized spherical coordinates and their Jacobian. We have

$$\|g\|_2^2 = \int_{\mathbb{R}^n} \frac{dx_1x_2\cdots x_n}{(1 + x_1^2 + x_2^2 + \cdots + x_n^2)^2} = c\int_0^\infty \frac{r^{n-1}dr}{(1+r^2)^2} = \infty$$

precisely when $n \ge 4$.

(4) We only give the proof for $n = 3$. Write $V = V_1 + V_2$ where $V_1 \in L^2(\mathbb{R}^3)$ and $V_2 \in L^\infty(\mathbb{R}^3)$. Let $f \in H^2(\mathbb{R}^3)$. Then

$$\begin{aligned}
\|Mf\|_2 &= \|Vf\|_2 \\
&= \|(V_1 + V_2)f\|_2 \\
&\leq \|V_1 f\|_2 + \|V_2 f\|_2 \\
&\leq \|V_1\|_2 \|f\|_\infty + \|V_2\|_\infty \|f\|_2 \\
&\leq \|V_1\|_2 (a\|Af\|_2 + b\|f\|_2) + \|V_2\|_\infty \|f\|_2 \\
&\leq a\|V_1\|_2 \|Af\|_2 + (\|V_2\|_\infty + b\|V_1\|_2)\|f\|_2.
\end{aligned}$$

By Question 2), we can make a small enough so that $a\|V_1\|_2 < 1$. Theorem 10.1.22 then implies that $A+M$ is closed.

Solution 10.3.20.

(1) Since A is injective, it makes sense to define $A^{-1} : \operatorname{ran}(A) \to D(A)$. Writing $D(A^{-1}) = \operatorname{ran}(A)$, we then have

$$G(A^{-1}) = \{(y, A^{-1}y) : \ y \in D(A^{-1})\} = \{(Ax, x) : \ x \in D(A)\}.$$

Hence A is closed iff A^{-1} is closed.

(2) By assumption $\lambda - A$ is injective. Besides, $D(\lambda - A) = D(A)$. Since A is closed, so is $-A$ and so $\lambda - A$ too is closed (by Exercise 10.3.13). By the previous question, $R(\lambda, A)$ is closed. Now, since $\lambda \in \rho(A)$, we have

$$D(R(\lambda, A)) = \operatorname{ran}(\lambda - A) = H,$$

that is, $R(\lambda, A)$ is closed and defined everywhere on a Hilbert space. By the Closed Graph Theorem, $R(\lambda, A) \in B(H)$.

(3) To show that $\sigma(A)$ is closed we may equivalently show that $\rho(A)$ is open. If $\rho(A) = \varnothing$, then $\rho(A)$ is open and there is nothing to prove. Assume then that $\rho(A) \neq \varnothing$. Let $\lambda_0 \in \rho(A)$, i.e. $R(\lambda_0, A) \in B(H)$.

We need to show that for some $\alpha > 0$, $B(\lambda_0, \alpha) \subset \rho(A)$. Choose $\alpha = \frac{1}{\|R(\lambda_0, A)\|}$. Let $\lambda \in B(\lambda_0, \alpha)$, i.e. $|\lambda - \lambda_0| < \alpha$ (remember that we have to show that $R(\lambda, A) \in B(H)$). We may write

$$\lambda I - A = (\lambda - \lambda_0)I + \lambda_0 I - A$$

so that

$$\lambda I - A = [(\lambda - \lambda_0)R(\lambda_0, A) + I](\lambda_0 I - A).$$

Since $|\lambda - \lambda_0| < \frac{1}{\|R(\lambda_0, A)\|}$, we have $\|(\lambda - \lambda_0)R(\lambda_0, A)\| < 1$. It follows that the bounded $(\lambda - \lambda_0)R(\lambda_0, A) + I$ is invertible. Thus $\lambda I - A$ is invertible too, i.e. $\lambda \in \rho(A)$.

SOLUTION 10.3.21. Since $\rho(A)$ is not empty, it contains at least one complex λ. Hence $A - \lambda I$ is invertible. By Proposition 10.1.13, $A - \lambda I$ is closed. By Exercise 10.3.13 $(A - \lambda I) + \lambda I$ is closed, that is, A is closed on $D(A)$.

SOLUTION 10.3.22. Since $\sigma(A) \neq \mathbb{C}$, A is closed by Proposition 10.1.16. Let $\lambda \in \rho(A)$. Since $A - \lambda I$ is surjective, we have $\operatorname{ran}(A - \lambda I) = H$. Finally, as $\operatorname{ran}(A) \subset D(A)$, we have

$$H = \operatorname{ran}(A - \lambda I) \subset D(A) \subset H,$$

i.e. $D(A) = H$. Since A is closed, the Closed Graph Theorem yields the boundedness of A.

REMARKS.

(1) In fact, the result of the previous exercise first appeared in **[162]** under the extra condition that A is closed. We have voluntarily omitted this assumption since it is tacitely assumed whenever we write $\sigma(A) \neq \mathbb{C}$ (by Exercise 10.3.21!). Besides, the proof we have given here is shorter.

(2) It is not always true that a closed linear operator A such that $\operatorname{ran}(A) \subset D(A)$ implies that $A \in B(H)$. A counterexample may be found in Exercise 10.5.4. See Exercise 10.5.8 for a similar result.

SOLUTION 10.3.23.

(1) Let $f \in L^2(\mathbb{R})$. We may write

$$\|Af\|_2^2 \leq \sup_{x \in \mathbb{R}}(e^{-x^2}) \int_{\mathbb{R}} |f(x-1)|^2 dx = 1 \times \int_{\mathbb{R}} |f(x)|^2 dx = \|f\|_2^2,$$

so that A is bounded on $L^2(\mathbb{R})$.

(2) To find A^*, let $f, g \in L^2(\mathbb{R})$. Then we have

$$\begin{aligned} < Af, g > &= \int_{\mathbb{R}} Af(x)\overline{g(x)}dx \\ &= \int_{\mathbb{R}} e^{-x^2} f(x-1)\overline{g(x)}dx \\ &= \int_{\mathbb{R}} f(x)e^{-(x+1)^2}\overline{g(x+1)}dx \\ &= \int_{\mathbb{R}} f(x)\overline{e^{-(x+1)^2}g(x+1)}dx \\ &= < f, A^*g > . \end{aligned}$$

Accordingly, for $f \in L^2(\mathbb{R})$

$$(A^*f)(x) = e^{-(x+1)^2} f(x+1).$$

(3) First, it is easy to see that

$$(A^n f)(x) = e^{-x^2-(x-1)^2-\cdots-[x-(n-1)]^2} f(x-n).$$

Hence, by Exercise 10.3.10

$$\|A^n\| = \sup_{x\in\mathbb{R}} e^{\varphi(x)}$$

where $\varphi(x) = -x^2 - (x-1)^2 - \cdots - [x-(n-1)]^2$.

But φ attains its maximum at $\frac{n-1}{2}$. Hence we may easily compute and check that

$$\begin{aligned}\varphi(\frac{n-1}{2}) &= -\frac{n(n-1)^2}{4} + \frac{n(n-1)^2}{2} - \frac{n(n-1)(2n-1)}{6} \\ &= -\frac{(n-1)n(n+1)}{12}.\end{aligned}$$

Finally,

$$\|A^n\| = e^{-\frac{(n-1)n(n+1)}{12}}.$$

(4) By the method of Exercise 10.3.9, A is clearly injective. By a similar argument, A^* too is injective. As a deduction

$$(\text{ran}A)^\perp = \ker A^* = \{0\} \Longrightarrow \overline{\text{ran}A} = (\ker A^*)^\perp = \{0\}^\perp = L^2(\mathbb{R}),$$

i.e. ran$A$ is dense in $L^2(\mathbb{R})$.

(5) We have on the one hand

$$r(A) = \lim_{n\to\infty} \|A^n\|^{\frac{1}{n}} = \lim_{n\to\infty} e^{-\frac{(n-1)(n+1)}{12}} = 0.$$

On the other hand,

$$0 = r(A) = \sup\{|\lambda| : \lambda \in \sigma(A)\}$$

which forces us to have

$$\sigma(A) = \{0\}.$$

(6) We start by showing that B is closed on $D(B)$. Let (f_n) be in $D(B)$ such that $f_n \to f$ and $Bf_n \to g$ (in L^2!). Since $f_n \in D(B) = \text{ran}A$, we may write $f_n = Ag_n$ for some $g_n \in L^2(\mathbb{R})$. Hence (as $BAf = f$ for all $f \in L^2(\mathbb{R})$)

$$Ag_n \longrightarrow f \text{ and } BAg_n = g_n \longrightarrow g.$$

Thus $g \in L^2(\mathbb{R})$ and since A is continuous, we deduce that $f = Ag \in \text{ran}(A) = D(B)$, i.e. $f \in D(B)$. Besides $Bf = BAg = g$ and this shows the closedness of B.

Now, we show that $\sigma(B) = \varnothing$. We need to investigate two cases:

(a) Let $\lambda = 0$. We need to show that $0 \times I - B$, i.e. B, is bijective. First, B is injective. If $Bf = 0$ where $f \in D(B) = \text{ran}A$, then as $f = Ag$ for some g, we must have $BAg = 0$. But $BA = I$ and so

$$0 = BAg = g.$$

Hence $g = 0$ and so $f = 0$. Thus, B is one-to-one.
Moreover, as $BA = I$, it is clear that B is onto. Therefore, B is bijective and so invertible by Proposition 10.1.15. Consequently, $0 \in \rho(B)$.

(b) Let $\lambda \neq 0$. We must show that $\lambda I - B$ is bijective. We start by injectivity. Let $f \in D(B)$. Since $\lambda \neq 0$, we may write

$$\begin{aligned}(\lambda I - B)f = 0 &\Longrightarrow \lambda Af - BAf = 0\\ &\Longrightarrow \lambda Af - f = 0\\ &\Longrightarrow \left(A - \frac{1}{\lambda}I\right) f = 0.\end{aligned}$$

Since $\sigma(A) = \{0\}$, $\rho(A) = \mathbb{C} \setminus \{0\}$, i.e. $A - \mu I$ is bijective for all $\mu \neq 0$. This implies that $A - \mu I$ is injective for all $\mu \neq 0$, in particular for $\mu = \frac{1}{\lambda}$ too. Hence $f = 0$, showing that $\lambda I - B$ is one-to-one.
Let's turn now to surjectivity. Let $g \in L^2(\mathbb{R})$. Since $A - \frac{1}{\lambda}I$ is surjective and $g/\lambda \in L^2(\mathbb{R})$, we have

$$\exists h \in D(A) : \left(A - \frac{1}{\lambda}I\right) h = \frac{g}{\lambda}.$$

Now, let $f = Ah$. Then

$$\begin{aligned}(\lambda I - B)f &= (\lambda I - B)Ah\\ &= \lambda Ah - BAh\\ &= \lambda Ah - h\\ &= \lambda \left(A - \frac{1}{\lambda}I\right) h\\ &= \lambda \times \frac{g}{\lambda}\\ &= g,\end{aligned}$$

proving that $\lambda I - B$ is onto for all $\lambda \neq 0$. Hence $\lambda I - B$ is bijective for all $\lambda \neq 0$, that is, invertible $\lambda I - B$ for all $\lambda \neq 0$ (Proposition 10.1.15), i.e. $\mathbb{C} \setminus \{0\} \subset \rho(B)$.

Accordingly, $\rho(B) = \mathbb{C}$ or $\sigma(B) = \varnothing$.

SOLUTION 10.3.24. Let $f \in D(A)$. In order to find the spectrum of A, we need to solve the following differential equation

$$\lambda f(x) - f'(x) = g(x),$$

where $g \in L^2[0,1]$ is given and $\lambda \in \mathbb{C}$. The way of solving the previous differential equation is known. We find that

$$f(x) = \int_0^x e^{\lambda(x-t)} g(t) dt$$

and it is the *unique* solution which belongs to $D(A)$ (do the details!). This means that $\lambda - A$ is invertible (why?) for all λ, that is,

$$\rho(A) = \mathbb{C} \text{ or } \sigma(A) = \varnothing.$$

10.4. Hints/Answers to Tests

SOLUTION 10.4.1. No, it is not! Since $C_0^\infty(\mathbb{R})$ is dense in $L^2(\mathbb{R})$, pick a sequence f_n in $C_0^\infty(\mathbb{R})$ which converges in $L^2(\mathbb{R})$ to $f(x) := \mathbb{1}_{[0,1]}(x)$ say. Observe that $f \notin C_0^\infty(\mathbb{R})$. Now, you need to show that $xf_n \to xf$ w.r.t. the $L^2(\mathbb{R})$ norm. This proves that A is not closed on $C_0^\infty(\mathbb{R})$.

SOLUTION 10.4.2. Take z in $\overline{\ker A}$. Write down what this means in terms of sequences, hence get $z \in \ker A$ by the closedness of A...

SOLUTION 10.4.3. If A is closed, then Theorem 10.1.10 says that $(D(A), \|\cdot\|_{G(A)})$ is complete. By the assumption $(*)$, the graph norms of A and B coincide. Accordingly, $(D(B), \|\cdot\|_{G(B)})$ too is complete, i.e. B is closed...

SOLUTION 10.4.4. Yes! Use Exercise 10.3.13 to the closed $A + B$ and the bounded $(-B)$!...

SOLUTION 10.4.5.

(1) **First Method:** Let $x_n \to x$ and $BAx_n \to y$. Since BB^* is invertible, the normality of B guarantees that B^*B be invertible too. Hence, by the continuity of B^*, $B^*BAx_n \to B^*y$. Therefore,

$$Ax_n \longrightarrow (B^*B)^{-1}B^*y.$$

But A is closed, hence $x \in D(A) = D(BA)$ and $Ax = (B^*B)^{-1}B^*y$. This implies that

$$B^*BAx = B^*y \text{ and hence } BB^*BAx = BB^*y$$

which, thanks to the invertibility of BB^*, clearly yields $BAx = y$, proving the closedness of BA.

(2) **Second Method:** Just apply Exercise 4.3.23 to get that B and also B^* are both invertible, then use the second question of Exercise 10.3.13. As a bonus, we also have that B^*A is closed.

CHAPTER 11

Functional Calculi

11.2. True or False: Answers

ANSWERS.

(1) True. To see this, let $x \in H$. Then we easily see that

$$A^2x = A(Ax) = A(\lambda x) = \lambda Ax = \lambda^2 x$$

and so by induction

$$A^n x = \lambda^n x,\ n = 1, 2, \cdots$$

Therefore,

$$p_n(A)x = p_n(\lambda)x$$

for any polynomial function p_n. Since f is continuous, we know by Theorem 11.1.6, $p_n(A)$ converges to $f(A)$ in norm, which implies that $p_n(A)$ converges to $f(A)$ strongly as well. Therefore,

$$f(A)x = f(\lambda)x,$$

as needed.

(2) False in general. Let A be such that $\sigma_p(A) = \varnothing$, e.g. if A is the multiplication operator by x on $L^2[0,1]$ (which is self-adjoint). Let f be identically 0. Then clearly

$$f(\sigma_p(A)) = f(\varnothing) = \varnothing \neq \sigma_p(f(A)) = \sigma_p(0) = \{0\}.$$

REMARK. See [**190**] for related results in Theorem 10.33; or on Page 349 of [**48**], where the following result is given: *If $A \in B(H)$ and f is an analytic function in a neighborhood of $\sigma(A)$, then*

$$f(\sigma_a(A)) = \sigma_a(f(A)).$$

(3) True. We already know from Proposition 7.1.19 that for all $\lambda \notin \sigma(A)$, we have

$$\|(A-\lambda)^{-1}\| \geq \frac{1}{d(\lambda, \sigma(A))}.$$

If A is self-adjoint, then the reverse inequality holds. In fact, in case of self-adjointness we may show the equality directly.

By Theorem 11.1.1, and taking the *continuous* $f(x) = \frac{1}{x-\lambda}$ on $\sigma(A)$ (whichever $\lambda \notin \sigma(A)$), $\|f(A)\|_{B(H)} = \|f\|_\infty$, that is,

$$\|(A-\lambda)^{-1}\|_{B(H)} = \|f\|_\infty = \sup_{x\in\sigma(A)}\left|\frac{1}{x-\lambda}\right| = \frac{1}{\inf\limits_{x\in\sigma(A)}|x-\lambda|} = \frac{1}{d(\lambda,\sigma(A))}.$$

Remark. We can reach the same conclusion with a normal operator A by exactly the same method. In fact, this holds even for "hyponormal" operators. See Exercise 12.3.17.

(4) False! For instance, $f(x) = x^2$ is increasing and continuous on $\mathbb{R}^+$ but it is not operator monotone on $\mathbb{R}^+$...

(5) True! Since $\Delta \subset \Delta'$, plainly $\Delta \cap \Delta' = \Delta$. Hence

$$E(\Delta)E(\Delta') = E(\Delta) \text{ or } E(\Delta) \leq E(\Delta')$$

by Proposition 5.1.9.

(6) True! We already know from Proposition 11.1.30 that $\mu_{x,x}$ is a measure (and we will confirm that in Exercise 11.3.19). The positiveness then follows from the fact that $E(\Delta)$ is an *orthogonal projection*. Indeed,

$$\mu_{x,x} =< E(\Delta)x, x >=< E(\Delta)^2x, x >=< E(\Delta)x, E(\Delta)x >= \|E(\Delta)x\|^2 \geq 0$$

for each $x \in H$.

(7) True. Clearly,

$$I = 1(A) = \int_{\sigma(A)} 1dE = E(\sigma(A))...$$

(8) False! The positivity and invertibility of A are not sufficient in general for the positivity of $\log A$. Indeed, let $A = \frac{1}{2}I$ (which is clearly invertible and positive!). Then

$$\log(\frac{1}{2}I) = \int_{\sigma(\frac{1}{2}I)} \log(\lambda)dE = \int_{\{\frac{1}{2}\}} \log\lambda dE = -\log(2)\int_{\{\frac{1}{2}\}} dE = -\log(2)I$$

is not positive.

Remark. It is clear now that if $\alpha > 0$, then

$$\log(\alpha I) = (\log\alpha)I.$$

(9) True (even for normal operators). To see that, let A be normal and let $x \in H$. Then

$$\begin{aligned} < e^A x, x > &= \int\limits_{\sigma(A)} e^\lambda d\mu_{x,x} \\ &= \int\limits_{\sigma(A)} \sum_{n=0}^{\infty} \frac{\lambda^n}{n!} d\mu_{x,x} \\ &= \sum_{n=0}^{\infty} \frac{1}{n!} \int\limits_{\sigma(A)} \lambda^n d\mu_{x,x} \text{ (to be justified below)} \\ &= \sum_{n=0}^{\infty} \frac{1}{n!} < A^n x, x > \\ &= < \sum_{n=0}^{\infty} \frac{1}{n!} A^n x, x > \text{ (why?)} \end{aligned}$$

and so

$$e^A = \sum_{n=0}^{\infty} \frac{A^n}{n!},$$

as needed. As mentioned, we need to justify the interchange of the integral and the summation above. By Theorem 1.38 of [**189**], say, we are done if we show that

$$\sum_{n=0}^{\infty} \int\limits_{\sigma(A)} \left| \frac{\lambda^n}{n!} \right| d\mu_{x,x} < \infty.$$

This, however, follows from the fact that $|\lambda| \leq \|A\|$ whenever $\lambda \in \sigma(A)$...

(10) The answer is yes! Let $A \in B(H)$ be self-adjoint. By the functional calculus, we could take $B = f(A)$ where $f(x) = \sqrt[3]{x}$ for all $x \in \mathbb{R}$. Hence $B^3 = A$ (by Exercise 11.3.2 say!).

To show the uniqueness of such a cube root, consider "another" cube root which we denote by C, that is, $C^3 = A$. Proposition 11.1.20 then yields $B = C$.

(11) True. This follows from the observation

$$[f(A)]^* = \overline{f}(A) = f(A).$$

(12) True. We have

$$[f(A)]^* f(A) = \overline{f}(A) f(A) = |f|^2(A) = f(A)\overline{f}(A) = f(A)[f(A)]^*.$$

(13) A normal operator has always a square root! This square root is, however, not unique (e.g. what are the self-adjoint square roots of I?).

Let us see why a normal operator admits a square root. Let $A \in B(H)$ be normal. Then, by invoking the Spectral Theorem, we know that there exists a unique spectral measure E on the Borel subsets of $\sigma(A)$ such that:

$$A = \int_{\sigma(A)} \lambda dE.$$

Now, set

$$B = \int_{\sigma(A)} \sqrt{\lambda} dE$$

where $\sqrt{\lambda}$ designates any of the two complex square roots of λ (we all know that a non-zero complex number has two square roots). It is also clear that $B^2 = A$.

11.3. Solutions to Exercises

Solution 11.3.1. Let $\mathcal{P}$ be the set of polynomials restricted to $\sigma(A)$. Then $\mathcal{P} \subset C(\sigma(A))$. By Corollary 7.1.60, the mapping $\mathcal{P} \to B(H)$ given by $p \mapsto p(A)$ is a well-defined isometry (we prefer not give this mapping any symbol to emphasize the dependence on A). Theorem 1.1.32 and Theorem 2.1.104 then allow us to extend it uniquely to an isometry on $C(\sigma(A))$. The new mapping $f \mapsto f(A)$ clearly satisfies (where $g \in C(\sigma(A))$):

(1) $(f+g)(A) = f(A) + g(A)$,
(2) $(\alpha f)(A) = \alpha f(A)$ for any complex α,
(3) $f(A)g(A) = fg(A)$,
(4) $[f(A)]^* = \overline{f}(A)$,
(5) $1(A) = I$

(Reason: Since these properties already hold for polynomials, just pass to the uniform limit...). By the same idea, we plainly see that

$$\|f(A)\|_{B(H)} = \|f\|_\infty$$

holds.

It only remains to prove the spectral mapping theorem (the following is mainly borrowed from [**105**]): If $\lambda \notin f(\sigma(A))$, then the function

$\frac{1}{\lambda - f}$ is continuous on $\sigma(A)$ and so (by the property $f(A)g(A) = fg(A)$)

$$(\lambda I - f(A))^{-1} = \left(\frac{1}{\lambda - f}\right)(A),$$

i.e. $\lambda \notin \sigma(f(A))$. Therefore, we have shown that $\sigma(f(A)) \subset f(\sigma(A))$. To prove the reverse inclusion, we need an intermediate step:

Suppose that f is continuous on $\sigma(A)$, real-valued, $f \geq 0$ and $f(A)$ is invertible. As $\sigma(f(A)) \subset f(\sigma(A)) \subset \mathbb{R}^+$, we clearly have $-\frac{1}{n} \in \rho(f(A))$ for all $n \geq 1$ and

$$R(-\frac{1}{n}, f(A)) = (-\frac{1}{n}I - f(A))^{-1} = \left(\frac{1}{-\frac{1}{n} - f}\right)(A).$$

Since $\lambda \mapsto R(\lambda, f(A))$ is continuous on $\rho(A)$ (why?), we have

$$\lim_{n\to\infty} R(-\frac{1}{n}, f(A)) = R(0, f(A)) = -(f(A))^{-1}.$$

On the other hand, by the property $\|f(A)\|_{B(H)} = \|f\|_\infty$, we have

$$\left\|R(-\frac{1}{n}, f(A))\right\| = \left\|\frac{1}{-\frac{1}{n} - f}\right\|.$$

If f vanished somewhere on $\sigma(A)$, then we would get $R(-\frac{1}{n}, f(A)) \to \infty$! This contradiction tells us that f does not vanish on $\sigma(A)$, that is, to say that $0 \notin f(\sigma(A))$.

In the end, let $f : \sigma(A) \to \mathbb{C}$ be continuous and let $\lambda \notin \sigma(f(A))$, i.e. $\lambda I - f(A)$ is invertible and hence so is its adjoint $\overline{\lambda}I - \overline{f}(A)$. Their product

$$[\lambda I - f(A)][\overline{\lambda}I - \overline{f}(A)] = |\lambda - f|^2(A)$$

is therefore invertible as well. By the "italic part", we know, by the positivity and the continuity of $|\lambda - f|^2$, that $|\lambda - f|^2$ does not vanish on $\sigma(A)$. This implies that $\lambda - f$ does not vanish on $\sigma(A)$ either. Thus, $\lambda \notin f(\sigma(A))$ and so $f(\sigma(A)) \subset \sigma(f(A))$.

SOLUTION 11.3.2. Relation (1) is true for $g(x) = x^2$. To see this, just go back to Theorem 11.1.1, and set $f = g$ in the first property to see that $(f(A))^2 = f^2(A)$. Hence

$$(f(A))^3 = (f(A))^2 f(A) = f^2(A)f(A) = f^3(A).$$

This shows that Relation (1) holds for $g(x) = x^3$ as well. Using a proof by induction, we may easily see that Relation (1) does hold for polynomials.

Now, let g be a continuous real-valued function defined on $f(\sigma(A))$. Since $f(\sigma(A))$ is compact (why?), there exists a sequence of polynomials (g_n) converging uniformly to g. Since $f(A)$ is self-adjoint, we may then write

$$\|g_n(f(A)) - g(f(A))\| = \|g_n - g\|_\infty = \max_{x\in\sigma(f(A))} |g_n(x) - g(x)|,$$

and as $\sigma(f(A)) = f(\sigma(A))$, then

$$\max_{x\in\sigma(f(A))} |g_n(x) - g(x)| = \max_{x\in\sigma(A)} |g_n(f(x)) - g(f(x))|.$$

Therefore

$$\begin{aligned}\|g_n(f(A)) - g(f(A))\| &= \max_{x\in\sigma(A)} |g_n(f(x)) - g(f(x))|\\ &= \max_{x\in\sigma(A)} |(g_n \circ f)(x) - (g \circ f)(x)|.\end{aligned}$$

Now, since $g \circ f$ and $g_n \circ f$ are all continuous, we also have

$$\|(g_n \circ f)(A) - (g \circ f)(A)\| = \max_{x\in\sigma(A)} |(g_n \circ f)(x) - (g \circ f)(x)|.$$

We do not know yet what $g\circ f(A)$ is (this is the question of the exercise), but we do know $(g_n \circ f)(A)$ for g_n are polynomials, and we have

$$(g_n \circ f)(A) = g_n(f(A)).$$

Therefore,

$$\|g_n(f(A))-g(f(A))\| = \|g_n(f(A))-(g\circ f)(A)\| = \max_{x\in\sigma(A)} |(g_n\circ f)(x)-(g\circ f)(x)|.$$

Sending n to infinity, and using the uniqueness of the limit, we finally obtain

$$(g \circ f)(A) = g(f(A)).$$

Solution 11.3.3.

(1) Since f is a homeomorphism, f is continuous so that $f(A)$ is well-defined. Since its inverse, g say, is continuous, using Proposition 11.1.47 we have

$$A = (g \circ f)(A) = g(f(A)) = g(f(B)) = (g \circ f)(B) = B.$$

(2) Just apply the previous result with $g(x) = x^{\frac{1}{n}}$.

Solution 11.3.4.

(1) We already know from Exercise 7.3.23 that $\sigma(A) = \{-1, 3\}$ (and so $\|A\| = 3$). Since the exponential function is continuous, and A is self-adjoint (and e^A is bounded!), we have

$$\sigma(e^A) = e^{\sigma(A)} = \{e^\lambda : \lambda \in \sigma(A)\} = \{e^{-1}, e^3\}.$$

(2) Since e^A is self-adjoint, we have by the Spectral Radius Theorem:

$$\|e^A\| = \sup\{\lambda : \ \lambda \in \sigma(e^A)\} = e^3 = e^{\|A\|}.$$

Remark. If we did not have the Spectral Mapping Theorem on our hands, then we would have first computed e^A (which takes us quite some time). We would find

$$e^A = \frac{1}{2}\begin{pmatrix} e^3 + e^{-1} & e^3 - e^{-1} \\ e^3 - e^{-1} & e^3 + e^{-1} \end{pmatrix}.$$

Solution 11.3.5.

(1) Simply consider

$$B = \begin{pmatrix} -2 & 0 \\ 0 & 1 \end{pmatrix}.$$

Then $\|B\| = 2$. Moreover,

$$e^B = \begin{pmatrix} e^{-2} & 0 \\ 0 & e \end{pmatrix}.$$

Hence, $\|e^B\| = e$. Therefore,

$$\|e^B\| = e \neq e^2 = e^{\|B\|}$$

(2) We already know that $\|e^A\| \leq e^{\|A\|}$ holds (by Exercise 2.3.32). To prove the other inequality, first remember that $\|A\| \in \sigma(A)$ (by Corollary 7.1.13). Then, by the Spectral Radius and Mapping Theorems, we have

$$\|e^A\| = \sup\{|\lambda| : \lambda \in \sigma(e^A)\} = \sup\{e^\mu : \ \mu \in \sigma(A)\} \geq e^{\|A\|},$$

as required.

(3) Since A is positive, so is A^n (by Exercise 5.3.17). Since A is also self-adjoint, it is normaloid, that is, $\|A^n\| = \|A\|^n$. Therefore, the previous question gives

$$\|e^{A^n}\| = e^{\|A^n\|} = e^{\|A\|^n}.$$

To establish the equality involving the self-adjoint B, notice that B^{2n} is positive (why?). Then, apply the previous result to get the desired result

$$\|e^{B^{2n}}\| = e^{\|B\|^{2n}}.$$

(4) By Exercise 9.3.21, $V + V^*$ or $\operatorname{Re} V$ is positive. Hence

$$\|e^{\operatorname{Re} V}\| = e^{\|\operatorname{Re} V\|}.$$

By Test 2.4.1, the norm of $V+V^*$ is just the norm of the operator A in that test, i.e. $\|\operatorname{Re} V\| = \frac{1}{2}$. Consequently,

$$\|e^{\operatorname{Re} V}\| = \sqrt{e},$$

as desired.

Solution 11.3.6. Since A^α is positive (hence self-adjoint!), the Spectral Radius Theorem combined with Spectral Mapping Theorem give:

$$\|A^\alpha\| = \sup\{\lambda^\alpha : \lambda \in \sigma(A)\} = (\sup\{\lambda : \lambda \in \sigma(A)\})^\alpha = \|A\|^\alpha,$$

as required.

Solution 11.3.7. Since $AB \geq 0$, we know by Lemma 5.1.40 that $\sqrt{A} = K\sqrt{B}$ for some positive contraction $K \in B(H)$ and $K\sqrt{B} = \sqrt{B}K$. Hence

$$A = K\sqrt{B}K\sqrt{B} = K^2B = BK^2.$$

Ergo

$$A^\alpha = (BK^2)^\alpha = B^\alpha K^{2\alpha} \text{ (because } K^2B = BK^2).$$

Finally, for any $x \in H$, the Reid Inequality (and a glance at Theorem 11.1.13) allows us to write:

$$< A^\alpha x, x >=< B^\alpha K^{2\alpha} x, x > \leq \|K^{2\alpha}\| < B^\alpha x, x >= \|K\|^{2\alpha} < B^\alpha x, x >$$

Since $\|K\| \leq 1$, we finally obtain:

$$< A^\alpha x, x > \leq < B^\alpha x, x >$$

for all x, as desired.

Solution 11.3.8. Define a function $f : \sigma(A) \to \mathbb{R}$ by

$$f(x) = \begin{cases} 1, & x = \lambda, \\ 0, & x \neq \lambda, \end{cases}$$

where λ is an isolated point in $\sigma(A)$. Then f is continuous (why?) and bounded so that $f(A) \in B(H)$. Besides $f(A)$ is self-adjoint. Since f is not identically null on $\sigma(A)$, $f(A) \neq 0$.

Now, set $p(x) = x - \lambda$. Hence

$$(A - \lambda I)f(A) = p(A)f(A) = (pf)(A).$$

But if $x \in \sigma(A)$, then $(pf)(x) = 0$. Thereupon $(A - \lambda I)f(A)$ vanishes too. As $f(A) \neq 0$, there exists a $y \in H$ such that $z = f(A)y \neq 0$. But $(A - \lambda I)f(A) = 0$ would then implies that $Az = \lambda z$, that is, $\lambda \in \sigma_p(A)$.

Solution 11.3.9. Let $A, B \in B(H)$ be self-adjoint. It is clear that e^{iB} is unitary so that we can do the following (as also $AB = BA$)

$$\|e^{iA} - e^{iB}\| = \|(e^{iA}e^{-iB} - I)e^{iB}\| = \|e^{iA}e^{-iB} - I\| = \|e^{i(A-B)} - I\|.$$

It becomes apparent (since $A - B$ is self-adjoint) that if we want prove the result of the exercise, then it suffices to do so with the following inequality

$$\|e^{iS} - I\| \leq \|S\|$$

with $S \in B(H)$ being self-adjoint.

Therefore, let's proceed to show the previous inequality. First, observe thanks to the Spectral Mapping Theorem,

$$\lambda \in \sigma(e^{iS} - I) \Longleftrightarrow \lambda = e^{i\mu} - 1, \ \mu \in \sigma(S).$$

Second, it is easy to see that $e^{iS} - I$ is normal. Hence by the Spectral Radius Theorem

$$\|e^{iS} - I\| = \sup\{|\lambda| : \ \lambda \in \sigma(e^{iS} - I)\} = \sup\{|e^{i\mu} - 1| : \ \mu \in \sigma(S)\}.$$

But, we know that for any $x \in \mathbb{R}$, $|e^{ix} - 1| \leq |x|$, and so

$$\sup\{|e^{i\mu} - 1| : \ \mu \in \sigma(S)\} \leq \sup\{|\mu| : \ \mu \in \sigma(S)\} = \|S\|$$

by calling on again the Spectral Radius Theorem, and this marks the end of the proof.

Solution 11.3.10. Suppose that $AB = BC$. Then

$$A^2B = AAB = ABC = BCC = BC^2.$$

Hence, for all n, we have

$$A^nB = BC^n.$$

If p is a polynomial of some degree, then clearly

$$p(A)B = Bp(C).$$

Theorem 11.1.6 then implies that (why?)

$$f(A)B = Bf(C).$$

Solution 11.3.11.

(1) Since f is continuous on $\sigma(B)$, there exists a sequence (p_n) of polynomials converging uniformly to f so that

$$\lim_{n\to\infty} \|p_n(B) - f(B)\|_{B(H)} = 0.$$

Since $f(B) \in B(H)$, $D(f(B)A) = D(A)$. So, let $x \in D(A)$ and consider $y = f(B)x$ ($\in H$) and $y_n = p_n(B)x$. By Exercise 10.3.4 and the above observation, we see that

$$Ay_n = Ap_n(B)x = p_n(B)Ax \longrightarrow f(B)Ax$$

as for all n

$$0 \leq \|p_n(B)Ax - f(B)Ax\|_H \leq \|p_n(B) - f(B)\|_{B(H)}\|Ax\|_H.$$

Since $y_n \to y$ (why?), by the closedness of A, we get

$$y = f(B)x \in D(A) \text{ and } Ay = Af(B)x = f(B)Ax,$$

that is, $x \in D(Af(B))$ and $Af(B)x = f(B)Ax$, which precisely means that $f(B)A \subset Af(B)$, as required.

(2) As for the last claim, $f(x) = \sqrt{x}$ is well defined and continuous on $\sigma(B)$ as $B \geq 0$. Accordingly,

$$\sqrt{B}A \subset A\sqrt{B}.$$

Solution 11.3.12.

(1) False! Consider

$$A = \begin{pmatrix} e^{\frac{2i\pi}{3}} & 0 \\ 0 & e^{\frac{2i\pi}{3}} \end{pmatrix} = e^{\frac{2i\pi}{3}} I,$$

where I designates the identity matrix. Then A is not positive (for instance since it is not self-adjoint) while

$$A^3 = [e^{\frac{2i\pi}{3}} I]^3 = e^{2i\pi} I^3 = I,$$

that is, A^3 is positive.

(2) Since A is self-adjoint and $A^3 \geq 0$, a glance at Theorem 11.1.15 gives $A \geq 0$...

Remark. Another proof of $A \geq 0$ is: Let $\lambda \in \sigma(A)$. Then $\lambda^3 \in (\sigma(A))^3 = \sigma(A^3)$. Since $A^3 \geq 0$, it follows that $\lambda^3 \geq 0$. Since A is self-adjoint, $\lambda \in \mathbb{R}$ and so $\lambda \geq 0$, as needed.

(3) Let $A, B \in B(H)$ be self-adjoint such that $ABA = A^2$ and $BAB = B^2$. Then

$$A^3 = AA^2 = A^2BA = ABABA = AB^2A = ABBA = (AB)(AB)^* \geq 0$$

so that by the previous question $A \geq 0$.

In a similar manner, we may show that $B \geq 0$.

Solution 11.3.13. Let $f(x) = \ln x$ be defined for $x > 0$.

(1) We already proved "$\Rightarrow$" in Exercise 2.3.32. Therefore, we only prove "$\Leftarrow$". Assume that $e^Ae^B = e^{A+B}$. Since $A + B$ is self-adjoint, e^{A+B} too is self-adjoint. Whence so is e^Ae^B, i.e.

$$e^Ae^B = (e^Ae^B)^* = (e^B)^*(e^A)^* = e^Be^A.$$

Since e^A is positive, we may write

$$f(e^A)e^B = e^Bf(e^A) \text{ or merely } \ln(e^A)e^B = e^B\ln(e^A)$$

and so

$$Ae^B = e^B A.$$

Since B is self-adjoint too, the same reasoning mutatis mutandis gives

$$AB = BA.$$

(2) As in the first answer!

(3) It is just obvious that

$$e^A = e^B \Longrightarrow e^A e^B = e^B e^A \Longrightarrow AB = BA.$$

(4) Define on $H \oplus H$ the operators

$$\tilde{A} = \begin{pmatrix} 0 & A \\ A & 0 \end{pmatrix} \text{ and } \tilde{B} = \begin{pmatrix} B & 0 \\ 0 & C \end{pmatrix}.$$

One has

$$\tilde{A}^2 = \begin{pmatrix} A^2 & 0 \\ 0 & A^2 \end{pmatrix}, \ \tilde{A}^3 = \begin{pmatrix} 0 & A^3 \\ A^3 & 0 \end{pmatrix}, \cdots$$

Hence

$$e^{\tilde{A}} = \begin{pmatrix} I + \frac{A^2}{2!} + \frac{A^4}{4!} + \cdots & A + \frac{A^3}{3!} + \frac{A^5}{5!} + \cdots \\ A + \frac{A^3}{3!} + \frac{A^5}{5!} + \cdots & I + \frac{A^2}{2!} + \frac{A^4}{4!} + \cdots \end{pmatrix} = \begin{pmatrix} \cosh A & \sinh A \\ \sinh A & \cosh A \end{pmatrix}.$$

Similarly, we can find that

$$e^{\tilde{B}} = \begin{pmatrix} e^B & 0 \\ 0 & e^C \end{pmatrix}.$$

Our hypotheses guarantee that $e^{\tilde{A}} e^{\tilde{B}} = e^{\tilde{B}} e^{\tilde{A}}$ and since $\tilde{A}$ and $\tilde{B}$ are both self-adjoint, Question (2) then implies that $\tilde{A}\tilde{B} = \tilde{B}\tilde{A}$, i.e.

$$\begin{pmatrix} 0 & AC \\ AB & 0 \end{pmatrix} = \begin{pmatrix} 0 & BA \\ CA & 0 \end{pmatrix}$$

Comparing the entries in the previous matrices, we see (by remembering also that A, B, C are all self-adjoint) that $AC = BA$, quod erat demonstrandum.

SOLUTION 11.3.14. Assume $e^A = I$ and let $\lambda \in \sigma(A)$. Then $e^\lambda = 1$. Since A is self-adjoint, λ is real and so $e^\lambda = 1$ implies $\lambda = 0$ only. Since $\sigma(A)$ is never empty, $\sigma(A) = \{0\}$. Again, since A is self-adjoint, by the Spectral Radius Theorem, we have

$$\|A\| = r(A) = \sup\{|\lambda| : \lambda \in \sigma(A)\} = 0 \Longrightarrow A = 0.$$

SOLUTION 11.3.15.

(1) We only prove "$\Rightarrow$". Since $e^A = e^B$ and both A and B are self-adjoint, Exercise 11.3.13 yields $AB = BA$. This implies

$$I = e^A e^{-A} = e^A e^{-B} = e^{A-B}.$$

But $A - B$ is obviously self-adjoint, so Exercise 11.3.14 gives $A = B$.

(2) Since A is self-adjoint, by the functional calculus

$$(e^{iA})^* = e^{-iA}$$

so that

$$(e^{iA})^* e^{iA} = e^{-iA} e^{iA} = e^{-iA+iA} = e^0 = I.$$

Similarly, we have

$$e^{iA}(e^{iA})^* = I,$$

i.e. e^{iA} is unitary whenever A is self-adjoint.

(3) The answer is no! According to the next question, the counterexample we are after must not be normal. Consider

$$A = \begin{pmatrix} -i\pi & 2i\pi \\ -i\pi & i\pi \end{pmatrix} \text{ so that } iA = \begin{pmatrix} \pi & -2\pi \\ \pi & -\pi \end{pmatrix} = \pi \underbrace{\begin{pmatrix} 1 & -2 \\ 1 & -1 \end{pmatrix}}_{\text{call it } B}.$$

Then it may easily be checked that

$$AA^* = -i\pi \begin{pmatrix} 1 & -2 \\ 1 & -1 \end{pmatrix} \times i\pi \begin{pmatrix} 1 & 1 \\ -2 & -1 \end{pmatrix} = \pi^2 \begin{pmatrix} 5 & 3 \\ 3 & 2 \end{pmatrix}$$

and

$$A^*A = i\pi \begin{pmatrix} 1 & 1 \\ -2 & -1 \end{pmatrix} \times (-i\pi) \begin{pmatrix} 1 & -2 \\ 1 & -1 \end{pmatrix} = \pi^2 \begin{pmatrix} 2 & -3 \\ -3 & 5 \end{pmatrix},$$

so that A is not normal and so it cannot be self-adjoint.

Next, we claim that $e^{iA} = -I$. We have

$$e^{iA} = e^{i(-iB)} = e^{\pi B}$$

(we have introduced B to get rid of the "i"). We need to compute integer powers of πB. We have

$$\pi^2 B^2 = \pi^2 \begin{pmatrix} 1 & -2 \\ 1 & -1 \end{pmatrix} \begin{pmatrix} 1 & -2 \\ 1 & -1 \end{pmatrix} = \pi^2 \begin{pmatrix} -1 & 0 \\ 0 & -1 \end{pmatrix} = -\pi^2 I$$

(good! Remaining calculations will be easy!). Hence

$$\pi^3 B^3 = \pi B(-\pi^2 I) = -\pi^3 B,$$

$$\pi^4 B^4 = (-\pi^2 I)(-\pi^2 I) = \pi^4 I, \text{ and so on...}$$

Therefore,

$$\begin{aligned}
e^{iA} &= e^{\pi B} \\
&= I + \pi B + \frac{\pi^2}{2!}B^2 + \frac{\pi^3}{3!}B^3 + \frac{\pi^4}{4!}B^4 + \frac{\pi^5}{5!}B^5 + \cdots \\
&= I + \pi B - \frac{\pi^2}{2!}I - \frac{\pi^3}{3!}B + \frac{\pi^4}{4!}I + \frac{\pi^5}{5!}B + \cdots \\
&= \underbrace{\left(1 - \frac{\pi^2}{2!} + \frac{\pi^4}{4!} + \cdots\right)}_{\cos\pi=-1} I + \underbrace{\left(\pi - \frac{\pi^3}{3!} + \frac{\pi^5}{5!} + \cdots\right)}_{\sin\pi=0} B \\
&= -I.
\end{aligned}$$

Hence

$$e^{iA}(e^{iA})^* = (e^{iA})^* e^{iA} = (-I)(-I) = I,$$

that is, e^{iA} is unitary whereas A is not self-adjoint.

(4) By the normality of A, we have

$$e^{iA-iA^*} = e^{iA}e^{-iA^*} = e^{iA}(e^{iA})^* = I.$$

Since $iA - iA^*$ is self-adjoint, Exercise 11.3.14 gives $A = A^*$, which completes the proof.

(5) Since A is normal, $(e^A)^* = e^{A^*}$. Hence if $A = -A^*$, then we have

$$e^A e^{A^*} = e^{-A^*}e^{A^*} = e^{-A^*+A^*} = e^0 = I = e^{A^*-A^*} = e^{A^*}e^{-A^*} = e^{A^*}e^A,$$

i.e. e^A is unitary.

Conversely, assume that e^A is unitary. We have, by remembering that A is normal,

$$e^{A+A^*} = e^A e^{A^*} = e^A(e^A)^* = I = e^0.$$

Since $A+A^*$ is self-adjoint, Exercise 11.3.14 does the remaining job, i.e. it yields $A + A^* = 0$.

(6) By Question (2), e^{iA} is unitary. We have

$$e^{B^*} = e^{-iA} \text{ and } e^{-B} = e^{-iA}.$$

Thus

$$e^{-B} = e^{B^*} \text{ so that } e^{B+B^*} = I$$

because B is normal. However, $B + B^*$ is always self-adjoint, whence $B^* = -B$ by calling on again Exercise 11.3.14.

Solution 11.3.16. First, AB is positive as by assumption $A, B \geq 0$ and $AB = BA$. Since A and B are invertible, it follows that AB too is invertible. Therefore, the quantities $\log(AB)$, $\log A$ and $\log B$ are all well defined. Notice also that

$$AB = BA \Longrightarrow \log A \log B = \log B \log A.$$

We are ready to prove the desired equality. We have

$$e^{\log A + \log B} = e^{\log A} e^{\log B} = AB.$$

Therefore (and since $\log A + \log B$ is self-adjoint),

$$\log(AB) = \log(e^{\log A + \log B}) = \log A + \log B,$$

as desired.

Finally, since A is invertible, $AA^{-1} = A^{-1}A = I$. Besides, A^{-1} is positive and invertible. Hence

$$0 = \log I = \log A + \log A^{-1}$$

and so

$$\log(A^{-1}) = -\log A,$$

as required.

Solution 11.3.17.

(1) It is clear from Proposition 11.1.25 that

$$A \geq I \Longrightarrow \log A \geq \log I = 0.$$

Since $\log A$ is positive, Exercise 11.3.5 yields

$$\|e^{\log A}\| = e^{\|\log A\|} \text{ and so } \|A\| = e^{\|\log A\|}.$$

Passing to the "real logarithm" (as $\|A\| \geq 1$) gives

$$\log \|A\| = \|\log A\|,$$

as desired.

(2) Using the previous question, we may write

$$\|A\| = 1 \Longrightarrow \|\log A\| = \log 1 = 0 \Longrightarrow \log A = 0 \Longrightarrow A = e^{\log A} = e^0 = I,$$

as needed.

(3) We have

$$e^B = I \Longrightarrow B = \log(e^B) = \log I = 0.$$

Solution 11.3.18. By assumption, $f(A)$ is a positive operator. From the third property of Theorem 11.1.1, we have

$$\|f(A)\|_{B(H)} = \sup_{x \in \sigma(A)} |f(x)| = \sup_{x \in \sigma(A)} f(x).$$

Since $A \geq 0$, Corollary 7.1.13 informs us that $\|A\| \in \sigma(A)$ and so

$$\sup_{x \in \sigma(A)} f(x) \geq f(\|A\|)$$

and so

$$\|f(A)\| \geq f(\|A\|)...(1).$$

If g is the continuous inverse of f which is also increasing (by hypothesis), then

$$g(\|f(A)\|) \geq g(f(\|A\|)) \text{ or } \|A\| \leq g(\|f(A)\|).$$

By a glance at Inequality (1), we have

$$\|A\| \leq g(\|f(A)\|) \leq \|g(f(A))\| = \|A\|.$$

Hence

$$\|A\| = g(\|f(A)\|).$$

Accordingly,

$$f(\|A\|) = f[g(\|f(A)\|)] = \|f(A)\|,$$

as suggested.

Solution 11.3.19. Let $x, y \in H$. Let us show that $\mu_{x,y}$ is an ordinary measure. Let $(\Delta_n)_n$ be a disjoint countable collection of elements of Σ. *We must show that*

$$\mu_{x,y}(\bigcup_{n\in\mathbb{N}} \Delta_n) = \sum_{n=1}^{\infty} \mu_{x,y}(\Delta_n).$$

Since $\Delta_n \cap \Delta_m = \varnothing$ for all $n \neq m$, By Definition 11.1.29, we have

$$E(\Delta_n)E(\Delta_m) = 0, \ \forall n \neq m.$$

Since $(E(\Delta_n))_n$ are mutually orthogonal, we can easily establish the convergence of the series $\sum_{n\in\mathbb{N}} \|E(\Delta_n)x\|^2$ (how?).

By Theorem 3.1.21, $\sum_{n\in\mathbb{N}} E(\Delta_n)x$ too converges. By Proposition 3.1.22, $\sum_{n\in\mathbb{N}} < E(\Delta_n)x, y >$ converges for all $y \in H$. Therefore, we can write

$$\begin{aligned}
\mu_{x,y}(\bigcup_{n=1}^{\infty} \Delta_n) &=< E(\bigcup_{n=1}^{\infty} \Delta_n)x, y > \\
&=< \sum_{n=1}^{\infty} E(\Delta_n)x, y > \\
&= \sum_{n=1}^{\infty} < E(\Delta_n)x, y > \\
&= \sum_{n=1}^{\infty} \mu_{x,y}(\Delta_n),
\end{aligned}$$

as needed.

Solution 11.3.20. First, it is clear that $E(\Delta) \in B(H)$ for all Δ. Let us now check the remaining properties of a spectral measure.

(1) $E(\varnothing) = 0$ as
$$E(\varnothing)f(x) = \mathbb{1}_{\varnothing}(x)f(x) = 0$$
and $E(X) = I$ because
$$E(X)f(x) = \mathbb{1}_X(x)f(x) = 1f(x) = f(x).$$

(2) Let $\Delta \in \Sigma$. Then $E(\Delta)$ is an orthogonal projection. Indeed, since $\mathbb{1}_\Delta(x)$ is real-valued, $E(\Delta)$ is self-adjoint (Exercise 10.3.9). Also, since $(\mathbb{1}_\Delta)^2 = \mathbb{1}_\Delta$,
$$(E(\Delta))^2 = E(\Delta).$$

(3) Let $\Delta_1, \Delta_2 \in \Sigma$. Then
$$E(\Delta_1 \cap \Delta_2) = E(\Delta_1)E(\Delta_2)$$
for $\mathbb{1}_{\Delta_1}\mathbb{1}_{\Delta_2} = \mathbb{1}_{\Delta_1 \cap \Delta_2}$.

(4) Let $(\Delta_n)_{n\in\mathbb{N}} \in \Sigma$ be pairwise disjoint. Set
$$\bigcup_{n=1}^{\infty} \Delta_n = \Delta.$$
Also, let
$$f_N = \sum_{n=1}^{N} \mathbb{1}_{\Delta_n} f \text{ where } f \in L^2(\mu).$$

Then since $(\Delta_n)_{n\in\mathbb{N}} \in \Sigma$ are pairwise disjoint, we obtain (pointwise)
$$\lim_{N\to\infty} |\mathbb{1}_\Delta f - f_N|^2 = 0.$$
Since also
$$|\mathbb{1}_\Delta f - f_N|^2 \leq 4|f|^2$$
for all N, and $|f|^2 \in L^1(\mu)$, the Dominated Convergence Theorem yields
$$\lim_{N\to\infty} \|\mathbb{1}_\Delta f - f_N\|_2^2 = 0.$$
Hence, we have w.r.t. the Strong Operator Topology
$$\mathbb{1}_\Delta f = \lim_{N\to\infty} f_N = \lim_{N\to\infty} \sum_{n=1}^{N} \mathbb{1}_{\Delta_n} f$$

and this means that (for all x)

$$E(\Delta)f(x) = E\left(\bigcup_{n=1}^{\infty}\Delta_n\right)f(x) = \sum_{n=1}^{\infty}E(\Delta_n)f(x),$$

as required. Thus E is a spectral measure.

SOLUTION 11.3.21. Write $A = \int \lambda dE$. Since $E(\Delta) = \int_{\sigma(A)} \mathbb{1}_\Delta(\lambda)dE$, we have

$$AE(\Delta) = \int_{\sigma(A)} \lambda \mathbb{1}_\Delta(\lambda)dE = \int_{\sigma(A)} \mathbb{1}_\Delta(\lambda)\lambda dE = E(\Delta)A.$$

Finally, $\operatorname{ran}(E(\Delta))$ reduces A by Theorem 4.1.42.

SOLUTION 11.3.22. Let $A = \int \lambda dE$. Then $BE(\Delta) = E(\Delta)B$ implies that $\{\operatorname{ran}(E(\Delta))\}_\Delta$ is a family of reducing subspaces for every operator which commutes with A. In case A is non scalar, then as already seen in Exercise 8.3.3, $\sigma(A)$ must contain at least two *distinct* points, λ and μ, say. Set $r = \frac{1}{2}|\lambda - \mu|$ (and so $r > 0$). Consider now the open ball centered at λ and with radius r:

$$B(\lambda, r) = \{z \in \mathbb{C} : \ |z - \lambda| < r\}$$

and set

$$\Delta = \sigma(A) \cap B(\lambda, r) \text{ and } \Delta' = \sigma(A) \setminus B(\lambda, r).$$

Then clearly,

$$\sigma(A) = \Delta \cup \Delta' \text{ and } \Delta \cap \Delta' = \varnothing.$$

Observe also that Δ is relatively open in $\sigma(A)$ and so is $\sigma(A) \setminus \overline{B(\lambda, r)}$. The latter plainly satisfies $\sigma(A) \setminus \overline{B(\lambda, r)} \subset \Delta'$. Hence, by the Spectral Theorem for normal operators, we have that

$$E(\Delta) \neq 0 \text{ and } E(\Delta') \neq 0$$

(indeed $E(\Delta') = 0$ would yield $E(\sigma(A) \setminus \overline{B(\lambda, r)}) = 0$!). Therefore,

$$I = E(\sigma(A)) = E(\Delta \cup \Delta') = E(\Delta) + E(\Delta'),$$

i.e.

$$E(\Delta) = I - E(\Delta') \neq I.$$

Accordingly, $0 \neq E(\Delta) \neq I$ which finally leads to

$$\{0\} \neq \operatorname{ran}(E(\Delta)) \neq H.$$

SOLUTION 11.3.23. It is clear that if $[A^*, A]A = 0$, then either $([A^*, A] = 0)$ or $([A^*, A] \neq 0$ and $[A^*, A]A = 0)$.

(1) If $[A^*, A] = 0$, then A is normal, and normal operators have always nontrivial invariant subspaces.

(2) If $[A^*, A] \neq 0$ and $[A^*, A]A = 0$, then Exercise 4.3.43 informs us that $\overline{\text{ran}(A)}$ and $\ker([A^*, A])$ are nontrivial subspaces for A.

SOLUTION 11.3.24. Since A is normal we may write $A = \int \lambda dE$ (for some unique spectral measure E). By Theorem 11.1.56, we have

$$AB = BA \Longrightarrow E(\Delta)B = BE(\Delta).$$

So, if we use the the ordinary measure $\mu_{x,y}$ which appeared in Proposition 11.1.30, then we may write for all $x, y \in H$

$$\mu_{x,B^*y} =< E(\Delta)x, B^*y >=< BE(\Delta)x, y >=< E(\Delta)Bx, y >= \mu_{Bx,y}.$$

Therefore,

$$\begin{aligned} < BA^*x, y > &=< A^*x, B^*y > \\ &= \int_{\sigma(A)} \overline{\lambda} d\mu_{x,B^*y} \\ &= \int_{\sigma(A)} \overline{\lambda} d\mu_{Bx,y} \\ &=< A^*Bx, y >, \end{aligned}$$

and so $BA^* = A^*B$, as needed. To prove the other implication, just apply the first one to the normal A^*. We get $BA^{**} = A^{**}B$ or merely $BA = AB$, as required.

SOLUTION 11.3.25. Assume that A is self-adjoint with a positive spectrum. The Spectral Theorem allows us to write for all $x \in H$

$$< Ax, x >= \int_{\sigma(A)} \lambda d\mu_{x,x}.$$

Since $\mu_{x,x}$ is a positive measure, since each $\lambda \geq 0$ (on $\sigma(A)$), we obtain

$$\forall x \in H, \ < Ax, x > \geq 0,$$

i.e. A is positive.

SOLUTION 11.3.26. Let $A \in B(H)$ be normal. Set

$$p(z) = |z| \text{ and } u(z) = \begin{cases} \frac{z}{|z|}, & z \neq 0, \\ 1, & z = 0. \end{cases}$$

where $z \in \sigma(A)$. Then p is continuous and u is Borel and bounded. Hence it is legitimate to define $p(A)$ and $u(A)$, denoted respectively by P and U. Then P is positive ($P = |A|$). Also, since

$$\overline{u}(z)u(z) = u(z)\overline{u}(z) = 1$$

for all z, we deduce that

$$U^*U = UU^* = I,$$

i.e. U is unitary. We then conclude by observing

$$z = u(z)p(z) = p(z)u(z)$$

for each z, that

$$A = UP = PU.$$

Therefore,

$$AP = UPP = PUP = PA$$

and

$$AU = UPU = UA.$$

Finally, if A is self-adjoint, then $\sigma(A) \subset \mathbb{R}$ and so the function u defined above becomes real-valued and so U is self-adjoint!

Solution 11.3.27.

(1) By the polar decomposition, $A = UP$, where U is unitary and P is positive (and invertible!). Hence $\sigma(P) \subset \mathbb{R}_+^*$. Let $f(x) = \ln x$ be defined on $\sigma(P)$. Then f is real-valued and continuous so that $B = f(P)$ is self-adjoint. Hence $P = e^B$ by Exercise 11.3.2.

Since U is unitary, $\sigma(U) \subset \{\lambda \in \mathbb{C} : |\lambda| = 1\}$. Now, define $g : \sigma(U) \to (0, 2\pi]$ such that $e^{ig(\lambda)} = \lambda$ (g is just the argument function and it is real-valued, Borel and bounded). Now, put $g(U) = C$. Then (as U is normal!)

$$C^* = [g(U)]^* = \overline{g}(U) = g(U) = C,$$

i.e. C is self-adjoint. Therefore, using the functional calculus, we infer that $U = e^{iC}$ so that finally we see that

$$A = e^{iC}e^B.$$

(2) Since A is normal, we may write $A = UP = PU$. Using the same functions f and g as before, and by properties of the functional calculus we obtain that

$$f(P)g(U) = g(U)f(P) \text{ or } BC = CB.$$

Hence, as B and C are self-adjoint, we know that $B+iC$ is then normal, call it N. Therefore (using again the commutativity of B and C)

$$A = e^{iC}e^B = e^{B+iC} = e^N.$$

Solution 11.3.28. Since A is normal, recall that

$$A = \int_{\sigma(A)} \lambda dE$$

for some unique spectral measure E. Also,

$$A^* = \int_{\sigma(A)} \overline{\lambda} dE \text{ and } AA^* = \int_{\sigma(A)} |\lambda|^2 dE \ (= A^*A).$$

Thanks to Exercises 7.3.3 & 7.3.24, we are only concerned with proving "right-to-left" implications.

(1) It is clear that if $\sigma(A)$ is real, and if $\lambda \in \sigma(A)$, then $\lambda = \overline{\lambda}$ and so $A = A^*$.
(2) Assume that $\sigma(A) \subset \{\lambda \in \mathbb{C} : |\lambda| = 1\}$. So clearly, if $\lambda \in \sigma(A)$, then $|\lambda| = 1$. Hence

$$AA^* = \int_{\sigma(A)} dE = E(\sigma(A)) = I.$$

Therefore as A is normal, then A is unitary.
(3) Let $\lambda \in \sigma(A)$, i.e. $\lambda \geq \alpha$. Hence for all $x \in H$ (and using $\mu_{x,x}$ of Proposition 11.1.30):

$$< Ax, x >= \int_{\sigma(A)} \lambda d\mu_{x,x} \geq \alpha \int_{\sigma(A)} d\mu_{x,x} = \alpha < E(\sigma(A))x, x >= \alpha < Ix, x >,$$

i.e. $A \geq \alpha I$, as desired.
(4) Assume that $\sigma(A) \subset \{0, 1\}$. Then A is self-adjoint. Moreover, for all $\lambda \in \sigma(A)$, $\lambda^2 = \lambda$. Therefore,

$$A^2 = \int_{\sigma(A)} \lambda^2 dE = \int_{\sigma(A)} \lambda dE = A,$$

i.e. A is also idempotent. Accordingly, A is an orthogonal projection.

Solution 11.3.29.

(1) Recall that $\|A\|_F = \sqrt{\operatorname{tr}(A^*A)}$. Using the linearity and the cyclic property of the trace of a product of matrices, we get

$$\begin{aligned}\|AB - BA\|_F^2 &= \operatorname{tr}[(AB - BA)^*(AB - BA)]\\ &= \operatorname{tr}(B^*A^*AB - B^*A^*BA - A^*B^*AB + A^*B^*BA)\\ &= \operatorname{tr}(B^*A^*AB) - \operatorname{tr}(B^*A^*BA) - \operatorname{tr}(A^*B^*AB) + \operatorname{tr}(A^*B^*BA)\\ &= \operatorname{tr}(A^*ABB^*) - \operatorname{tr}(AB^*A^*B) - \operatorname{tr}(BA^*B^*A) + \operatorname{tr}(AA^*B^*B)\\ &= \operatorname{tr}(A^*ABB^*) - \operatorname{tr}(AB^*A^*B) - \operatorname{tr}(ABA^*B^*) + \operatorname{tr}(AA^*B^*B)\\ &= \operatorname{tr}(A^*ABB^* - AB^*A^*B - ABA^*B^* + AA^*B^*B).\end{aligned}$$

By replacing each B by its adjoint, we obtain

$$\|AB^* - B^*A\|_F^2 = \operatorname{tr}(A^*AB^*B - ABA^*B^* - AB^*A^*B + AA^*BB^*).$$

Calling on again the linearity of the trace, we obtain

$$\begin{aligned}&\|AB - BA\|_F^2 - \|AB^* - B^*A\|_F^2\\ &= \operatorname{tr}(-A^*AB^*B + A^*ABB^* + AA^*B^*B - AA^*BB^*)\\ &= \operatorname{tr}[-(A^*A - AA^*)(B^*B - BB^*)]\\ &= -\operatorname{tr}[(A^*A - AA^*)(B^*B - BB^*)],\end{aligned}$$

as desired.

(2) If B (or A) is normal, then the previous equation clearly yields

$$\|AB - BA\|_F = \|AB^* - B^*A\|_F$$

from which we get

$$AB = BA \Longleftrightarrow AB^* = B^*A,$$

as suggested.

SOLUTION 11.3.30.

(1) Since $TA = BT$, we clearly have

$$TA^2 = (TA)A = (BT)A = B(TA) = B(BT) = B^2T,$$

so that it easily follows by induction that

$$TA^n = B^nT \text{ for } n = 1, 2, \cdots.$$

Hence

$$Tp(A) = p(B)T,$$

where $p(z)$ is a polynomial in z and of degree n, say.

(2) By the Normal Functional Calculus, we know that the bounded Borel functions $e^{i\overline{z}A}$ and $e^{i\overline{z}B}$ are limits of polynomials in A and B respectively. Thus

$$Te^{i\overline{z}A} = e^{i\overline{z}B}T,\ \forall z \in \mathbb{C}.$$

(3) Let $f(z) = e^{-izB^*}Te^{izA^*}$, which defines a function from $\mathbb{C}$ into $B(H)$. Since A commutes with A^*, and B commutes with B^*, Exercise 2.3.32 gives

$$e^Ae^{A^*} = e^{A+A^*} \text{ and } e^Be^{B^*} = e^{B+B^*}.$$

The previous, combined with $e^{-i\overline{z}B}Te^{i\overline{z}A} = T$ imply that

$$\begin{aligned}f(z) =& e^{-izB^*}Te^{izA^*}\\ =& e^{-izB^*}e^{-i\overline{z}B}Te^{i\overline{z}A}e^{izA^*}\\ =& e^{-izB^*-i\overline{z}B}Te^{i\overline{z}A+izA^*}\\ =& e^{-i(zB^*+\overline{z}B)}Te^{i(zA^*+\overline{z}A)}.\end{aligned}$$

Now, for any $z \in \mathbb{C}$

$$(zB^* + \overline{z}B)^* = \overline{z}B^{**} + \overline{\overline{z}}B^* = \overline{z}B + zB^* = zB^* + \overline{z}B,$$

that is, $zB^* + \overline{z}B$ is self-adjoint and so is $zA^* + \overline{z}A$. Hence, $e^{-i(zB^*+\overline{z}B)}$ and $e^{i(zA^*+\overline{z}A)}$ are both unitary. So, by Exercise 4.3.6

$$\|f(z)\| = \|e^{-i(zB^*+\overline{z}B)}Te^{-i(zA^*+\overline{z}A)}\| = \|T\|,$$

for all $z \in \mathbb{C}$. Therefore, f is bounded. It is also known that f is entire (cf. Proposition 2.1.97) so that by the Liouville Theorem (Theorem 2.1.98), f is constant.

(4) Since f is constant on $\mathbb{C}$, $f'(z) = 0$ for all $z \in \mathbb{C}$. Hence

$$0 = f'(z) = -iB^*e^{-izB^*}Te^{izA^*} + ie^{-izB^*}TA^*e^{izA^*},$$

and letting $z = 0$ gives

$$0 = -iB^*T + iTA^* \text{ or } TA^* = B^*T,$$

as desired.

(5) Let

$$T = \begin{pmatrix} 1 & 1 \\ -1 & 1 \end{pmatrix},\ A = \begin{pmatrix} 0 & 1 \\ 1 & 0 \end{pmatrix} \text{ and } B = \begin{pmatrix} 1 & 0 \\ 0 & -1 \end{pmatrix}.$$

Then T, A and B are all normal (in fact, A and B are even self-adjoint and unitary!). Now

$$TA = BT = \begin{pmatrix} 1 & 1 \\ 1 & -1 \end{pmatrix}$$

whereas

$$T^*A = \begin{pmatrix} -1 & * \\ * & * \end{pmatrix} \neq \begin{pmatrix} 1 & * \\ * & * \end{pmatrix} = BT^*.$$

SOLUTION 11.3.31. Of course, we are only concerned with establishing "Fuglede $\Rightarrow$ Putnam".

Let T, A, B be all in $B(H)$ such that both A and B are normal. Assume that $TA = BT$. We need to show that $TA^* = B^*T$ (which is the Putnam's version) by using the Fuglede version only. Consider

$$\tilde{A} = \begin{pmatrix} B & 0 \\ 0 & A \end{pmatrix} \text{ and } \tilde{T} = \begin{pmatrix} 0 & T \\ 0 & 0 \end{pmatrix},$$

which are two matrices of operators defined on $H \times H$, and where 0 stands for the zero operator on H. Hence

$$\tilde{A}^* = \begin{pmatrix} B^* & 0^* \\ 0^* & A^* \end{pmatrix} = \begin{pmatrix} B^* & 0 \\ 0 & A^* \end{pmatrix}.$$

Then we have

$$\tilde{A}^*\tilde{A} = \begin{pmatrix} B^*B & 0 \\ 0 & A^*A \end{pmatrix} = \begin{pmatrix} BB^* & 0 \\ 0 & AA^* \end{pmatrix} = \tilde{A}\tilde{A}^*$$

because A and B are normal and therefore $\tilde{A}$ too is normal.

Furthermore, and as $TA = BT$, we can easily check that $\tilde{T}\tilde{A} = \tilde{A}\tilde{T}$. Applying the Fuglede version to the previous equation yields $\tilde{T}\tilde{A}^* = \tilde{A}^*\tilde{T}$, i.e.

$$\begin{pmatrix} 0 & TA^* \\ 0 & 0 \end{pmatrix} = \begin{pmatrix} 0 & B^*T \\ 0 & 0 \end{pmatrix}.$$

Comparing the entries in the matrices intervening in the previous equation gives $TA^* = B^*T$, as required.

SOLUTION 11.3.32.

(1) False! Take

$$N = \begin{pmatrix} 1 & 1 \\ -1 & -1 \end{pmatrix}, \; M = \begin{pmatrix} -1 & -1 \\ 1 & 1 \end{pmatrix}, \; A = \begin{pmatrix} 0 & 1 \\ 1 & 0 \end{pmatrix} \text{ and } B = \begin{pmatrix} 1 & 0 \\ 0 & 1 \end{pmatrix}.$$

Then M and N are normal, and A and B are self-adjoint. We have

$$AN = \begin{pmatrix} -1 & -1 \\ 1 & 1 \end{pmatrix} = MB$$

but

$$AN^* = \begin{pmatrix} * & -1 \\ * & * \end{pmatrix} \neq \begin{pmatrix} * & 1 \\ * & * \end{pmatrix} = M^*B.$$

(2) False again! Take

$$M = \begin{pmatrix} 0 & 1 \\ -1 & 0 \end{pmatrix}; \; N = \begin{pmatrix} 1 & 0 \\ 0 & -1 \end{pmatrix}; \; A = \begin{pmatrix} 0 & 1 \\ 1 & 0 \end{pmatrix} \text{ and } B = \begin{pmatrix} -1 & 0 \\ 0 & -1 \end{pmatrix}.$$

Then all operators involved are unitary where, in addition, A and B are self-adjoint. We do have $AN = MB$ but we *do not* have $AN^* = M^*B$.

(3) We have

$$A = ANN^* = (AN)N^* = (MB)N^* = M(BN^*).$$

Then

$$M^*A = M^*MBN^* = BN^*.$$

(4) If one takes S to be the unilateral shift defined on ℓ^2, then by setting

$$M = N = A = B = S \text{ (and hence } N \text{ is not a co-isometry)},$$

one sees that $AN = MB$ while $BN^* \neq M^*A$.

And if one sets

$$M = N = A = B = S^* \text{ (and hence } M \text{ is not an isometry)},$$

then $AN = MB$ whereas $BN^* \neq M^*A$.

Solution 11.3.33. Since A and B are self-adjoint, we may write

$$B(AB) = BAB = (AB)^*B.$$

Since AB and $(AB)^*$ are normal, the Fuglede-Putnam Theorem gives

$$B(AB)^* = (AB)^{**}B \text{ or merely } B^2A = AB^2.$$

Consequently,

$$B^2A^2 = AB^2A = A^2B^2.$$

On the other hand, we easily see that

$$|AB|^2 = (AB)^*AB = AB(AB)^* = AB^2A = A^2B^2$$

and so by Exercise 5.3.28

$$|AB| = \sqrt{A^2B^2} = \sqrt{A^2}\sqrt{B^2} = |A||B|,$$

as required.

Let us pass to counterexamples:

- Let

$$A = \begin{pmatrix} 2 & 0 \\ 0 & -1 \end{pmatrix} \text{ and } B = \begin{pmatrix} 0 & 1 \\ 1 & 0 \end{pmatrix}.$$

Then each of A and B is self-adjoint. Also

$$AB = \begin{pmatrix} 0 & 2 \\ -1 & 0 \end{pmatrix}$$

and so AB is not normal. We can easily check that

$$|AB| = \begin{pmatrix} 1 & 0 \\ 0 & 2 \end{pmatrix}, \ |A| = \begin{pmatrix} 2 & 0 \\ 0 & 1 \end{pmatrix} \text{ and } |B| = \begin{pmatrix} 1 & 0 \\ 0 & 1 \end{pmatrix},$$

i.e.

$$|AB| \neq |A||B|.$$

- Let
$$A = \begin{pmatrix} 0 & 1 \\ 2 & 0 \end{pmatrix} \text{ and } B = \begin{pmatrix} 0 & 2 \\ 1 & 0 \end{pmatrix}.$$
Then, neither A nor B is normal. Also,
$$AB = \begin{pmatrix} 1 & 0 \\ 0 & 2 \end{pmatrix}$$
which is even positive. We may easily compute their absolute values. We obtain
$$|A| = \begin{pmatrix} 2 & 0 \\ 0 & 1 \end{pmatrix}, \ |B| = \begin{pmatrix} 1 & 0 \\ 0 & 2 \end{pmatrix} \text{ and } |AB| = \begin{pmatrix} 1 & 0 \\ 0 & 2 \end{pmatrix}.$$
Accordingly,
$$|AB| = \begin{pmatrix} 1 & 0 \\ 0 & 2 \end{pmatrix} \neq \begin{pmatrix} 2 & 0 \\ 0 & 2 \end{pmatrix} = |A||B|.$$

Solution 11.3.34. Let $A, B \in B(H)$ be self-adjoint such that $B \geq 0$, say. Then by Corollary 7.1.37, $\sigma(AB)$ is a subset of $\mathbb{R}$. Since AB is normal, with real spectrum, it is self-adjoint (Proposition 11.1.39).

Solution 11.3.35. Let us assume that A is positive (the proof for B positive is just identical). By the Fuglede-Putnam Theorem, and since both AB and $(AB)^*$ are normal, we can write

$$A(BA) = (AB)A \iff A(AB)^* = (AB)A \iff A(AB)^{**} = (AB)^*A,$$

that is,

$$A^2B = BA^2 \text{ or merely } AB = BA$$

by Theorem 5.1.28. Thus, AB is self-adjoint.

Solution 11.3.36. Let T be an invertible operator such that $T^{-1}AT = B$, where A and B are normal. The aim is to show that there exists a unitary operator U such that $U^{-1}AU = B$.

Since T is invertible, it has the unique polar decomposition $T = UP$, where U is unitary and P $(= \sqrt{T^*T})$ is positive and invertible (see Proposition 6.1.32). Since $T^{-1}AT = B$, $TB = AT$, and since A and B are normal, the Fuglede-Putnam Theorem gives $TB^* = A^*T$. Passing to adjoints yields

$$BT^* = (TB^*)^* = (A^*T)^* = T^*A.$$

Hence

$$BT^*T = T^*AT = T^*TB,$$

that is,

$$BP^2 = P^2B \text{ or } BP = PB$$

(by Theorem 5.1.28). Finally, we obtain

$$TB = AT \iff (UP)B = A(UP) \iff UBP = AUP \iff UB = AU.$$

Therefore, $U^{-1}AU = B$ and this completes the proof.

SOLUTION 11.3.37.

(1) "$\Longrightarrow$"([**94**]): Let $A, B \in \mathcal{N}$. Since $\mathcal{N}$ is a vector space, $A+B$ and $A+iB$ ($i^2 = -1$) are in $\mathcal{N}$, i.e. $A+B$ and $A+iB$ are both normal. This means that if

$$C = (A+B)(A+B)^* - (A+B)^*(A+B)$$

and

$$D = (A+iB)(A+iB)^* - (A+iB)^*(A+iB),$$

then $C = D = 0$. Hence $C + iD = 0$. Working this out (with $AA^* = A^*A$ and $BB^* = B^*B$ in mind) yields

$$AB^* + BA^* - A^*B - B^*A + i(-iAB^* + iBA^* - iA^*B + iB^*A) = 0,$$

so that after simplification we are only left with

$$2(AB^* - B^*A) = 0 \text{ or } AB^* = B^*A.$$

Applying the Fuglede Theorem to the normal B finally gives $AB = BA$.

(2) "$\Longleftarrow$": The set $\mathcal{N}$ is not empty as 0 is normal. Now, let $A, B \in \mathcal{N}$ and let $\alpha, \beta \in \mathbb{C}$.

Since A and B are normal, so are αA and βB. Since $AB = BA$, $(\alpha A)(\beta B) = (\beta B)(\alpha A)$ so that Exercise 4.3.11 is applicable and yields the normality of $\alpha A + \beta B$, that is, $\mathcal{N}$ is a vector space.

SOLUTION 11.3.38.

(1) First, we have

$$A^2 = (B+N)^2 = B^2 + NB + BN + N^2 = B^2 + 2BN$$

for N commutes with B and $N^2 = 0$. Hence $A^2 - B^2 = 2BN$.

Second, we may write

$$\begin{aligned} AB = BA \Longrightarrow AB^2 &= BAB = B^2A \\ \Longrightarrow A^2B^2 &= AB^2A = B^2A^2 \\ \Longrightarrow A^2(-B^2) &= (-B^2)A^2, \end{aligned}$$

so that Exercise 4.3.11 guarantees the normality of $A^2 - B^2$ or that of $2BN$ or simply that of BN.

Now, by Exercise 7.3.29, we have $\sigma(A) = \sigma(B)$ so that B is invertible because A is. Since B is normal and invertible, B^{-1} is normal. Hence

$$NB = BN \Longrightarrow N = BNB^{-1} \; (= B^{-1}BN).$$

Since BN is normal (and commutes with the normal B^{-1}), we infer by Theorem 4.1.16 that N is normal. Finally, since $N^2 = 0$, we obtain $N = 0$. Thus $A = B$, that is, A is normal.

(2) If the invertibility of A is dropped, the result may fail to hold. Assume that $\dim H = 2$. Then every $A \in B(H)$ may be written as $B+N$, where B is normal, $N^2 = 0$ and $BN = NB$ (this is more commonly known as the **Dunford Decomposition**). If the result of the foregoing question were to hold by forgetting about the invertibility of A, then this would amount to say that A is normal iff A^2 is normal, something not always true as there are non-normal matrices having normal squares.

SOLUTION 11.3.39.

(1) "$\Longrightarrow$": Since A is normal, it may be written as $A = UP$, where U is unitary, P is positive (recall that $P = \sqrt{A^*A} = |A|$) and P commutes with U. By hypothesis, B commutes with $AA^* = A^*A$ and so B commutes with P. Hence

$$U^*ABU = U^*UPBU = BPU = BA.$$

Therefore, BA is unitarily equivalent to AB, which is normal. By Exercise 4.3.17, BA is normal too.

(2) "$\Longleftarrow$": We may write $A(BA) = (AB)A$. But both AB and BA are normal, the Fuglede-Putnam Theorem then implies that

$$A(BA)^* = (AB)^*A \text{ or } AA^*B^* = B^*A^*A.$$

Taking adjoints yields $BAA^* = A^*AB$, which by the normality of A, just means that B commutes with AA^*.

SOLUTION 11.3.40.

(1) Since A and B are both normal, so are λA and λB. So by the Fuglede-Putnam Theorem, we have

$$AB = \lambda BA \Longrightarrow AB^* = \overline{\lambda}B^*A \text{ and } A^*B = \overline{\lambda}BA^*.$$

We have on the one hand

$$(AB)^*AB = B^*A^*AB = B^*AA^*B = \overline{\lambda}B^*ABA^* = |\lambda|^2B^*BAA^*,$$

and on the other hand,

$$AB(AB)^* = ABB^*A^* = AB^*BA^* = \overline{\lambda}B^*ABA^* = |\lambda|^2B^*BAA^*.$$

Thus AB is normal.

(2) Assume for example that A is positive. By the first question, both AB and BA are normal. Since $A \geq 0$, Exercise 11.3.35 applies and yields the self-adjointness of both AB and BA. Hence

$$BA = B^*A^* = (AB)^* = AB = \lambda BA,$$

yielding $\lambda = 1$.

(3) We have

$$A^2B = \lambda ABA = \lambda^2BA^2.$$

Since A is self-adjoint, A^2 is positive so that by the foregoing question $\lambda^2 = 1$ or merely $\lambda \in \{-1, 1\}$.

(4) By Exercise 4.3.13, we have

$$\|AB\| = \|BA\|.$$

Since $\|AB\| \neq 0$, we immediately obtain $|\lambda| = 1$.

Remark. The approach of the solution (in the first three questions) appeared in [**41**]. The last one is due to [**44**]. Other approaches can be found in [**35**] and [**215**].

Solution 11.3.41. Let $A, M, N \in B(H)$ where N and M are also assumed to be normal. Assume also that $AN = MA$.

Define on $H \oplus H$ the following operators

$$\widetilde{N} = \begin{pmatrix} M^* & 0 \\ 0 & N \end{pmatrix} \text{ and } \widetilde{A} = \begin{pmatrix} 0 & A \\ 0 & 0 \end{pmatrix}$$

(remember that $\sigma(\widetilde{N}) = \sigma(M^*) \cup \sigma(N)$). Since M and N are normal, so is $\widetilde{N}$ (as one can easily check). Besides one has

$$\widetilde{A}\widetilde{N} = \begin{pmatrix} 0 & A \\ 0 & 0 \end{pmatrix}\begin{pmatrix} M^* & 0 \\ 0 & N \end{pmatrix} = \begin{pmatrix} 0 & AN \\ 0 & 0 \end{pmatrix}$$

and

$$\widetilde{N}^*\widetilde{A} = \begin{pmatrix} M & 0 \\ 0 & N^* \end{pmatrix}\begin{pmatrix} 0 & A \\ 0 & 0 \end{pmatrix} = \begin{pmatrix} 0 & MA \\ 0 & 0 \end{pmatrix}.$$

Since $AN = MA$, it follows that

$$\widetilde{A}\widetilde{N} = \widetilde{N}^*\widetilde{A}.$$

By hypothesis whenever z is not real either $z \notin \sigma(M^*) \cup \sigma(N)$ or $\overline{z} \notin \sigma(M^*) \cup \sigma(N)$, that is, whenever z is not real either $z \notin \sigma(\widetilde{N})$ or $\overline{z} \notin \sigma(\widetilde{N})$. Consequently, Theorem 11.1.61 applies and gives

$$\widetilde{A}\widetilde{N} = \widetilde{N}\widetilde{A} \text{ or } AN = M^*A,$$

as needed.

11.4. Hints/Answers to Tests

Solution 11.4.1.

(1) Since $[f(A)]^* = \overline{f}(A)$, we have

$$\|f(A) - [f(A)]^*\| = \|f(A) - \overline{f}(A)\| = \|f - \overline{f}\|_\infty$$

and so clearly

$$f(A) \text{ is self-adjoint} \Longleftrightarrow f \text{ is real-valued.}$$

(2) Calling on Corollary 7.1.13 and the result $\sigma(f(A)) = f(\sigma(A))$ proves the desired equivalence.

Solution 11.4.2. Yes!

$$AB = BA \Longrightarrow B^{-1}A \subset AB^{-1}...$$

Then use Exercise 11.3.11...

Solution 11.4.3. The answer is evident if one knows the answer to the following:

(1) If $T = A + iB$, then when is T normal?
(2) What happens to the normal AB if A and B are self-adjoint, where one of them is positive?

Solution 11.4.4.

(1) Write $B(AB) = (BA)B$...Then apply the Fuglede-Putnam Theorem version of Exercise 11.3.32 (Question 3). Finally proceed as in Exercise 11.3.35.
(2) Under the same hypothesis, we have (haven't we?)

$$AB \text{ isometry} \Longleftrightarrow AB \text{ unitary.}$$

Prove it!

SOLUTION 11.4.5. Let

$$A = \begin{pmatrix} 0 & \pi \\ -\pi & 0 \end{pmatrix} \text{ and } B = \begin{pmatrix} \pi & -2\pi \\ \pi & -\pi \end{pmatrix}.$$

Then

$$e^A = e^B = -I \text{ while } AB \neq BA...$$

SOLUTION 11.4.6. First, set

$$\eta_{x,y} =< F(\Delta)x, y > .$$

Then write

$$\int_{f(\sigma(A))} \lambda d\eta_{x,y}(\lambda) = ... = \int_{\sigma(A)} f(z) d\mu_{x,y}(z) =< f(A)x, y >$$

by the *abstract change of variable formula* (see e.g. **[174]**)...

Then apply the Spectral Mapping Theorem (of Theorem 11.1.41!) and the uniqueness of the spectral measure to get the desired result...

SOLUTION 11.4.7. True even for normal operators. Indeed a normal operator is unitarily equivalent to a multiplication operator in some setting. Then use Exercise 4.3.6...

SOLUTION 11.4.8. WLOG, we assume that $|A| \leq I$ and $|B| \leq I$. We can always find an $n \in \mathbb{N}$ such that $p, q \leq 2^n$. Then by Example 11.1.33, we have

$$|A|^{\frac{p}{2^n}} \geq |A| \text{ and } |B|^{\frac{q}{2^n}} \geq |B|.$$

Therefore (why?),

$$|A|^p \geq |A|^{2^n} \text{ and } |B|^q \geq |B|^{2^n}.$$

Thus,

$$|A|^p + |B|^q \geq |A|^{2^n} + |B|^{2^n}.$$

Since $|A|^{2^n} + |B|^{2^n}$ is already invertible (Exercise 6.3.5), we obtain the invertibility of $|A|^p + |B|^q$.

SOLUTION 11.4.9. By Exercise 11.3.27, we know that if A is invertible then

$$A = e^{iC} e^B$$

where B and C are self-adjoint. The mapping $f : [0, 1] \to G$ defined by

$$f(x) = e^{ixC} e^{xB}$$

is continuous with $f(0) = I$ and $f(1) = A$. Therefore, G is path-connected and so it is connected.

SOLUTION 11.4.10. An isometry is quasinormal...

Solution 11.4.11. Since A is invertible, $0 \notin \sigma(A)$. Hence

$$|A|^{-1} = \int_{\sigma(A)} \frac{1}{|\lambda|} dE = \int_{\sigma(A)} |\lambda^{-1}| dE = |A^{-1}|...$$

CHAPTER 12

Hyponormal Operators

12.2. True or False: Answers

ANSWERS.

(1) The reason is that in a finite dimensional space, normal and hyponormal operators coincide. For a proof, see Exercise 12.3.1.
(2) No! In fact, if A^* is hyponormal, then we say that A is **co-hyponormal**. If both A and A^* are hyponormal, then this simply means that A is normal. So, for a counterexample of a hyponormal whose adjoint is not hyponormal, it suffices to consider a hyponormal operator which is not normal.
(3) No again! For a counterexample, see Exercise 12.3.1.

REMARK. It is worth noticing that if we denote the class of hyponormal operators whose squares are not hyponormal by Q, and the set of hyponormal operators by H_0, then

$$\overline{Q} = H_0,$$

where the topology involved is that of the operator norm. See **[72]**.

(4) No! In fact, we have just seen that the square of a hyponormal operator need not be hyponormal. So how are we going to define $f(A)$ if just A^2 is not always hyponormal?!
(5) False! Let A be any hyponormal operator ($A^*A \geq AA^*$) which is not normal. If $-A^*$ were hyponormal, then we would have

$$(-A^{**})(-A^*) \geq (-A^*)(-A^{**}) \text{ or } AA^* \geq A^*A$$

which would lead to the normality of A, and this is the sought contradiction.
(6) False in the event of left-invertibility! Let S be the shift operator on ℓ^2. Then S is hyponormal (see Exercise 12.3.3) and left invertible but it is not invertible. In the case of right invertibility, this result is true. See Exercise 12.3.16.

(7) False again! As before, let S be the shift operator on ℓ^2. Then S is hyponormal and

$$\sigma(S^*S) = \{1\} \neq \sigma(SS^*) = \{0, 1\}.$$

(8) False! We know that the shift operator is hyponormal. Does it have a square root?...

(9) False (cf. Exercise 12.3.8) even for a larger class of operators. Indeed, let B be self-adjoint and V be an isometry. Then B and V are hyponormal. If they were commuting in norm, then we would have

$$\|VB\| = \|BV\|.$$

Now, consider the example of Exercise 4.3.6, which is:

$$V(x_1, x_2, \cdots) = (x_1, 0, x_2, 0, \cdots)$$

and

$$B(x_1, x_2, \cdots) = (0, x_2, 0, x_4, 0, \cdots),$$

both defined on ℓ^2. Then V is an isometry and B is self-adjoint. Besides, we have found that

$$\|BV\| = 0 \neq 1 = \|VB\|.$$

(10) Only the left-to-right implication is valid. Indeed, since A is hyponormal, we get

$$AA^* \leq A^*A \text{ or } |A^*| = \sqrt{AA^*} \leq \sqrt{A^*A} = |A|$$

by Theorem 5.1.28.

For the reverse implication to hold, we need for example the commutativity of A and A^* (which is not necessarily the case!). Counterexamples, however, are not easy to find. We refer readers to **[214]** (the counterexample is in the form of "singular integral operator"). We can do it differently but first, we have

Definition 12.2.1. *An operator $A \in B(H)$ such that $|A^*| \leq |A|$, that is, $(AA^*)^{\frac{1}{2}} \leq (A^*A)^{\frac{1}{2}}$ is called **semi-hyponormal**. More generally, an $A \in B(H)$ such that*

$$(AA^*)^p \leq (A^*A)^p,$$

*for some positive p, is called p-**hyponormal**.*

Remark. By the first implication above, we can say that a hyponormal operator is semi-hyponormal. In fact, a hyponormal operator is p-hyponormal for any $p \in (0, 1]$ thanks to the Löwner-Heinz Inequality.

One interesting result (though not strictly within the context of this book) is:

PROPOSITION 12.2.2. *(see Page 183 of* **[81]***) Let A be p-hyponormal with $p \in (0,1]$. Then A^n is $\frac{p}{n}$-hyponormal for any $n \in \mathbb{N}$.*

We have already mentioned a hyponormal operator A whose square A^2 *is not hyponormal* (details to be given in Exercise 12.3.1). Since this A is hyponormal, that is, A is 1-hyponormal, by the previous proposition A^2 is always is $\frac{1}{2}$-hyponormal, i.e. A^2 is always semi-hyponormal. So the sought counterexample is just $B = A^2$. Then B is semi-hyponormal but it is not hyponormal!

12.3. Solutions to Exercises

SOLUTION 12.3.1.

(1) Let $x \in H$. Then

$$\begin{aligned} AA^* \leq A^*A &\Longleftrightarrow < AA^*x, x > \leq < A^*Ax, x >, \ \forall x \in H \\ &\Longleftrightarrow < A^*x, A^*x > \leq < Ax, A^{**}x >, \ \forall x \in H \\ &\Longleftrightarrow < A^*x, A^*x > \leq < Ax, Ax >, \ \forall x \in H \\ &\Longleftrightarrow \|A^*x\|^2 \leq \|Ax\|^2, \ \forall x \in H \\ &\Longleftrightarrow \|A^*x\| \leq \|Ax\|, \ \forall x \in H. \end{aligned}$$

(2) Let S be the usual shift. Set

$$A = S^* + 2S.$$

Then

$$A^* = S^{**} + 2S^* = S + 2S^*.$$

Therefore

$$AA^* = (S^* + 2S)(S + 2S^*) = S^*S + 2S^2 + 2S^{*2} + 4SS^*$$

so that

$$AA^* = I + 2S^2 + 2S^{*2} + 4SS^*.$$

We also have

$$A^*A = (S + 2S^*)(S^* + 2S) = SS^* + 2S^2 + 2S^{*2} + 4S^*S$$

so that

$$A^*A = SS^* + 2S^2 + 2S^{*2} + 4I.$$

Now

$$A^*A - AA^* = 3I - 3SS^* = 3(I - SS^*).$$

It is clear that A is not normal. However, by Exercise 5.3.2, $I - SS^* \geq 0$, and so A is hyponormal.

Another example is the operator A defined in Exercise 4.3.22, i.e.

$$A(\cdots, x_{-1}, \boldsymbol{x_0}, x_1, \cdots) = (\cdots, \alpha_{-1}x_{-2}, \boldsymbol{\alpha_0 x_{-1}}, \alpha_1 x_0, \cdots).$$

Then

$$AA^*(\cdots, x_{-1}, \boldsymbol{x_0}, x_1, \cdots) \leq A^*A(\cdots, x_{-1}, \boldsymbol{x_0}, x_1, \cdots)$$

if and only if

$$|\alpha_n| \leq |\alpha_{n+1}|, \ \forall n \in \mathbb{Z}.$$

Accordingly, to find the required example, take $\alpha_n \in \mathbb{R}^+$, increasing and such that $\alpha_n \neq \alpha_0$.

(3) Remember that $\operatorname{tr}(ST) = \operatorname{tr}(TS)$ for any square (finite) matrices S and T. Therefore

$$\operatorname{tr}(A^*A - AA^*) = 0.$$

Since A is hyponormal, $A^*A - AA^*$ is a positive matrix. But a positive matrix with trace zero is the zero matrix (see Exercise 5.3.7). Thus, $A^*A = AA^*$, that is, A is normal.

(4) ([**109**]) Not necessarily! Take again the operator $A = S^* + 2S$ introduced above. Then we have already seen that A is hyponormal. So, we need only check that A^2 is not hyponormal. It suffices therefore to find an $x_0 \in \ell^2(\mathbb{N})$ such that

$$\|(A^2)^*x_0\| = \|(A^*)^2x_0\| \not\leq \|A^2x_0\|, \text{ that is, } \|(A^*)^2x_0\| > \|A^2x_0\|.$$

The reader may check that one possible choice is to take $x_0 = (1, 0, -2, 0, \cdots)$, and this finally gives us

$$\|(A^*)^2x_0\| = \sqrt{89} > \sqrt{80} = \|A^2x_0\|.$$

Solution 12.3.2.

(1) Evident! Just write the definition of a hyponormal operator in terms of norms, then use Theorem 3.1.69...

(2) Since A is hyponormal, we now know that $A^* = KA$ for some contraction $K \in B(H)$. But A is invertible, and so

$$K = A^*A^{-1} \text{ or } K^* = (A^{-1})^*A.$$

Therefore,

$$(A^{-1})^* = K^*A^{-1},$$

which establishes the hyponormality of A^{-1} by the previous question for K^* is a contraction.

Solution 12.3.3.

(1) Let $A \in B(H)$ be hyponormal and $\alpha \in \mathbb{C}$. Then for all $x \in H$

$$\|(\alpha A)^* x\| = \|\overline{\alpha} A^* x\| = |\overline{\alpha}|\|A^* x\| = |\alpha|\|A^* x\| \leq |\alpha|\|Ax\| = \|\alpha Ax\|,$$

proving the hyponormality of αA.

(2) Let $A \in B(H)$ be an isometry, i.e. $A^*A = I$. We also know that $\|A\| = 1 = \|A^*\|$. Ergo, for all $x \in H$

$$\|A^*x\| \leq \|A^*\|\|x\| = \|x\| = \|A^*Ax\| \leq \|A^*\|\|Ax\| = \|Ax\|,$$

and we have proved that A is hyponormal.

(3) By assumption,

$$\|A\| = \|A^{-1}\| = \|(A^{-1})^*\| = \|(A^*)^{-1}\| \leq 1.$$

So for all $x \in H$, we have

$$\|Ax\| = \|A(A^*)^{-1}A^*x\| \leq \|A\|\|(A^*)^{-1}\|\|A^*x\| \leq \|A^*x\|,$$

i.e. A^* is hyponormal. Since A is hyponormal by assumption, we finally infer that A is normal.

Solution 12.3.4. By assumption, there exists a *unitary* operator $U \in B(H)$ such that

$$U^{-1}AU = U^*AU = B.$$

Then $A = UBU^*$ and also $A^* = UB^*U^*$. So if B is hyponormal (that is $BB^* \leq B^*B$), then

$$AA^* = UBU^*UB^*U^* = UBB^*U^* \leq UB^*BU^* \text{ (cf. Exercise 5.3.8).}$$

Since

$$UB^*BU^* = UB^*U^*UBU^* = A^*A,$$

we immediately obtain

$$AA^* \leq A^*A,$$

i.e. A is hyponormal and this proves half of the equivalence. To prove the other half, the interested reader may reason similarly.

Solution 12.3.5. Let (A_n) be a sequence of hyponormal operators acting on a Hilbert space H (i.e. $\|A_n^*x\| \leq \|A_nx\|$ for all x and for each $n \in \mathbb{N}$) such that $\|A_nx - Ax\| \to 0$ as $n \to \infty$ (for all $x \in H$). Then by Exercise 1.3.2, we have that (for all $x \in H$)

$$\lim_{n\to\infty} \|A_nx\| = \|Ax\|.$$

Strong convergence also implies weak convergence so that (for all $x, y \in H$)

$$\lim_{n\to\infty} < A_n x, y >=< Ax, y >$$

which, in turn, is known to hold if and only if

$$\lim_{n\to\infty} < A_n^* x, y >=< A^* x, y >$$

(for all $x, y \in H$).

Now, we have by the Cauchy-Schwarz Inequality (and the hyponormality of (A_n))

$$| < A_n^* x, y > | \leq \|A_n^* x\| \|y\| \leq \|A_n x\| \|y\|$$

for all natural n and all $x, y \in H$.

Passing to the limit as n goes to infinity, we obtain

$$| < A^* x, y > | \leq \|Ax\| \|y\|$$

for all x and all y. Setting $y = A^* x$ finally yields (for all $x \in H$)

$$\|A^* x\|^2 \leq \|Ax\| \|A^* x\|$$

or simply (why?)

$$\|A^* x\| \leq \|Ax\|, \ \forall x \in H,$$

that is, A is hyponormal and this finishes the proof.

Solution 12.3.6.

(1) Let $n \in \mathbb{N}$ and $x \in H$. First, we know from Exercise 2.3.7 that $\|A^n\| \leq \|A\|^n$ for all n. So, it only remains to show the reverse inequality. Let us do that using a proof by induction. Obviously, the statement is true for $n = 1$, hence assume that $\|A^n\| \geq \|A\|^n$ and let us show that $\|A^{n+1}\| \geq \|A\|^{n+1}$. Let $x \in H$. We have

$$\begin{aligned}
\|A^n x\|^2 &=< A^n x, A^n x > \\
&=< A^* A^n x, A^{n-1} x > \\
&\leq \|A^* A^n x\| \|A^{n-1} x\| \\
&\leq \|A^{n+1} x\| \|A^{n-1} x\| \text{ (for } A \text{ is hyponormal)} \\
&\leq \|A^{n+1}\| \|A^{n-1}\| \|x\|^2
\end{aligned}$$

so that we obtain

$$\|A^n\|^2 \leq \|A^{n+1}\| \|A^{n-1}\| \leq \|A^{n+1}\| \|A\|^{n-1}.$$

But, by hypothesis $\|A^n\| \geq \|A\|^n$, hence

$\|A\|^{2n} \leq \|A^{n+1}\| \|A\|^{n-1}$ or simply $\|A\|^{n+1} \leq \|A^{n+1}\|$, establishing the result.

(2) We easily see that

$$r(A) = \lim_{n\to\infty} \|A^n\|^{\frac{1}{n}} = \|A\|,$$

that is, hyponormal operators are normaloid.

Solution 12.3.7. By hypothesis, $\|A^*x\| \leq \|Ax\|$ and $\|B^*x\| \leq \|Bx\|$ for all $x \in H$. Also, since $AB^* = B^*A$, we have $A^*B = BA^*$. It then follows that for each x

$$\|(AB)^*x\| = \|B^*A^*x\| \leq \|BA^*x\| = \|A^*Bx\| \leq \|ABx\|,$$

that is, AB is hyponormal.

Solution 12.3.8.

(1) Since A^* is hyponormal, we have

$$\|ABx\| \leq \|A^*Bx\|$$

for all $x \in H$. Consequently,

$$\|AB\| \leq \|A^*B\|.$$

(2) Since $A = (A^*)^*$ is hyponormal, applying the previous result gives

$$\|A^*B\| \leq \|AB\|.$$

To answer the other question, notice that if A is hyponormal, then

$$\|A^*A\| \leq \|AA\| \leq \|A\|^2.$$

Since we already know that $\|A^*A\| = \|A\|^2$, we get

$$\|A^2\| = \|A\|^2.$$

(3) As in the first answer...

(4) The inequality here may be shown as in the previous cases...

(5) Since A and B^* are hyponormal, we know that

$$\|BA\| = \|(BA)^*\| = \|A^*B^*\| \leq \|A^*B\| \leq \|AB\|,$$

as required.

(6) Interchanging the roles of A and B in the previous question leads to the desired inequality.

Solution 12.3.9.

(1) Since A is normal, we know that

$$A = PU = UP$$

where P is positive and U is unitary. Hence

$$AA^*B = BAA^* \Longrightarrow P^2B = BP^2 \Longrightarrow PB = BP$$

so that

$$U^*ABU = U^*UPBU = PBU = BPU = BA.$$

Since AB is hyponormal, it follows by Exercise 12.3.4 that BA is hyponormal.

(2) Let A and B be acting on the standard basis (e_n) of ℓ^2 by:

$$Ae_n = \alpha_n e_n \text{ and } Be_n = e_{n+1},\ \forall n \geq 1$$

respectively. Assume further that α_n is bounded, *real-valued* and *positive*, for all n. Hence A is self-adjoint (hence normal!) and positive. Then

$$ABe_n = \alpha_n e_{n+1},\ \forall n \geq 1.$$

For convenience, let us carry out the calculations as infinite matrices. Then

$$AB = \begin{bmatrix} 0 & 0 & & & & 0 \\ \alpha_1 & 0 & 0 & & & \\ 0 & \alpha_2 & 0 & 0 & & \\ & 0 & \alpha_3 & 0 & \ddots & \\ & & 0 & \ddots & 0 & \\ 0 & & & \ddots & \ddots & \ddots \end{bmatrix}$$

so that

$$(AB)^* = \begin{bmatrix} 0 & \alpha_1 & & & & 0 \\ 0 & 0 & \alpha_2 & & & \\ 0 & 0 & 0 & \alpha_3 & & \\ & 0 & 0 & 0 & \ddots & \\ & & 0 & \ddots & 0 & \\ 0 & & & \ddots & \ddots & \ddots \end{bmatrix}.$$

Hence

$$AB(AB)^* = \begin{bmatrix} 0 & 0 & & & & 0 \\ 0 & \alpha_1^2 & 0 & & & \\ 0 & 0 & \alpha_2^2 & 0 & & \\ & 0 & 0 & \alpha_3^2 & \ddots & \\ & & 0 & \ddots & \ddots & \\ 0 & & & \ddots & \ddots & \ddots \end{bmatrix}$$

and

$$(AB)^*AB = \begin{bmatrix} \alpha_1^2 & 0 & & & & 0 \\ 0 & \alpha_2^2 & 0 & & & \\ 0 & 0 & \alpha_3^2 & 0 & & \\ & 0 & 0 & \ddots & \ddots & \\ & & 0 & \ddots & \ddots & \\ 0 & & & \ddots & \ddots & \ddots \end{bmatrix}.$$

Therefore,

$$AB \text{ is hyponormal} \iff \alpha_n \leq \alpha_{n+1}.$$

Similarly

$$BAe_n = \alpha_{n+1}e_{n+1}, \ \forall n \geq 1.$$

Whence the matrix representing BA is given by:

$$BA = \begin{bmatrix} 0 & 0 & & & & 0 \\ \alpha_2 & 0 & 0 & & & \\ 0 & \alpha_3 & 0 & 0 & & \\ & 0 & \alpha_4 & 0 & \ddots & \\ & & 0 & \ddots & 0 & \\ 0 & & & \ddots & \ddots & \ddots \end{bmatrix} \text{ so that } (BA)^* = \begin{bmatrix} 0 & \alpha_2 & & & & 0 \\ 0 & 0 & \alpha_3 & & & \\ 0 & 0 & 0 & \alpha_4 & & \\ & 0 & 0 & 0 & \ddots & \\ & & 0 & \ddots & 0 & \\ 0 & & & \ddots & \ddots & \ddots \end{bmatrix}.$$

Therefore,

$$BA(BA)^* = \begin{bmatrix} 0 & 0 & & & & 0 \\ 0 & \alpha_2^2 & 0 & & & \\ 0 & 0 & \alpha_3^2 & 0 & & \\ & 0 & 0 & \alpha_4^2 & \ddots & \\ & & 0 & \ddots & \ddots & \\ 0 & & & \ddots & \ddots & \ddots \end{bmatrix}$$

and

$$(BA)^*BA = \begin{bmatrix} \alpha_2^2 & 0 & & & & 0 \\ 0 & \alpha_3^2 & 0 & & & \\ 0 & 0 & \alpha_4^2 & 0 & & \\ & 0 & 0 & \ddots & \ddots & \\ & & 0 & \ddots & \ddots & \\ 0 & & & \ddots & \ddots & \ddots \end{bmatrix}.$$

Accordingly, BA is hyponormal iff $\alpha_n \leq \alpha_{n+1}$ (thankfully, this is the same condition for the hyponormality of AB).

Finally,

$$BA^2 = \begin{bmatrix} 0 & 0 & & & & 0 \\ \alpha_1^2 & 0 & 0 & & & \\ 0 & \alpha_2^2 & 0 & 0 & & \\ & 0 & \alpha_3^2 & 0 & \ddots & \\ & & 0 & \ddots & 0 & \\ 0 & & & \ddots & \ddots & \ddots \end{bmatrix} \neq A^2B = \begin{bmatrix} 0 & 0 & & & & 0 \\ \alpha_2^2 & 0 & 0 & & & \\ 0 & \alpha_3^2 & 0 & 0 & & \\ & 0 & \alpha_4^2 & 0 & \ddots & \\ & & 0 & \ddots & 0 & \\ 0 & & & \ddots & \ddots & \ddots \end{bmatrix}.$$

Remark. An explicit example of such an (α_n) verifying the required hypotheses would be to take:

$$\begin{cases} \alpha_1 = 0 \\ \alpha_{n+1} = \sqrt{2 + \alpha_n} \end{cases}$$

Then (α_n) is bounded (in fact, $0 \leq \alpha_n < 2$, for all n), increasing and such that $\alpha_1 = 0 \neq \alpha_2 = \sqrt{2}$.

(3)

(a) "$\Longrightarrow$": Since A is normal, we have $A = UP = PU$ where U is unitary and P is positive. Since $AA^*B = BA^*A$, we obtain $P^2B = BP^2$ or just $BP = PB$ by the positivity of P.
Therefore, we may write

$$U^*ABU = BA \text{ or } U^*(AB)^*U = (BA)^*.$$

Since $(BA)^*$ is hyponormal, so is $(AB)^*$ by Exercise 12.3.4. It immediately follows that AB is normal.
To establish the normality of BA it suffices to show that BA is hyponormal given that $(BA)^*$ is hyponormal by hypothesis. A similar idea as before may be applied after observing that BA is unitarily equivalent to the hyponormal AB.

(b) "$\Longleftarrow$": To prove the the reverse implication, we use the Fuglede-Putnam Theorem. We have:

$$\begin{aligned} ABA = ABA &\Longrightarrow A(BA) = (AB)A \\ &\Longrightarrow A(BA)^* = (AB)^*A \\ &\Longrightarrow AA^*B^* = B^*A^*A \\ &\Longrightarrow BAA^* = A^*AB. \end{aligned}$$

This completes the proof.

SOLUTION 12.3.10.

(1) Let $x \in H$. Then we have

$$\|Ax\|^2 = < Ax, Ax > = < A^*Ax, x > \leq \|A^*Ax\|\|x\| \leq \|A^2x\|\|x\|,$$

and it is clear where we have used the hyponormality of A and the Cauchy-Schwarz Inequality.

(2) Assume that A^2 is compact and let (x_n) be a sequence in H such that $\|x_n\| = 1$. We must show that (Ax_n) has a convergent subsequence. Since A^2 is compact and $\|x_n\| = 1$, we know that (A^2x_n) has a convergent subsequence, denoted by $(A^2x_{n_k})$. On that account, this subsequence is Cauchy. But it clear from the first question that for any *unit* vectors $x_n, x_m \in H$

$$\|Ax_n - Ax_m\|^2 \leq \|A^2x_n - A^2x_m\|\|x_n - x_m\| \leq 2\|A^2x_n - A^2x_m\|,$$

so that if (A^2x_n) is Cauchy, then so is (Ax_n). Hence (Ax_{n_k}) is Cauchy in H as well. Since H is complete, (Ax_{n_k}) converges. Accordingly, A is compact.

REMARKS.

(1) A fortiori, if A is self-adjoint or normal, then A^2 compact does imply that A is compact.

(2) We have already met an example of a non-compact operator whose square is compact (cf. Exercise 9.3.7).

(3) There is nothing special about A^2 in the proof in the second answer. In fact, the same proof works if we have A^n compact for some $n \geq 3$.

SOLUTION 12.3.11. We only do the first case and we leave the second one to interested readers.

It is clear that $A - A^*$ is anti-symmetric and so $(A - A^*)^2 \leq 0$. Hence

$$(A - A^*)^2 \leq 0 \Longleftrightarrow A^2 + A^{*^2} - AA^* - A^*A \leq 0.$$

But, A is hyponormal and so $-AA^* - A^*A \geq -2A^*A$. So,

$$A^2 + A^{*^2} - 2A^*A \leq 0$$

or

$$A^2 + A^{*^2} + AA^* + A^*A \leq A^2 + A^{*^2} + 2A^*A \leq 4A^*A.$$

Therefore,

$$(A + A^*)^2 \leq 4A^*A \text{ or } |\operatorname{Re} A| \leq |A|$$

by Theorem 5.1.39.

Remembering that $B^- = \frac{1}{2}(|B| - B) \geq 0$ whenever B is self-adjoint, we conclude that

$$\operatorname{Re} A \leq |\operatorname{Re} A| \leq |A|,$$

as required.

SOLUTION 12.3.12.

(1) First, we have

$$(\text{Re}A)^2 = \frac{A^2 + A^{*2} + AA^* + A^*A}{4}$$

and

$$(\text{Im}A)^2 = -\frac{A^2 + A^{*2} - AA^* - A^*A}{4}.$$

Hence

$$(\text{Re}A)^2 + (\text{Im}A)^2 = \frac{AA^* + A^*A}{2}.$$

Now A is hyponormal iff $AA^* \leq A^*A$, that is, iff

$$|A|^2 = A^*A \geq \frac{AA^* + A^*A}{2},$$

and this completes the proof.

(2) Since A is hyponormal, by the previous question (or by Exercise 12.3.11), we have

$$(\text{Re}A)^2 \leq |A|^2 \text{ or } |\text{Re}A| \leq |A|.$$

By our assumption, we then obtain

$$|\text{Re}A| = |A| \text{ or } (\text{Re}A)^2 = |A|^2.$$

Thus

$$(A - A^*)^2 = 0.$$

Since $A - A^*$ is normal, we finally infer that

$$A = A^*.$$

SOLUTION 12.3.13.

(1) Let S be the shift operator. Set $B = S + 3S^*$. Then we have

$$BB^* = (S + 3S^*)(S^* + 3S) = SS^* + 3S^{*2} + 3S^2 + 9S^*S$$

and

$$B^*B = (S^* + 3S)(S + 3S^*) = S^*S + 3S^2 + 3S^{*2} + 9SS^*.$$

Remembering that $S^*S = I$, we obtain after simplification

$$BB^* - B^*B = 8I - 8SS^* \geq 0$$

(by Exercise 5.3.2). Accordingly, B is not hyponormal.

(2) Put $B = A + \lambda A^*$ so that $B^* = A^* + \overline{\lambda} A$. Hence

$$BB^* = AA^* + \lambda A^{*2} + \overline{\lambda} A^2 + |\lambda|^2 A^* A$$

and

$$B^* B = A^* A + \lambda A^{*2} + \overline{\lambda} A^2 + |\lambda|^2 AA^*.$$

It follows that

$$\begin{aligned} BB^* - B^* B &= AA^* - A^* A + |\lambda|^2 (A^* A - AA^*) \\ &\leq AA^* - A^* A + A^* A - AA^* \text{ (since } |\lambda| \leq 1) \\ &= 0, \end{aligned}$$

i.e. $A + \lambda A^*$ hyponormal.

Solution 12.3.14. Let AK be co-hyponormal, i.e. $(AK)^*$ is hyponormal and let $A \geq 0$. The inequality is evident when $K = 0$. So, assume that $K \neq 0$. It is then clear that $\frac{K}{\|K\|}$ satisfies

$$KK^* \leq \|K\|^2 I.$$

Hence

$$|(AK)^*|^2 = AKK^* A \leq \|K\|^2 A^2$$

or simply $|(AK)^*| \leq \|K\| A$.

Now, for all $x \in H$

$$\begin{aligned} | < AKx, x > | = | < x, (AK)^* x > | &= |\overline{< (AK)^* x, x >}| \\ &= | < (AK)^* x, x > |. \end{aligned}$$

Since $(AK)^*$ is hyponormal, invoking Proposition 12.1.6 and $|(AK)^*| \leq \|K\| A$ give

$$\begin{aligned} | < AKx, x > | &= | < (AK)^* x, x > | \\ &\leq < |(AK)^*| x, x > \leq \|K\| < Ax, x >, \end{aligned}$$

completing the proof.

Solution 12.3.15.

(1) Let λ be in $\mathbb{C}$ and let x be in H. Since $(A-\lambda I)^* = A^* - \overline{\lambda} I$, we may write

$$\begin{aligned}
\|(A-\lambda I)x\|^2 &= < (A-\lambda I)x, (A-\lambda I)x > \\
&= < (A-\lambda I)^*(A-\lambda I)x, x > \\
&= < (A^*-\overline{\lambda} I)(A-\lambda I)x, x > \\
&= < (A^*A-\overline{\lambda}A-\lambda A^*+|\lambda|^2 I)x, x > \\
&\geq < (AA^*-\overline{\lambda}A-\lambda A^*+|\lambda|^2 I)x, x > \text{ (as } A \text{ is hyponormal)} \\
&= < (A-\lambda I)(A^*-\overline{\lambda} I)x, x > \\
&= < (A-\lambda I)^*x, (A-\lambda I)^*x > \\
&= \|(A-\lambda I)^*x\|^2,
\end{aligned}$$

establishing the hyponormality of $A-\lambda I$.

Remark. Another way of seeing this is to show that

$$(\lambda I - A)^*(\lambda I - A) - (\lambda I - A)(\lambda I - A)^* = A^*A - AA^*$$

for any complex λ. Thereupon, A is hyponormal iff $\lambda I - A$ is hyponormal.

(2) By the previous question, we have

$$\|(A^* - \overline{\lambda} I)x\| = \|(A-\lambda I)^*x\| \leq \|(A-\lambda I)x\|$$

for all $x \in H$. Therefore, we deduce that if $\lambda \in \sigma_p(A)$, then $\overline{\lambda} \in \sigma_p(A^*)$.

(3) Let $x, y \neq 0$ be such that $Ax = \lambda x$ and $Ay = \mu y$ where $\lambda \neq \mu$. Hence by the previous question $A^*y = \overline{\mu} y$ and so

$$\lambda < x, y > = < \lambda x, y > = < Ax, y > = < x, A^*y > = < x, \overline{\mu} y > = \mu < x, y >$$

an so $< x, y > = 0$.

(4) $\sigma_r(A^*)$ is always empty for any hyponormal operator A. If $\sigma_r(A^*)$ were not empty, it would contain at least a certain λ. Hence

$$\ker(\lambda - A^*) = \{0\} \text{ and } \overline{\operatorname{ran}(\lambda - A^*)} \neq H.$$

But by Question (2),

$$\ker(\lambda - A^*) = \{0\} \Longrightarrow \ker(\overline{\lambda} - A) = \{0\}$$

and so $\overline{\operatorname{ran}(\lambda - A^*)} = H$.

This means that we *cannot* have $\ker(\lambda - A^*) = \{0\}$ and $\overline{\operatorname{ran}(\lambda - A^*)} \neq H$ simultaneously, that is, $\sigma_r(A^*)$ is always empty.

SOLUTION 12.3.16. Assume that A is right invertible, that is, for some $B \in B(H)$: $AB = I$. Hence, A is onto and so

$$\operatorname{ran} A = H \Longrightarrow \ker(A^*) = \{0\}.$$

The hyponormality of A then gives $\ker(A) = \{0\}$. Therefore, A is a bijection. Thus, the Banach Isomorphism Theorem finally implies the invertibility of A.

SOLUTION 12.3.17.

(1) Since A is hyponormal, by Exercise 12.3.15, $A-\lambda I$ is hyponormal for any $\lambda \in \mathbb{C}$. So to prove $(A-\lambda I)^{-1}$ is hyponormal, it then suffices to prove that the inverse of a hyponormal operator is hyponormal (we have already seen a simple proof of this result, here we give an algebraic proof). Let $B \in B(H)$ be an invertible hyponormal operator.

Since B is hyponormal, $B^*B - BB^* \geq 0$. Hence by Exercise 5.3.8

$$0 \leq B^{-1}(B^*B - BB^*)(B^{-1})^* = B^{-1}(B^*B - BB^*)(B^*)^{-1}$$

which, after simplification, yields

$$B^{-1}B^*BB^{*-1} - I \geq 0 \text{ or } B^{-1}B^*BB^{*-1} \geq I.$$

Remembering that if $C \geq I$, then $C^{-1} \leq I$, leads to

$$[B^{-1}B^*BB^{*-1}]^{-1} = B^*B^{-1}B^{*-1}B \leq I,$$

that is,

$$I - B^*B^{-1}B^{*-1}B \geq 0.$$

Calling on Exercise 5.3.8, we have

$$B^{*-1}(I - B^*B^{-1}B^{*-1}B)B^{-1} \geq 0,$$

which, after simplifications, gives us

$$B^{*-1}B^{-1} - B^{-1}B^{*-1} \geq 0,$$

which is exactly what B^{-1} being hyponormal means.

(2) Since $(A-\lambda I)^{-1}$ is hyponormal, by Exercise 12.3.6 we have

$$\|(A-\lambda I)^{-1}\| = r((A-\lambda I)^{-1}).$$

Consequently, for all $x \in H$ such that $\|x\| = 1$, we have

$$\begin{aligned}\|(A-\lambda I)^{-1}x\| &\leq \|(A-\lambda I)^{-1}\|\\ &= \sup\{|\mu| : \mu \in \sigma[(A-\lambda I)^{-1}]\}\\ &= \frac{1}{\inf\{|\mu| : \mu \in \sigma(A-\lambda I)\}}\\ &= \frac{1}{\inf\{|\mu-\lambda| : \mu \in \sigma(A)\}}\\ &= \frac{1}{d(\lambda,\sigma(A))}.\end{aligned}$$

Solution 12.3.18. By Exercise 7.3.10, we know that $\sigma_a(A^*) \subset \sigma(A^*)$ for any $A^* \in B(H)$. To prove the reverse inequality, let $\lambda \notin \sigma_a(A^*)$. Then $A^* - \lambda I$ is bounded below, that is, $(A-\overline{\lambda}I)^*$ is bounded below. Ergo,

$$\exists \alpha > 0,\ \forall x \in H :\ \|(A-\overline{\lambda}I)^*x\| \geq \alpha\|x\|.$$

Since $A - \overline{\lambda}I$ is hyponormal, we have for all $x \in H$,

$$\|(A-\overline{\lambda}I)x\| \geq \|(A-\overline{\lambda}I)^*x\|$$

so that $A-\overline{\lambda}I$ too is bounded below. Since its adjoint $(A-\overline{\lambda}I)^*$ is clearly injective, we immediately conclude that $(A-\overline{\lambda}I)^*$ is invertible (by Proposition 5.1.23). Hence $A^*-\lambda I$ is also invertible, that is, $\lambda \notin \sigma(A^*)$ and this completes the proof.

Solution 12.3.19.

(1) Recall that

$$\operatorname{Re} A = \frac{A+A^*}{2} \text{ and } \operatorname{Im} A = \frac{A-A^*}{2i}.$$

Hence

$$\begin{aligned}A^*A &= |A|^2\\ &\leq (\operatorname{Re} A)^2 + (\operatorname{Im} A)^2\\ &= \frac{(A+A^*)^2}{4} - \frac{(A-A^*)^2}{4}\\ &= \frac{A^2+A^{*2}+AA^*+A^*A-(A^2+A^{*2}-AA^*-A^*A)}{4}\\ &= \frac{AA^*+A^*A}{2},\end{aligned}$$

that is, iff $A^*A \leq AA^*$, i.e. iff A^* is hyponormal.

(2) By the previous question, A^* is hyponormal. To show that A is self-adjoint, and since A^* is hyponormal, it suffices by virtue of Corollary 12.1.10 to show that A (or A^*!) has real spectrum.

Calling on Proposition 12.1.12, we know that $\sigma(A) = \sigma_a(A)$ (as A^* is hyponormal). So let $\lambda \in \sigma_a(A)$. Then

$$\exists x_n \in H, \ \|x_n\| = 1 \text{ and } \|(A-\lambda)x_n\| \longrightarrow 0$$

and hence $\|Ax_n\| \to |\lambda|$ as n tends to infinity. Now, we can easily deduce (do it!), using the hyponormality of A^*, that when $n \to \infty$

$$\|(\operatorname{Re} A)x_n\| \longrightarrow |\operatorname{Re}\lambda| \text{ and } \|(\operatorname{Im} A)x_n\| \longrightarrow |\operatorname{Im}\lambda|$$

Since $|A|^2 \leq (\operatorname{Re} A)^2 + \alpha(\operatorname{Im} A)^2$, by Proposition 6.1.3, we get that for all $n \in \mathbb{N}$

$$\|Ax_n\|^2 \leq \|(\operatorname{Re} A)x_n\|^2 + \alpha\|(\operatorname{Im} A)x_n\|^2.$$

Letting $n \to \infty$ yields

$$|\lambda|^2 \leq |\operatorname{Re}\lambda|^2 + \alpha|\operatorname{Im}\lambda|^2$$

and since $\lambda = \operatorname{Re}\lambda + i\operatorname{Im}\lambda$, we finally obtain

$$(\alpha - 1)|\operatorname{Im}\lambda|^2 \geq 0$$

which, because $\alpha < 1$, forces $\operatorname{Im}\lambda = 0$, that is $\lambda \in \mathbb{R}$.

Remark. The proof given here is an improved version of the one which first appeared in [**75**].

(3) It is clear that if A is self-adjoint, then

$$|A|^2 = A^2 = \left(\frac{A+A}{2}\right)^2 = (\operatorname{Re} A)^2.$$

Conversely, by assumption $|A|^2 \leq (\operatorname{Re} A)^2$ and so

$$|A|^2 \leq (\operatorname{Re} A)^2 + \alpha(\operatorname{Im} A)^2$$

for all $\alpha > 0$, in particular for those in $(0,1)$. Therefore, by the previous question, A is self-adjoint.

Solution 12.3.20. Since T is hyponormal, we have (cf. Exercise 4.3.36)

$$T^*T - TT^* = 2i(AB - BA) = 2i[AB - (AB)^*] \geq 0.$$

Since A and B are self-adjoint, we may write

$$(AB)A = A(BA) = A(AB)^*.$$

As AB (and $(AB)^{**}$!) is hyponormal, Theorem 12.1.14 gives

$$(AB)^*A = A(AB)^{**} = A(AB).$$

Hence

$$[AB + (AB)^*]A = A[AB + (AB)^*]$$

and

$$[AB - (AB)^*]A = A[(AB)^* - AB].$$

Thus, by setting $C = i[AB - (AB)^*]$ (which is positive!)

$$CA = -AC$$

which clearly implies that

$$C^2A = AC^2.$$

Since $C \geq 0$, we obtain by Theorem 5.1.28 that $CA = AC$, that is, $(AB)^* - AB$ commutes with A, and so A commutes with AB as we already know that $(AB)^* + AB$ commutes with A.

Therefore,

$$A(AB - BA) = (AB - BA)A = 0.$$

Theorem 8.1.10 now tells us that $\sigma(AB - BA) = \{0\}$. Finally, since $AB - BA$ is skew symmetric, it is normal and it follows that

$$AB - BA = 0,$$

proving the normality of T.

Solution 12.3.21. Let S be the shift operator on ℓ^2. Then S is hyponormal and $S^2 = S^2$, that is, $SS = SS$. If the Fuglede-Putnam Theorem held for hyponormal operators, then we would have $SS^* = S^*S$, i.e. we would obtain $SS^* = I$ which is a contradiction!

Solution 12.3.22. WLOG we may assume that $\|A\| = 1$.

(1) We claim that $A^*Ax = x$ iff $\|Ax\| = \|x\|$: If $A^*Ax = x$, then

$$\|Ax\|^2 =< A^*Ax, x >=< x, x >= \|x\|^2.$$

Conversely, if $\|Ax\| = \|x\|$, then

$$\begin{aligned}\|A^*Ax - x\|^2 &= \|A^*Ax\|^2 - 2\operatorname{Re} < A^*Ax, x > + \|x\|^2 \\ &\leq \|Ax\|^2 - 2\|Ax\|^2 + \|x\|^2 \\ &\leq 0.\end{aligned}$$

Therefore, $A^*Ax = x$ as needed. Using what we have just shown, we can re-write M as follows:

$$M = \{x \in H : A^*Ax = x\} = \ker(A^*A - I)$$

which is clearly a closed subspace since $A^*A - I \in B(H)$.

(2) We need to show that $AM \subset M$. Let $x \in M$. Then

$$\|A(Ax)\| \leq \|A\|\|Ax\| = \|Ax\| = \|x\| = \|A^*Ax\| \leq \|A(Ax)\|,$$

and hence $\|A(Ax)\| = \|Ax\| = \|A\|\|Ax\|$, i.e. $Ax \in M$.

(3) Since A is normal, $AA^* = A^*A$. Also, A and A^* are both hyponormal. Hence $M = \ker(A^*A - I) = \ker(AA^* - I)$ is invariant for A and A^*. Accordingly, M is reducing for A.

12.4. Hints/Answers to Tests

SOLUTION 12.4.1. Since A is invertible, it follows that AA^* too is invertible. Hence $0 \leq AA^* \leq A^*A$ with Theorem 5.1.39 give

$$A^{-1}(A^{-1})^* = (A^*A)^{-1} \leq (AA^*)^{-1} = (A^{-1})^*A^{-1}...$$

SOLUTION 12.4.2. The solution is almost an identical copy of the proof of Theorem 6.1.22 which appeared in Exercise 6.3.14. The only difference lies in using Exercise 12.3.7 and Exercise 12.3.11...

SOLUTION 12.4.3. No! If it were, we would have (wouldn't we?)

$$\frac{2}{\pi} = \|V\| = r(V) = 0!$$

REMARK. Another proof may be found in Exercise 12.5.1.

SOLUTION 12.4.4. Just routine...

SOLUTION 12.4.5. Yes, it does! Observe that if $B \geq 0$ with P as its unique positive square root, then

$$\sigma(AB) = \sigma(BA) = \sigma(PAP)...$$

CHAPTER 13

Similarities of Operators

13.2. Solutions to Exercises

SOLUTION 13.2.1.

(1) Since $\sigma(B)$ is compact in $\mathbb{C}$, to show that $\sigma(B) \subset \mathbb{R}$, it then suffices to show that its boundary $\partial\sigma(B)$ is a subset of $\mathbb{R}$. Since the latter is a subset of $\sigma_a(B)$, it is sufficient to show that $\sigma_a(B) \subset \mathbb{R}$. So let $\lambda \in \sigma_a(B)$. Then there exists a sequence of unit vectors (x_n) in H such that $\|(B-\lambda)x_n\| \to 0$. Now we have

$$|(\overline{\lambda}-\lambda) < A^{-1}x_n, x_n > | = |(\overline{\lambda}-\lambda) < A^{-1}x_n, x_n > + < (B^*-B^*)A^{-1}x_n, x_n > |$$

which is equal to

$$| < (B^* - \lambda)A^{-1}x_n, x_n > - < (B^* - \overline{\lambda})A^{-1}x_n, x_n > |$$

and the latter, using the Cauchy-Schwarz Inequality (and the fact that $\|x_n\| = 1$ for all n), is less than

$$\|(B^* - \lambda)A^{-1}x_n\| + \|A^{-1}\|\|(B^* - \overline{\lambda})x_n\|.$$

Since $A^{-1}BA = B^*$ and $B - \lambda$ is hyponormal, we obtain that

$$\begin{aligned}|(\overline{\lambda} - \lambda) < A^{-1}x_n, x_n > | &\leq \|A^{-1}(B - \lambda)x_n\| + \|A^{-1}\|\|(B^* - \overline{\lambda})x_n\| \\ &\leq \|A^{-1}(B - \lambda)x_n\| + \|A^{-1}\|\|(B - \lambda)x_n\| \\ &\leq \|A^{-1}\|\|(B - \lambda)x_n\| + \|A^{-1}\|\|(B - \lambda)x_n\| \\ &= 2\|A^{-1}\|\|(B - \lambda)x_n\|.\end{aligned}$$

Now, the hypothesis $0 \notin \overline{W(A)}$ entails that $0 \notin \overline{W(A^{-1})}$ (by Test 8.4.5). Hence, by passing to the limit in the previous inequalities, we obtain

$$\lim_{n\to\infty} |(\overline{\lambda} - \lambda) < A^{-1}x_n, x_n > | = 0$$

forcing $\lambda = \overline{\lambda}$, i.e. $\lambda \in \mathbb{R}$, as required.

(2) This follows from the fact that a hyponormal operator with a real spectrum is self-adjoint (Corollary 12.1.10).

SOLUTION 13.2.2.

(1) Let $N = A+iB$ where A and B are two commuting self-adjoint operators. Hence $e^A e^{iB} = e^{iB} e^A$. Consequently,

$$\begin{aligned} e^S e^N = e^N e^S &\Longleftrightarrow e^S e^A e^{iB} = e^A e^{iB} e^S \\ &\Longleftrightarrow e^S e^A e^{iB} = e^{iB} e^A e^S \\ &\Longleftrightarrow e^S e^A e^{iB} = e^{iB} (e^S e^A)^*. \end{aligned}$$

(2) Since B is self-adjoint, e^{iB} is unitary. It is also cramped by the spectral hypothesis on B. Now, Theorem 13.1.3 implies that $e^S e^A$ is self-adjoint, i.e.

$$e^S e^A = e^A e^S.$$

Exercise 11.3.13 then gives us $AS = SA$.

(3) It only remains to show that $BS = SB$. Since $e^S e^A = e^A e^S$, we immediately obtain

$$e^S e^N = e^N e^S \Longrightarrow e^S e^A e^{iB} = e^A e^{iB} e^S \text{ or } e^A e^S e^{iB} = e^A e^{iB} e^S$$

and so

$$e^S e^{iB} = e^{iB} e^S$$

by the invertibility of e^A.

Now, using Lemmas 11.1.53 & 11.1.54 we immediately see that

$$e^S e^{-B} = e^{-B} e^S \text{ or } e^S e^B = e^B e^S.$$

Exercise 11.3.13 then yields $BS = SB$ so that finally

$$SN = S(A+iB) = (A+iB)S = NS.$$

Remark. By inspecting the previous solution, we see that we may have taken $(-\frac{\pi}{2}, \frac{\pi}{2})$ in lieu of $(0, \pi)$ without any problem. Hence the same results hold with this new interval. Thus any self-adjoint operator (remember that its imaginary part then must vanish) obeys the given condition on the spectrum.

Solution 13.2.3.

(1) Since M is normal, e^M too is normal. Hence

$$e^M e^N = e^N e^M \Longrightarrow e^{M^*} e^N = e^N e^{M^*}$$

by the Fuglede Theorem because e^M is normal. Using again the normality of M

$$e^{M^*} e^M e^N = e^{M^*} e^N e^M \Longrightarrow e^{M^*} e^M e^N = e^N e^{M^*} e^M$$

or

$$e^{M^*+M} e^N = e^N e^{M^*+M}$$

Since $M^* + M$ is self-adjoint, Exercise 13.2.2 is applicable and yields

$$(M^* + M)N = N(M^* + M) \text{ or just } CN = NC.$$

(2) The previous implies that $N^*C = CN^*$ and hence

$$(N + N^*)C = C(N + N^*).$$

Therefore, we have

$$AC = CA \text{ and hence } BC = CB.$$

Doing the same work for N in lieu of M, very similar arguments and Exercise 13.2.2 all yield

$$AM = MA \text{ and hence } AD = DA.$$

(3) Going back to the equation $e^N e^M = e^M e^N$, by the commutativity of B and C and by that of A and D, we then obtain

$$e^A e^{iB} e^C e^{iD} = e^C e^{iD} e^A e^{iB} \Longleftrightarrow e^A e^C e^{iB} e^{iD} = e^C e^A e^{iD} e^{iB}.$$

Since A and C commute and since $e^A e^C$ is invertible, we are left with

$$e^{iB} e^{iD} = e^{iD} e^{iB}.$$

Lemmas 11.1.53 & 11.1.54 then yield

$$e^{-B} e^{-D} = e^{-D} e^{-B} \text{ or } e^B e^D = e^D e^B$$

which leads to $BD = DB$ by Exercise 11.3.13. Therefore,

$$BM = MB.$$

(4) Finally, we have

$$NM = (A+iB)M = AM+iBM = MA+iMB = M(A+iB) = MN,$$

completing the proof.

SOLUTION 13.2.4.

(1) First, A is clearly normal. It is easy to see that

$$AB = \pi^2 \begin{pmatrix} 1 & * \\ * & * \end{pmatrix} \neq \pi^2 \begin{pmatrix} 2 & * \\ * & * \end{pmatrix} = BA.$$

(2) We have

$$\text{Im} A = \frac{A - A^*}{2i} = \begin{pmatrix} 0 & -i\pi \\ i\pi & 0 \end{pmatrix}$$

and hence $\sigma(\text{Im}\, A) = \{\pi, -\pi\}$ (which signifies that $\sigma(\text{Im}\, A)$ cannot be inside an open interval of length equals to π).

(3) By computing integer powers of A we may easily check that
$$e^A = \begin{pmatrix} \cos\pi & \sin\pi \\ -\sin\pi & \cos\pi \end{pmatrix} = \begin{pmatrix} -1 & 0 \\ 0 & -1 \end{pmatrix} = -I.$$
Similarly, we find that $e^B = -I$.

Solution 13.2.5.

(1) The normality of N implies that of e^N, and so by the Fuglede Theorem
$$Ae^N = e^N A \Longleftrightarrow Ae^{N^*} = e^{N^*}A$$
or
$$A^*e^N = e^N A^*.$$
Hence
$$(A+A^*)e^N = e^N(A+A^*) \text{ or } (\mathrm{Re}A)e^N = e^N(\mathrm{Re}A)$$
so that
$$e^{\mathrm{Re}A}e^N = e^N e^{\mathrm{Re}A}.$$
But, ReA is self-adjoint, so Exercise 13.2.2 applies and then gives
$$(\mathrm{Re}A)N = N(\mathrm{Re}A).$$

(2) Similarly, we find that
$$(\mathrm{Im}A)e^N = e^N(\mathrm{Im}A)$$
and as ImA is self-adjoint, similar arguments to those applied before yield
$$(\mathrm{Im}A)N = N(\mathrm{Im}A).$$
Therefore clearly $AN = NA$.

(3) The hypothesis $\sigma(\mathrm{Im}N) \subset (0,\pi)$ cannot merely be dropped. Take N to be the operator A in Exercise 13.2.4, and take A (of our exercise!) to be any operator which does not commute with N. Then
$$Ae^N = -A = e^N A \text{ but } AN \neq NA.$$

Solution 13.2.6. We have
$$e^{A+B}e^A = e^Be^Ae^A = e^Ae^Be^A = e^Ae^{A+B}.$$
Since $A+B$ is normal and $\sigma(\mathrm{Im}(A+B)) \subset (0,\pi)$, Exercise 13.2.5 gives
$$(A+B)e^A = e^A(A+B) \text{ or just } Be^A = e^AB$$
for A commutes with e^A. Now, right multiplying both sides of the previous equation by e^B leads to
$$Be^Ae^B = e^ABe^B = e^Ae^BB$$

or

$$Be^{A+B} = e^{A+B}B.$$

Applying Exercise 13.2.5 once more, we see that

$$B(A+B) = (A+B)B \text{ or } AB = BA,$$

establishing the result.

SOLUTION 13.2.7.

(1) We need only prove the implication "$\Longrightarrow$". It is plain that $A + A^*$ is self-adjoint. Hence the remark below the solution of Exercise 13.2.2 combined with Exercise 13.2.6 give us

$$AA^* = A^*A.$$

(2) Obviously

$$Be^B = e^BB.$$

Hence

$$Be^A = e^AB.$$

Exercise 13.2.5 then does the remaining job, i.e. it gives the commutativity of A and B.

(3) By the example of Exercise 13.2.4,

$$e^A = e^B = -I,$$

while $AB \neq BA$. This shows (again!) the importance of the assumption $\sigma(\text{Im}A) \subset (0,\pi)$.

SOLUTION 13.2.8. Since $0 \notin \overline{W(A)}$, A is invertible. So, let $A = UP$ be its polar decomposition. Remember that $P = (A^*A)^{\frac{1}{2}}$ is positive and $U = A(A^*A)^{-\frac{1}{2}}$ is unitary. By Theorem 13.1.2, U is even cramped.

Since $A^*AB = BA^*A$, we have

$$P^2B = BP^2 \text{ or } PB = BP.$$

Hence we may write

$$\begin{aligned}
&A^{-1}B^*A = B\\
\Longleftrightarrow &P^{-1}U^*B^*UP = B\\
\Longleftrightarrow &U^*B^*U = PBP^{-1}\\
\Longleftrightarrow &U^*B^*U = BPP^{-1}\\
\Longleftrightarrow &U^*B^*U = B\\
\Longleftrightarrow &B^* = UBU^*.
\end{aligned}$$

As U is cramped, Theorem 13.1.3 applies and yields the self-adjointness of B, establishing the result.

Solution 13.2.9. Let $A = UP$ where U is unitary and P is positive (where $P = (A^*A)^{\frac{1}{2}}$). We then have

$$BA^*A = A^*AB \Longrightarrow A^*AB^{-1} = B^{-1}A^*A,$$

hence

$$P^2B^{-1} = B^{-1}P^2 \text{ so that } PB^{-1} = B^{-1}P.$$

Therefore,

$$\begin{aligned} &A^{-1}B^*A = B^{-1} \\ \Longleftrightarrow &P^{-1}U^*B^*UP = B^{-1} \\ \Longleftrightarrow &U^*B^*U = PB^{-1}P^{-1} \\ \Longleftrightarrow &U^*B^*U = B^{-1}PP^{-1} \\ \Longleftrightarrow &U^*B^*U = B^{-1} \\ \Longleftrightarrow &B^* = UB^{-1}U^*. \end{aligned}$$

Since $0 \notin \overline{W(A)}$, U is cramped so that Theorem 13.1.5 applies and gives us $B^* = B^{-1}$, completing the proof.

Solution 13.2.10.

(1) Since A is normal, so is $-A$. They clearly commute so that Theorem 13.1.6 guarantees that $A = -A$, i.e. $A = 0$. If T is normal, then the Fuglede-Putnam Theorem yields

$$TA = -AT \Longrightarrow -AT^* = T^*A \Longrightarrow -TA^* = A^*T.$$

Hence

$$T(A - A^*) = -(A - A^*)T.$$

Since $A - A^*$ is normal, Theorem 13.1.6 ensures the self-adjointness of A.

(2) As $TA = A^*T$ and $TA^* = AT$, then

$$T(A - A^*) = -(A - A^*)T.$$

The previous question tells us that $A = A^*$.

(3) Since $TA = A^*T$, whether A is normal or T is unitary we always have (why?)

$$TA^* = AT.$$

Now apply the foregoing result.

Solution 13.2.11.

(1) Since $TA = BT$, and A and B are normal, the Fuglede-Putnam Theorem gives us $TA^* = B^*T$. Taking adjoints in the previous equation and by the self-adjointness of T we obtain $AT = TB$. Thus,

$$T(A-B) = -(A-B)T.$$

Since T is self-adjoint and $0 \not\in W(T)$, e.g. Corollary 13.1.10 gives us the desired result, i.e. $A = B$.

(2) Since A is normal, A^* is normal too. Hence, the previous question gives $A = A^*$, that is, A is self-adjoint.

(3) Since A and B are self-adjoint, we may write

$$B(AB) = BAB = (AB)^*B.$$

Since AB is assumed to be normal and $0 \not\in W(B)$, the previous question yields the self-adjointness of AB.

Remark. We would have had reached the same conclusion had we assumed that $0 \not\in W(A)$.

Solution 13.2.12.

(1) Let

$$A = \begin{pmatrix} 2 & 0 \\ 0 & 1 \end{pmatrix},\ B = \begin{pmatrix} 1 & 0 \\ 0 & 2 \end{pmatrix} \text{ and } T = \begin{pmatrix} 0 & 1 \\ 1 & 0 \end{pmatrix}.$$

Then A and B are normal (even self-adjoint!) and $A \neq B$. We may also easily check that A commutes with B and that $TA = BT$. But $0 \in W(A)$ because if $X = \begin{pmatrix} 0 \\ 1 \end{pmatrix}$, then $\|X\| = 1$ and

$$< TX, X >= 1 \times 0 + 0 \times 1 = 0.$$

(2) Let

$$A = \begin{pmatrix} 1 & i \\ -i & 2 \end{pmatrix},\ B = \begin{pmatrix} 1 & 1 \\ 1 & 2 \end{pmatrix} \text{ and } T = \begin{pmatrix} 1 & 0 \\ 0 & i \end{pmatrix}$$

(i being the usual complex number). Then we observe that A and B are self-adjoint. The reader may also check that $TA = BT$ and that A does not commute with B. If 0 were in $W(T)$, then we would have for $\|(x,y)\| = 1$

$$< T\begin{pmatrix} x \\ y \end{pmatrix}, \begin{pmatrix} x \\ y \end{pmatrix} >= |x|^2 + i|y|^2 = 0,$$

which would force $x = y = 0$, and this is not consistent with $\|(x,y)\| = 1$!

Solution 13.2.13. Obviously,

$$AA^2 = A^2A.$$

Since A^2 is normal, the Fuglede Theorem yields

$$AA^{2*} = A^{2*}A \text{ (or } AA^{*2} = A^{*2}A) \text{ or } A^*A^2 = A^2A^*.$$

Hence

$$A^*A^2A^* = A^2A^*A^* = AAA^{*2} = AA^{*2}A.$$

Therefore,

$$(A^*A)(AA^*) = (AA^*)(A^*A),$$

that is A^*A and AA^* commute. Everything now is ready to apply Corollary 13.1.8. The latter gives immediately the normality of A, establishing the result.

Solution 13.2.14.

(1) First, we have

$$UA = A^*U \Longrightarrow A^*U^* = U^*A.$$

We may also write

$$\begin{aligned} U^2AU^{*2} =& U(UAU^*)U^* \\ =& UA^*U^* \\ =& UU^*A \\ =& A, \end{aligned}$$

giving $U^2A = AU^2$ or $AU^{*2} = U^{*2}A$ or $U^2A^* = A^*U^2$.

(2) We have

$$\begin{aligned} AU =& U^*A^*UU \\ =& U^*A^*U^2 \\ =& U^*UUA^* \\ =& UA^*. \end{aligned}$$

Hence, $U^*A^* = AU^*$.

(3) Set $S = \frac{1}{2}(U + U^*)$. Then $S > 0$. First, we will show that $SAA^* = A^*AS$. We have

$$UAA^* = A^*UA^* = A^*AU$$

and

$$U^*AA^* = A^*U^*A^* = A^*AU^*.$$

Hence

$$\begin{aligned}SAA^* &= \frac{1}{2}(U+U^*)AA^*\\ &= \frac{1}{2}UAA^* + \frac{1}{2}U^*AA^*\\ &= \frac{1}{2}A^*AU + \frac{1}{2}A^*AU^*\\ &= A^*AS.\end{aligned}$$

Since S is self-adjoint and $0 \notin W(S)$, we obtain by Exercise 13.2.11 that $AA^* = A^*A$, that is, A is normal.

(4) Since A is normal, by Exercise 13.2.10 we finally get

$$UA = A^*U \Longrightarrow A = A^*$$

as $0 \notin \overline{W(U)}$.

13.3. Hints/Answers to Tests

Solution 13.3.1. If $A > 0$, write

$$ABA = A(AB)^* = ABA.$$

Then apply Corollary 13.1.7 to get $AB = (AB)^*$...

Solution 13.3.2. Easy! Just write

$$A(A^*A) = (AA^*)A...$$

Bibliography

1. M. B. Abrahamse, Commuting subnormal operators, *Illinois J. Math.*, **22/1** (1978), 171-176.
2. N. I. Akhiezer, I. M. Glazman, Theory of linear operators in Hilbert space. Translated from the Russian and with a preface by Merlynd Nestell. Reprint of the 1961 and 1963 translations. Two volumes bound as one. *Dover Publications*, Inc., New York, 1993.
3. C. D. Aliprantis, K. C. Border, *Infinite dimensional analysis. A hitchhiker's guide*, Third edition. Springer, Berlin, 2006.
4. C. D. Aliprantis, O. Burkinshaw, *Problems in real analysis*, A workbook with solutions. Second edition. Academic Press, Inc., San Diego, CA, 1999.
5. G. R. Allan, Power-bounded elements and radical Banach algebras, Linear operators (Warsaw, 1994), 9-16, Banach Center Publ., **38**, Polish Acad. Sci. Inst. Math., Warsaw, 1997.
6. T. Ando, On Hyponormal Operators, *Proc. Amer. Math. Soc.*, **14** (1963) 290-291.
7. T. Ando, Concavity of certain maps on positive definite matrices and applications to Hadamard products, *Linear Algebra Appl.*, **26** (1979), 203-241.
8. W. Arendt, F. Räbiger, A. Sourour, Spectral properties of the operator equation $AX+XB=Y$, *Quart. J. Math. Oxford Ser. (2)*, **45/178** (1994) 133-149.
9. W. Arveson, An invitation to C^*-algebras. Graduate Texts in Mathematics, No. **39**. *Springer-Verlag*, New York-Heidelberg, 1976.
10. S. Axler, *Linear algebra done right.* Third edition. Undergraduate Texts in Mathematics. Springer, Cham, 2015.
11. G. Bachman, Elements of abstract harmonic analysis, with the assistance of Lawrence Narici, *Academic Press*, New York-London 1964. xi+256 pp.
12. G. Bachman, L. Narici, Functional analysis. Reprint of the 1966 original. *Dover Publications, Inc.*, Mineola, NY, 2000.
13. S. Banach, Théorie des opérations linéaires (French) [Theory of linear operators] Reprint of the 1932 original. *Éditions Jacques Gabay*, Sceaux, 1993.
14. G. de Barra, J. R. Giles, B. Sims, On the numerical range of compact operators on Hilbert spaces, *J. London Math. Soc.*, **2/5** (1972) 704-706.
15. M. Barraa, M. Boumazghour, Numerical range submultiplicity, *Linear Multilinear Algebra.* DOI:10.1080/03081087.2015.1005567
16. B. Beauzamy, Un opérateur sans sous-espace invariant: simplification de l'exemple de P. Enflo. (French) [An operator with no invariant subspace: simplification of the example of P. Enflo]. *Integral Equations Operator Theory,* 8/3 (1985) 314-384.
17. W. A. Beck, C. R. Putnam, A Note on Normal Operators and Their Adjoints, *J. London Math. Soc.*, **31** (1956), 213-216.

18. W. Beckner, Inequalities in Fourier analysis on $\mathbb{R}^n$, *Proc. Nat. Acad. Sci. U.S.A.*, **72** (1975), 638-641.
19. A. Benali, M. H. Mortad, Generalizations of Kaplansky's theorem involving unbounded linear operators, *Bull. Pol. Acad. Sci. Math.*, **62/2** (2014), 181-186.
20. S. K. Berberian, Note on a Theorem of Fuglede and Putnam, *Proc. Amer. Math. Soc.,* **10** (1959) 175-182.
21. S. K. Berberian, A note on operators unitarily equivalent to their adjoints, *J. London Math. Soc.*, **37** (1962) 403-404.
22. S. K. Berberian, A note on hyponormal operators, Pacific J. Math., **12** (1962) 1171-1175.
23. S. K. Berberian, The spectral mapping theorem for a Hermitian operator, *Amer. Math. Monthly,* **70** (1963) 1049-1051.
24. S. K. Berberian, The numerical range of a normal operator, *Duke Math. J.*, **31**, (1964) 479-483.
25. S. K. Berberian, Introduction to Hilbert space, Reprinting of the 1961 original. With an addendum to the original. *Chelsea Publishing Co.*, New York, 1976.
26. S. K. Berberian, Extensions of a theorem of Fuglede and Putnam, *Proc. Amer. Math. Soc.*, **71/1** (1978) 113-114.
27. R. Bhatia, Matrix analysis, Graduate Texts in Mathematics, **169**. *Springer-Verlag,* New York, 1997.
28. R. Bhatia, P. Rosenthal, How and why to solve the operator equation $AX - XB = Y$. *Bull. London Math. Soc.*, **29/1** (1997) 1-21.
29. A. M. Bikchentaev, On invertibility of some operator sums, *Lobachevskii J. Math.*, **33/3** (2012), 216-222.
30. A. M. Bikchentaev, Tripotents in algebras: invertibility and hyponormality, *Lobachevskii J. Math.*, **35/3** (2014) 281-285.
31. M. Sh. Birman, M. Z. Solomjak, Spectral theory of selfadjoint operators in Hilbert space. Translated from the 1980 Russian original by S. Khrushchëv and V. Peller. Mathematics and its Applications (Soviet Series). *D. Reidel Publishing Co.*, Dordrecht, 1987.
32. F. F. Bonsall, J. Duncan, Numerical ranges of operators on normed spaces and of elements of normed algebras. London Mathematical Society Lecture Note Series, 2 *Cambridge University Press,* London-New York 1971.
33. F. F. Bonsall, J. Duncan, Complete normed algebras. Ergebnisse der Mathematik und ihrer Grenzgebiete, Band **80**. *Springer-Verlag,* New York-Heidelberg, 1973.
34. H. Brezis, Analyse fonctionnelle. (French) [Functional analysis] Théorie et applications. [Theory and applications] Collection Mathématiques Appliquées pour la Maîtrise. [Collection of Applied Mathematics for the Master's Degree] Masson, Paris, 1983.
35. J. A. Brooke, P. Busch, D. B. Pearson, *Commutativity up to a factor of bounded operators in complex Hilbert space*, R. Soc. Lond. Proc. Ser. A Math. Phys. Eng. Sci., **458/2017** (2002) 109-118.
36. S. L. Campbell, Linear operators for which T^*T and $T+T^*$ commute, *Pacific J. Math.* **61/1** (1975) 53-57.
37. A. Chaban, M. H. Mortad, *Global Space-Time L^p-Estimates for the Airy Operator on $L^2(\mathbb{R}^2)$ and Some Applications*, Glas. Mat. Ser. III, **47/67** (2012) 373-379.

38. A. Chaban, M. H. Mortad, *Exponentials of Bounded Normal Operators*, Colloq. Math., **133/2** (2013) 237-244.
39. I. Chalendar, J. R. Partington, *Modern approaches to the invariant-subspace problem*, Cambridge Tracts in Mathematics, **188**, Cambridge University Press, Cambridge, 2011.
40. J. Charles, M. Mbekhta, H. Queffélec, Analyse fonctionnelle et théorie des opérateurs (French), Dunod, Paris, 2010.
41. Ch. Chellali, M. H. Mortad, Commutativity up to a Factor for Bounded and Unbounded Operators, *J. Math. Anal. Appl.*, **419/1** (2014), 114-122.
42. W. Cheney, Analysis for applied mathematics. Graduate Texts in Mathematics, **208**. *Springer-Verlag*, New York, 2001.
43. P. R. Chernoff, A Semibounded Closed Symmetric Operator Whose Square Has Trivial Domain, *Proc. Amer. Math. Soc.*, **89/2** (1983) 289-290.
44. M. Cho, J. I. Lee, T. Yamazaki, *On the operator equation $AB = zBA$*, Sci. Math. Jpn., **69/2** (2009), 257-263.
45. W. F. Chuan, *The unitary equivalence of compact operators*, Glasgow Math. J., **26/2** (1985), 145-149.
46. P. J. Cohen, A counterexample to the closed graph theorem for bilinear maps, *J. Functional Analysis*, **16** (1974) 235-240.
47. J. B. Conway, Functions of one complex variable, Second edition. Graduate Texts in Mathematics, **11**. *Springer-Verlag*, New York-Berlin, 1978. xiii+317 pp.
48. J. B. Conway, A Course in Functional Analysis, *Springer*, 1990 (2nd edition).
49. J. B. Conway, A course in operator theory, Graduate Studies in Mathematics, **21**. *American Mathematical Society*, Providence, RI, 2000.
50. C. Costara, D. Popa, Exercises in Functional Analysis, Kluwer Texts in the Mathematical Sciences, **26**, *Kluwer Academic Publishers Group, Dordrecht*, 2003.
51. E. B. Davies, *Quantum theory of open systems*, Academic Press, London-New York, 1976.
52. E. B. Davies, *Linear operators and their spectra*, Cambridge Studies in Advanced Mathematics, **106**. Cambridge University Press, Cambridge, 2007.
53. D. Deckard, C. Pearcy, Another class of invertible operators without square roots, Proc. Amer. Math. Soc., **14** (1963) 445-449.
54. S. Dehimi and M. H. Mortad, *Bounded and Unbounded Operators Similar to Their Adjoints*, Bull. Korean Math. Soc., **54/1 (2017)** 215-223.
55. S. Dehimi and M. H. Mortad, *Right (Or Left) Invertibility of Bounded and Unbounded Operators and Applications to the Spectrum of Products*, Complex Anal. Oper. Theory, **12/3** (2018) 589-597.
56. S. Dehimi, M. H. Mortad, *Generalizations of Reid Inequality*, Mathematica Slovaca (to appear).
57. A. Devinatz, A. E. Nussbaum, J. von Neumann, *On the Permutability of Self-adjoint Operators*, Ann. of Math. (2), **62** (1955) 199-203.
58. T. Diagana, Schrödinger Operators with a Singular Potential, *Int. J. Math. Math. Sci.*, **29/6** (2002) 371-373.
59. T. Diagana, A Generalization Related to Schrödinger Operators with a Singular Potential, *Int. J. Math. Math. Sci.*, **29/10** (2002) 609-611.
60. W. F. Donoghue, Jr., The lattice of invariant subspaces of a completely continuous quasi-nilpotent transformation, *Pacific J. Math.* **7** (1957) 1031-1035.

61. R. G. Douglas, On majorization, factorization, and range inclusion of operators on Hilbert space, *Proc. Amer. Math. Soc.*, **17** (1966) 413-415.
62. N. Dunford, Spectral operators. *Pacific J. Math.*, **4** (1954) 321-354.
63. N. Dunford, J. T. Schwartz, Linear operators. Part I. General theory. With the assistance of William G. Bade and Robert G. Bartle. Reprint of the 1958 original. Wiley Classics Library. A Wiley-Interscience Publication. *John Wiley & Sons*, Inc., New York, 1988.
64. N. Dunford, J. T. Schwartz, Linear operators. Part II. Spectral theory. Self-adjoint operators in Hilbert space. With the assistance of William G. Bade and Robert G. Bartle. Reprint of the 1963 original. Wiley Classics Library. A Wiley-Interscience Publication. *John Wiley & Sons*, Inc., New York, 1988.
65. J. Duoandikoetxea, *Fourier Analysis*, American Mathematical Society, G.S.M. Vol. 29, 2001.
66. T. Eisner, A "typical" contraction is unitary, *Enseign. Math.* (2) **56/3-4** (2010) 403-410.
67. M. R. Embry, Conditions implying normality in Hilbert space, *Pacific J. Math.*, **18** (1966) 457-460.
68. M. R. Embry, Similarities Involving Normal Operators on Hilbert Space, *Pacific J. Math.*, **35** (1970) 331-336.
69. M. R. Embry, A connection between commutativity and separation of spectra of operators. *Acta Sci. Math. (Szeged)*, **32** (1971) 235-237.
70. P. Enflo, A counterexample to the approximation problem in Banach spaces, *Acta Math.*, **130** (1973) 309-317.
71. P. Enflo, On the invariant subspace problem for Banach spaces, *Acta Math.*, **158/3-4** (1987), 213-313.
72. P. Fan, J. Stampfli, On the density of hyponormal operators, *Israel J. Math.*, **45/2-3**, (1983) 255-256.
73. C. Foiaş, J. P. Williams, Some remarks on the Volterra operator, *Proc. Amer. Math. Soc.*, **31**, (1972) 177-184
74. G. B. Folland, A course in abstract harmonic analysis. Second edition. Textbooks in Mathematics. *CRC Press, Boca Raton*, FL, 2016.
75. C. K. Fong, V. I. Istrăţescu, Some characterizations of Hermitian operators and related classes of operators, *Proc. Amer. Math. Soc.*, **76**, (1979) 107-112.
76. C. K. Fong, S. K. Tsui, A note on positive operators, *J. Operator Theory*, **5/1**, (1981) 73-76.
77. F. G. Friedlander, Introduction to the theory of distributions. Second edition. With additional material by M. Joshi. *Cambridge University Press*, Cambridge, 1998.
78. B. Fuglede, A Commutativity Theorem for Normal Operators, *Proc. Nati. Acad. Sci.*, **36** (1950) 35-40.
79. T. Furuta, A simplified proof of Heinz inequality and scrutiny of its equality, *Proc. Amer. Math. Soc.*, **97/4** (1986) 751-753.
80. T. Furuta, $A \geq B \geq 0$ assures $(B^r A^p B^r)^{1/q} \geq B^{(p+2r)/q}$ for $r \geq 0$, $p \geq 0$, $q \geq 1$ with $(1+2r)q \geq p+2r$, *Proc. Amer. Math. Soc.*, **101/1** (1987) 85-88.
81. T. Furuta, *Invitation to Linear Operators: From Matrices to Bounded Linear Operators on a Hilbert Space*, Taylor & Francis, Ltd., London, 2001.
82. I. Gelfand, Normierte Ringe (German), Rec. Math. [Mat. Sbornik] N. S., **51/9** (1941) 3-24.

83. I. Gohberg, S. Goldberg, Basic operator theory. Reprint of the 1981 original. *Birkhäuser Boston*, Inc., Boston, MA, 2001.
84. I. Gohberg, S. Goldberg, M. A. Kaashoek, *Basic classes of linear operators*. Birkhäuser Verlag, Basel, 2003.
85. L. Golinskii, V. Totik, Orthogonal polynomials: from Jacobi to Simon. Spectral theory and mathematical physics: a Festschrift in honor of Barry Simon's 60th birthday, 821-874, Proc. Sympos. Pure Math., **76**, Part **2**, *Amer. Math. Soc.*, Providence, RI, 2007.
86. R. Grone, C. R. Johnson, E. M. Sa, H. Wolkowicz, Normal matrices, *Linear Algebra Appl.*, **87** (1987), 213-225.
87. K. Gustafson, M. H. Mortad, *Conditions Implying Commutativity of Unbounded Self-adjoint Operators and Related Topics*, J. Operator Theory, **76/1** (2016) 159-169.
88. K. Gustafson, D. K. M. Rao, *Numerical range. The field of values of linear operators and matrices*, Universitext. Springer-Verlag, New York, 1997.
89. S. J. Gustafson, I. M. Sigal, Mathematical concepts of quantum mechanics. Second edition. Universitext. *Springer*, Heidelberg, 2011.
90. B. C. Hall, Quantum theory for mathematicians. Graduate Texts in Mathematics, **267**. *Springer*, New York, 2013.
91. P. R. Halmos, Commutativity and spectral properties of normal operators. *Acta Sci. Math. Szeged*, **12**, (1950). Leopoldo Fejér Frederico Riesz LXX annos natis dedicatus, Pars B, 153-156.
92. P. R. Halmos, What does the spectral theorem say?, *Amer. Math. Monthly*, **70** (1963) 241-247.
93. P. R. Halmos, Ten problems in Hilbert space, Bull. Amer. Math. Soc., **76** (1970) 887-933.
94. P. R. Halmos, *A Hilbert Space Problem Book*, Springer, 1982 (2nd edition).
95. P. R. Halmos, *Linear algebra problem book*, The Dolciani Mathematical Expositions, **16**. Mathematical Association of America, Washington, DC, 1995.
96. P. R. Halmos, Introduction to Hilbert space and the theory of spectral multiplicity. Reprint of the second (1957) edition. *AMS Chelsea Publishing*, Providence, RI, 1998.
97. P. R. Halmos, G. Lumer, J. J. Schäffer, Square roots of operators, *Proc. Amer. Math. Soc.*, **4** (1953) 142-149.
98. F. Hansen, An operator inequality, *Math. Ann.*, **246/3** (1979/80) 249-250.
99. F. Hansen, G. K. Pedersen, Jensen's inequality for operators and Löwner's theorem, *Math. Ann.*, **258/3** (1981/82) 229-241.
100. V. Hardt, A. Konstantinov, R. Mennicken, On the spectrum of the product of closed operators, *Math. Nachr.*, **215**, (2000) 91-102.
101. G. H. Hardy, J. E. Littlewood, G. Pólya, Inequalities. 2d ed. *Cambridge, at the University Press*, 1952.
102. S. Hassi, Z. Sebestyén, H. S. V. de Snoo, On the nonnegativity of operator products, *Acta Math. Hungar.*, **109/1-2**, (2005) 1-14.
103. E. Hille, On roots and logarithms of elements of a complex Banach algebra, *Math. Ann.*, **136** (1958) 46-57.
104. E. Hille, R. S. Phillips, Functional analysis and semi-groups, revised. American Mathematical Society Colloquium Publications, vol. **31**. *American Mathematical Society, Providence, R. I.*, 1957.

105. F. Hirsch, G. Lacombe, *Elements of functional analysis*, Translated from the 1997 French original by Silvio Levy. Graduate Texts in Mathematics, **192**. Springer-Verlag, New York, 1999.
106. M. Hladnik, M. Omladič, *Spectrum of the Product of Operators*, Proc. Amer. Math. Soc., **102/2**, (1988) 300-302.
107. R. A. Horn, C. R. Johnson, Topics in matrix analysis. Corrected reprint of the 1991 original. *Cambridge University Press*, Cambridge, 1994.
108. Ch. Horowitz, An elementary counterexample to the open mapping principle for bilinear maps, *Proc. Amer. Math. Soc.*, **53 /2** (1975) 293-294.
109. T. Ito, T. K. Wong, Subnormality and Quasinormality of Toeplitz Operators, *Proc. Amer. Math. Soc.*, **34**, (1972) 157-164.
110. Z. J. Jabłoński, Il B. Jung, J. Stochel, Unbounded quasinormal operators revisited. *Integral Equations Operator Theory*, **79/1** (2014) 135-149.
111. J. Janas, On unbounded hyponormal operators. II, *Integral Equations Operator Theory*, **15/3**, (1992) 470-478.
112. R. Kadison, J. R. Ringrose, Fundamentals of the theory of operator algebras, Vol. I. Elementary theory. Reprint of the 1983 original, G.S.M., **15**, American Mathematical Society, Providence, RI, 1997.
113. I. Kaplansky, Products of normal operators, *Duke Math. J.*, **20/2** (1953) 257-260.
114. T. Kato, Perturbation Theory for Linear Operators, *Springer*, 1980 (2nd edition).
115. J. L. Kelley, Decomposition and representation theorems in measure theory, Math. Ann., **163** (1966) 89-94.
116. F. Kittaneh, On generalized Fuglede-Putnam theorems of Hilbert-Schmidt type, *Proc. Amer. Math. Soc.*, **88/2** (1983) 293-298.
117. F. Kittaneh, On normality of operators, *Rev. Roumaine Math. Pures Appl.*, **29/8** (1984) 703-705.
118. F. Kittaneh, *Notes on some inequalities for Hilbert space operators*, Publ. Res. Inst. Math. Sci., 24/2 (1988), 283-293.
119. F. Kittaneh, Spectral radius inequalities for Hilbert space operators, *Proc. Amer. Math. Soc.*, **134/2** (2006), 385-390 (electronic).
120. F. Kittaneh, Norm inequalities for commutators of positive operators and applications, *Math. Z.*, **258/4** (2008), 845-849.
121. H. Kosaki, On Intersections of Domains of Unbounded Positive Operators, *Kyushu J. Math.*, **60/1** (2006) 3-25.
122. S. G. Kreĭn, Linear differential equations in Banach space. Translated from the Russian by J. M. Danskin. Translations of Mathematical Monographs, Vol. **29**. American Mathematical Society, Providence, R.I., 1971.
123. E. Kreyszig, Introductory functional analysis with applications. Wiley Classics Library. *John Wiley & Sons, Inc.*, New York, 1989.
124. C. S. Kubrusly, Hilbert space operators, A problem solving approach, *Birkhäuser* Boston, Inc., Boston, MA, 2003.
125. C. S. Kubrusly, The elements of operator theory, Second edition, *Birkhäuser/Springer*, New York, 2011.
126. S. H. Kulkarni, M. T. Nair, G. Ramesh, Some properties of unbounded operators with closed range, *Proc. Indian Acad. Sci. Math. Sci.*, **118/4** (2008) 613-625.

127. B. W. Levinger, The square root of a 2×2 matrix, *Math. Mag.*, **53/4** (1980) 222-224.
128. C.K. Li, Y.T. Poon, Spectrum, numerical range and Davis-Wielandt shell of a normal operator, *Glasg. Math. J.*, **51/1, (2009)** 91-100.
129. E. H. Lieb, M. Loss, Analysis. Second edition. Graduate Studies in Mathematics, **14**. *American Mathematical Society*, Providence, RI, 2001.
130. B. V. Limaye, Linear functional analysis for scientists and engineers. *Springer*, Singapore, 2016.
131. C.-S. Lin, Inequalities of Reid type and Furuta, *Proc. Amer. Math. Soc.*, **129/3** (2001) 855-859.
132. V. I. Lomonosov, Invariant subspaces of the family of operators that commute with a completely continuous operator (Russian). *Funkcional. Anal. i Priložen.*, **7/3** (1973) 55-56.
133. G. Lumer, M. Rosenblum, Linear operator equations. *Proc. Amer. Math. Soc.*, **10** (1959) 32-41.
134. I. J. Maddox, The norm of a linear functional, *Amer. Math. Monthly*, **96/5** (1989) 434-436.
135. A. Mansour. Résolution de deux types d'équations opératorielles et interactions. Équations aux dérivées partielles [math.AP]. Université de Lyon, 2016. Français (French). <NNT : 2016LYSE1151>. <tel-01409645>
136. R. A. Martínez-Avendaño, P. Rosenthal, An introduction to operators on the Hardy-Hilbert space. Graduate Texts in Mathematics, **237**. *Springer*, New York, 2007.
137. M. Mbekhta, Partial isometries and generalized inverses, *Acta Sci. Math. (Szeged)*, **70/3-4** (2004) 767-781.
138. R. Meise, D. Vogt, Introduction to Functional Analysis, Oxford G.T.M. **2**, *Oxford University Press* 1997.
139. A. Montes-Rodriguez, S. A. Shkarin, *New results on a classical operator*, Recent advances in operator-related function theory, 139-157, Contemp. Math., **393**, Amer. Math. Soc., Providence, RI, 2006.
140. R. L. Moore, D. D. Rogers, T. T. Trent, A note on intertwining M-hyponormal operators, *Proc. Amer. Math. Soc.*, **83/3** (1981) 514-516.
141. M. H. Mortad, Normal products of self-adjoint operators and self-adjointness of the perturbed wave operator on $L^2(\mathbb{R}^n)$. Thesis (Ph.D.)-The University of Edinburgh (United Kingdom). *ProQuest LLC, Ann Arbor, MI,* 2003.
142. M. H. Mortad, An Application of the Putnam-Fuglede Theorem to Normal Products of Self-adjoint Operators, *Proc. Amer. Math. Soc.*, **131/10**, (2003) 3135-3141.
143. M. H. Mortad, Self-adjointness of the Perturbed Wave Operator on $L^2(\mathbb{R}^n)$, $n \geq 2$, *Proc. Amer. Math. Soc.*, **133/2**, (2005) 455-464.
144. M. H. Mortad, On L^p-Estimates for the Time-dependent Schrödinger Operator on L^2, *J. Ineq. Pure Appl. Math.*, **8/3**, (2007) Art. 80, 8pp.
145. M. H. Mortad, Yet More Versions of The Fuglede-Putnam Theorem, *Glasg. Math. J.*, **51/3**, (2009) 473-480.
146. M. H. Mortad, *On a Beck-Putnam-Rehder Theorem*, Bull. Belg. Math. Soc. Simon Stevin, **17/4** (2010), 737-740.
147. M. H. Mortad, *Similarities Involving Unbounded Normal Operators*, Tsukuba J. Math., **34/1**, (2010) 129-136.

148. M. H. Mortad, *Exponentials of Normal Operators and Commutativity of Operators: A New Approach*, Colloq. Math., **125/1** (2011) 1-6.
149. M. H. Mortad, *Products and Sums of Bounded and Unbounded Normal Operators: Fuglede-Putnam Versus Embry*, Rev. Roumaine Math. Pures Appl., **56/3** (2011), 195-205.
150. M. H. Mortad, *An all-unbounded-operator version of the Fuglede-Putnam theorem*, Complex Anal. Oper. Theory, **6/6** (2012) 1269-1273.
151. M. H. Mortad, *On the Closedness, the Self-adjointness and the Normality of the Product of Two Unbounded Operators*, Demonstratio Math., **45/1** (2012), 161-167.
152. M. H. Mortad, *Commutativity of Unbounded Normal and Self-adjoint Operators and Applications*, Operators and Matrices, **8/2** (2014), 563-571.
153. M. H. Mortad, Introductory topology. Exercises and solutions. 2nd edition. (English). *Hackensack, NJ: World Scientific* (ISBN 978-981-3146-93-8/hbk; 978-981-3148-02-4/pbk). xvii, 356 p. (2017).
154. M. H. Mortad, *A Contribution to the Fong-Tsui Conjecture Related to Self-adjoint Operators*. arXiv:1208.4346.
155. M. H. Mortad, *On The Absolute Value of The Product and the Sum of Linear Operators*, Rend. Circ. Mat. Palermo, II. Ser (to appear). DOI: 10.1007/s12215-018-0356-8.
156. M. H. Mortad, *On the Invertibility of the Sum of Operators*. arXiv:1804.07288v1.
157. M. H. Mortad, *On the Existence of Normal Square and Nth Roots of Operators*. arXiv:1801.06884
158. B. Sz.-Nagy, *Perturbations des Transformations Linéaires Fermées (French)*, Acta Sci. Math. Szeged, **14** (1951) 125-137.
159. M. Naimark, On the Square of a Closed Symmetric Operator, *Dokl. Akad. Nauk SSSR*, **26** (1940) 866-870; ibid. **28** (1940), 207-208.
160. L. Narici, E. Beckenstein, Topological vector spaces. Second edition. Pure and Applied Mathematics (Boca Raton), 296. *CRC Press, Boca Raton*, FL, 2011.
161. J. von Neumann, Approximative properties of matrices of high finite order, *Portugaliae Math.*, **3** (1942) 1-62.
162. Y. Okazaki, *Boundedness of closed linear operator T satisfying $R(T) \subset D(T)$*, Proc. Japan Acad. Ser. A Math. Sci., **62/8** (1986) 294-296.
163. S. Ôta, *Closed linear operators with domain containing their range*, Proc. Edinburgh Math. Soc., (2) **27/2** (1984) 229-233.
164. F. C. Paliogiannis, *A Generalization of the Fuglede-Putnam Theorem to Unbounded Operators*, J. Oper., 2015, Art. ID 804353, 3 pp.
165. A. B. Patel, S. J. Bhatt, *On unbounded subnormal operators*, Proc. Indian Acad. Sci. Math. Sci., **99/1** (1989) 85-92.
166. A. B. Patel, P. B. Ramanujan, *On Sum and Product of Normal Operators*, Indian J. Pure Appl. Math., **12/10** (1981) 1213-1218.
167. G. K. Pedersen, *Some operator monotone functions*, Proc. Amer. Math. Soc., **36** (1972) 309-310.
168. C. R. Putnam, *On Normal Operators in Hilbert Space*, Amer. J. Math., **73** (1951) 357-362.
169. C. R. Putnam, *Commutation properties of Hilbert space operators and related topics*, Springer-Verlag, New York, 1967.

170. Qoqosz, http://math.stackexchange.com/questions/155899/norm-of-integral-operator-in-l-2.
171. H. Radjavi, P. Rosenthal, Hyperinvariant subspaces for spectral and n-normal operators, *Acta Sci. Math. (Szeged)*, **32** (1971) 121-126.
172. H. Radjavi, P. Rosenthal, Invariant subspaces for products of Hermitian operators, *Proc. Amer. Math. Soc.*, **43** (1974) 483-484.
173. H. Radjavi, P. Rosenthal, Invariant Subspaces, *Dover*, 2003 (2nd edition).
174. I. K. Rana, An introduction to measure and integration. Second edition. Graduate Studies in Mathematics, **45**. American Mathematical Society, Providence, RI, 2002.
175. C. J. Read, A solution to the invariant subspace problem, *Bull. London Math. Soc.*, **16/4** (1984) 337-401.
176. C. J. Read, A solution to the invariant subspace problem on the space ℓ^1, *Bull, London Math. Soc.*, **17/4** (1985) 305-317.
177. C. J. Read, Quasinilpotent operators and the invariant subspace problem, *J. London Math. Soc. (2)*, **56/3** (1997) 595-606.
178. M. Reed, B. Simon, Methods of Modern Mathematical Physics, Vol. **1**: Functional Analysis, *Academic Press*, 1972.
179. M. Reed, B. Simon, Methods of Modern Mathematical Physics, Vol. **2**: Fourier Analysis, Self-Adjointness, *Academic Press*, 1975.
180. W. Rehder, On the Adjoints of Normal Operators, *Arch. Math. (Basel)*, **37/2** (1981) 169-172.
181. W. T. Reid, *Symmetrizable completely continuous linear transformations in Hilbert space,* Duke Math. J., **18** (1951) 41-56.
182. N. M. Rice, On nth roots of positive operators, *Amer. Math. Monthly*, **89/5** (1982) 313-314.
183. M. Rosenblum, On the operator equation $BX - XA = Q$, *Duke Math. J.*, **23** (1956) 263-269.
184. M. Rosenblum, On a theorem of Fuglede and Putnam, *J. London Math. Soc.*, **33**, (1958) 376-377.
185. M. Rosenblum, The operator equation $BX - XA = Q$ with self-adjoint A and B. *Proc. Amer. Math. Soc.*, **20** (1969) 115-120.
186. W. E. Roth, The equations $AX - YB = C$ and $AX - XB = C$ in matrices. *Proc. Amer. Math. Soc.*, **3** (1952) 392-396.
187. W. Rudin, Function theory in polydiscs. *W. A. Benjamin, Inc.*, New York-Amsterdam 1969.
188. W. Rudin, Principles of mathematical analysis. Third edition. International Series in Pure and Applied Mathematics. *McGraw-Hill Book Co.*, New York-Auckland-Düsseldorf, 1976.
189. W. Rudin, Real and Complex Analysis, Third edition, *McGraw-Hill Book Co.*, New York, 1987.
190. W. Rudin, Functional Analysis, *McGraw-Hill Book Co.*, Second edition, International Series in Pure and Applied Mathematics, McGraw-Hill, Inc., New York, 1991.
191. B. P. Rynne, M. A. Youngson, Linear functional analysis. Second edition. Springer Undergraduate Mathematics Series. *Springer-Verlag London*, Ltd., London, 2008.

192. T. Saitô, T. Yoshino, On a conjecture of Berberian, *Tôhoku Math. J.*, (2) **17** (1965) 147-149.
193. Ch. Schmoeger, A note on logarithms of self-adjoint operators. http://www.math.us.edu.pl/smdk/SCHMOEG1.pdf.
194. K. Schmüdgen, On Domains of Powers of Closed Symmetric Operators, *J. Operator Theory*, **9/1** (1983) 53-75.
195. K. Schmüdgen, *Unbounded Self-adjoint Operators on Hilbert Space,* Springer GTM **265** (2012).
196. A. Schweinsberg, The operator equation $AX - XB = C$ with normal A and B. *Pacific J. Math.*, **102/2** (1982) 447-453.
197. Z. Sebestyén, Positivity of operator products. *Acta Sci. Math. (Szeged)*, **66/1-2** (2000) 287-294.
198. J. H. Shapiro, "Notes on the Numerical Range", 2017. http://www.joelshapiro.org
199. B. Simon, Operator theory. A Comprehensive Course in Analysis, Part **4**. *American Mathematical Society, Providence, RI*, 2015.
200. U. N. Singh, K. Mangla, Operators with inverses similar to their adjoints, *Proc. Amer. Math. Soc.*, **38** (1973), 258-260.
201. G. Sirotkin, Infinite matrices with "few" non-zero entries and without non-trivial invariant subspaces, *J. Funct. Anal.*, **256/6** (2009) 1865-1874.
202. J. G. Stampfli, *Hyponormal operators and spectral density*, Trans. Amer. Math. Soc., **117** (1965) 469-476.
203. E. M. Stein, R. Shakarchi, Fourier analysis. An introduction. Princeton Lectures in Analysis, 1. Princeton University Press, Princeton, NJ, 2003.
204. E. M. Stein, G. Weiss, Introduction to Fourier analysis on Euclidean spaces. Princeton Mathematical Series, No. 32. Princeton University Press, Princeton, N.J., 1971.
205. D. Sullivan, The square roots of 2×2 matrices, *Math. Mag.*, 66/5 (1993) 314-316.
206. J. J. Sylvester, Sur l'équation en matrices $px = xq$ (French). *C. R. Acad. Sci. Paris* **99**, (1884) 67-71, 115-116
207. G. Teschl, Mathematical methods in quantum mechanics with applications to Schrödinger operators. Second edition. Graduate Studies in Mathematics, **157**. *American Mathematical Society*, Providence, RI, 2014.
208. M. Uchiyama, *Commutativity of selfadjoint operators*, Pacific J. Math., **161/2** (1993) 385-392.
209. I. Vidav, *On idempotent operators in a Hilbert space*, Publ. Inst. Math., (Beograd) (N.S.) **(4) 18** (1964) 157-163.
210. J. Weidmann, Linear operators in Hilbert spaces (translated from the German by J. Szücs), Srpinger-Verlag, GTM **68** (1980).
211. R. L. Wheeden, A. Zygmund, Measure and integral, An introduction to real analysis. Pure and Applied Mathematics, Vol. **43**. Marcel Dekker, Inc., New York-Basel, 1977.
212. R. Whitley, The spectral theorem for a normal operator, Amer. Math. Monthly, **75** (1968) 856-861.
213. J. P. Williams, Operators Similar to Their Adjoints, *Proc. Amer. Math. Soc.*, **20**, (1969) 121-123.

214. D. X. Xia, On the nonnormal operators-semihyponormal operators. Sci. Sinica, **23/6** (1980) 700-713.
215. J. Yang, Hong-Ke Du, *A Note on Commutativity up to a Factor of Bounded Operators*, Proc. Amer. Math. Soc., **132/6** (2004), 1713-1720.
216. N. Young, An Introduction to Hilbert Space, *Cambridge University Press*, 1988.
217. R. Zeng, Young's inequality in compact operators-the case of equality, *JIPAM. J. Inequal. Pure Appl. Math.*, **6/4** (2005), Article 110, 10 pp.
218. X. Zhan, Matrix inequalities, Lecture Notes in Mathematics, 1790. *Springer-Verlag*, Berlin, 2002.
219. M. Zima, *A theorem on the spectral radius of the sum of two operators and its application*, Bull. Austral. Math. Soc., **48/3** (1993), 427-434.
220. https://math.berkeley.edu/sites/default/files/pages/Spring86.pdf
221. (A bunch of authors) Elementary operators & applications. In memory of Domingo A. Herrero. Proceedings of the International Workshop held in Blaubeuren, June 9-12, 1991. Edited by Martin Mathieu. *World Scientific Publishing Co.*, Inc., River Edge, NJ, 1992.

Index

Printed in the United States
By Bookmasters